AN ILLUSTRATED GUIDE TO DINOSAUR FEEDING BIOLOGY

AN ILLUSTRATED GUIDE TO

Dinosaur Feeding Biology

Ali Nabavizadeh and David B. Weishampel

 JOHNS HOPKINS UNIVERSITY PRESS BALTIMORE

© 2023 Johns Hopkins University Press
All rights reserved. Published 2023
Printed in Canada on acid-free paper
9 8 7 6 5 4 3 2 1

Johns Hopkins University Press
2715 North Charles Street
Baltimore, Maryland 21218
www.press.jhu.edu

Library of Congress Cataloging-in-Publication Data

Names: Nabavizadeh, Ali, 1987– author. | Weishampel, David B., 1952–
    author.
Title: An illustrated guide to dinosaur feeding biology / Ali Nabavizadeh,
    David B. Weishampel.
Description: Baltimore, Maryland : Johns Hopkins University Press, 2023. |
    Includes bibliographical references and index.
Identifiers: LCCN 2022030764 | ISBN 9781421413532 (hardcover ; acid-free
    paper) | ISBN 9781421445878 (ebook)
Subjects: LCSH: Dinosaurs—Food. | Dinosaurs—Adaptation. | Paleoecology. |
    BISAC: SCIENCE / Paleontology | SCIENCE / Life Sciences / Biology
Classification: LCC QE861.6.B44 N33 2023 | DDC 567.9—dc23/eng20221202
LC record available at https://lccn.loc.gov/2022030764

A catalog record for this book is available from the British Library.

Special discounts are available for bulk purchases of this book. For more information,
please contact Special Sales at specialsales@jh.edu.

To my wife Nora

—Ali

To my daughters Sarah and Amy

—David

Contents

Acknowledgments

First and foremost, this book would not have been possible without decades of detailed work done by countless researchers across multiple generations. While conducting our research, we were amazed by just how much work had been done over the years and how well each generation of scientists connected with and improved upon past studies with new ideas, additional fossil material, and more advanced modeling techniques.

We would like to give a special thanks to Johns Hopkins University Press, in particular Ezra Rodriguez, Tiffany Gasbarrini, Ashleigh McKown, Juliana McCarthy, Kyle Kretzer, and Vincent Burke, as well as all other editors and reviewers for their help, guidance, and incredible patience with us as we finished this manuscript. The book would not have been possible without them and their willingness to allow us to explore all the possible ways we could make this book as great and effective as it could be, both in writing and illustrations. Thank you from the bottom of our hearts.

We especially thank Dr. Peter Dodson for his undying support throughout this process and for believing in our ability to write a great book (that we hope will prove useful to paleontologists and paleontology enthusiasts alike for years to come). Peter's mentorship and incredible optimism regarding our work has been an immense help, and for that we are eternally grateful. Additionally, we would like to extend our deepest gratitude to our many colleagues in paleontology and anatomy with whom invaluable discussions, insights, and access to museum specimens and other resources helped expand our knowledge and improve our approach to writing this volume. These wonderful people include Dr. Karen Poole, Dr. Paul Barrett, Dr. David Button, Dr. Catherine Forster, Dr. Joy Reidenberg, Dr. Jeffrey Laitman, Dr. Eugenia Gold, Dr. Daniel Barta, Dr. Ashley Morhardt, Dr. Victoria Arbour, Dr. Thomas Holtz, Dr. Jack Horner, Ashley Hall, Lee Hall, Evan Johnson-Ransom, Dr. Amanda Falk, Dr. Larry Witmer, Dr. Darren Naish, Dr. David Evans, Dr. Matthew Carrano, Dr. Hans-Dieter Sues, Dr. Michael Brett-Surman, Amy Henrici, Carl Mehling, Dr. Peter Makovicky, Dr. François Therrien, Dr. Annelise Folie, Dr. Pascal Godefroit, Dr. Pascaline Lauters, Dr. Jordan Mallon, Dr. Frank Varriale, Dr. David Burnham, Dr. Desui Miao, and so many others. Ali would especially like to also thank the late, great Dr. Larry Martin for being a wonderful mentor during his undergraduate years at the University of Kansas and for initially sparking his interest in dinosaur feeding biology.

We are also eternally grateful to the faculty and students in the Center for Functional Anatomy and Evolution at Johns Hopkins University School of Medicine, particularly those present during Ali's PhD years, for their encouragement and support. We extend a special thanks to Rachel Frigot for her invaluable help with biomechanical discussions and with the initial planning of this book. We also thank Dr. Valerie DeLeon, Dr. Kenneth Rose, Dr. Christopher Ruff, Dr. Jonathan Perry, Dr. Mark Teaford, Dr. Katrina Jones, Dr. Francois Gould, Catherine Sartin, Dr. Heather Ahrens, Dr. Nicky Squyres, Dr. M. Loring Burgess, Dr. Heather Kristjanson, Dr. Ellen Fricano, Dr. Kaya Zelazny, and many others.

The remainder of these acknowledgments are specifically from Ali Nabavizadeh.

I am tremendously thankful to my colleagues at the University of Chicago during my time as a postdoc there for their encouragement and wonderful, incredibly helpful discussions, including Dr. Julia Schultz, Dr. David Grossnickle, Dr. Bhart-Anjan Bhullar, Dr. Aaron Olsen, Dr. Courtney Orsbon, Dr. Jacqueline Lungmus, April Isch Neander, Dr. Robert Burroughs, Dr. Kenneth Angielczyk, Dr. Callum Ross, Dr. Mark Westneat, Dr. Zhe-Xi Luo, Dr. Richard Madden, Dr. Regan Dunn, and many others.

I am also thankful to the faculty (Gary Lees, Timothy Phelps, David Rini, Jennifer Fairman) and students of the Art as Applied to Medicine Department at Johns Hopkins University School of Medicine. I learned so much about scientific illustration from each and every one of them, and I feel extremely lucky to have been able to work with such talented and kind people who helped me improve my illustration skill so that I could include it in this book. I would also like to thank my current colleagues at the University of Pennsylvania School of Veterinary Medicine for their continued encouragement throughout the final stretch of writing this book, including Dr. Paul Orsini, Dr. Barbara Grandstaff, Dr. Elizabeth Woodward, Adelaide Paul, and Dr. Peter Hand. Also, the encouragement of my good friends Dr. Sangita Phadtare, Dr. Ehsan Basafa, Dr. Nick Boire, and Vajini Atukorale has also been monumentally helpful in my journey in becoming an anatomist and paleontologist.

I would like to thank my esteemed graduate mentor and coauthor on this book, Dr. David Weishampel, for his endless knowledgeable input, his ever-helpful discussions, his uplifting encouragement, his gracious support, and of course for reading my dissertation and everything else I have written since. If it weren't for him, I would not be anywhere near where I am right now, and for that I am eternally grateful to him. I am lucky to have worked with him as my mentor and friend, and I am in awe that I am lucky enough to coauthor a book with him.

Thanks also go to my amazing parents, Shohreh and Vahid, and my incredible brother, Omid, for their endless support, encouragement, and love. They have always been there supporting me and cheering me on through every success and obstacle I have run into in life. I love them so much, and I am so very thankful for having such a wonderful family in my life.

I would like to thank my sweet little mini goldendoodle, Abigail, for her caring nature, unconditional love, and limitless ability to keep a smile on my face as I write for hours on end every day.

And finally, I am the most grateful to the love of my life—my wife, Nora. Her undying support and years and years of love, encouragement, determination, and faith in my ability to finish this book is the only reason I was able to keep going. Without her, the book would not have been completed. Words cannot express my love for her. I am a better man today because of her, and I am so fortunate to have her in my life.

AN ILLUSTRATED GUIDE TO DINOSAUR FEEDING BIOLOGY

CHAPTER 1

Early Dinosaur Feeding Studies, Mesozoic Landscapes, and Dietary Ecology

Dinosaurs: What Are They?

When you think of a dinosaur, what comes to mind? Many people think of dinosaurs as humongous reptiles that are long dead, leaving no descendants—that they were largely ferocious, brainless meat-eaters. Yet that has not stopped dinosaurs from being the stars of (at last count) six blockbuster cinematic creations, complete with some of the best computer-generated animations to hit movie screens. Covering how much of this is true and how much is false would take a large volume or more on dinosaurs and their close vertebrate relatives, although many authors have attempted such a feat (e.g., Colbert, 1961; Bakker, 1986; Weishampel et al., 1990, 2004; Farlow and Brett-Surman, 1997; Brett-Surman et al., 2012; Brusatte, 2018; among countless others). These reptiles are now known from all earthly continents—even Antarctica (Hammer and Hickerson, 1994)—but they are known from a relatively small sliver of geologic time. Earth is thought to be 4.6 billion years old, and life on this planet dates (relatively) not too long thereafter (4 billion years ago). For 3 billion years, life on earth consisted primarily of single-celled organisms of different kinds. It was not until the advent of the Phanerozoic, some 540 million years ago, that complex life arose. Of the Phanerozoic, the middle portion is called the Mesozoic, and in particular from the Late Triassic Period through the end of the Cretaceous Period (229 to 66 million years ago), we have the age of dinosaurs.

But what are these dinosaurs, these terribly great reptiles? Originally, Richard Owen (1804–92; fig. 1.1),

a comparative anatomist and vertebrate paleontologist, recognized three extinct reptiles named *Iguanodon*, *Megalosaurus*, and *Hylaeosaurus* from the Mesozoic of England as the first members of Dinosauria in 1842 (Owen, 1842). Owen made no attempt to discern the internal classification of these three dinosaurs, but it did not take too long before more dinosaurian discoveries and further studies of existing material inspired vertebrate paleontologists to establish relationships among dinosaurs and with other reptile groups.

Early vertebrate paleontologists, such as Thomas Henry Huxley (1825–95), Edward Drinker Cope (1840–97),

Figure 1.1. Portrait of Richard Owen (1804–92), a British comparative anatomist and vertebrate paleontologist who also coined the name Dinosauria.

Figure 1.2. Portrait of Thomas Henry Huxley (1825–95), a British vertebrate paleontologist and comparative anatomist who, among many other things, conducted research on the evolutionary relationship of birds and theropod dinosaurs.

Figure 1.3. Portraits of paleontological rivals (a) Edward Drinker Cope and (b) Othniel Charles Marsh. Edward Drinker Cope (1840–97) was an American vertebrate paleontologist and comparative anatomist whose systematic research on dinosaurs divided the group into four categories, none of which have proved to be useful today. Othniel Charles Marsh (1831–99) was an American vertebrate paleontologist who divided Dinosauria into four categories, most of which we still use today.

Figure 1.4. Portrait of Harry Govier Seeley (1839–1909), a British vertebrate paleontologist who recognized the two-fold division of Dinosauria into Saurischia (lizard-hipped dinosaurs) and Ornithischia (bird-hipped dinosaurs).

and Othniel Charles Marsh (1831–99), attempted to subdivide Dinosauria to represent the evolutionary development of the group. Huxley (fig. 1.2), a comparative anatomist and evolutionary biologist in England, had an interest in dinosaur systematics focused on the relationship between one group of dinosaurs called theropods and birds (Huxley, 1868), particularly in the newly discovered *Archaeopteryx* and *Compsognathus* from the Solnhofen limestone (Late Jurassic) of southern Germany.

Cope (1883; fig. 1.3a) and Marsh (1878, 1884; fig. 1.3b) were by far more interested in how Dinosauria was subdivided into various groups, however. Cope, a vertebrate paleontologist from Philadelphia, principally used variation in the ankle and hindlimb in his classification of dinosaurs and gave us Orthopoda (not Ornithopoda!), Goniopoda, Hallopoda, and Opisthicoela. In contrast, Marsh, a Yale University vertebrate paleontologist and rival to Cope, looked to anatomical similarities and differences from throughout the skeleton. On the basis of these features, Marsh's schema included Ornithopoda, Stegosauria, Theropoda, and Sauropoda. Marsh's dino-

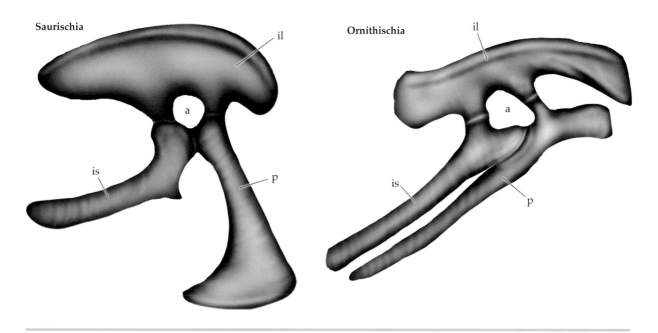

Figure 1.5. Comparison of (*left*) generalized saurischian dinosaur pelvis and (*right*) ornithischian dinosaur pelvis. Abbreviations: a, acetabulum; il, ilium; is; ischium; p, pubis.

saur groups have stood the test of time, while Cope's names have not.

But neither Cope nor Marsh was to establish the relationships among these groups. That happened in 1887, when Harry Govier Seeley (1839–1909; fig. 1.4) hit upon a scheme by which all dinosaurs could be divided into two major groups, which were thought to represent distinct evolutionary lines. Seeley was a vertebrate paleontologist from King's College at the University of London whose specialties were European dinosaurs and pterosaurs.

Based ultimately on specimen hip structure, Seeley gave us ornithischian and saurischian dinosaurs (fig. 1.5). Ornithischians were those forms, like *Iguanodon* and *Hylaeosaurus*, whose pubis bone of the pelvis has rotated backward to lie parallel to the shaft of the ischium. In contrast, the pubis of *Megalosaurus* (a saurischian) projects rostrally (forward), lying at a distinct angle to the shaft of the ischium. Seeley's basic classification into Ornithischia and Saurischia has been the organizing principal in systematics for the past 130 years of dinosaur research. With the advent of phylogenetic systemat-

ics, the discovery of close relationship among clades (groups of organisms more closely related to each other than to any other groups of organisms; Hennig, 1966; Wiley and Lieberman, 2011), many scientists have tried to fit character support to Seeley's twofold dinosaurian clades, assuming each had a single origin (monophyly) and formed a treelike branching pattern that indicated closeness of evolutionary relation.

Jacques A. Gauthier, a Yale University systematist who focused principally on dinosaurs and lizards, was the first to analyze Dinosauria cladistically, emphasizing a variety of theropods and their relationship with birds (Gauthier, 1986). He identified an increase in length of the scapula, an increase in asymmetry of the manus, and various aspects of the pelvis and hindlimb as unique dinosaurian features. He further cladistically identified several theropod groups and recognized their phylogenetic relationships: Ceratosauria, Tetanurae, Carnosauria, Coelurosauria, and Maniraptora among several others. Almost simultaneously, Paul C. Sereno, an archosaur systematist at the University of Chicago, analyzed the phylogenetic relationships among ornithischian clades

(Sereno, 1986). His work established a number of previously known clades: Thyreophora (including Stegosauria and Ankylosauria), Marginocephalia (including Ceratopsia and Pachycephalosauria), and Ornithopoda, but even more significantly, he posited the relationships among these ornithischians.

That is where things have stood over the past few decades: scientists discovering new taxa and characters. Very recently, these basic dinosaurian relationships have been challenged. Matthew G. Baron, David B. Norman, and Paul M. Barrett, vertebrate paleontologists at Cambridge University and the Natural History Museum, sampled the earliest members of Saurischia, Ornithischia, and Theropoda, as well as a variety of non-dinosaurian dinosauromorphs, and cladistically reanalyzed them to see what pattern arose (Baron et al., 2017a). Gone is the standard grouping of sauropodomorphs and theropods into saurischians, but rather Saurischia became the clade of sauropodomorphs and herrerasaurs, a small group of bipedal dinosaurs formerly associated with early theropods. Theropods themselves are no longer associated with sauropodomorphs; instead they form a separate, much larger clade as a sister group to ornithischian dinosaurs—a clade the authors called Ornithoscelida. These new ideas present a rich source of inquiry about the evolution of dinosaurian taxa and their characters. In our opinion, however, it is far too soon to switch allegiances prior to these explorations of Baron et al. (2017a). Therefore we will keep to the older schema of dinosaur classification, keeping in mind that some, but certainly not all, evolutionary trajectories of feeding strategies may change if their conclusions hold true.

A Short History of Early Dinosaur Feeding Studies

Historically, the study of dinosaur feeding largely focused on the geometries of the jaws and how they might have worked during mastication. Evidence of dinosaur feeding can also be gleaned from other sources of information, however, such as general aspects of dinosaurian

biomes (predator–prey relationships, plant–animal associations, etc.), coprolites (fossilized feces; see below), and other aspects of gut contents, among others. Most of these studies are based on single genera or species (or single specimens). Few constitute whole clade comparisons, although Weishampel and Norman (1989) qualitatively described chewing in most Mesozoic herbivorous vertebrates, from procolophonids and dicynodonts to dinosaurs, as transverse grinders (ornithopods), orthal (up-and-down) pulpers (pachycephalosaurs, ankylosaurs, and stegosaurs), orthal slicers (ceratopsians), and gut processors (sauropodomorphs). As will be discovered later in this book, however, numerous studies of their feeding mechanisms have since found them to be much more complex.

This section is not meant to be a comprehensive or complete discussion of the history of ideas about dinosaurian feeding. It is intended to emphasize the major historical aspects of dinosaurian feeding studies and to highlight the kind of approaches brought to bear on the feeding structures in organisms that no longer feed (or even behave in any appreciable way; Weishampel, 1995). We begin by outlining the earlier history of studies of ornithopod dinosaurs because they have arguably the most written about their jaw mechanics and because these investigations constitute the most elaborate among dinosaur clades. Thereafter, we discuss early studies in the jaw systems of ceratopsians, ankylosaurs, sauropodomorphs, and finally theropods, including birds. Unfortunately, there are no detailed early studies of stegosaurs and pachycephalosaurs, but their feeding has been studied in much more detail recently (as has the feeding in all other aforementioned dinosaur groups), and we will be covering this subject in more detail in later chapters.

The first to examine the jaw mechanics in any dinosaur was Gideon Mantell (1790–1852; fig. 1.6). In 1848, he published a description of an upper and lower jaw, complete with replacement teeth, of *Iguanodon*. He described the distinct differences between this dentition and that of both extinct and extant reptiles but noted that these features were more like herbivorous mammals. On the basis of dentary tooth wear, he determined

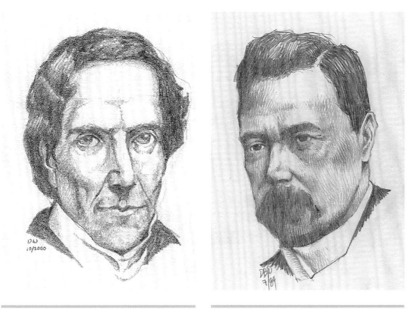

Figure 1.6. Portrait of Gideon Mantell (1790–1852), a British vertebrate paleontologist whose work on the second dinosaur to be discovered—*Iguanodon*—led him to make the first claims about the anatomical basis of dinosaur feeding.

Figure 1.7. Portrait of Louis Dollo (1857–1931), a Belgian vertebrate paleontologist and engineer who wrote extensively on the abundant remains of *Iguanodon* collected at the small village of Bernissart.

that *Iguanodon* chewed transversely, foraging on contemporary conifers, ferns, and cycads.

In his elaborate and meticulous studies of the abundant material of *Iguanodon* from Bernissart, Belgium, Louis Dollo (1857–1931; fig. 1.7) discussed their jaw mechanics in general and comparative ways. In contrast to Mantell (1848), Dollo's work on feeding in *Iguanodon*, in keeping with the preservation of more or less complete skulls, involves skull morphology and the reconstruction of the cranial musculature. Two studies in particular (Dollo, 1883, 1884) show his methods that combined engineering and paleobiology to understand the jaw systems in *Iguanodon*. In the former paper, he described bone by bone seven skulls of *Iguanodon*, including information on the masticatory muscles and on the complex replacement of teeth, comparing this ornithopod with crocodilians, iguanas, and tuataras (Dollo, 1883). In contrast, Dollo's (1884) paper explores how diet and skull construction are interwoven by exploring the jaw adductor musculature and the way food was dis-

membered in the mouth. He first compares the skulls and musculature of living carnivorous and herbivorous mammals, crocodilians, and lizards to evaluate which muscle groups were better developed for particular feeding groups. He then attempted to identify the ways in which four dinosaurs—*Iguanodon*, a hadrosaurid (*Edmontosaurus*), a sauropod (*Diplodocus*), and a theropod (*Ceratosaurus*)—corresponded to the different jaw systems he had identified for the modern vertebrates. From his comparisons, *Iguanodon* and the *Edmontosaurus* were found to be more like a lizard, with a temporal-muscle-dominated jaw system, while the theropod and sauropod were more like herbivorous mammals and modern crocodilians, with the jaw apparatus dominated by the palatal (pterygoid) musculature.

By 1869, hadrosaurids were recognized as having teeth that were organized into a dental battery consisting of two to three functional teeth and up to five replacement teeth per tooth position, and these were distributed over 60 tooth positions or families in the upper

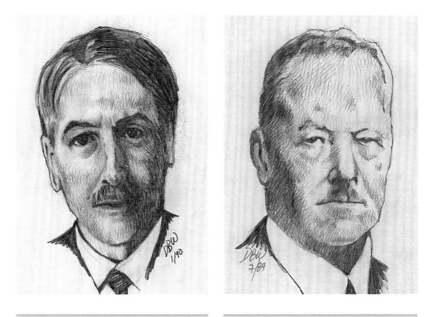

Figure 1.8. Portrait of Franz Baron Nopcsa (1877–1933), a Hungarian vertebrate paleontologist who wrote extensively on the dwarfed dinosaurs from Transylvania, including the ornithopod *Telmatosaurus*, which was the first dinosaur to have its jaw musculature reconstructed and its jaw mechanics analyzed.

Figure 1.9. Portrait of Jan Versluys (1873–1939), a German comparative anatomist and vertebrate paleontologist who was the first to investigate the details of movement of the quadrate against the skull roof in dinosaurs in general (streptostyly).

and lower jaws. Marsh seems to be among the first to examine the jaw apparatus of hadrosaurids in 1895, through his brief comments on mobility of the quadrate-squamosal joint called streptostyly. But it was Franz Baron Nopcsa (1877–1933; fig. 1.8) who first addressed broad issues in hadrosaurid jaw mechanics in his description of *Telmatosaurus* in 1900. He suggested that *Telmatosaurus* (and thereby all hadrosaurs) chewed its food by movement of the lower jaws at the jaw joint and by fore-and-aft motion of the lower jaws through mobility of the quadrate-squamosal joint, the same that Marsh (1895) hypothesized (Nopcsa, 1900).

Jan Versluys (1873–1939; fig. 1.9) was the first to investigate the details of movement of the quadrate against the skull roof in dinosaurs in general (Versluys, 1910, 1923). In his most detailed study, he suggested that the lower jaws of *Edmontosaurus* (formerly *Anatosaurus*) moved in a fore-aft motion, associated with streptostyly

of the quadrates, combined with an independent outward rotation of the mandibles along their long axis to produce an overall transverse grinding movement between opposing teeth during mastication.

Dominik von Kripp (1907–67) based his interpretation of jaw mechanics in hadrosaurids on a biomechanical analysis of the same skull of *Edmontosaurus* that Versluys had used (Kripp, 1933), and Kripp's hypothesis also included a peculiar kind of streptostyly. In addition, Kripp proposed that the mandibles rotated inwardly along their long axes to function in fracturing vegetation before swallowing it.

Up until the 1960s, virtually all ideas about ornithopod chewing involved at least some degree of streptostyly, and these were largely of European origin. Only Lawrence M. Lambe (1863–1919; fig. 1.10a), Richard S. Lull (1867–1957; fig. 1.10b), and Nelda E. Wright (1901–92) considered mastication in these dinosaurs, but only

Figure 1.10. Portraits of (a) Lawrence Lambe and (b) Richard Swann Lull.

in passing as ersatz mammals, through the outcome of shear between occluding teeth (Lambe, 1920; Lull and Wright, 1942).

It was John H. Ostrom (1928–2005) who broke with this tradition of treating streptostyly as part of the jaw mechanics in hadrosaurids (fig. 1.11a) in his 1961 monograph of the cranial anatomy of these dinosaurs (Ostrom, 1961). In contrast to Nopcsa, Versluys, and Kripp, he rejected streptostyly and long-axis rotation of the mandibles, hypothesizing instead that the lower jaws moved only in the fore-aft direction, while the remainder of the skull remained rigid. In a similar fashion, Peter M. Galton (1942–) advocated a similar jaw mechanism for *Hypsilophodon* in his 1974 monograph (Galton, 1974).

In 1984, David B. Norman (1952–) presented a new style of jaw mechanism in ornithopod dinosaurs, which he called pleurokinesis (Norman, 1984). Based initially on his osteological and functional studies, he reconstructed the jaw musculature of *Iguanodon* and examined how these muscles would produce the kinds of wear seen in the same skulls. Surprisingly, the upper jaws were co-opted to swing slightly laterally to produce the transverse chewing as indicated by wear of the dentition of *Iguanodon*.

Also in 1984, David B. Weishampel (1952–) presented three-dimensional computer models of pleuroki-

nesis as well as many other jaw mechanisms that have been postulated for ornithopods. These models were used to virtually generate tooth wear, seen as the test of preexisting mechanics from the literature (Weishampel, 1984a). Together, Norman and Weishampel went on to investigate the ways that pleurokinesis may have co-evolved with contemporary floras (Norman and Weishampel, 1985; Weishampel and Norman, 1989; see also Weishampel and Jianu, 2000). For a deeper, updated look at the many studies regarding the evolution of ornithopod feeding mechanisms, see chapter 12.

The jaw mechanics of ceratopsians (fig. 1.11b) have also been thoroughly studied, although not quite as much as that of ornithopods early on. In the first monographic treatment of ceratopsians by John Bell Hatcher (1861–1904), Othniel Charles Marsh, and Richard Swann Lull (Hatcher et al., 1907) and in Lull's subsequent revisionary work on these same dinosaurs (Lull, 1933), the authors treated these animals as exclusively herbivorous with relatively simple orthal, slicing, or cutting jaw mechanics. Georg Haas (1905–81) reconstructed the muscles of mastication in *Protoceratops*, but he did not postulate a particular style of mastication for ceratopsians (Haas, 1955).

Ostrom (1964a, 1966) similarly reconstructed the jaw adductor musculature, but in many more taxa, and

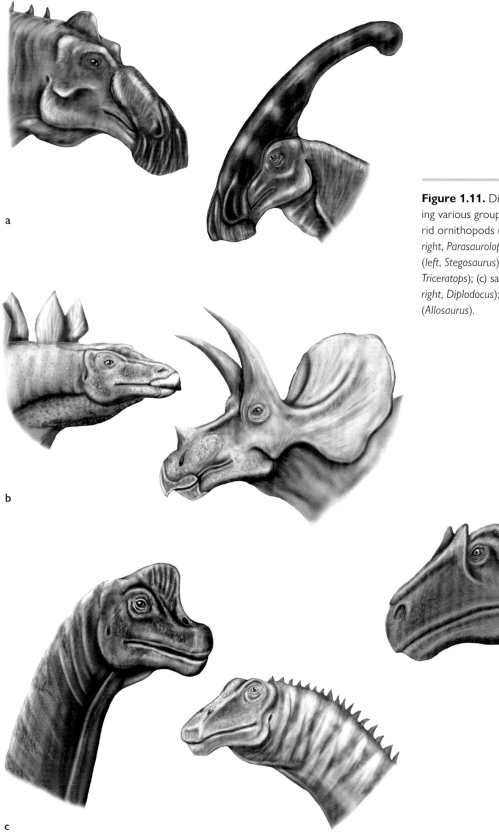

Figure 1.11. Dinosaur heads representing various groups, including (a) hadrosaurid ornithopods (*left*, *Edmontosaurus*; *right*, *Parasaurolophus*); (b) stegosaurs (*left*, *Stegosaurus*) and ceratopsians (*right*, *Triceratops*); (c) sauropods (*left*, *Giraffatitan*; *right*, *Diplodocus*); and (d) theropods (*Allosaurus*).

used the resulting muscle vectors to determine how the upper and lower teeth moved with respect to each other during chewing. When the occluding teeth in the upper and lower jaws of forms like *Triceratops*, *Centrosaurus*, *Pentaceratops*, and many other ceratopsians slid past each other, they created a vertical plane of wear between them. As in hadrosaurids, the teeth were organized into a dental battery, but only a single tooth in each tooth position came into wear at a time. It is likely that ceratopsians fed on fibrous vegetation in a scissorlike fashion. Much of Ostrom's work on ceratopsian jaw mechanics was later supported by Peter Dodson (1946–) in his comprehensive book on the paleobiology of these horned dinosaurs (Dodson, 1996). He objected to Ostrom's reconstruction of the jaw adductor muscles anchored to virtually all of the dorsum of the frill, however, instead hypothesizing that the adductor attachment was restricted to the lower portion of the parietal fenestra. For a deeper, updated look at the many studies regarding the evolution of ceratopsian (as well as pachycephalosaur) feeding mechanisms, see chapter 13.

Early on, little was written about ankylosaur feeding, in large part because of their simple teeth and relatively uncomplicated, rigid skulls. Haas (1969) reconstructed the cranial muscles in *Euoplocephalus*, noting that the entire muscular system of the head in these animals was generally weak and that they fed on relatively soft vegetation. For a deeper, updated look at the comparatively numerous studies regarding the evolution of ankylosaur (as well as stegosaur; fig. 1.11b) feeding mechanisms, see chapter 11.

Sauropodomorphs (fig. 1.11c) were long thought to lack any sort of oral processing, instead using their jaws to ingest foliage and then deliver this bolus of food without mastication, and therefore without tooth wear, to the immense gastrointestinal tract for further breakdown via gastroliths (swallowed stones used in digestion) and chemical digestion. Indeed, gastroliths lodged in a muscular gizzard have previously been suggested as the default mechanism for fracturing vegetation (Bakker, 1986; Bonaparte and Mateus, 1999; Sanders et al. 2001), although the interpretation of pebbles preserved in the

abdomen as gastroliths has been a controversial one (Calvo, 1994b; Wings and Sander, 2007). Following the short remarks of *Diplodocus* cranial musculature by Dollo (1884), Haas (1963) was the first to reconstruct the cranial musculature of a sauropodomorph, in this case also of *Diplodocus*; beyond that, however, sauropod feeding mechanisms and ecology were rarely studied until the 1990s (e.g., Fiorillo, 1991; Calvo, 1994a; Barrett and Upchurch, 1994). For a detailed look at these and all of the many other studies regarding feeding mechanisms in sauropodomorphs, see chapters 8 and 9.

Theropods (fig. 1.11d), particularly many of the nonavian varieties, are largely carnivorous, having little or no oral processing following ingestion of their animal food, be it a fresh kill or scavenged carcass. Most had narrow, recurved, and serrated teeth; possessed sharp claws on the manus and pes; and had bipedal stance, often with long tails. Theropods also possessed an intramandibular joint between the dentary and surangular, which has been argued by some to possibly aid in increasing the gape of the mouth. A vast majority of feeding studies about predatory theropods are from the 1990s onward, but select early works have focused on broadscale anatomy of the skull and teeth and made general inferences about function and ecology in theropods (e.g., Dollo, 1884; Gilmore, 1915; Gregory and Adams, 1915; Lambe, 1917; Adams, 1919; Gregory, 1920; Ostrom, 1969, 1976; Molnar, 1973; Barsbold, 1983; Welles, 1984). At least three theropod groups exhibit features that reflect diets other than carnivory: oviraptorosaurs, therizinosaurs, and ornithomimosaurs. Of the three, the edentulous oviraptorosaurs depart the most from the pattern of theropod feeding as judged from skull morphology. Through comparative anatomy, muscle reconstruction, and kinematic modeling, Rinchen Barsbold (1936–) suggested that this unusual group of theropods may have used their powerful toothless beaks to crush mollusk shells (Barsbold, 1977), while work by others indicate a diet of eggs and/or plants (Smith, 1992; Currie et al., 1993). For a much more in-depth, updated look at the many theropod feeding studies in the literature, see chapters 4 through 7.

There has been an abundance of feeding studies on birds (Aves), both extinct and particularly extant. Much of the latter comes from Leiden in the Netherlands, particularly the Dullemeijer research group (see Dullemeijer, 1974), with later examples from Zweers (1991) on feeding mechanics in the mallard and Heidweiller and Zweers (1990, 1992) on the drinking mechanics of finches and chickens. In addition, Walter Bock's (1964) study of intracranial mobility of modern birds models and reviews avian cranial kinesis. For a discussion of bird feeding studies, see chapter 7.

Mesozoic Earth and Vegetative Landscapes

Before we delve into the intricate details of the different kinds of feeding and feeding studies regarding all dinosaurs and what these mean in terms of dinosaur biology, we need to touch on a number of terms, concepts, and other things that you will encounter in this book. These include geologic time scales, plate tectonics, climate, vegetative landscapes, and the types of direct evidence of diet and digestion. Furthermore, the study of bones, teeth, muscle, and functional morphology will be covered in chapter 2, followed by an overview of dinosaur anatomy in chapter 3.

Earth throughout the Mesozoic

Earth is roughly 4.6 billion years old, and complex multicellular life has been here for only approximately one-tenth of that time, during the Phanerozoic (542 million years ago to present). Of that Phanerozoic time interval (Walker et al., 2018), dinosaur existence began in the Mesozoic, some 230 million years ago. At the very beginning of the Mesozoic, during the Early and Middle Triassic, dinosaurs had not yet evolved. Instead, their archosaurian ancestors and other preceding vertebrates dominated the terrestrial realms, including pseudosuchian archosaurs (the group leading to the evolution of modern-day crocodilians), nonmammalian synapsids (predecessors to mammals), and other early tetrapods. It was not until the Late Triassic that dinosaurs first crept onto the scene, and it is the rest of the divisions of the Mesozoic and the Cenozoic that are used to determine when different groups of dinosaurs (and all other animals and groups of organisms, including plants and fungi) had their stay on this planet.

The Late Triassic (237 to 201 million years ago; fig. 1.12) began with all the continents united as the supercontinent called Pangaea. The global climate was dry—arid to semiarid. It was at the end of the Triassic and into the Jurassic that Pangaea began to rift apart to eventually form two major continental landmasses, Laurasia and Gondwana (to the north and to the south, respectively).

The Early Jurassic (201 to 174 million years ago) followed a mass extinction, separating it from the Late Triassic. It was a time of somewhat warmer climate with marked seasonality from the northern realm to the South Pole, which characterized great swaths of terrestrial environments. The North Atlantic Ocean continued to open as North America moved to the west, and Europe and Northern Africa moved to the east.

The Middle Jurassic (174 to 163 million years ago) was a time of somewhat cooler, temperate climate (compared to the rest of the Jurassic) as well as reduced deposition of terrestrial sediments. Consequently, although we know that various dinosaur groups must have existed during the Middle Jurassic, their remains were not well preserved. The dinosaur record over this time interval is exceedingly poorly known.

The Late Jurassic (163 to 145 million years ago; fig. 1.12), especially toward the end of the interval, was a time marked by seasonality and aridity. The extensive epeiric seas present then had a stabilizing effect on climate. This is particularly true during the terrestrial deposition of the Morrison Formation of western North America, a time when the dinosaur faunas were dominated by diverse and large-bodied dinosaurs.

The Early Cretaceous (145 to 100 million years ago) and the Late Cretaceous (100 to 66 million years ago; fig. 1.12) saw the North Atlantic Ocean increasing in size as North America continued to separate from northern Africa and Europe. Furthermore, the southern continents, which collectively originated from Gondwana, began rifting apart, especially between Africa and South America. Considerable mountain-building occurred on a global scale to produce what eventually would become the Rocky Mountains of North America and the Andes Mountains of South America. Climate was uniform because of inundation of shallow seaways on the continental landmasses. The Late Cretaceous was especially hot and humid. India separated from Australia, Africa, and Antarctica in its long travel northward to form the Himalayas with southern Asia.

Earth experienced a worldwide extinction approximately 66 million years ago, marking the end of the Cretaceous and the beginning of the Tertiary. This extinction event was relatively short and apocalyptic, most probably due to the collision of a large (10-km diameter) asteroid with earth just off the coast of the Yucatan Peninsula. It killed off all non-avian dinosaurs, many mammal and bird groups, and a variety of marine animals such as plesiosaurs, seagoing lizards called mosasaurs (which underwent total extinction), as well as multichambered and often coiled cephalopod molluscs and single-celled animals called foraminifera, both of which suffered partial but extreme mass extinctions. Finally, the fossil record of plants indicates that a major extinction occurred geologically instantaneously at the boundary between the Cretaceous and Tertiary, the same time as the extinction of the other aforementioned groups. The Tertiary (66 to 2.6 million years ago) was a time of continued breakup of the global continents, when they came to resemble their modern configurations and positions. The climate consisted of relatively long periods of hot global temperatures.

Paleobotany and Vegetative Landscapes in the Mesozoic

Like modern-day mammals, most dinosaurs were herbivores, feeding on contemporary plants, their leaves, and fruits. The classification of this vegetation, then and now, indicates that there are two major groups.

The first group consists of non-seed-bearing plants including ferns, lycopods, and sphenopsids (horsetails) (fig. 1.13). All tend to be low-growing and shrubby, vascular plants, possessing specialized tissues that conduct water and nutrients throughout the plant.

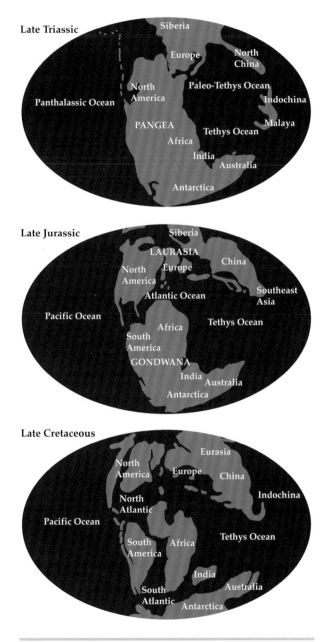

Figure 1.12. Global landmass distribution in three different Mesozoic time periods: (*top*) Late Triassic Period (237 to 201 million years ago), (*middle*) Late Jurassic (163 to 145 million years ago), and (*bottom*) Late Cretaceous Period (100 to 66 million years ago).

Figure 1.13. Non-seed-bearing plants, including (*right*) ferns and (*far right*) horsetails (sphenopsids).

Figure 1.14. Seed-bearing plants, including (*clockwise from top left*) cycads, ginkgos, angiosperms, and conifers.

The second group of plants (known as spermatophytes) consists of gymnosperms and angiosperms, uniquely united by possessing a seed developed for the survival of gametes (fig. 1.14). The best-known gymnosperms of present today are conifers (e.g., pines and cypress). Additionally, ginkgophytes (ginkgos) and cycadophytes (cycads)—shrubby plants with a large pineapple-like stem and bunches of tough leaves sprouting from the top—were much more prevalent during the Mesozoic but are less common today. Finally, angiosperms are known today from magnolias, grasses, maples, orchids, roses, and many more. All of these angiosperms comprise the so-called flowering plants because of their development in close association with insect pollinators early in their evolution. These also thrived because of their relatively short generation time, permitting rapid population growth and easier colonization of disturbed habitats over that of gymnosperms.

Seen in total, the world's florae of the Mesozoic were decidedly different than those of today (Wing and Sues, 1992). In the Late Triassic, lycopods, sphenopsids, ferns, and seed ferns decreased in abundance across the globe. From then until the end of the Mesozoic, these groups represented an approximate constant proportion of the world's floras. In contrast, gymnosperms, especially the conifers, dramatically increased their representation in the world's floras thereafter. Shrubby and arborescent ferns and conifers dominated terrestrial environments in the Late Triassic, and there were no flowering plants, as angiosperms were not to evolve until some 50 million years later. Otherwise, Late Triassic floras consisted of sphenopsids and a few kinds of non-coniferin gymnosperms.

For the Jurassic, plant communities were composed of ginkgos, conifers, ferns, cycads, and cycadeoids. Pteridosperms (seed ferns), sphenopsids, and lycopsids were in decline in the Early Jurassic, while woody gymnosperms and herbaceous ferns dominated the flora. Conifers, particularly members of Cheirolepidaceae (a clade that appears to have been adapted for semiarid and coastal settings), formed much of the diversity of large trees. Some fossil floras known from the Middle Jurassic indicate that conifers and gingkoes made up the tall foliage of the times, while cycadophytes, seed ferns, ferns, and lycopods made up the low foliage. This floral makeup persisted throughout the Cretaceous as well.

Rapid diversification of angiosperms (flowering plants) of shrubby habitats led to dominance, including arborescent forms, in the Late Cretaceous. Diversification of angiosperms continued to the present. Furthermore, grasses (also angiosperms) did not appear any time throughout the vast majority of the Mesozoic, so most dinosaurs did not feed on grasses. But evidence of microscopic structures called silica phytoliths (many of which are specific to grasses) found within some sauropod coprolites (fossilized feces; see below) from the Late Cretaceous of India suggests that some of the latest-surviving non-avian dinosaurs may have potentially grazed on grasses, at least in passing (Prasad et al., 2005; see also Piperno and Sues, 2005).

Tracing Dietary Ecology

When assessing the interaction between dinosaurs and their environment throughout the Mesozoic (e.g., herbivores feeding on available plant life), some of the clearest evidence of these interactions when it comes to feeding ecology comes from factors or contents outside of the anatomy of the specimens at hand. Dinosaurs are known to us not only from their bones and teeth (which you will see from here on out), their footprints and skin impres-

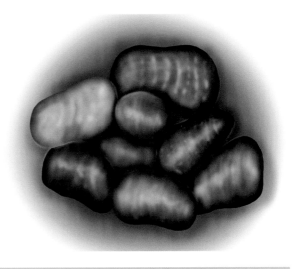

Figure 1.15. Gastroliths of various shapes and sizes.

sions, and their eggs and embryos, but also from other associated structures that pertain to their feeding ecology.

For instance, there are aggregates of smoothly polished stones lodged in the muscular part of the stomach. These stones are called gastroliths (fig. 1.15) and were probably used at the very least for grinding and processing consumed food items (Wieland, 1906; Christiansen, 1996; Currie and Padian, 1997; Wings, 2007), much like naturally occurring, nonorganic "teeth" in the crop immediately prior to the stomach proper in birds. Gastroliths, together creating a mechanism sometimes referred to as a "gastric mill," are mainly seen in herbivorous birds for triturating (grinding up) tough fibrous plant material and seeds, among other things (Wings, 2007). In cases where an alleged gastric mill is discovered with a specimen of a non-avian dinosaur, it is mainly used as a signal for herbivory in that animal (Zanno and Makovicky, 2011). Still, other possible uses of gastroliths include using it as a source of mineral supply and storage, cleaning the stomach, maintenance of microbial gut flora, or alleviating hunger (Wings, 2007), so these factors should also be considered when assessing diet in dinosaurs. Furthermore, the rarity of gastroliths as well as the slightly different textures of some of these smooth stones present caution in identifying these structures as gastric mills in some cases.

Cololites (fossilized gut contents) are also sometimes associated with some generally smaller dinosaur specimens, are usually found within the abdomen or nearby, and can also be an indicator of diet. Cololites can be seen as the animal's final meal before death. This can tell a lot about an animal's diet in and of itself; however, it is important to keep in mind that these cololites only represent one meal and likely do not represent the full range of its diet. Examples of cololites range from bones of other animals within predatory dinosaur specimens to the rarer cases of plant material found within the body cavity of herbivores.

Another commonly found fossilized structure known as a coprolite represents the ultimate fossil disposition of the physical remains of the undigested food—in other words, fossilized feces (Behrensmeyer and Hook, 1992). Distasteful as these kinds of fossils may seem, they could be important in linking herbivores with plant matter as well as predators with crushed-up bone from prey items. Depending on the geological age of the coprolite, its geographical location, its size, and its undissolved contents, it is sometimes possible to determine what type of plant matter an animal has consumed or, in the case of predators versus prey, to determine approximately who ate whom (Thulborn, 1991; Chin et al., 1998, 2003). The diverse shapes and sizes of coprolites are often highly variable, leading to a lot of uncertainty as to which animal actually produced it.

Bite marks (e.g., tooth punctures and scratches) and breakages seen in the bones of prey items also give clear evidence of predatory behavior. Shed teeth are also sometimes associated with prey, and in some cases these shed teeth are found still embedded within the bone. A good indication of whether the prey survived attacks like this is whether the wounded bone has been clearly healed over with remodeled bone. In other cases, the bone becomes diseased, which is another condition that is sometimes seen. Identification of which predators made these attacks is largely done by matching the tooth puncture size and shape to the carnivorous theropods that lived within that environment to see if they match. Another thing to keep in mind when examining these bite marks is whether it might have been made in an attack or merely a scavenging action—a distinction that is often difficult to assess.

Other signs of dietary ecology include taphonomic assemblages of dinosaurs (i.e., how different dinosaurs are found associated with one another) as well as the dietary and environmental information gained from studying stable isotopes in teeth (see chap. 2 for a brief discussion). All of these types of dietary evidence are referenced throughout this book in association with the various types of dinosaurs discussed in the literature on dietary feeding ecology. It is no surprise, however, that by far the most cited source of information regarding feeding ecology in dinosaurs is their anatomy, a realm that we explore in chapters 2 and 3.

CHAPTER 2

Bones, Teeth, Muscle, and the Study of Functional Morphology

Studying Dinosaurs

Functional morphology, or the study of how anatomical structures relate to an organism in a functional context, continues to be an ever-growing field in the realms of comparative anatomy and paleontology. It is a discipline that has been explored for practically all vertebrate groups in the scientific literature and is especially prevalent in dinosaur paleontology. Dinosaur functional morphologists have used both qualitative and quantitative methods in biomechanical studies. Qualitative studies include close examination and description of bone structure and joint surfaces and architecture, as well as reconstructed musculature to infer broadscale functional adaptations and evolutionary trends. Quantitative biomechanical and statistical studies delve deeper into the detailed minutiae of biomechanical performance and significance of individual and combined anatomical structures in certain specimens of interest. The rigor of quantitative biomechanical studies is especially helpful in dinosaur paleontology because it methodically builds upon our understanding of taxa, of which we may only have a select few specimens.

Functional morphology research of the musculoskeletal system is of major importance if we as paleontologists want to grasp the significance and magnitude of dinosaur evolution, diversification, and paleoecology. When performing these studies, one must consider that any given function (whether it be feeding, locomotion, defense, and so on) is a product of all anatomical features working on that particular system. Incidentally, in

no other functional system is the consideration of such anatomical integration more important than in the feeding system. Feeding function as a whole includes a combination of studies in bone and joint biology (external morphology, surface texture, internal microstructure, joint surface size and structure), tooth biology (external morphology, internal microstructure, dental macro- and microwear patterns), and muscle biology (overall anatomy, fiber types and lengths, biomechanical performance).

Of course, it would be an incredible feat to integrate all these disciplines into one study. The reality, however, is that functional morphology studies—especially ones that relate to feeding mechanics in particular—are driven by individual questions about a certain aspect of a given functional system (especially given the paucity of specimens available), which then require post hoc holistic thinking and discussion that integrate and (ideally) either verify or falsify what has been hypothesized in previously published studies. Hypotheses, which by nature must be falsifiable, are built on the basis of knowledge provided by in-depth morphological analysis of the specimens at hand, comparing them with other specimens and other taxa, as well as—and most especially—knowledge of the functional morphology or biomechanics of either related or functionally analogous living animals that would help give context and verification for hypotheses about extinct animals. The functional and biomechanical testing methodologies often use sophisticated (both two-dimensional, 2D, and three-dimensional, 3D) techniques and are also best when verified previously with extant animal studies. Here we present the general biological concepts and methods integrated in studying dinosaur anatomy and function, including osteology, dentition, soft tissue reconstruction, and biomechanical analyses.

Bones and Joints

There is no denying the awe-inspiring beauty and complexity of the vertebrate skull. The endless morphological diversity of the many bones making up such an essential structural entity is truly one of the most fascinating enigmas of vertebrate evolution. Both inside and out, its functions are countless, including but not limited to feeding, vision, olfaction (smelling), respiration (breathing), hearing, balance, and, most important of all, protecting the brain. The rest of skeleton is no different. Postcranial elements are packed full of intricate functional systems with so many key components to any vertebrate lifestyle, such as locomotion and housing organs (just to name a couple). It is no wonder why many paleontologists and anatomists love studying vertebrate skeletons so much! The nature of the challenges inherent in studying the diversity of vertebrate skeletal anatomy—especially in extinct animals—keeps us all guessing about the functions that each bone shape, bump, groove, and hole provides, or would have provided, in life.

Before we can even begin to think about feeding function in the extinct, non-avian dinosaurs, we first need to be confident in our understanding of what living vertebrates do with their skulls (especially the living dinosaurs—birds) and the rest of their bodies. Understanding the functional implications of the bones and joints in extant vertebrate heads, for instance, is essential in speculating about feeding function in extinct animal groups with similar traits. Various osteological (bony), arthrological (joint-related), and associated myological (muscular) complexes that relate to feeding system function in vertebrates include the mechanics of simple lower jaw elevation and depression; forward, backward, and rotational lower jaw motions; many forms of cranial kinesis (coordinated movements of individual cranial elements that aid in feeding); and intramandibular kinesis (movements at the mandibular symphysis—the chin—and other possible joints within the mandible). Morphological characteristics in dinosaur feeding systems account for much of their taxonomic and phenotypic diversity, and many aspects of them show functional similarities with living vertebrates. Most known feeding preferences are presented throughout Dinosauria (carnivory, piscivory, insectivory, omnivory, and herbivory), and each has its own set of adaptations presented in their bony anatomy.

Studying Dinosaur Bone Morphology

Dinosaur bones are remarkable things to work with for so many reasons, whether it be digging them up out in the field or studying them in museum collections (fig. 2.1), as they can provide clues about how dinosaurs lived their lives. The unfortunate reality we face as dinosaur anatomists (or, really, anatomists studying any other extinct animal group) is that because most dinosaur groups are long extinct, the only frames of reference for functional implications of their bones are the highly derived descendants of the theropod dinosaur lineage (birds), their close archosaurian relatives (crocodilians), other more distantly related reptiles (lizards), or any vertebrate megafauna that could be thought of as possible ecological analogues (megaherbivores like elephants or rhinoceros, for example). Our understanding of bone biology and joint function gives additional information to piece together functional and ecological adaptations dinosaurs may have acquired for their specific feeding, locomotor, or other habits.

Whatever the question may be, however, the fact remains that the fossilized bone we have to work with only gives us part of the story. This is where having a real knack for detective work comes in handy as a paleontologist. Every subtle texture in the bone—every bump, ridge, concavity, nook, and cranny—gives us another piece of the puzzle. From a purely morphological standpoint, studying the comparative anatomy of dinosaur bones not only gives us information about their taxo-

Figure 2.1. Skulls of *Triceratops* in a museum collection, surrounded by plaster jackets of other fossils.

nomic status (to determine whether a specimen represents a new taxon or one that already exists in the literature), but also allows us to understand the taxon's phylogenetic position (when comparing and coding all traits seen in the bones across dinosaur taxa). Although early dinosaur paleontologists primarily used comparative anatomical descriptions alone to assess relationships between dinosaurs, more recent morphological phylogenetic analyses improved this process. These phylogenetic analyses code all anatomical characters in data matrices that then calculate how similar or different various taxa are to one another. Together with those comparisons, researchers assess evolutionary relationships using parsimony (by interpreting the fewest steps it takes to evolve each combination of characters). The presence of convergently evolved characters (or analogous characters shared between distantly related taxa) may skew these results, however, and is also taken into account on a case-by-case basis.

Comparative anatomy lets us dive even deeper to understand the functional adaptations dinosaurs acquired throughout their evolution and, in turn, the potential behaviors they may have exhibited. Both comparative functional morphology and comparative behavior are what we are primarily interested in when it comes to

feeding. Functional morphology interprets the resulting biomechanical manifestations of the anatomy of an animal and is of particular interest to many vertebrate paleontologists. The anatomical diversity seen specifically in dinosaurs has led paleontologists to investigate the interconnected functional adaptations of bones, teeth, ligaments, joints, muscles, vasculature, nerves, integument, and other soft tissues to understand more and more about their feeding function, locomotion, senses, vocalization, and other physiological processes of the body. Drawing comparisons between anatomical structures also gives insight into the evolution of functional adaptations as well as the impact paleoecological pressures had on their evolution. Behavioral implications can be deduced from comparative functional morphological studies, which are necessary because we obviously are not able to observe extinct non-avian dinosaurs in the wild. The fossil record also provides at least some direct evidence of behaviors, allowing us to determine function, including tooth wear that indicates dietary preferences and jaw motion, and fossil tracks, indicating locomotor behavior.

In the past couple of decades, studies in dinosaur morphology and function have greatly benefited from computed tomography (CT). CT scanning has fundamentally transformed the way paleontologists are able to analyze bones, fossilized or extant, by creating high-resolution visualization of the internal anatomy of bones (including that of the brain cavity and sinuses within the skull) as well as 3D volumetric data of the outer surface of bone. If internal anatomy is not important to a particular study and one is only concerned about the outer surface anatomy, another option is 3D surface scanning of the outer surface of bone or a more recent technique known as photogrammetry, where many photographs are taken of a specimen from different angles then put together as a 3D digitization. Three-dimensional bone data are essential for quantifying the morphology of bone in the most accurate way possible.

A useful technique used for quantifying bone morphology itself and comparing it across taxa is geometric morphometrics (GMM). GMM is a highly sophisticated technique used to quantify and compare the external or internal morphology of bones with great accuracy. Using many landmarks indicative of informative characteristics of bone, this method helps in our understanding of the evolution of morphological variation, both inter- and intraspecifically. This approach can be done with 2D and 3D data sets, depending on the question asked as well as the resources at hand. GMM uses these biologically meaningful landmarks of features throughout the skeleton, standardizes them using a technique called Generalized Procrustes analysis (GPA) to minimize the impact of body size and specimen orientation of a specimen, and compares their relative distances from one another to quantify relative size and morphology of various characters across a range of taxa. It can measure ratios in size and volume as well as fluctuations in relative morphology both between taxa (to measure interspecific variation) as well within a particular taxon (to measure intraspecific variation) using multivariate statistics. GMM has more recently been applied specifically to feeding studies, such as in comparisons of ornithischian beak morphology (Mallon and Anderson, 2014a) and of skull morphologies across three distantly related herbivorous taxa (Lautenschlager et al., 2016). Further functional analyses of dinosaur bone morphology have

also been performed to elucidate mechanical advantages of jaw apparatuses with 2D lever arm techniques as well as 3D computer biomechanical modeling techniques, such as finite element analysis (FEA) and multibody dynamics analysis (MDA), both of which have been widely used in recent dinosaur feeding function studies and are discussed in more depth below.

As fascinating and important as the macroanatomy of dinosaur bones is to study, their microanatomy is just as fascinating and important. Paleohistology is the study of fossilized bone and tooth microstructure through thin-sectioning and microscopy. Dinosaur paleobiological studies have benefited greatly from paleohistological analysis since the very beginnings of dinosaur research (Mantell, 1841; Owen, 1840). Paleohistology addresses major paleobiological questions of physiological function and growth of these animals and provides more depth to anatomical descriptions of new dinosaur taxa. Patterns of bone matrix and vascularity can be easily detected histologically, as can remodeling and replacement of bone and lines of arrested growth (LAGs, which document growth processes in the bone throughout life). Paleohistology can also be informative of soft tissues that might have attached to bone surfaces, such as signs of cartilage types in joints and Sharpey's fibers that indicate keratinous attachment of ornithischian beaks or, at times, even muscle attachment sites. Furthermore, dinosaur feeding studies have benefited from paleohistological studies that describe tooth microstructure and provide implications for feeding anatomy and function (Sander, 1999; Hwang, 2011; Erickson et al., 2012; Brink et al., 2015, 2016; Button et al., 2017a).

Biology of Bones, Cartilage, and Joints

Bones are mineralized, structurally rigid connective tissues that together form the anatomical framework and support system of (most) vertebrates—the bony skeleton. Think of the steel frame of beams within skyscrapers, and you generally have a basic understanding of the overarching purpose of the skeleton. Other biomechanical functions of bones include the creation of joint surfaces with cartilage and ligaments as well as muscle attachments and anchoring to a much larger extent. Without bones, you would not be able to do most, if not all, of the things you do on a daily basis. The pivotal role of bones in our lifestyle, in turn, makes it easy to take such an amazing structure and its uses for granted.

Bone primarily consists of deposits of minerals called hydroxyapatite (a form of calcium phosphate) that give bone its incredibly rigid structure resistant to compressive forces. This calcium phosphate is incorporated into a layered matrix (lamellae) of the more flexible collagen protein as well as other proteins, salts, water, and neurovasculature. As bone is laid down, a network of woven bone stretches out to create an initial structural framework, followed by lamellae filling in the gaps. This interweaving of woven and lamellar bone creates what is known as fibrolamellar bone, and as a result, structured cells called osteons are formed. The first of these are called primary osteons, but as bone matures and remodels throughout life, secondary osteons start forming, arranged in concentric rings around neurovascular canals and creating what are called Haversian systems. Osteons, when bundled together in a dense mass, form compact bone, which reinforces the structural integrity of the bone as a whole by absorbing forces that are acting upon it. In contrast, the more porous regions within bone are called spongy bone (also known as cancellous bone), and they allow a bit more pliability in certain regions.

Cortical bone is what makes up the outer circumference of the bony element, while the medullary cavity lies within the core. During bone development and growth, bones consistently go through cycles of osteogenesis (with bone-creating osteoblasts) and bone cell removal (with bone-destroying osteoclasts) in order to continually grow and be modified in response to environmental influences. When bone cells created by osteoblasts are matured and embedded within spaces called lacunae in the matrix of bone, they are known as osteocytes.

There are two categories of bone types, which are named on their mode of formation: endochondral bone and intramembranous bone. Developmentally, both types are ultimately formed from a network of cells from me-

soderm—the middle of the three embryonic germ layers of vertebrates—called mesenchyme. But whereas endochondral bone is formed within cartilage originally produced in mesenchyme, intramembranous bone is formed directly from the mesenchyme itself. In addition to endochondral bone, the subcategory of intramembranous bone we focus on here is dermal bone (ossified mesenchyme within skin layers), as they are what make up most of the bones in the skull. Other intramembranous bones include sesamoid bones (bones formed within a muscle tendon, such as the patella, or kneecap), and perichondral and periosteal bones (formed by deeper cell layers).

Although bones are what make up a majority of the skeleton, there are other key structures that come into play functionally as well. The second most important kind, you will recall, is called cartilage. Cartilage, in contrast to bone, is a much more flexible connective tissue made of cells called chondrocytes, also within lacunar spaces. It is mostly derived from mesoderm, although some cartilage in the head is also derived from neural crest cells created in ectoderm—the outermost layer of the three embryonic germ layers of vertebrates. Unlike bone, cartilage is avascular. It is seen in joints and the junctions between bones as well as in the form of extensions from plates of bone. Hyaline cartilage is typically the most prevalent type of cartilage in the body. It is seen at articular ends of bones or even just as structural support (as in our nose) and has a smooth outer surface. Synovial joints typically possess hyaline cartilage boundaries within a capsule filled with a lubricant called synovium, allowing motion to occur between elements. Fibrocartilage consists of collagen fibers used to absorb compressive and tensile forces between two bones. Lastly, elastic cartilage is a much more flexible, typically seen to support structures such as the pinna of the ear in mammals (the entire framework of the external ear).

As mentioned previously—and as you will repeatedly see throughout this and future chapters—bones are connected both to each other as well as to muscles by different types of tissues. When bone is connected to bone, it forms a joint. Joints can come in all shapes and

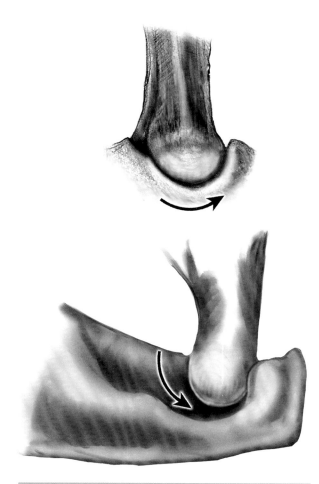

Figure 2.2. Jaw joints of two different dinosaurs showing representations of (*top*) a hinge (pivot) joint (in an early thyreophoran) and (*bottom*) a spheroidal (ball-in-socket) joint (in a hadrosaurid).

as well as peg-in-socket joints, both with the freedom to rotate in a circular manner in all directions. Saddle-shaped joints and ellipsoid joints also allow motion in multiple axes, but these are more constrained when moving in each direction owing to guiding ridges in different planes. Lastly, planar joints are those with one element sliding across a flat surface of another element.

Rigid connective tissues called ligaments act in stabilization between elements. Ligaments can come in fibrous, cartilaginous, as well as combined "fibrocartilaginous" forms, and each of these helps in allowing a certain amount of motion between bones (or restricting any motion from happening). Ligaments can be seen in mobile and nonmobile joints as well as sutures. Much of the structural support in joints depends highly on the extent and orientations of ligaments throughout the body.

Dentition

Teeth, when present, hold some of the most important clues into the diets of most vertebrates. Morphological diversity, microstructure, and occlusal wear patterns of teeth are all indicative of their taxonomic, functional, and paleoecological significance and are key anatomical structures for deciphering where an animal is along the trophic spectrum (from carnivore to omnivore to herbivore). The word "spectrum" is key here, as there are never any definite lines distinguishing an animal as, for example, an absolute obligate herbivore or carnivore. Still, the clues that dinosaur teeth hold lead us to a generally confident understanding of its trophic level (in most cases). Teeth are still great indicators of the ecology of most non-avian dinosaurs (and most toothed vertebrates as a whole) in any given ecosystem.

Teeth evolve largely to adapt to ecological pressures in an ecosystem, be it the types of prey available to a predator, the type of plant life available to an herbivore, or really anything in between. Carnivores tend to have sharper, more bladelike dentition ideal for slicing meat, while herbivores tend to have either "leaf-shaped" dentition or flat occlusal surfaces that are better for oral

sizes, each type serving a specific function. For instance, joints involving the junction between two flat sheets of bone include butt joints (with both elements joined against each other evenly), lap joints (with one element overlapping the other), scarf joints (with the edges of both elements wedged against each other diagonally), and serrate, or interdigitate, joints (with both elements interwoven at their junction, as in cranial sutures). When an element is allowed rotation about one axis against another element (by means of a synovial joint), it is what is known as a hinge or pivot joint, as seen in many animals with an orthal (up-and-down) motion of the jaws (fig. 2.2). There are also spheroidal (or ball-in-socket) joints grinding of different vegetation (depending on its toughness). Omnivory (eating both meat and vegetation) is more difficult to identify in vertebrate dentition, but it broadly presents with characteristics of both carnivory and herbivory within each tooth. The incredible diversity of dinosaur teeth shows all of these characteristics and tells a fascinating story about the evolution of feeding behavior across Dinosauria.

Studying Dentition

Vertebrate paleontologists have used teeth to decipher ancient vertebrate feeding behavior and paleoecology since the beginnings of the field. In the case of extinct non-avian dinosaurs, it is especially difficult to know feeding strategies given the fact that one cannot just observe these creatures in the wild—meaning paleontologists have to depend on what is known about the shapes of dentition in living animals (measuring and comparing how different types of teeth are used across various taxa). These comparisons then aid in the detective work necessary to understand what types of food different dinosaurs ate based on tooth morphology. Whether built for slicing meat with bladed and serrated ("ziphodont") teeth or built for grinding tough vegetation with so-called leaf-shaped ("phyllodont") or occlusally flattened teeth, comparative morphology of teeth and how they are arranged within the jaw are what paleontologists depend on to deduce diet and feeding strategies. Statistical analysis is also especially useful in this regard, such as in studies using tooth characters or even geometric morphometrics to compare the shapes of different teeth. In these analyses, teeth can group together in various morphospaces that can tell us whether an animal was more of a carnivore, herbivore, or generalized omnivore, especially if compared to the teeth of vertebrates with known diets.

In addition to overall anatomical morphology, paleohistological studies of dinosaur teeth are also incredibly useful for a number of reasons. By cutting miniscule slices of teeth and observing and measuring them through high-powered microscopes, paleohistological tooth studies can distinguish tooth tissue types and

their distribution throughout a specific tooth, giving insights into taxonomic and evolutionary trajectories (Sander, 1999; Hwang, 2005, 2010, 2011; Stokosa, 2005; Wang et al., 2015; Brink et al., 2016; Button et al., 2017a) as well functionalities given the various complexities of the teeth (Erickson et al., 2012, 2015). Paleohistology is also used to determine tooth growth and tooth replacement rates by observing daily growth lines (lines of von Ebner) created by the daily development of both enamel and dentine. Specifically, because dinosaur teeth replaced continuously throughout life, comparisons of von Ebner line counts are made between aged and newer teeth to estimate tooth replacement rates in various dinosaur taxa (Erickson, 1996).

In addition to microscopic anatomical studies, isotope geochemistry of dinosaur teeth has also been a significant methodology in elucidating the paleoenvironmental conditions that different dinosaurs may have adapted to and migrated through. Variable carbon isotopes of surrounding plant life (mainly that of C_3 plants—plants where carbon dioxide is fixed into a three-carbon compound prior to photosynthesis) and oxygen isotope ratios of surrounding surface water are recorded within teeth after an animal has ingested organic material from a specific region (Fricke and Pearson, 2008). This information leads to inferences of diet, paleoenvironmental adaptation, niche partitioning, migration, and more (Straight et al., 2004; Fricke and Pearson, 2008; Tütken, 2011; Goedert et al., 2016; Suarez et al., 2017; Cullen et al., 2020).

One of the most useful methods for inferring diet is through tooth wear studies—both microwear and mesowear (fig. 2.3). Dental microwear studies involve observing the occlusal surface of teeth at the microscopic level and inferring diet and feeding motions based on the morphology and orientation of the scratches and pits that are visible. Scratches are typically formed by harder plant materials an herbivore had been eating, such as those where the silica phytoliths within the plants would have left a clear mark on the tooth surface. Pits are usually formed through a grittier diet, such as in herbivorous browsers eating bark or consuming large amounts of dirt in the process (Walker, 1981; Teaford, 1988, 1994). The ratio of the number of pits versus scratches on a tooth surface indicate whether an animal is more of a browser, feeding on more woody plant material, or more of ground-level feeder on brush or other low plant material. In some cases, these results are also linked to changes in snout morphology, connecting wider snouts to more generalist feeders and narrower snouts to more selective feeders (Solounias and Moelleken, 1993; Whitlock, 2011; Mallon and Anderson, 2014a). Dental microwear studies can also use orientation of tooth wear to assess the directionality of the movement of the jaw during oral processing to determine the feeding mechanisms used by a particular animal (Weishampel, 1984a; Fiorillo, 1998; Rybczynski and Vickaryous, 2001; Williams et al., 2009; Varriale, 2011, 2016; Whitlock, 2011; Mallon and Anderson, 2014b). Mesowear, or macrowear in general (tooth wear

Figure 2.3. Illustrations of (*top*) dental microwear (with tooth scratches) on a hadrosaur tooth occlusal surface and (*bottom*) differential mesowear across ankylosaur teeth in a jaw line.

visible without a microscope), is also helpful in assessing feeding mechanisms, as it clearly indicates more forceful power strokes in focused orientations (Thulborn, 1974; Galton, 1986; Barrett, 2001).

Although teeth are such small anatomical structures, the amount of information they can convey to vertebrate paleontologists is almost limitless. From diet and feeding function to paleoenvironmental inferences, teeth hold many clues about the lives of extinct animals—especially dinosaurs. But what are teeth exactly, and what is it about them that makes them so unique in the clues they hold?

Tooth Structure

From the outside, the teeth of most tetrapods are composed of the crown, which projects out of the gumline (or gingiva), and a root, which lies lodged within the tooth socket (or alveolus) and is anchored by thickened collagenous tissues know as periodontal ligaments. The pulp is a connective tissue located within the tooth (the pulp cavity) that consists of neurovasculature that enters via an apical foramen beneath the root. This neurovasculature contributes to the growth and nourishment of the teeth.

On a more microscopic scale, three hard tissues make up a tooth. The first and hardest of all three is enamel, which is composed of acellular mineralized tissue that envelops and forms the outer platform of the crown. Calcium salt deposits create the enamel before eruption within the oral epithelium. The enamel-formed crown is the surface that comes in direct contact with the opposing teeth and processes food, so increased enamel hardness is key to tooth function. Dentin, the second hard tissue, is similar in structure to bone and lies deep to the enamel, forming the outer boundary of the pulp cavity and a large amount of the inner structural integrity of the tooth. Dentin is constantly deposited by odontoblasts within each tooth, and the resulting microscopic layers of dentin deposits are called "lines of von Ebner," which paleohistologists use to further understand the biology and life span of each tooth they investigate (see above). The third hard tissue in a tooth is known as cementum, which also grows in layers and coats ("cements") the surrounding surface of the dentin within the embedded root of the tooth. In some cases, as in many herbivores, cementum can even contribute to higher-crowned teeth for added structural integrity.

Developmentally, teeth form within the skin (epidermis and dermis) and erupt in bands of oral epithelium and dental lamina along the oral margin, with the crown developing first, followed by the root, and lastly the periodontal ligament and cementum. The enamel is formed in the epidermis by way of ameloblasts of the enamel organ, while the neural crest mesenchymal cells of the dermis form the dental papilla, which in turn produces odontoblasts that secrete the dentin. Reptilian teeth (including those of dinosaurs) are continuously replaced in alternating patterns throughout life—a condition known as "polyphyodonty" that is different from the mammalian "diphyodont" condition, in which there are only two sets of teeth grown throughout the life of the animal. The advantage of polyphyodonty over diphyodonty is that more, continuously replacing teeth are available throughout the life of an animal owing to excessive wear, age of the tooth, or breakage.

Among dinosaurs, the number of teeth is typically more or less conserved along the premaxilla and maxilla in the upper jaw and the dentary in the lower jaw. Although many reptiles also possess palatal dentition, this is not the case in any known dinosaur. The extent to which the dentition is retained or lost along any of the aforementioned elements is highly variable depending on phylogeny, function, or ecology. In many ornithischian dinosaurs, for instance, the premaxillary dentition is independently lost in multiple subclades. Moreover, the dentition can be lost or retained from either direction along the jawline, whether it be mesial to distal (front to back) or distal to mesial (back to front), and this can also contribute to paleoecological and functional variation in feeding mechanisms.

Three main modes of tooth implantation in the bone are prevalent across reptiles. Acrodonty refers to when the base of a tooth is fused to the oral margin of

the jawbone without being incorporated or anchored into an alveolus—a condition seen in only a few extinct reptilian groups, such as captorhinids and rhynchosaurs. Pleurodonty, a more prevalent condition in lepidosaurs (lizards and snakes), refers to when a tooth is anchored only by the labial (outer) wall of a dental groove because the lingual (inner) wall is nonexistent. In this case, the alveolus and its contents are visible on the medial surface of the tooth-bearing bone because its root is not bound completely. Of most relevance to this book is the condition known as thecodonty, seen in archosaurs (including all dinosaurs and crocodilians) and mammals. Thecodonty (also known as "gomphodonty") consists of a complete bony socket into which the root of a tooth is firmly affixed by a periodontal membrane and cementum both labially and lingually and with no exposed part of the root. This morphology adds extra anchoring support for teeth to withstand increased bite force and different motions of the jaw, which is ideal for dinosaurian feeding mechanisms (Fong et al., 2016; LeBlanc et al., 2017). Some dinosaurs, such as hadrosaurids and ceratopsids, add even more structural support of the dentition by creating a dental battery of up to hundreds of tightly packed functional and replacement teeth. The functional teeth in the dental battery form a single large occlusal surface for extra resistance to forces from processing tough vegetation (Erickson et al., 2012, 2015; LeBlanc et al., 2016; Bramble et al., 2017).

The external morphology of teeth is incredibly variable across all vertebrates and tells us a lot about dietary preference and function. In mammals, the crown is made up of many cusps, ridges, and basins that correspond to the morphology of opposing teeth. This is what constitutes precise occlusion, which is when the upper and lower teeth fit perfectly with each other in order to process food in certain directions. Reptiles, and specifically dinosaurs, for the most part do not show precise occlusion. Generally, teeth slide past each other, with the premaxillary and maxillary teeth occluding external to the dentary teeth of the lower jaw. A few minor exceptions to this pattern include the more direct occlusion of dental batteries of ceratopsids and hadrosaurids.

Dinosaur tooth rows are sometimes homodont (possessing similarly shaped teeth all along the oral margin) and other times heterodont or pseudoheterodont (teeth of varying specialized morphology along different regions of the oral margin for various feeding modes). Heterodonty, at least as it is broadly defined, is not as common in dinosaurs; however, it does exist in some cases (such as the extreme case in the aptly named heterodontosaurids). Many dinosaurs can be characterized by possessing a pseudoheterodont dentition, which changes only slightly along the tooth row (Hendrickx et al., 2019).

Muscles and Other Soft Tissues

When it comes to thinking about the biology of dinosaurs—and in particular from the perspective of the heads of dinosaurs—one great passion we (both authors of this book) share is the love of using illustration to investigate and visualize anatomy and function. In fact, for both of us, having an excuse to draw dinosaurs all the time is one of the most enjoyable parts of doing this type of research. The fusion of science and art in research is an excellent way to blend the intellectual thrill of the science you investigate with its overall aesthetic beauty in nature. This beauty, especially in the field of paleontology, is largely the reason we come to admire enigmatic ancient animals like dinosaurs and why we grow curious about them in the first place. As dinosaur paleontologists who also illustrate, we have realized that drawing the bones that are sitting in front of us is an exceptional way to learn how to pay close attention to their minute detail. Each bump, groove, ridge, and hole is important to document and measure, both when writing a description of a specimen as well as when trying to assess its complex anatomy in its entirety and when trying to reconstruct the soft tissue anatomy surrounding the bones.

As seen above, the bones and teeth that are actually preserved in the fossil record reveal a lot about the paleobiology and evolutionary relationships of dinosaurs.

Without all the soft tissues surrounding the hard tissues, however, one has scratched only the surface when thinking about the functional anatomy of these dinosaurs as living, breathing beasts. This is unfortunately the part of paleontology in which researchers are at the utmost disadvantage—muscles, ligaments, arteries, veins, nerves, and organs don't typically fossilize (save for a select few exceptions). This is where paleontologists need to do some serious, creative detective work using the fossil bones that are provided.

Soft Tissue Reconstruction

Bones oftentimes leave hints of soft tissue attachment or placement along their surfaces, which in turn can allow inference of presence or absence of soft tissues (muscles, joint structures, ligaments, neurovasculature) as well as breadth of attachment along a particular region of the bone (McGowan, 1979, 1982; Bryant and Seymour, 1990; Bryant and Russell, 1992; Witmer, 1995; Hutchinson, 2001a; Holliday, 2009). These clues are known as osteological correlates—visible, tangible manifestations of soft tissue attachment or placement marked along the morphology of a given bone (Bryant and Russell, 1992; Witmer, 1995). Some of the most common osteological correlates are those informative of muscle attachment sites. Depending on the nature and size of the muscle attached to a particular part of a bone, various muscle attachment types can be represented through a broad range of characteristics, including but not necessarily limited to striations or rugosities (identifying a more tendinous attachment; fig. 2.4) and ridges and smooth depressions in bone called fossae (implying the presence of larger fleshy muscle bodies). Additionally, muscles do not always leave a marking on bone and instead might have been attached to an already smooth surface. These absences of osteological correlates can create more difficulty in reconstructing and measuring exact muscle attachment size and, in turn, overall muscle architecture with accuracy, so extra precautions and considerations are necessary.

In addition to osteological correlates informing us

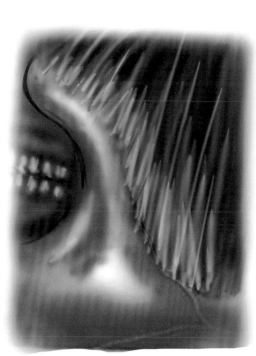

Figure 2.4. Example of a muscle scar on the medial aspect of a hadrosaur coronoid process with an illustration of how the deeper jaw muscles would have attached onto the hadrosaur jaw in life.

about individual muscular attachment sites, they are also vastly informative of the overall trajectory of vertebrate evolution. Extant phylogenetic bracketing (EPB; Witmer, 1995) is one of the most commonly used methods for reconstructing muscle anatomy as well as the anatomy of other soft tissues. EPB uses what we already know about the soft tissue anatomy of living animals most closely related to the extinct animal in question through dissection and imaging techniques. The evolutionary relationships among the living and extinct taxa are then used as context to deduce the presence or absence and placement of particular anatomical structures. For non-avian dinosaurs, this includes crocodilians, which are the closest living sister group to non-avian dinosaurs and are archosaurs themselves, and birds, which are derived forms of theropod dinosaurs themselves. These two create a "bracket" (albeit a very wide bracket) that provides initial insights into the anat-

omy. Anatomical information from other reptiles like the more distantly related lizards and turtles are also used in some cases where they share more similar osteological correlates with non-avian dinosaurs than some living archosaurs (Witmer, 1995).

The EPB brings confidence in soft tissue structure presence and architecture through examination of anatomical structures in the examined fossil bones and identifying homologous structures in the bones of phylogenetically closely related extant animals where the corresponding soft tissue is already known. Phylogenetic confidence in presence and architecture of soft tissue structures are classified from Level I through Level III. A Level I inference indicates the utmost phylogenetic confidence that a soft tissue structure is likely present because it, along with its osteological correlate, is present in both an extant sister taxon and the next more distantly (or "ancestrally") related taxon along the same clade. A Level II inference indicates that a soft tissue structure and its osteological correlate is present in an extant sister taxon, but it is not present in the next more distantly related extant taxon along the same clade. Finally, a Level III inference indicates the least phylogenetic confidence, with the soft tissue structure inferred without it present in the most immediate extant relatives. Additionally, in each of these levels, a prime (') designation might be added to indicate the absence of an osteological correlate. For instance, a fossil taxon with a designation of Level I' indicates that, although all related extant relatives of a fossil taxon may possess a soft tissue structure, the osteological correlate might be missing in the fossil itself (Witmer, 1995).

In addition to dissection, anatomists and paleontologists in recent years have developed new ways to view fine-scale soft tissue architecture (including individual muscle fibers and how they are arranged and attached) by iodine staining extant vertebrate specimens, including those of living archosaurs and lepidosaurs, and CT scanning at extremely high definition—a technique that has gained much traction and continues to produce some of the most beautiful and vastly informative 3D volumetric data sets (Gignac et al., 2016).

Given the fact that we consider what we know about the anatomy of living relatives of extinct dinosaurs and that we use that knowledge to elucidate dinosaur soft tissue anatomy, many details often lead to differences in opinion regarding the presence or exact placement of individual soft tissue structures like muscles. Certain bony structures, for instance, may either be homologous or convergent between two taxa, so it takes special care to deduce whether bony structures seen in the fossil indicate similarities in assumed surrounding soft tissues or whether they might have served a completely different purpose. Paying attention to surrounding anatomical landmarks that may correspond to the bony structure in question would also help one consider whether, for example, a muscle may be present either as an analog or a homolog to a known muscle in a living taxon. All possibilities must be considered in reconstructing soft tissue structures before determining what is the most likely reconstruction given

surrounding anatomy and whatever might be known of the functionality of any given body part.

Both 2D and 3D methodologies are essential to our knowledge of dinosaur paleobiology, and both complement each other beautifully. Whereas 2D studies have (qualitatively and quantitatively) compared muscular anatomy across wide ranges of phylogenetically related taxa, 3D modeling studies provide considerable depth in quantifying the fine details of individual specimens. By comparing muscular and other soft tissue anatomy (like that of joints, integument, sensory systems, and others) using all of these modeling techniques, paleontologists continue to grow closer to figuring what dinosaurs may have looked like in life.

Skeletal Muscle Structure

Muscular reconstruction is the most prevalent of all forms of soft tissue reconstruction in fossils because of its essential role in functional studies. Without muscles, a dinosaur would have been no more active in life than it is when on display at a museum. Skeletal muscles are striated and typically voluntary. They contract and shorten to act upon bones (although sometimes they act upon neighboring tissues such as integument) to create motion in a desired orientation about a particular joint (or multiple joints). Muscle bundles possess fibers composed of microscopic structures called actin and myosin protein complexes that link into each other to shorten the muscle during an entire contraction.

Muscles have attachment sites on bones, with each muscle typically showing at least one origin and at least one insertion site (although many have more than one of each). The insertion site mainly moves toward the origin during muscle contraction, but there are some cases where there is ambiguity in this distinction of muscle attachments. Sometimes muscles show fleshy attachments to bone, but many times they attach to bone with tendons. A tendon is a macroscopic fibrous connective tissue (which is often cordlike in structure) that runs within muscle to hold muscle fibers and creates an attachment of muscle to bone to act upon it. Sometimes tendinous fibers of muscle can be more sheetlike in morphology to attach along a much larger surface area of one or more bones, creating what is known as an aponeurosis.

Skeletal muscle bodies come in multiple forms. Pennate muscles have fibers that are oriented at an oblique angle relative to the tendon's line of action, allowing for more fibers and, in turn, a much greater amount of force, even within a smaller range. There are different types of pennate muscle. Unipennate muscle is when all of the angled muscle fibers are oriented in one direction on only one side of the tendon (much like a quill pen). Bipennate muscle is when the angled muscle fibers are seen on both sides of a midline tendon (with each side at an opposite angle to the other). When the tendon itself branches within the muscle, it is considered multipennate. Outside of pennate muscles, other types of muscles include fusiform (a spindle-shaped muscle with muscle fibers running parallel to the muscle's length), parallel (a more rectangular, elongate muscle with all muscle fibers running parallel to one another with consistent thickness throughout), convergent (a fan-shaped muscle with all of its fibers converging at a common muscle attachment), and circular (with concentric muscle bodies—sometimes forming a sphincter—surrounding an opening or recess). Each muscular form serves a specific function that acts in accordance with the integral muscle fiber orientations.

Most of the tendinous or fleshy attachment sites leave distinct scars on the bone, which are well recorded in the dinosaur fossil record. Muscle scarring can consist of either a smooth depression in bone, indicating the attachment of large muscle bundles, or striations or rugosities on bone, which are indicative of more tendinous (tendon-based) attachments. In some cases, no clear indication of muscular attachment sites is visible on fossilized bone, particularly in craniomandibular bone material. Therefore many of the attachments are estimated using the EPB. Holliday (2009), for instance, used knowledge of muscle attachment sites in living archosaurs and other reptiles to infer dinosaur cranial muscle anatomy, as discussed in chapter 3.

Biomechanical Analysis

Biomechanical modeling of dinosaurs, and in particular dinosaur skulls, takes a highly integrated knowledge of all of the anatomical factors discussed above in making a feeding (or any other) system work the way it might have all of those millions of years ago. Engineering techniques have been integrated in functional morphological studies for many decades, and it is within the principles of engineering that quantitative functional morphological studies are carried out. Three major engineering concepts help define the overall nature and function of vertebrate skulls and feeding systems: statics, kinematics, and dynamics (Weishampel, 1993).

Statics refers to the capacity of the skull or other bones to transmit and absorb forces running throughout their overall structure. This concept mostly depends on beam theory (modeling structures as a beam, applying loads perpendicular to its long axis to make it bend, and calculating characteristics of its deflection and ability to carry loads). The facial skeleton and mandible of dinosaurs have been modeled as statically loaded beams in numerous recent feeding studies and have also been subject to numerous studies of force transmission with lever systems (see below; Weishampel, 1993). Kinematics in feeding systems refers to the idea that many vertebrate skulls are inherently mobile and can be modeled as movable segments part of a network of bones. For instance, mobility of bones against one another within the skull creates a mechanism known as cranial kinesis and includes various kinetic mechanisms such as streptostyly and prokinesis seen in lizards and birds, respectively, as well as pleurokinesis, which has been proposed for ornithopod dinosaurs (Norman, 1984; Weishampel, 1984a; Norman and Weishampel, 1985; see chap. 12). Kinematics investigates the displacement patterns of various components within a system (Weishampel, 1993). Finally, dynamics models help analyze forces and movements of skull elements (the dynamic nature of the bones) and integrating moments of force acting on joints with displacement pathways into the model, thereby making it possible to assess dynamic performance of feeding (or any other) systems (Weishampel, 1993). All three of these approaches have set the standards for many quantitative biomechanical feeding studies in dinosaurs to various capacities.

Sophisticated quantitative biomechanical methodologies have substantially improved our understanding of dinosaur function and paleoecology, especially in the past few decades (e.g., Ostrom, 1966; Rayfield et al., 2001; Lautenschlager, 2020; Sakamoto, 2021; Cost et al., 2022; and many more listed throughout this book). Biomechanical analyses can be performed both using 2D and 3D methodologies. Two-dimensional methods are useful for looking into evolutionary trends and processes at a much grander scale, as seen below. Still, 2D methods have their limitations. The reality, of course, is that in life, animals work in 3D, with fluctuating structures of bone that can potentially alter how a system works. Three-dimensional methods can give a clearer picture of the animal itself, especially when it comes to elucidating the interplay between size and shape, and its effects on various functions, such as true bite force in feeding studies. These issues call for 3D computerized approaches in quantifying form and function in indi-

vidual specimens that are studied in greater depth, especially with 3D surface scans and CT scans (Cost et al., 2022). For all of these reasons, it is important to consider what the question is for a particular study as well as what resources are available to use in performing the study.

Photographs of specimens in various orientations are used to perform 2D biomechanical analyses. To study any musculoskeletal system, for instance, 2D lever arm mechanics methodology can be implemented (fig. 2.5). The mechanical advantage, or leverage, of a reconstructed muscle can be calculated from bone morphology and orientation of the corresponding muscle. To calculate the mechanical advantage, one needs to determine the length of the moment arm (or the perpendicular length from the fulcrum to the vector of the muscle, known as orientation of force) and divide that number by the length of the output lever (or the distance between the fulcrum to the point of force). One unfortunate downside to this methodology is that, when looking at fossil skulls in two dimensions, there is no way to get a clear-cut number for the size of the muscle (and, by association, its strength) because there is no way

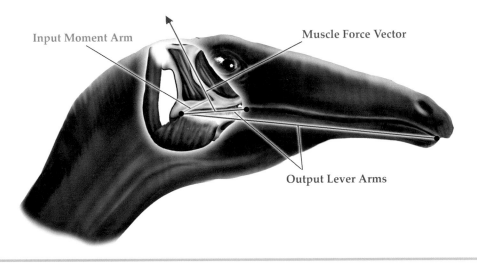

Figure 2.5. *Coelophysis* head with a portrayal of two-dimensional mandibular leverage. The red bar represents the adductor muscle force vector, the green bar represents the input moment arm, and the two blue bars represent the distance from the jaw joint to the most mesial (front) and most distal (back) bite points. The long distance to the most mesial bite point suggests that the long snout of *Coelophysis* was built more for speed rather than sustained higher bite force.

to measure the physiological cross-sectional area of the muscle body. The muscle just is not there anymore! For this reason, the results of many 2D lever arm mechanics analyses are based on a constant musculature unit for all specimens used in a particular study—a method that takes the size of the animal out of the equation. An upside to this method, however, is that biomechanical performance of a functional unit can be traced through evolutionary changes in dinosaur skull shape.

Two-dimensional lever arm mechanics methodology has been used in dinosaur feeding studies for decades (Ostrom, 1964a, 1966; Tanoue et al., 2009b; Sakamoto, 2010; Molnar, 2013; Mallon and Anderson, 2015; Nabavizadeh, 2016). In a given feeding system viewed in profile (or lateral view), the craniomandibular joint (or, more precisely, the joint between the quadrate bone of the cranium and the articular joint of the mandible) can act as the fulcrum, with various points along the tooth row where output levers are directed from that fulcrum, since the bite point is where all of the force is implemented. In many of the aforementioned studies, the m. adductor mandibulae externus (mAME; see chap. 3) is the muscle group reconstructed to calculate the muscle vector with its corresponding attachment sites. Two-dimensional lever arm analyses with these coordinates result in the mechanical advantage of the jaw, or what some refer to as the relative bite force (RBF) when put in an evolutionary context.

Mandibular mechanical advantage has also been calculated in a different way with an overhead or dorsal view in ceratopsian skulls by Tanoue et al. (2009b) using a method from Greaves (1978) to understand the relationship of tooth placement and bite force in horses, giving a different perspective on mechanical advantage. Additionally, modeling a mandible as though it acted as a beam under bending loads is another way paleontologists have attempted to get a more 3D perspective with 2D views of a vertebrate jaw. Researchers can calculate the cross-sectional geometry of a given region of a bone to see how much stress it can take in different directions before breaking. For instance, in the case of the mandibular symphysis, loading stresses include those created

from wishboning (with both ends of the two hemimandibles pulled away from each other), shear (with the two dentaries moving against each other at the symphysis), and torsional forces (with at least one side rotating around its long axis) (Therrien et al., 2005).

Because 2D views of specimens limit the ability to analyze possible cranial kinesis, Weishampel (1984a) pioneered a computerized methodology using advanced linkage systems. These linkage systems modeled various joint structures within the skull in multiple views to get a 3D perspective in testing hypothesized cranial kinesis in ornithopods (the pleurokinetic model). This method brought new light to the idea that movements of bones against each other can be quantified, making it possible to test cranial kinesis in a variety of vertebrates. Kinetic linkage systems and associated mechanical advantages of different joints have since been analyzed in much more depth in many vertebrates, including modern birds (Olsen and Westneat, 2016; Olsen, 2019).

Since the early 2000s (Rayfield et al., 2001), dinosaur paleontologists have increasingly been performing biomechanical analyses using a method known as finite element analysis (FEA). FEA is a method used by engineers to test how well different structures perform under various forces, specifically looking into the stresses (how forces are distributed across a structure), strains (how much a structure is stretched or pulled), and deformations (distortion through movement of a structure) that these structures go through with a given external force (fig. 2.6). FEA is a method that has been used in a large range of fields, from engineering and industry to medicine (Rayfield, 2007). Paleontologists have used it in testing the function of fossilized (and extant; e.g., Porro et al., 2011) bone, with truly enlightening results.

The complexity of any given shape can vastly complicate how much stress, strain, and distortion it can go through with external forces. With FEA, one can take a given structure and break it down into much smaller components of equal size and shape, the connections of which form what is known as a mesh. A force is applied to the mesh, and depending on how it is transferred throughout the individual components of the mesh,

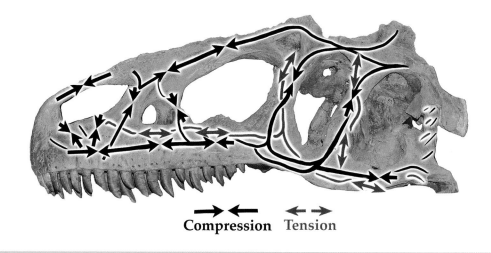

Compression   Tension

Figure 2.6. Illustration of stresses and strains (compression and tension) through an *Allosaurus* skull, based on finite element analysis by Rayfield et al. (2001).

stresses, strains, and deformations are quantified. FEA of bony anatomy can be performed both in two and three dimensions. Analysis of 2D meshes, although not nearly as comprehensive, can be done much more efficiently and with a much greater number of specimens with photographs and drawings because there are fewer data points the computer needs to solve. Three-dimensional meshes can be created with surface laser scans, photogrammetry, and CT scans. As one might imagine, these kinds of data are more involved, with exponentially higher complexity and many more components working against each other in various ways. Because of this complexity, computerized 3D analyses tend to take much longer than 2D analyses.

Once a mesh is formed, it has specific material properties—internal structural properties that result in varying stiffness of an object (from rigid to elastic)—that will determine how a structure will change with external force. The more rigid an object is, the more it is capable of deformation with sufficient force. The more elastic an object is, the more flexible a structure will be, and in turn the less likely it is to be deformed with external force. In analyses of fossilized bone, the material properties are not preserved, leading paleontologists to use the properties of the most closely related extant animal. In the case of dinosaurs, the material properties of croco-

dilians and birds are mostly implemented for analysis (with some exceptions), which itself is problematic owing to the different ecologies of extinct dinosaurs. Regardless, just like in any soft tissue reconstruction, paleontologists need to work with what is available to them.

Once constraints are set up on the model based on the reconstructed muscle forces, external forces are applied to calculate the stresses and strains distributed across all the components throughout the mesh. Numerical values are obtained for stresses and strains in the model, and the resulting stresses and strains within the bone are visible as a representative color scheme. Regions with high stresses and strains turn to warmer colors (from yellow to red at the warmest), whereas low stresses and strains turn to cooler colors (from green to blue at the coolest). With these data, paleontologists can elucidate what kinds of feeding behavior (or any other behavior, for that matter) could be plausible in different taxa and what areas of the skull are under compression or tension with different feeding mechanisms. The more stresses or strains throughout a skull with a particular modeled behavior, the less likely it is that the animal actually made those specific actions. Although there is clearly nothing simple about the use of FEA, the intuitive nature of the final output results, both visually and numerically, has allowed dinosaur paleontologists to explore feeding behavior in all types of dinosaurs (Rayfield et al., 2001, 2007; Rayfield, 2005a, 2005b, 2011; Bell et al., 2009; Young et al., 2012; Lautenschlager et al., 2013, 2016; Cuff and Rayfield, 2015; Lautenschlager, 2017).

FEA is not the only highly sophisticated engineering computational modeling strategy used for functional studies in dinosaurs. Multibody dynamics analysis (MDA) takes modeling dinosaur function a step further by allowing paleontologists to simulate movements of 3D volumetric bone models induced by reconstructed muscle forces (Lautenschlager, 2020). MDA allows multifaceted functional systems (bones, joints, and muscles) to be linked and work together in three dimensions to simulate how a functional system (like the craniomandibular feeding system) would work with different bone and joint structures as well as different muscle forces acting on it in real time. Additionally, forces at any given bite point, joint reaction forces, and even range of motion can be calculated on the basis of all the coordinates applied to the model (Lautenschlager, 2020).

One of the advantages of MDA is that one needs only volumetric data of a particular specimen in order to perform the analysis (as opposed to FEA, which also requires input about the biology of bone itself). MDA software provides the possibility of re-creating external soft tissues as needed and then simulating what happens when the system starts to work with all of the external components acting on it. Biologically accurate movements and flexibility of muscles and joints as well as other properties are all incorporated into the model (Lautenschlager, 2020). These components make MDA a complex yet vital method that paleontologists have recently started implementing and will continue to use for years to come in feeding studies (Bates and Falkingham, 2012; Snively et al., 2013; Lautenschlager et al., 2016) and locomotion studies (Sellers et al., 2009, 2017; Bates et al., 2012; Maidment et al., 2014a).

The possibilities are seemingly endless when it comes to the resources available for exploring functional morphology in dinosaurs. The studies of the past few decades have given paleontologists so much to work with, allowing them to get closer to understanding dinosaurs as living, functioning organisms rather than just fossils. The interplay between form and function is a phenomenon that has intrigued anatomists from the very beginnings of the field. The more sophisticated techniques are used to elucidate this interplay, especially with the constant advances in computer modeling technology, the more we can learn about how these animals lived their everyday life.

CHAPTER 3

An Overview of Dinosaur Anatomy

Anatomical Orientations

In this section, we explore the general anatomy of dinosaurs, particularly the osteology and soft tissues of the heads of dinosaurs (with key focus on biomechanical components that aid in the feeding system) in addition to important aspects of their postcranial anatomy. This overview introduces the basic structure and terminology included throughout this book.

Before delving into descriptions, it is essential to understand anatomical terminology. Using terms like "up," "down," "above," "below," "forward," "backward," "right," "left," and others is typically frowned upon in the anatomical sciences. Although some of these terms may show up from time to time in future pages for the sake of ease, we mainly use formal anatomical terminology throughout. The following is an overview of these terms:

Directionality throughout the Body

Rostral: toward the nose, when discussing orientation within the head ("forward").

Cranial/Anterior: toward the head, when discussing orientation throughout the body ("forward").

Caudal/Posterior: toward the tail, when discussing both the head and rest of the body ("backward").

Dorsal/Superior: toward the skull roof or spinal column of an animal ("above").

Ventral: toward the underside of an animal ("below").

Medial: toward the midline or inner part of the subject matter.

Lateral: toward the sides relative to the subject matter.

Proximal: toward the center of the torso of an animal.

Distal: toward the outer extremities relative to the torso.

Superficial: a layer of tissue closer to the external surface of the animal.

Deep: a layer of tissue that is more internal.

Directionality within Teeth

Mesial: toward the rostral-most extent of a tooth or tooth row.

Distal: toward the caudal-most extent of a tooth or tooth row.

Labial/Buccal: the lateral side of dentition (toward the side with extra-oral tissue).

Lingual: the medial side of dentition (toward the inner oral cavity, where the tongue is located).

Apical: toward the tip of the tooth.

Basal: toward the base of the tooth.

Skull Morphology, Teeth, and Feeding Mechanisms

Cranial Fenestrae

All dinosaurs, which include living birds, possess what is known as a diapsid skull, meaning there are two fenestrae (windowlike holes; singular is "fenestra") caudal to the orbit, inside of which the brain, cranial musculature (including mainly jaw musculature), and other soft tissue structures are protected within the cranium. Di-

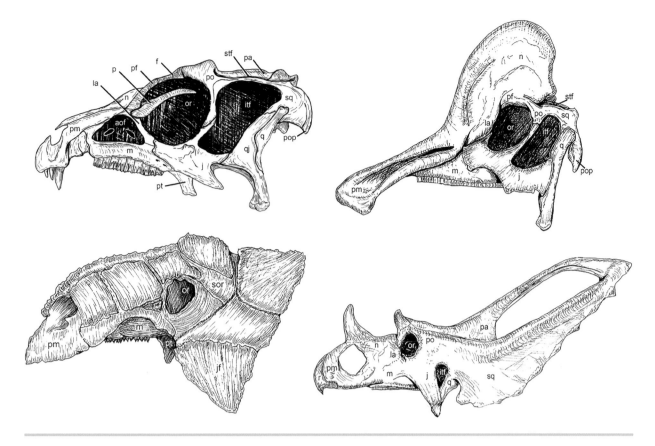

Figure 3.1. Bones of the dinosaur cranium seen in four different ornithischians (*clockwise from top left*): *Heterodontosaurus*, *Corythosaurus*, *Chasmosaurus*, and *Euoplocephalus*. Abbreviations: aof, antorbital fenestra; f, frontal; itf, infratemporal fenestra; j, jugal; jf, jugal flange; la, lacrimal; m, maxilla; n, nasal; or, orbit; p, palpebral; pa, parietal; pf, prefrontal; pm, premaxilla; po, postorbital; pop, paroccipital process; pt, pterygoid; q, quadrate; qj, quadratojugal; r, rostral; sor, supraorbital; sq, squamosal; stf, supratemporal fenestra.

apsid skulls are also seen in all other living archosaurs (crocodilians and birds) as well as lepidosaurs (lizards, snakes, and tuataras). In contrast, the other major amniote condition is the possession of a synapsid skull, with only one fenestra behind the orbit, as in the common ancestor of the mammalian line. This distinction in cranial morphology, along with being a key feature in the identification of vertebrate groups, marks a huge divide in the function of the jaws and jaw musculature in feeding mechanics between most diapsids (including dinosaurs) and synapsids (mammals).

At first glance, dinosaur skulls in lateral view are instantly recognized by the fenestrae composed of the bounding of complex articulations of skull elements (fig. 3.1). The pair of fenestrae that give dinosaurs their diapsid nature, both the supratemporal and infratempo-

ral fenestrae, are placed most caudally in the cranium for placement of the jaw musculature (allowing room for bulging muscles during contraction) and are separated by the supratemporal bar. The supratemporal fenestra is located dorsally, and the infratemporal fenestra is positioned just ventral to it on the lateral side of the skull, mostly closed off ventrally by a bony bar known as an infratemporal arch. The infratemporal fenestra is often either vertical or obliquely angled with a caudodorsal orientation and is usually either quadrangular or ovoid (as is the supratemporal fenestra).

There are several other cranial fenestrae in diapsids. Perhaps the most recognizable fenestra is the orbit, placed just caudal to the muzzle. Present in all vertebrates, the orbits house the eyes and are usually larger and circular or ovoid in shape. Typically, in many cases, such as in ceratopsians, the orbits tend to decrease in relative size as genera become larger. Dinosaurs also possess an antorbital fenestra rostral to the orbit, which is usually somewhat triangular. This is more prominent in theropod and sauropodomorph dinosaurs and is typically much more reduced in early-diverging ornithischians or absent in most other ornithischian dinosaur species (Osmólska, 1985; Witmer, 1987, 1997). Some theropods variably also possess other smaller cranial fenestrae rostral to the antorbital fenestra known as the maxillary fenestra and, smallest of all, the premaxillary fenestra. The external naris is a fenestra for entry into the nasal passage toward the rostral tip of the snout (especially large in ceratopsids). The lower jaw commonly contains fenestrae as well, including a mandibular fenestra (external and internal), seen in many taxa, and, less commonly, a surangular foramen.

Embryological Components of the Skull

As mentioned, the skull has many functions, and as such, structural integrity and morphological accommodations for these functions are crucial. As in all vertebrates, the majority of skull elements are paired, with a few exceptions that will be mentioned accordingly.

Bony elements of the skull are often easily categorized in terms of three main embryological components that produce multiple modular complexes, both between and within them. The first is the neurocranium (also known as the chondrocranium), which is made up of endochondral bone. The neurocranium is primarily the bony hub that houses the brain and major special sensory organs. The second component discussed here is known as the dermatocranium, which is made of intramembranous dermal bone and creates the majority of the exterior of the cranium and elements of the upper muzzle and lower jaw that hold dentition. The third component is known as the splanchnocranium, which is also made up of endochondral bone and is what, embryologically, creates the jaw joint, the auditory apparatus, the hyoid, and other components originally made up of the branchial gill arches.

Braincase and Palate

The braincase is highly variable across Dinosauria, both in relative size and shape. It is made up of a collection of bones that together form a chamber for the brain to occupy. These bones include the basioccipital, sphenethmoid, basisphenoid, laterosphenoid, presphenoid, orbitosphenoid, prootic, opisthotic, supraoccipital, and exoccipital. Much of the occipital region has been found to be evolutionarily strongly correlated throughout Dinosauria and archosaurs as a whole, which speaks considerably to their biomechanical necessity (Watanabe et al., 2019). As it is not pertinent to our current understanding of feeding function, most details of anatomical variation of each of these bones in general will not be discussed here.

There are a few key aspects of the braincase, however, that are important to ongoing discussions of muscle attachments and possible cranial kinesis and jaw function in general. Cranial kinesis is the movement of intracranial elements against each other, a mechanism seen in birds (avian dinosaurs; see chap. 7) and lizards. For instance, the basisphenoid articulates with the pterygoid bones of the palatal region to form a synovial joint known as the basal or basipterygoid joint, although some have argued that this joint was likely mainly immobile in non-avian dinosaurs (Holliday and Witmer, 2008). The laterosphenoid, along the dorsomedial wall of the orbit, articulates with the postorbital bone, a joint that is agreed upon as being synovial (Norman, 1984; Weishampel, 1984a; Norman and Weishampel, 1985; Holliday and Witmer, 2008; Bailleul et al., 2017), although it is now also suggested as being a mostly immobile joint, except to ensure that mobility of the braincase against the skull roof is minimal (Holliday and Witmer, 2008; Bailleul et al., 2017). In some taxa, the laterosphenoid possesses a joint with the epipterygoid, creating the epipterygolaterosphenoid joint (say that five times fast!). This joint seems to be lost in hadrosaurs and ceratopsids (and maybe even in sauropods and theropods) and possibly replaced with soft tissue (Holliday and Witmer, 2008). The exoccipital is also an important bone to note for jaw function, as it forms the caudolaterally oriented paroccipital process that is roughly triangular to squared-off in shape and is the site of origin of m. depressor mandibulae, an important muscle used in opening the mouth during feeding (see below).

Ventrally, the skull is composed of various elements that make up the palate. (Press your tongue up against the roof of your mouth, and you've found your palate.) Broadly speaking, the palate is the dividing cluster of bones between your nasal cavity and oral cavity. Centrally, they are formed from the palatine and pterygoid bones (with laterally attached ectopterygoids) and primarily incorporate the ventral aspects of the vomer and premaxillae rostrally and the single midline parasphenoid bone caudally. The pterygoid bones are especially important for origination and anchoring of the m. pterygoideus complex, both dorsally and ventrally (see below). In most dinosaurs, and most reptiles in general, the palate is slender and triangular, coming to a point rostromedially in ventral view. Of note, the oddballs in terms of the palate are the ankylosaurs, which possess a secondary palate as well, creating highly complex narial anatomy (Coombs, 1978; Witmer and Ridgely, 2008).

Snout, Skull Roof, and Postorbital Region

There are two cranial elements known to hold dentition. The premaxilla is the rostral-most cranial element in most dinosaurs, making up the tip of the snout or beak (aside from the additional rostral bone in ceratopsian beaks). It shares a border with the external naris and nasal dorsally. Caudally, it articulates with the maxilla, nasal, and vomer. The premaxilla contains the ventrally oriented dentition in many taxa, though multiple herbivorous ornithischian taxa secondarily lose this premaxillary dentition. The premaxilla takes on many shapes and sizes, from slender to broad. An extreme case of this is seen in the "duck-billed" hadrosaurs, with a proximally slender but distally broad bill. The premaxilla in ceratopsians, specifically, does not form the rostral-most edge of the cranium. Instead, the premaxilla caudally borders the rostral bone, an unpaired, median element unique to this group.

The other more prominently tooth-bearing cranial element is the maxilla, which is located just caudal to the premaxilla and creates the ventral border of the antorbital fenestra. In addition to the premaxilla (bordered rostrally and dorsally), the maxilla is also bordered by the palatine and vomer medially, the ectopterygoid caudally, and the lacrimal and jugal caudodorsally. A distinct buccal emargination of the ventral oral margin creates a medially inset tooth row in most ornithischians, which likely allowed room for soft tissues (but more on that later). The medially inset tooth row is relatively straight in nature and continues caudally medial to the ventral jugal process in most ornithischians but is largely continuous with the jugal in sauropodomorphs and theropods.

The nasal bones create the dorsal roof of the snout, bordering the dorsal edge of the maxilla. The dorsal edge of the nasal continues the rostral tapering of the skull in dorsal view and is situated between the paired premaxillae. Dorsally, the orbit is overlapped laterally by the caudodorsally oriented palpebral bone (Maidment and Porro, 2010), which arises from the lacrimal, a small element that forms the rostrodorsal margin of the orbit, separates the antorbital fenestra from the orbit, and is bordered ventrally by the jugal. The jugal is a large linking element that creates the ventral edge of the orbital region and a majority of the postdental extent of the ventral margin of the cranium. The jugal is highly variable among dinosaurs (Sullivan and Xu, 2017) and often exhibits a rounded to triangular projection extending caudoventrally from it. A smaller bone known as the quadratojugal often connects the jugal to the quadrate, the caudal-most element of the cranium. The prefrontal, frontal, and postfrontal bones form the rostral margin of the cranial roof caudal to the snout of the animal, with the supraorbital and postorbital bones articulating ventrolaterally, creating the dorsal margin of the orbit. The parietal articulates caudal to the frontal to form, along with the squamosal, most of the caudodorsal roof of the cranium. Of note is the caudally expanded nature of the ceratopsian parietosquamosal region, forming a "frill" at the back of the head.

The squamosal is ventrally joined with the dorsal articular surface of the quadrate bone through a junction known as the otic joint. This joint has been argued to be mobile in some dinosaurs such as ornithopods (Norman, 1984; Weishampel, 1984a; Norman and Weishampel, 1985), although morphological studies (Holliday and Witmer, 2008; Rybczynski et al., 2008) and a recent histological study of extant archosaurs (Bailleul et al., 2017) have suggested otherwise. The quadrate bone extends ventrally (or, at times, rostroventrally) to join with the glenoid surface of the articular bone of the mandible to create the synovial craniomandibular joint, which takes on many forms in accordance with individual functions. The quadrate also forms an internal wing that articulates with the pterygoid bones to form a continuous surface internally for further muscle attachments

(such as m. adductor mandibulae posterior; see below). The morphological diversification of the distal end of the quadrate is highly correlated with the evolutionary shaping of the jugal-quadratojugal region, indicative of ultimate infratemporal fenestra morphology and jaw articulation (Watanabe et al., 2019). Internal to the quadrate sits the stapes, a cranial element involved in transmitting auditory signals from the outside world to the side of the braincase (Colbert and Ostrom, 1958).

Mandible and Hyoid

The mandible, or lower jaw, is composed of multiple elements (fig. 3.2). The largest bone in the mandible is the

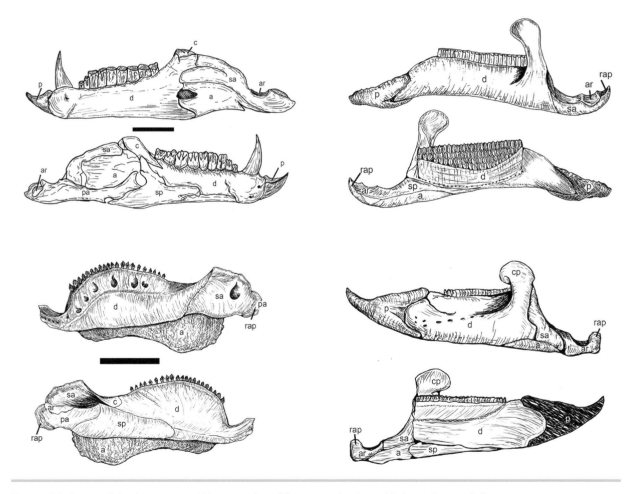

Figure 3.2. Bones of the dinosaur mandible seen in four different ornithischians (*clockwise from top left*): *Heterodontosaurus*, *Corythosaurus*, *Chasmosaurus*, and *Euoplocephalus*. Abbreviations: a, angular; ar, articular; c, coronoid; cp, coronoid process; d, dentary; p, predentary; pa, prearticular; rap, retroarticular process; sa, surangular; sp, splenial.

dentary, which is rostrocaudally elongate with a variably gracile to robust body (depending on the taxon) and articulates with its counterpart rostrally at the mandibular symphysis. It holds all of the mandibular dentition and articulates caudally with both the surangular (positioned dorsally) and angular (positioned ventrally)—an articulation that, in theropods, creates what is known as the intramandibular joint. The surangular often has a dorsally extended margin that is either rounded to triangular in shape (the coronoid eminence) or even, in the case of ceratopsids and hadrosaurids, possesses a tall, columnar process known as a coronoid process that bears an expanded apex. Both the coronoid eminence and coronoid process (which are also herein sometimes collectively called "coronoid elevations") act as insertion sites for m. adductor mandibulae externus (see below). Other elements include the coronoid bone (absent in some dinosaurs and typically seen at the dorsal margin of the coronoid elevation), the prearticular, and the articular, which is the bone that forms the main glenoid surface of the jaw joint articulating with the quadrate, as described above. Note that the articular comes from the splanchnocranium and not the dermatocranium as in the rest of the mandible (which makes sense, since it is part of the jaw joint).

Ornithischian dinosaurs are unique in possessing an additional unpaired, midline element known as the predentary, which is located at the rostral tip of the mandibular symphysis (articulating rostral to the tips of the dentaries), creating the ventral part of the beak. The predentary has been suggested to have important kinetic functions associated with the dentaries in feeding mechanics of various ornithischian clades (as discussed in later chapters; see also Nabavizadeh and Weishampel, 2016, for a full review).

Lastly, one set of elements that is rarely found in dinosaurs is the compilation of hyoid bones that come from the splanchnocranium (paraglossals, two ceratobranchials, as well as other elements such as basihyals and ceratohyals). Hyoid elements are suspended ventral to the skull by muscles that aid in tongue and throat function (Li and Clarke, 2015).

Dentition

Teeth can be one of the first indicators of dietary preference of a dinosaur. Distinguishing carnivores from herbivores or omnivores can vary in difficulty depending on the taxon. Although the earliest forms of reptiles possessed teeth that were conical, tooth shape became particularly more sophisticated in dinosaurs (fig. 3.3). Most of the carnivorous dinosaurs (all of which are members of Theropoda) largely possess labiolingually compressed teeth with serrated edges (carinae) and come to a sharp point apically. This bladelike, serrated tooth morphology is known as "ziphodont," and it is well designed for slicing the meat of prey (Holtz, 2003). In many cases, ziphodont teeth are variably hooked distally at the apex, making it an effective trap for prey struggling to escape. The serrations (denticles) of ziphodont dentition are variable in form, ranging from tall, pointed serrations to low, plateau-like serrations, implicating a broad range of functionalities (D'Amore, 2009; Brink et al., 2015; Torices et al., 2018). Many other dental features are also present throughout Theropoda, such as size variation along the tooth row, constriction at the base, various orientations and twisting of carinae and ridges, longitudinal grooves along the crown surface for extra grip (such as in "fluted teeth"), and even edentulism (loss of

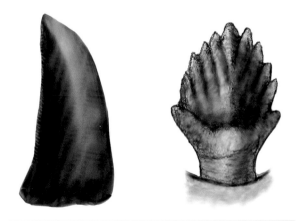

Figure 3.3. Dentition in dinosaurs, showing (*left*) one tooth that is ziphodont (from an allosauroid) and (*right*) one tooth that is phyllodont (from an ankylosaur).

teeth altogether). Some of these features and their combinations within a tooth give us clues about dietary preference and function (see Hendrickx et al., 2019). Some carnivorous theropods (tyrannosaurids and *Allosaurus*) go the extra mile, growing a "pachydont" dentition, which consists of elongate teeth that are much thicker labiolingually (roughly 60% of the mesiodistal length) and distally recurved. *Tyrannosaurus*, in particular, possessed some teeth that were deeply rooted and about one foot in length—able to withstand their powerful bite forces. Additionally, some other theropods, most notably spinosaurids, possessed thinner, more conical ("conidont") teeth with minimal to no denticles as well as a fluted crown, possibly indicating a change to a preference of fish in their diets (Holtz, 2003; Hendrickx et al., 2019).

Herbivorous dinosaurs had a much broader range of tooth morphologies, but generally, most possessed what is known as "phyllodont" or "folidont" maxillary and dentary teeth (or, more colloquially, "leaf-shaped" teeth) with denticles along the apical margin and sometimes an apicobasally oriented midline ridge on the outer surface. This morphology is ideal for slicing and cropping vegetation and is seen in early diverging sauropodomorphs as well as in some heterodontosaurids, thyreophorans, non-hadrosaurid ornithopods, pachycephalosaurs, and non-ceratopsid ceratopsians. In the cases of ceratopsids and hadrosaurids (and some derived heterodontosaurids), the teeth are occlusally more flattened and packed together, with ceratopsids and hadrosaurids showing extremely packed columns of dentition that form dental batteries made with a conglomeration of various tissues (Erickson et al., 2012, 2015; LeBlanc et al., 2016). Their physical properties make them well suited for applying powerful grinding forces during feeding. Additionally, sauropod dinosaurs show a range of tooth morphologies adapted for plant cropping, from the more cupped, spatulate teeth of macronarians to the pencil-like teeth of diplodocoids. All of these morphologies are highly variable in their presentation depending on the taxon, but each one serves to nip or process tough vegetation. Premaxillary teeth are generally smaller and suited for nipping in initial acquisition of the food, whereas cheek teeth are used more for the actual act of oral crushing or processing. Omnivorous dinosaurs are more difficult to pinpoint along the trophic spectrum because their teeth tend to show traits characterized for both carnivory and herbivory. It is this type of detective work that makes studying these animals ever the more fascinating.

The Basics of Feeding Mechanisms

With the help of dental microwear (discussed previously in chap. 2), orientation of jaw movements can be discerned based on orientation of scratches on the occlusal surface, leading to studies that try to understand how the feeding apparatus as a whole would have worked in order for it to move in those directions. These various feeding motions will be referred to throughout the rest of this book and include the following.

Orthal motion refers to the dorsoventral (up-and-down) movement of the teeth in occlusion. This motion is seen in most dinosaurs, at least acting as one part of the feeding mechanism, if not the only part (depending on the taxon).

Palinal motion refers to caudally oriented movement of the teeth while in occlusion. The term "palinal" is sometimes conflated with "propalinal," meaning "fore-aft" movement (with the additional "pro" coming from the word "proal," which is a rostral, or forward, motion of the teeth in occlusion—a feeding motion seen in elephants and rodents).

Transverse motion refers to the teeth grinding from side to side. Since dinosaurs likely did not move their entire lower jaws from side to side while feeding, transverse wear patterns have been attributed either to lateral motion of the maxillae in pleurokinetic mechanisms (see chap. 12) or to long-axis rotation of each individual side of the mandible, occluding with maxillary dentition (hemi-mandibular long axis rotation or "roll")—a mechanism enabled by the predentary acting as a brace, seen in ornithischians (see chaps. 10 through 14).

Owing to the restrictive nature of joints related to cranial kinesis in most non-avian dinosaurs, including that of the otic and basal joints (among others; see above), any specific hypotheses of cranial kinesis will be discussed along with the description of feeding mechanisms of the individual taxa with which they exist in the literature (see chaps. 4 through 13). A discussion of different kinds of cranial kinesis in birds can be found in chapter 7.

Musculature and Other Soft Tissues of the Head

Dinosaur skull bones and teeth are clearly complex with so many small features that, when combined, can mean many things related to their paleobiology. (And we haven't even scratched the surface yet on the actual meanings behind all this morphological diversity!) The inner workings of skull bones, joints, and teeth play a key role in our understanding of their intricate feeding system (as well as many other physiological processes that go on within the skull). But bones and teeth themselves tell only a fraction of the story without doing considerably more detective work. The complex and difficult task of reconstructing soft tissues like muscles and neurovasculature helps us build on our understanding of the functional and physiological significance of what we see in the bones. Only then can we fully understand how these mysterious animals lived their lives millions of years ago. In this section, we focus on the soft tissue anatomical details of the dinosaur head, including the musculature and other soft tissues that likely acted on the skull bones.

Cranial Musculature

Cranial muscle anatomy and orientation are key components in determining jaw function, as they are a large part of what determines exactly how and in what direction the jaws could move, which in turn determines the animal's preferred feeding mechanism. Cranial musculature in crocodilians and birds is well known (Lakjer, 1926; Hofer, 1950; Starck and Barnikol, 1954; Iordan-

sky, 1964; Schumacher, 1973; Holliday and Witmer, 2007), and as stated above, this knowledge has aided interpretation of dinosaur jaw musculature (Holliday, 2009), as crocodilians and birds are the closest living relatives of the extinct non-avian dinosaurs (Benton, 1985). Lepidosaurs also have been suggested to have similar musculature to select dinosaur taxa, although they are more distantly related phylogenetically. All together, dinosaur cranial musculature has been interpreted through numerous observations of muscle scars on skulls and jaws in a variety of taxa. We will now focus on muscle reconstruction in particular so we can appreciate just how important it is in our understanding of dinosaur biomechanics (fig. 3.4).

The following is an anatomical account of these muscles—based mainly on the works of Haas (1955, 1963, 1969), Ostrom (1961, 1964a, 1966), Molnar (1973, 2008), Galton (1974, 1985a), Weishampel (1984a), Sues and Galton (1987), Barrett and Upchurch (1994), Witmer (1997), Holliday (2009), Lautenschlager (2013), Button et al. (2014, 2016), and Nabavizadeh (2016, 2020a, 2020b)—and a discussion of their function.

First of all, the mouth needs to open to ingest appropriate types of food, whether it be meat or plant matter. This is mainly done with a muscle behind the head called m. depressor mandibulae (mDM). (Note that "m." appears in front of all muscle names mentioned here. This is the convention for naming individual muscles and stands for *musculus*, which is Latin for "muscle." The plural form is "mm." and stands for *musculi*.) The mDM, which is innervated by the seventh cranial nerve (CN) known as the facial nerve (CN VII), originates from the paroccipital process (on the caudal aspect of the exoccipital) and extends ventrally to attach to the retroarticular process just caudal to the jaw joint. When mDM is contracted (shortening the muscle body), it pulls the retroarticular process dorsally behind the skull, which in turn pulls the rostral, or front, end of the mouth downward to open as wide as it can. Granted, there is a bit of help from gravity in this process as well, but the physical effort to open the jaw is made by mDM. Once the jaw is open, the mouth is now ready to close down

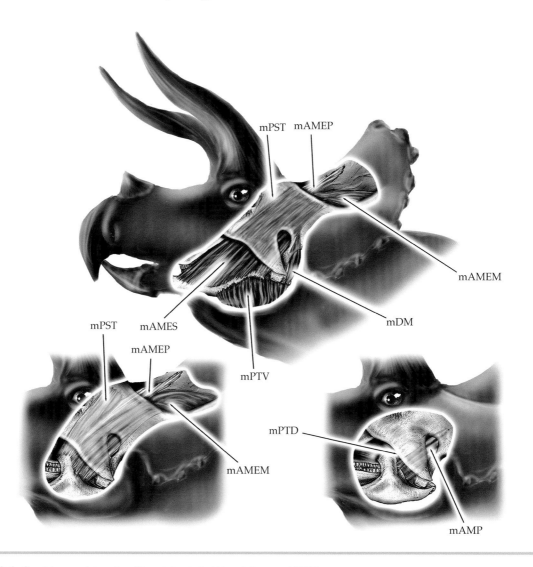

Figure 3.4. Cranial musculature in a *Triceratops* skull. Abbreviations: mAMEM, m. adductor mandibulae externus medialis; mAMEP, m. adductor mandibulae externus profundus; mAMES, m. adductor mandibulae externus superficialis; mAMP, m. adductor mandibulae posterior; mDM, m. depressor mandibulae; mPST, m. pseudotemporalis; mPTD, m. pterygoideus dorsalis; mPTV, m. pterygoideus ventralis.

on food. These closing muscles, however, are considerably more complex.

The main jaw closing (adducting) muscles—all innervated by the mandibular division of the trigeminal nerve (CN V3)—extend rostroventrally through the adductor chamber and attach onto the postdentary region of the lower jaw. These muscles act in lifting the jaw to close in an orthal motion and retract the jaw in a palinal feeding motion. As you will recall (see above), the ad-

ductor chamber is a large, contained space in the postorbital region, inside of which the adductor muscles originate along its superior rim (around the inner margins of the supratemporal fenestra). The muscular structure taking up a majority of the space within this chamber is the m. adductor mandibulae externus (mAME) complex. This muscle complex is a fan-shaped, pennate set of muscles that originate in the temporal region along the supratemporal fenestra and insert along the coronoid elevation (eminence or process) laterally, medially, and along its superior margin. In all, mAME is composed of three main sheetlike muscle bodies that are layered from deep to superficial. The deepest of all of these muscle layers is m. adductor mandibulae externus profundus (mAMEP), which originates on the caudomedial rim of the supratemporal fenestra (equivalent to the caudolateral aspect of the braincase) and extends ventrally to attach onto the apex of the coronoid eminence or process. The next layer just superficial to mAMEP is m. adductor mandibulae externus medialis (mAMEM), which originates along the caudal rim of the supratemporal fenestra and extends ventrally, likely mixing in with muscle fibers and tendinous sheet of the neighboring mAMEP in some cases, ultimately attaching along the middle portion of the caudal rim of the coronoid eminence or process.

Both mAMEP and mAMEM seem to be consistent among Dinosauria in their attachment sites, although the ceratopsids are a particularly interesting case. Previously, the mAME complex in ceratopsids like *Triceratops* has been proposed to have extended its origin all the way out to the caudodorsal rim of the frill, with an enormous muscle body that extends all the way down the frill, into the supratemporal fenestra, and down to insert onto the coronoid process (Ostrom, 1964a, 1966). Dodson (1996) argued against this in his book *The Horned Dinosaurs*, mentioning how unnecessary and disadvantageous it would have been to have an enormous muscle body attaching onto a relatively small coronoid process. With that, he proposed a more conservative origin of mAME at the base of the frill in ceratopsids. Ceratopsids possess a narrow supratemporal fenestra, allowing mAMEP and mAMEM to extend out to attach onto the base of their elaborate frill in a v shape (for more on this, see chap. 13).

Lastly, in all dinosaurs, the most superficial layer of the mAME complex is m. adductor mandibulae externus superficialis (mAMES). This muscle originates on the inner (medial) margin of the supratemporal bar, which is also the lateral border of the supratemporal fenestra. It extends rostroventrally with the deeper layers

of mAME and has been proposed to insert on the lateral surface of the coronoid eminence in theropods, sauropodomorphs, and many ornithischians (mainly on the lateral surface of the surangular, although it may even have extended onto the caudolateral surface of the dentary). The insertion site of mAMES has been more contentious among some advanced ornithischians like ceratopsids, hadrosaurids, and ankylosaurs, however, as the usual insertion on the lateral surface of the coronoid elevation was not quite visible in these taxa. As a result, its insertion was thought to have been mainly on the caudal rim of the coronoid process in these animals. Recently, Nabavizadeh (2020a) proposed that mAMES may have inserted further rostrolabially than previously thought in neoceratopsians especially, but also in hadrosaurids and ankylosaurs to a certain degree. This proposed insertion site is along a prominent labial dentary ridge seen in ornithischians that was previously thought to have been an insertion site of a separate, novel sheetlike muscle body extending from the labial maxillary ridge downward to the labial dentary ridge as "cheek" muscles (Galton, 1973; see below). With this alternative explanation, mAMES would have presented a much greater mechanical advantage and overall support system necessary in orthal and palinal feeding motions (Nabavizadeh, 2020a, 2020b). This reconstruction is also supported by the fact that the jugal flange flares out laterally in many ornithischians, especially ceratopsians, unlike in many sauropodomorphs and theropods, and this would have allowed a large muscle body to exit the adductor chamber and attach onto the labial dentary ridge. Still, adductor musculature in general may have attached along the rostral rim of the coronoid elevation in many other ornithischians as well as some herbivorous theropods and sauropodomorphs, depending on feeding mechanics and the amount of space provided within the skull (whether it be a part of mAME or another more rostrally positioned cranial muscle; Nabavizadeh, 2020b).

Another set of adductor muscles is the m. pseudotemporalis (mPST) complex, composed of m. pseudotemporalis superficialis (mPSTS), and a palatal muscle, m. pseudotemporalis profundus (mPSTP). The mPSTS originates along the lateral margin of the braincase at the rostral margin of the supratemporal fenestra. It extends ventrally and attaches medial to the coronoid eminence or process. In some extant birds, such as ratites, there is a tendinous sheet that continues as m. intramandibularis, which then inserts into the mandibular fossa (Holliday and Witmer, 2007). There has been speculation, however, that this muscle is actually just a ventral continuation of mPSTS (Holliday and Witmer, 2007). The mPSTP likely originated on the lateral surface of the epipterygoid in some non-avian dinosaurs. But because some derived sauropods, ornithopods, and ceratopsians have lost the epipterygoid altogether, there is a chance these taxa have also completely lost this muscle. The insertion of mPSTP for those taxa that retained an epipterygoid would have extended ventrally to also attach to the medial aspect of the coronoid region. Both of these muscles within this complex would have aided in adduction, or closure, of the lower jaw in conjunction with mAME.

Palatal muscles also act in powerful jaw closure. The m. adductor mandibulae posterior (mAMP), also known as m. adductor mandibulae caudalis (mAMC), innervated by the mandibular branch of the trigeminal nerve (CN V_3), is a palatal muscle that is more or less consistent among Dinosauria (and reptiles in general). It originates from a lateral fossa on the rostral wing of the quadrate bone (that attaches to the pterygoids) and then extends rostroventrally to attach along and inside the mandibular fossa on the medial aspect of the lower jaw. This muscle would have aided mAME and mPST in jaw closure as well as possibly slightly aided in any palinal feeding that occurred in some dinosaurs (iguanodontian ornithopods, neoceratopsians, and ankylosaurs).

The primary palatal muscle complex involved in the main feeding mechanisms of all dinosaurs is m. pterygoideus (mPT), which is also innervated by the mandibular branch of the trigeminal nerve (CN V_3). In all, mPT consisted of two muscle bodies: m. pterygoideus dorsalis (mPTD) and m. pterygoideus ventralis (mPTV). Although previously thought to originate more rostrally along the inner margins of the antorbital fenestrae, it is now accepted that mPTD originates mainly on the dorsal

surface of the palate and pterygoid flange, with variability as to how far rostrally it attaches depending on the taxon (Witmer, 1997). It extends caudoventrally to attach to the medial surface of the retroarticular process or caudomedial margin of the mandible itself. The mPTV originates on the ventral surface of the pterygoid flange, extending caudoventrally to wrap ventrally around the caudal margin of the mandible, to insert onto the lateral aspect of the retroarticular process or, at times, on the caudolateral surface of the postdentary mandible itself. These muscles would have helped in medial movement, restriction, or even stabilization of the lower jaw. Additionally, it would have facilitated orthal bite forces and provided a rostral vector for mandibular return in any dinosaurs that would have had a palinal feeding component in their jaw mechanisms. Holliday (2009) proposed mPTV may have even extended up to attach onto the lateral surface of the jugal in hadrosaurs and theropods, referring to striations and scarring on these jugal surfaces, possibly indicating extra muscle attachment. If this is true, it would give their feeding mechanisms especially even more support and power in orthal jaw closure.

A group of muscles called the constrictor dorsalis group has been described in sauropsids. They consist of three muscles that aid in any possible intracranial tissue function, including cranial kinesis. These three muscles include m. tensor periorbitae (mTP), or m. levator bulbi (mLB), m. levator pterygoideus (mLP), and m. protractor pterygoideus (mPP). The mTP originates from a buttress on the laterosphenoid and inserts onto the rostral border of the orbit. The mLP originates from the laterosphenoid and inserts onto the medial surface of the pterygoid and epipterygoid. Lastly, mPP originates from the ala basisphenoid and inserts onto the pterygoid and quadrate. These three muscles all play specific roles in the movements of intracranial bones to aid in their mobility (in taxa with kinetic intracranial joints—mainly birds), tighten the bones among their surrounding tissues, and assist in extra forces through feeding (Holliday and Witmer, 2008).

A number of small intracranial muscles have not been thoroughly investigated within Dinosauria, such as extraocular (eye) muscles and narial muscles that are in need of future study. The attachments of these muscles are, unfortunately, not easy to assess, as there is no clear demarcating muscle scar that coincide with these muscles, especially extraocular muscles, which originate from a tendinous ring within the orbit. Outside of intracranial musculature, there are a number of muscles that make up the floor of the mouth as well as the entirety of the tongue and associated muscles. The m. intermandibularis (mIM) can be thought of as the diaphragm of the lower jaw. It is innervated by the mandibular branch of the trigeminal nerve (CN V_3) and is thought to be synonymous with m. mylohyoideus in mammals, with lateral attachments along the inner margin of the mandibular rami that both attach at a midline raphe, forming a big sheet of muscle that spans the entire lower jaw ventrally.

Hyolingual musculature has not been thoroughly investigated in Dinosauria owing to the lack of much hyoid material in most dinosaurs. Certain individual hyoid elements have been found in various taxa, but there has been only one study (Hill et al., 2015) in which prominent hyoid elements of the ankylosaur *Pinacosaurus* have produced a general reconstruction of intrinsic hyolingual muscles (mm. hyoglossus et ceratohyoideus), extrinsic hyolingual protractors (mm. mandibulohyoideus et branchiomandibularis), and extrinsic hyolingual retractors (m. sternohyoideus). These muscles help in the various movements of the tongue for different feeding mechanisms. The tongue is often ignored in studies in feeding mechanisms, especially in fossil taxa, owing to lack of material. Still, the importance of the tongue should never be underestimated when it comes to feeding function. Whether all non-avian dinosaurs had very mobile tongues is debatable. Because many birds have extremely mobile tongues, however, it is worth investigating further possibilities of tongue use in feeding (Hill et al., 2015; Li and Clarke, 2015, 2016; Li et al., 2018).

Other Soft Tissues of the Head

Although muscles are the most commonly reconstructed, they are not the only soft tissue structures surrounding the bones we see in the fossil record. Neurovasculature (nerves and blood vessels) often leaves grooves on bone, which indicate their presence and pathways and can also be implicated with the presence of small neurovascular foramina, through which nerves and blood vessels can exit a bone (like the skull) and supply external features (George and Holliday, 2013; Barker et al., 2017; Holliday et al., 2020; Porter and Witmer, 2020). For instance, neurovascular foramina can be seen near the base of teeth, indicating nourishment to the gingiva surrounding the dentition.

Some have also hypothesized that these neurovascular foramina along the oral margin might be indication of dermal "lip" structures or even "cheek-like" skin flaps surrounding the oral cavity, with a caudally positioned rictus (skin fold on the caudal angle of the mouth), although speculation of the extent of these soft tissues has been the subject of numerous studies in buccal anatomy and is variable across taxa (Galton, 1973; Papp and Witmer, 1998; Czerkas and Gillette, 1999; Knoll, 2008; Morhardt, 2009; Carr et al., 2017; Nabavizadeh, 2020a). For instance, ornithischians were long thought to have possessed novel "cheek" muscles analogous to the buccinator in mammals owing to their buccal emargination (labial dentary ridge outside of the inset tooth row), which were thought to have helped them in feeding and in keeping food in their mouths during their complex feeding mechanisms. These "cheek" muscles were proposed to have derived from rostral fibers of adductor musculature—specifically either from m. levator anguli oris (an extension of mAMES, which is not present archosaurs; Holliday, 2009) or more likely from mAMES itself (Galton, 1973). Additional novel muscles including a "pseudomasseter" (like that seen in parrots) and "m. adductor mandibulae externus ventralis" have also been posited for psittacosaurids and heterodontosaurids (Sereno et al., 2010; Sereno, 2012). Because of the unparsimonious nature of these novel muscles and the variability in buccal emargination morphology, the existence of these muscles in non-avian dinosaurs is merely speculative (see Nabavizadeh, 2020a, for a synthesis of studies in ornithischian buccal anatomy).

A keratinous rhamphotheca, or beak, along the rostral oral margin in some dinosaur skulls is another example of cephalic dermal structures, including those seen in ornithischians and some maniraptoriforms (including all birds). Rhamphothecae (or impressions of them) have been found preserved in a select few dinosaurs, including hadrosaurs and ornithomimosaurs (Morris, 1970; Barrett, 2005; Farke et al., 2013) but can also be inferred with comparisons to extant birds and turtles (Morris, 1970; Czerkas and Gillette, 1999; Hieronymus et al., 2009; Lautenschlager et al., 2013; Button et al., 2017a). A good indicator of a keratinous beak is the presence of many grooves for attachment of the rhamphotheca as well as for blood vessels running down the rostral-most extent of the snout. These blood vessels would have been vital for nourishing the beak throughout life. A beak would have been ideal for cropping plant material, acting especially like the incisors of modern herbivorous mammals in nipping vegetation. The size and breadth of the rhamphotheca could also be an indicator of how much of a generalist or specialized feeder a particular taxon might have been (Mallon and Anderson, 2014a). Additional keratinous structures seen in dinosaur heads include the keratin-covered horns of ceratopsids and other keratinous cranial adornments across various other dinosaurs.

Reconstruction of sensory systems in the head is also crucial for our understanding of dinosaur biology. For instance, reconstruction of the brains in fossilized vertebrates dates back to the 1920s, with endocasts (internal molds) of cranial brain cavities created to broadly describe brain morphology (Edinger, 1929). Recent three-dimensional visualization, computed tomography imaging techniques, and iodine staining have vastly improved our understanding of the detailed anatomy of brain endocasts. Because of these innovations, we can determine the size and shape of dinosaurs' brains and their different regions to piece together more clues and make inferences regarding their paleoecology (sense of smell, or olfaction; visual acuity; and general cognitive capabilities). Imaging techniques also help in deciphering the anatomy (both hard and soft tissue) of other internal cranial regions, such as the morphology of the eyes and intraorbital soft tissues; the nasal cavity and its complexities of soft tissue anatomy in different taxa (and what it might mean for olfaction of different food types, whether prey or vegetation); and inner ear morphology, including that of the vestibular system and what it might suggest regarding how dinosaurs held their heads during feeding (Witmer, 1997, 2001; Witmer and Ridgley, 2008, 2009; Bourke et al., 2014, 2018; Gignac et al., 2016; Gold and Watanabe, 2018; Holliday et al., 2020; Porter et al., 2020). Each of these kinds of studies brings us closer to figuring out the diverse physiologies of these incredible organisms, and they will be discussed on a case-by-case basis among taxa throughout the rest of this book.

Postcranial Anatomy

The postcranial skeleton (everything but the skull) is composed of two main components: the axial skeleton (including the entire vertebral column and ribcage) and the appendicular skeleton (composed of the pectoral limb, or forelimb, and pelvic limb, or hindlimb), all with associated musculature and ligaments.

Overview of the Axial Skeleton

The vertebral column is composed of many individual elements (vertebrae) that articulate with one another, forming the "backbone" of the body (fig. 3.5). It is divided into four main sections of vertebrae of different morphologies: cervical vertebrae (the neck), dorsal or thoracolumbar vertebrae (the back), fused sacral vertebrae (the sacrum, composed of at least three or more vertebrae in dinosaurs), and caudal vertebrae (the tail).

A vertebra is made up of multiple components. The centrum (or body) is largely cylindrical and is the largest part of the vertebra. Just dorsal to the centrum is a neural arch for passage of the spinal cord through the neural canal. Extending dorsally is the neural spine (or spinous process) with two transverse processes extended laterally (one on either side of the vertebrae), for attachment of epaxial (back) musculature. Two sets of small bilateral processes called zygapophyses extend from the neural arch both cranially and caudally. Projecting cranially from the neural arch are the paired prezygapophyses, which have articular surfaces angled dorsomedially, and projecting caudally are the paired postzygapophyses, which have articular facets angled ventrolaterally. These zygapophyses articulate with their counterparts in successive vertebrae to help articulate the entire vertebral column and control how much movement is allowed between them. Additionally, thick intervertebral cartilages are present between the centra of the vertebrae to allow for extra mobility, cushioning, and support along the axial skeleton.

Articulations between vertebrae can also depend on the morphology of the cranial and caudal surfaces of the centra, which determine the type of mobility that would occur at the joint. These morphologies include amphiplatyan (with both sides flat), amphicoelous (with both ends concave), procoelous (with the cranial end concave and the caudal end convex and rounded), and opisthocoelous (with the cranial end convex and rounded and the caudal end concave). Other morphological structures sometimes occur in dinosaur vertebrae, including the hyposphene and hypantrum, which are separate small projections that extend from the caudal and cranial sides of the base of the neural spine in saurischians (sauropodomorphs and many non-avian theropods), among other extinct archosaurs. Small holes called pleurocoels also lead into small chambers or networks of chambers inside the vertebrae of saurischians.

Ribs are long, thin, bowed bones that articulate with ventrolateral margin of both sides of the neural arches as well as from the centra of dorsal vertebrae (with two articulation points on either side), altogether forming a cage (ribcage) around the heart, lungs, and other essential organs. Gastralia ("belly ribs") form the ventral aspect of the ribcage and belly in some dinosaurs (theropods and early sauropodomorphs), and these ribs also help in respiratory functions.

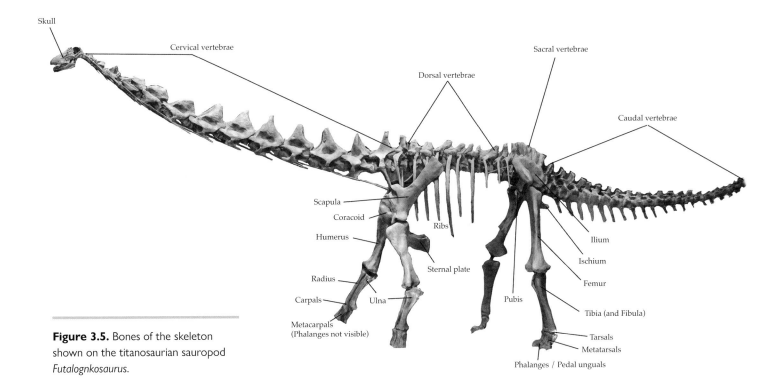

Figure 3.5. Bones of the skeleton shown on the titanosaurian sauropod *Futalognkosaurus*.

Several extraneous structures along the vertebrae help serve as stabilizers of the vertebral column. Many intervertebral ligaments hold the entire vertebral column together tightly along its entire length. One notable ligament is the nuchal ligament (or ligamentum nuchae), which some have postulated to have existed in many saurischians and ornithischians (Woodruff, 2017; Bertozzo et al., 2021), extending from the occiput at the back of the skull to neural spines of the dorsal vertebral column, with attachments along the entire cervical vertebral column along the way. This ligament would have allowed for immense storage of elastic energy for rebounding of the neck in head movements during feeding (as it does in modern herbivorous mammals, for instance). Cervical vertebrae also have ribs, which project caudoventrally from their respective vertebra, sometimes overlapping successive cervical ribs. Caudal vertebrae also have chevrons, which are ventrally extending riblike structures. Other structures involved in vertebral stability are ossified tendons seen in ornithischians, which help stiffen the vertebral columns in various taxa, as well as elongated prezygapophyses in some saurischians.

Additionally, dermal bones called osteoderms are sometimes present within regions of the skin and add an outer layer of protection for certain parts of the body, depending on the taxon. These osteoderms are especially prevalent in thyreophorans in the form of plates, spikes, and other armorlike structures but are also seen in some other dinosaurs, including some sauropodomorphs.

Neck, Trunk, and Tail Musculature

The musculature of the cervical, trunk, and caudal regions of the body are primarily made up of hypaxial muscles, which are innervated by ventral spinal nerves and attach to and surround a majority of the axial skeleton and ribs ventral to the vertebral column, and epaxial muscles, which are innervated by dorsal spinal nerves and are typically longitudinal and attach along the vertebral column between the neural spines and transverse processes.

Cervical (neck) muscles (fig. 3.6) are used to support the head and to move the neck in different directions. They typically originate at or near the occiput of the skull and extend caudally, attaching across various parts of the cervical vertebral series depending on the functionality of the muscle (Snively and Russell, 2007a). Cervical musculature is broadly made up of three main functional groups—dorsiflexors, ventroflexors, and lateroflexors—with many of these muscles being co-opted for more than one of these functionalities in various taxa. Dorsiflexors are more dorsally attached muscles that pull the head and neck upward; these can include m. transversospinalis capitis, m. complexus, m. splenius capitis, m. transversospinalis cervicis (or m. longus colli dorsalis), m. intercristales, and m. interspinales. Lateroflexors are laterally attached muscles that pull the head and neck to the sides; these can include m. complexus, m. splenius capitis, m. longissimus capitis superficialis, m. longissimus capitis profundus, m. iliocostalis capitis, and m. rectus capitis ventralis (which possibly includes both m. flexor lateralis and m. flexor medialis). Ventroflexors are ventrally positioned muscles that pull the head and neck downward; they include m. rectus capitis ventralis, m. longissimus capitis profundus, and sometimes m. longissimus capitis superficialis (in *Allosaurus*; Snively and Russell, 2007a).

The different types of neck movement are informative about feeding ecology in dinosaurs, no matter if it was a carnivore or an herbivore. Strength and leverage of neck musculature and flexibility in the neck help in allowing for a certain feeding envelope (or range of feeding in space while standing), in both horizontal and vertical directions. Typically, the longer the neck, the lighter the head, whereas a shorter neck was good for supporting a large head, especially those with larger or more complex dentition. For instance, allosaurid theropods had highly flexible necks and were able to quickly lunge at prey items (see chap. 5), whereas larger tyrannosaurids likely had relatively shorter, but girthier, necks able to more powerfully subdue larger prey (see chap. 6). Additionally, the gargantuan sauropod dinosaurs possessed extremely long necks, sometimes able to forage at

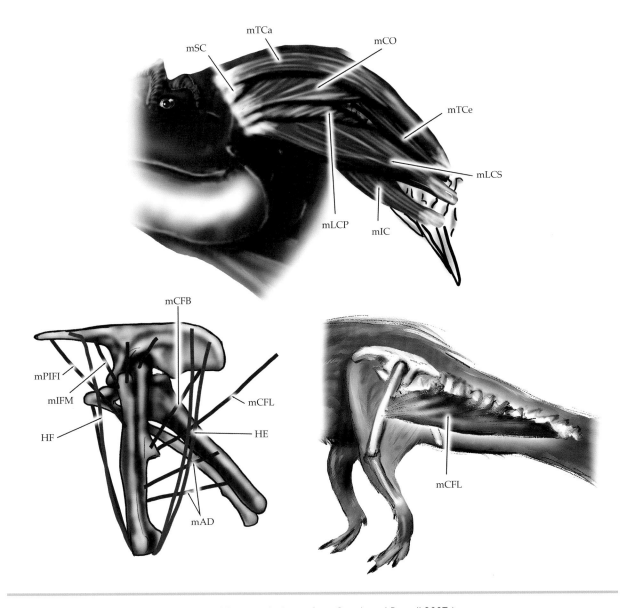

Figure 3.6. Musculature of (*top*) the neck in *Allosaurus* (redrawn from Snively and Russell 2007a) and (*bottom left*) the hindlimb in *Scelidosaurus*, showing muscle lines from origin to insertion (redrawn from Maidment and Barrett 2011; Maidment et al. 2014), and (*bottom right*) *Heterodontosaurus*, showing m. caudofemoralis longus. Abbreviations: mAD, m. adductor; HE, hip extensor muscles; HF, hip flexor muscles; mCFB, m. caudofemoralis brevis; mCFL, m. caudofemoralis longus; mCO, m. complexus; mIC, m. iliocostalis capitis; mIFM, m. iliofemoralis; mLCP, m. longissimus profundus; mLCS, m. longissimus capitis superficialis; mPIFI, m. puboischiofemoralis internus; mSC, m. splenius capitis; mTCa, m. transversospinalis capitis; mTCe, m. transversospinalis cervicis.

enormous heights or extensive horizontal ranges compared to other dinosaurs (see chaps. 8 and 9). But their heads were comparatively tiny with a less sophisticated feeding mechanism—many times smaller than the much smaller-bodied ceratopsids, which in turn had short, even

fused, cervical series supporting an enormous head (see chap. 13). Although many studies have used the articulation of the cervical vertebrae to assess neck flexibility, the extent of soft tissues (muscles, ligaments, cartilages, etc.) that surrounded these vertebrae would have added more complexity to those ranges of motion (Cobley et al., 2013) ecology and overall biology. Several hypotheses have been proposed, based primarily on osteological data, suggesting different degrees of neck flexibility. This study attempts to assess the effects of reconstructed soft tissues on sauropod neck flexibility through systematic removal of muscle groups and measures of flexibility of the neck in a living analogue, the ostrich (Struthio camelus.

The hypaxial muscles of the trunk were quite large compared to the epaxial musculature. Muscles surrounding the pectoral girdles and thoracic regions (including intercostal muscles and muscles associated with pectoral girdles, mm. trapezius et rhomboideus), muscles in the abdominal region (including rectus abdominus, or down the midline of the belly), obliques, and transverse abdominal muscles (wrapping around the flanks of the animals) all helped support the trunk and rib cage in various activities and ventilation (Dilkes, 2000; Schachner et al., 2009). Epaxial musculature of the trunk was composed of longitudinal muscles along the dorsal rim of the vertebral column, including m. transversospinalis (most medially), m. longissimus dorsi, and m. iliocostalis (most laterally), along with a number of other smaller muscles (Organ, 2006a; Schwarz-Wings et al., 2010). All of these muscles aided in dorsiflexion, ventroflexion, and lateroflexion of the vertebral column and were adapted for certain degrees of flexibility of the spine in these planes of motion.

Caudal (tail) musculature largely follows a similar pattern of hypaxial and epaxial musculature of the back. Epaxial musculature continues down longitudinally between the neural spines and transverse processes, becoming smaller and narrower at the distal-most end of the tail. Epaxial musculature of the tail is made up primarily of m. transversospinalis (or m. spinalis) dorsomedially and m. longissimus more laterally, and these muscles often helped movement and stiffening of the tail to various degrees. Hypaxial musculature extended from the caudal aspect of the pelvis (mainly the ischium) and would have helped mainly in lateral flexion, support, and stiffening of the tail as well, more so at the proximal end of it where

the muscle body is much larger. One large, elongate muscle called m. caudofemoralis longus extends from the fourth trochanter of the femur onto the ventral aspect of multiple caudal vertebrae and is an important muscle in powerful retraction of the femur in locomotion as well as powerful lateral flexion of the tail in locomotion. Another more superficial hypaxial muscle complex, m. ilio-ischiocaudalis (composed of both m. iliocaudalis and m. ischiocaudalis), originates at both the ilium and ischium and inserts onto the ventrolateral aspect of the caudal vertebrae, with a muscle body that is much thicker proximally than distally. It ultimately extends further distally along the tail than m. caudofemoralis longus (Arbour, 2009; Persons and Currie, 2011b). The morphology of the caudal vertebrae (including the centra) can vary in shape and thickness depending on the degree of hypaxial muscle attachment. Generally speaking, the centra are proximodistally shorter most proximally at the base of the tail (for controlled lateral movement of the whole tail), longer further down the tail (where m. caudofemoralis longus attached, for instance), and then shorter again distally (Hone et al., 2021).

Overview of the Appendicular Skeleton

As noted above, the appendicular skeleton is composed of the pectoral limbs (forelimbs) and pelvic limbs (hindlimbs) (fig. 3.5). The pectoral limb articulates with the cranial end of the trunk via the pectoral girdle. The pectoral girdle includes a large elongate bone called the scapula (making up the shoulder blade) as well as the smaller platelike coracoid bone articulating cranioventral to it. This articulation creates the glenoid fossa, a semicircular joint for articulation of the humerus. The sternum is made up of a pair of plate-like bones (sternals) in the chest region between the forelimbs. Many dinosaurs also have clavicles that connect the sternum to the rest of the pectoral girdle, with theropods fusing the clavicles to form a boomerang-shaped furcula (or wishbone).

The humerus is the largest, most robust bone of the forelimb outside the pectoral girdle, making up the arm portion of the forelimb skeleton. It is elongate and articulates with the glenoid fossa proximally and with the smaller ulna and radius distally. The elongate ulna and radius make up the forearm, with the ulna being the larger element of the two as well as being positioned more caudally. Distally to the forearm are the carpal bones, a series of small, flat bony elements making up the wrist and creating the beginning of the manus (or hand). Articulating distal to the carpals are up to five metacarpals, which are small, narrow, elongate elements making up the palm of the hand and creating the base of the digits. The medial-most digit is considered the first digit, or "thumb." Each metacarpal is associated with up to three phalanges (with each element referred to as a "phalanx"), and the distal-most phalanx (ungual) creates either a claw- or hooflike structure covered with keratin.

The pelvic limb articulates with the caudal end of the torso at the sacrum via the pelvic girdle (pelvis or hip bone). Each side of the pelvic girdle is made up of three bones: the ilium, pubis, and ischium. The ilium is the largest and most dorsally positioned of the three bones, with a craniocaudally elongate structure that is typically narrower in ornithischians and more arched and robust in saurischians. The ilium articulates with the sacrum along its medial surface. The pubis articulates cranioventral to the ilium, with the ischium articulating directly caudal to the pubis. The pubis of saurischian dinosaurs is typically oriented cranioventrally (with few exceptions), while that of ornithischians is oriented caudoventrally. In all cases, the ischium is directed caudoventrally. A large, circular hole is formed by the combined articulations of the ilium, pubis, and ischium, called the acetabulum, also known as the hip socket. Dinosaurs are unique in possessing what is known as an open acetabulum, referring to the fact that there is no sheet of bone closing off the medial aspect of the acetabulum (as in most other tetrapods). Articular cartilages that surrounded the hip joint in life would have acted in strong mechanical support of large compressive loads (and sometimes shear load) of the hindlimbs, especially in much larger dinosaurs (e.g., Tsai and Holliday, 2015; Tsai et al., 2018, 2020).

The femur (or thigh bone) is the largest bone of the lower limb, with a condyle that articulates with the acetabulum perpendicular to the shaft that extends ventrally. The femur possesses multiple protuberances (trochanters) for hip and thigh muscle attachment, including the greater and lesser trochanters dorsally and the fourth trochanter part way down the shaft (for attachment of m. caudofemoralis musculature; see below). Two femoral condyles articulate with the lower leg bones—the larger, medially positioned tibia and the narrower fibula laterally. Distal to these elements are the tarsal bones (analogous to the carpals of the forelimb) that create the ankle (and the beginning of the pes, or foot)—composed of the more proximal astragalus (medially) and calcaneum (laterally) as well as smaller distal tarsals. A straight hinge joint is formed between the proximal and distal tarsal rows in dinosaur ankles (compared to that of crocodylomorphs, which show a more complexly angled joint between the proximal tarsal bones). Articulating distal to the carpals are five metatarsals, which are small, narrow, elongate bones (analogous to the metacarpals of the forelimb) that form the bottom of the foot and create the base of the digits. The medial-most digit is, again, considered the first digit, and each metatarsal is associated with phalanges, the distal-most being referred to as an ungual, forming a claw- or hooflike structure covered with keratin.

Dinosaurs are known to have had more straight, erect limb posture compared to their sprawling crocodilian relatives. The extent of bipedality versus quadrupedality is highly variable among dinosaurs, however. Dinosaurs were likely ancestrally bipedal, with a vast majority of theropods showing clear signs of bipedality, including much shorter forelimbs than hindlimbs as well as grasping adaptations of the hands that would have precluded quadrupedal stance. Many early-diverging members of all other clades were bipedal as well. Typically, bipeds are generally considered cursors (or adapted for cursoriality), meaning they were adapted to running on the ground (although there are nuanced arguments for further distinctions among these taxa; Carrano, 1999). There are also intermediate forms that bridge the gap between bipedality and quadrupedality (aptly named facultative bipeds or facultative quadrupeds, depending on which they were more adapted for). There are four major instances in dinosaur evolution where a broadscale evolutionary transition from bipedality to obligate quadrupedality are apparent: within Sauropodomorpha, within Thyreophora, within Ornithopoda, and within Marginocephalia. Each of these four clades show small, bipedal early-diverging taxa, with later members transitioning to facultative bipeds/quadrupeds and later forms adapted for obligate quadrupedality. These obligate quadrupeds are typically herbivores of larger body size (typically considered graviportal locomotors) and show relatively larger forelimbs than bipeds (although many times their forelimbs are still shorter than their hindlimbs, but not always; Carrano, 1999). Limb musculature also adapted and co-opted functionalities accordingly, along with the transitions in locomotor strategy, depending on what was needed.

Appendicular Musculature

Vertebrate limb musculature is enormously complex, with many muscles providing a multitude of functions, including elevation, protraction, retraction, flexion, extension, abduction, adduction, and rotation of various parts of the limbs (depending on the functional need) as well as general functions in overall support of body weight on the limbs. Musculature is regionalized within both the forelimbs and hindlimbs.

The forelimb is suspended from the body by musculature that surrounds the pectoral girdle. This musculature includes the enlarged m. pectoralis ventrally, supporting the ribcage and sternum, and m. serratus ventralis, m. trapezius, and m. rhomboideus dorsally. Together, these muscles attach to the scapula, acting in various shoulder movements and forming a sling to hold the thorax up while supporting weight on the forelimbs. Muscles acting in extension of the humerus (upper arm) include the large fan-shaped m. latissimus dorsi, originating along the dorsal vertebral column, and m. deltoideus (a major humeral abductor at the shoulder girdle), among other

smaller muscles between the scapula, coracoid, and humerus (m. subcoracoscapularis and m. scapulohumeralis). The major forearm extensor is m. triceps brachii, acting in powerful elbow extension by attaching to the olecranon process of the ulna. Flexors and adductors of the arm include the large, fan-shaped m. pectoralis attaching along the sternum and ventral ribcage as well as m. coracobrachialis and m. supracoracoideus. The most prominent flexor of the forearm is m. biceps brachii, which acts along with other muscles like m. brachialis and m. humeroradialis. Many narrower muscles exist within the forearm, including flexors (m. flexor digitorum longus, m. flexor carpi radialis, and m. flexor carpi ulnaris) and extensors (m. extensor digitorum, m. extensor carpi radialis, and m. extensor carpi ulnaris), and most have tendons that lead toward the wrist and into individual digits to act in their various functions along with intrinsic hand muscles (Lull and Wright, 1942; Borsuk-Bialynicka, 1977; Coombs, 1978; Nicholls and Russell, 1985; Norman, 1986; Ostrom, 1995; Dilkes, 2000; Carpenter and Smith, 2001; Langer et al., 2007; Carpenter and Wilson, 2008; Maidment and Barrett, 2011; Burch, 2014, 2017).

The hindlimb is suspended from the body by muscles that surround the pelvic girdle (fig. 3.6). Major muscles originating from the pelvic girdle and acting on the femur (thigh) include m. puboischiofemoralis internus and m. iliofemoralis (mainly acting in femoral protraction), m. ischiotrochantericus and m. puboischiofemoralis externus (mainly acting in lateral femoral rotation), m. caudofemoralis brevis (inserting at the fourth trochanter of the femur and mainly acting in femoral retraction), and the adductor muscles. Major muscles that run along the femur and act on the knee include m. triceps femoris, which consists of three large muscle bodies that converge distally into a tendon acting in knee extension. The action of m. iliofibularis retracts (extends) the hip as well as flexes the knee. The flexor cruris muscles (m. flexor tibialis externus and internus) also perform these actions, much like our hamstrings. M. caudofemoralis longus shares an attachment with m. caudofemoralis brevis, but instead of stretching to the pelvic girdle, it extends much further for powerful femoral retraction and lateral movement of the tail, as noted above. The lower leg musculature consists of ventral plantarflexors (m. gastrocnemius as well as digital flexors) as well as dorsal dorsiflexors (m. extensor digitorum longus and m. tibialis cranialis), all of which have tendons that run past the ankle and individual digits of the foot for their respective functions, along with any intrinsic foot musculature (Romer, 1923, 1927; Lull and Wright, 1942; Galton, 1969; Norman, 1986; Dilkes, 2000; Hutchinson, 2001a, 2001b, 2002; Carrano and Hutchinson, 2002; Maidment and Barrett, 2011).

Gastrointestinal Tract (or the Gut System)

The gastrointestinal tract is perhaps the most important part of the dietary ecology of an animal. Unfortunately for dinosaur paleontologists, reconstructing the morphology of the gut system in non-avialan dinosaurs is incredibly difficult—practically impossible even—because there is next to no preservation of the gut system in a vast majority of these animals. This leaves researchers to mostly guess what dinosaur gut systems may have looked like based on the types of diets they had. In living mammals, carnivorans typically have much shorter, simpler gut systems than herbivorous mammals like ungulates, which possess some of the most elaborate gut systems known in any land vertebrate. This is because processing meat takes considerably less energy and microbial power than processing tough fibers of vegetation. Therefore the gut system of carnivorous dinosaurs was likely much shorter and much less complex than those of herbivorous dinosaurs. Typically, the larger herbivores have much larger gut cavities, which in turn are indicative of a much larger, more complex gut system, possibly including multiple stomach chambers like living foregut fermenters such as ruminants, or even an enlarged large caecum like living hindgut fermenters such as horses and rhinos. Ornithischians may have had more room for a larger gut cavity owing to the caudal retroversion of their pubic bones.

The enlargement of gut cavities, both in relative as well as absolute length and width, would have allowed more room for larger gastrointestinal tracts to digest the tough cellulose of plant cell walls, to be broken down by enzymes known as cellulases and general fermentation by microbes. The larger the gut, the more retention time that would have been allotted (Franz et al., 2011). A long retention time would have been especially helpful in the large sauropod dinosaurs, which were not able to break down food as well orally as some ornithischians with more sophisticated feeding apparatuses or in any herbivorous dinosaur that did not possess a gastric mill. Extremely few examples of gut preservation in non-avialan dinosaurs have been found (a *Scipionyx* liver and intestine; Dal Sasso and Signore, 1998), making speculation of gut morphology in dinosaurs difficult, with only derived gut morphology in living birds as our primary evidence of possible visceral morphology in non-avialan dinosaurs.

Brief Introduction to Phylogenetic Relationships

A few remarks are presented here to help you understand some of the phylogenetic terminology you will encounter throughout this text (see also Fastovsky and Weishampel, 2009). Perhaps the most important is taxon (taxa for plural), which is any group of organisms designated by a name, of any rank, within the ebb and flow of the hierarchy of life. So, for humans, we have *Homo sapiens*. For whales and their relatives, we have Cetacea. And for all dinosaurs, we have Dinosauria.

A special kind of taxon, particularly when viewed from a phylogenetic perspective, is the monophyletic taxon, a group of organisms that has a single origin (one ancestor) and contains all the descendants of this single ancestor (in contrast to the lesser used paraphyletic taxon, which does not contain all the descendants). These monophyletic groups are revealed by having homologous characters (isolated features found uniquely in a group of organisms) called synapomorphies, which are derived features shared among descendant taxa and found through analysis of parsimony (requiring the smallest number of evolutionary steps possible to get the morphology they present). A hierarchical branching diagram that shows the distribution of shared derived characters and thereby identifies monophyletic taxa is called a cladogram.

In the following 10 chapters, we discuss dinosaur feeding function and ecology in a phylogenetic context, with each taxon presented alongside their closest relatives to gain a better picture of how the feeding mechanisms observed in these impressive animals had evolved.

We start with the beginnings of Dinosauria and the evolution of feeding throughout Theropoda in four chapters (chaps. 4 through 7). Then we explore the evolution of feeding in the gargantuan, long-necked, and largely herbivorous sauropodomorph dinosaurs (chaps. 8 and 9), followed by the evolution of feeding throughout the diverse herbivorous Ornithischia (chaps. 10 through 13). Finally, in chapter 14, we discuss the overall paleoecology of all of these dinosaurs, including how they fit into their respective ecosystems.

CHAPTER 4

Early Dinosaurs and Non-Tetanuran Theropods

Dinosaur anatomy—particularly that of the skull—is enormously complex, and as such, evolutionary processes have resulted in a wide array of morphological variation across taxa. Dinosaurs have traditionally been thought to have evolved into two main groups: Saurischia (composed of Theropoda and Sauropodomorpha) and Ornithischia (everything else). But a recent phylogenetic analysis by Baron et al. (2017a) suggests that Ornithischia and Theropoda may have been more closely related to each other in a group they called Ornithoscelida, and Sauropodomorpha was a separate dinosaur clade sister to Herrerasauridae. Because of the standing uncertainty of this discrepancy, we herein embrace the traditional (former) hypothesized relationship, although we will discuss the anatomy and biology in Theropoda, Sauropodomorpha, and Ornithischia separately.

A typical early dinosaur skull shape can be described as roughly wedge-shaped or triangular, both in lateral and dorsal view, with a wider caudal end of the skull and a narrowed rostral end presented in varying lengths. This morphology can be seen in many theropods and early sauropodomorphs as well as basal members of ornithischian clades. A clade within Dinosauria that has received a significant amount of attention as far as qualitative and quantitative analysis of cranial mechanics and bite force in feeding, among many other factors, is Theropoda—a majority of which is represented by all the predatory dinosaurs known to have existed (see descriptions below for references).

Theropods were mainly obligate bipeds, retaining the ancestral body plan of an S-shaped neck; shorter, clawed forelimbs for grasping; long digitigrade hindlimbs; and a long tail to counterbalance the front half of the body. Many small- to medium-sized theropods were likely swift, agile predators (Holtz, 2003), while some of the largest theropods may not have been as fast for their body size. When dinosaurs started to diversify throughout the Triassic-Jurassic boundary, theropod body size saw a gradual increase as they became the dominant carnivores of their time (Griffin and Nesbitt, 2020). Nonavian theropods likely possessed a respiratory system with lungs constrained within the ribcage that would require air sacs to take in oxygen, as in their living avian descendants (Schachner et al., 2009). Their powerful hindlimbs were supported by strong hip joints supported by hyaline cartilage and other soft tissues and padding that were built to resist extensive compression and shear forces, with substantial clade-specific variation (Tsai et al., 2018, 2020).

The fascination for this group in feeding studies lies in the role of many of its members as top predators. There is general interest of how much force it would have taken the top predators to bite, kill, and dismember their prey, much of which consisted of large herbivorous dinosaurs (see, e.g., Erickson et al., 1996; Rayfield et al., 2001; Meers, 2002; Therrien et al., 2005; Bates and Falkingham, 2012; Gignac and Erickson, 2017). Theropod skulls varied widely in shape, from relatively triangular to rectangular. They also varied in size and robusticity (ranging from robust—e.g., tyrannosaurids, ceratosaurs, and allosauroids—to narrow—e.g., coelophysoids, dilophosaurids, and dromaeosaurids—and even, at times, somewhat "crocodile-like"—e.g., spinosaurids), with a variety of cranial ornamentations and variations of prominent bladelike, serrated dentition (known as "ziphodont" dentition in most cases) with characteristics specific to each group (Holtz, 2003; Hendrickx et al., 2019). Similar triangular cranial shapes are maintained in coelurosaurian theropods that turned to herbivory, such as therizinosaurs and ornithomimosaurs (and excluding oviraptorosaurs, with variably more box-shaped crania), but with alterations in dental shape or loss of dentition altogether. Variation in skull shape in theropods is widespread, mainly in relative rostrocaudal length and depth of the snout, and these variations have been suggested to be most strongly correlated with phylogeny, rather than biting function, although skull shapes of herbivorous theropods do show more overall variability than carnivores (Brusatte et al., 2012).

Specific characteristics of theropod skulls show widespread variability in and of themselves. Orbital shapes in

Figure 4.1. Phylogeny of early Dinosauria and Theropoda, largely based on compiled phylogenies in Hendrickx et al. (2015a).

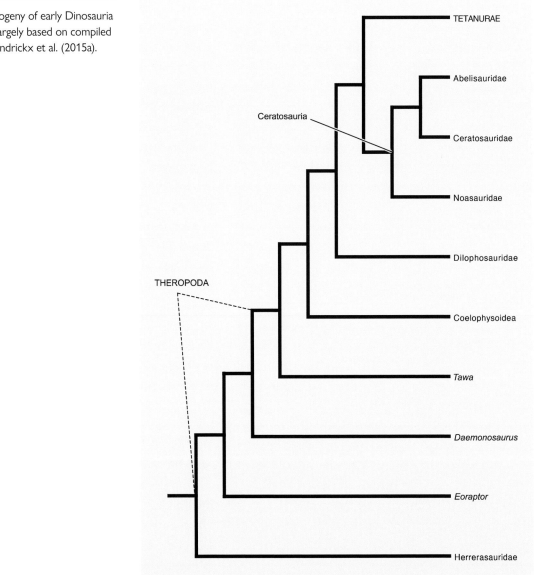

theropod skulls are distinctly variable (Chure, 1998) and have been suggested to be predictive of the amount of bite force resistance in the back of the skull, with larger carnivores exhibiting more elliptical, rostrocaudally compressed orbits (Henderson, 2002). Like their extant avian descendants, non-avian theropods possess many endocranial sinuses; in particular, the antorbital sinus pneumatized other cranial bones and has a diverticulum called the suborbital sinus, which helped keep air flowing through those sinuses (Witmer and Ridgely, 2008).

Table 4.1. Early dinosaurian and theropod genera organized by clade		
Major clade	**Subclade**	**Genus**
Early Dinosauria / Early Theropoda		
		Herrerasaurus
		Staurikosaurus
		Eoraptor
		Daemonosaurus
		Liliensternus
		Chilesaurus
		Tawa
Coelophysoidea		
		Coelophysis
Dilophosauridae		
		Dilophosaurus
Ceratosauria		
	Ceratosauridae	
		Ceratosaurus
	Abelisauridae	
		Abelisaurus
		Carnotaurus
		Majungasaurus
		Spectrovenator
		Rugops
	Noasauridae	
		Masiakasaurus
		Elaphrosaurus
		Limusaurus

Theropods also generally possess a dorsally extended lacrimal bone and a prominent intramandibular joint between the dentary and postdentary elements (potentially used as a sort of shock absorber for struggling prey; Holtz, 2003). Birds—which are highly derived maniraptoran theropods (Gauthier, 1986)—have unique, highly modified skulls with a keratinous beak.

A common method used for estimating bite performance over the past decade—particularly in theropods—has been finite element analysis (FEA) because it gives researchers a chance to visualize the stresses that occur throughout the skull and mandible once specific forces are applied to various parts of the jaw (as discussed in chap. 2). This method ultimately provides insight into why the cranial elements are shaped the way they are and their evolutionary importance to the success and feeding habits of theropods, such as *Tyrannosaurus*, *Allosaurus*, *Carnotaurus*, *Deinonychus*, and spinosaurids (Rayfield et al., 2001, 2007; Rayfield, 2004, 2005a, 2005b; Mazzetta et al., 2009). Researchers have also gone beyond cranial mechanics and have focused on mechanics of cervical spines of theropods as well, as cervical spines are hypothesized to have played a major role in head-striking to capture prey (Snively and Russell, 2007a, 2007b, 2007c; Snively et al., 2013). The combination of cranial and cervical mechanics as well as overall body plan ultimately gives a much more complete understanding of predatory behavior in these animals and the successful acquisition of prey. For reference in understanding relationships of theropods discussed in this chapter, see the phylogeny (fig. 4.1) and taxon list (table 4.1).

Early Dinosauria and the Rise of Theropoda

The bipedal, carnivorous Herrerasauridae (including *Herrerasaurus* and *Staurikosaurus*) is an isolated dinosaur clade found to have likely either been a basal saurischian, basal theropod, or, in Baron et al.'s (2017a) study, a sister group to Sauropodomorpha. Despite the discrepancy in its relation with other dinosaurs, herrerasaurids generally share many cranial features with theropod dinosaurs. As one of the earliest dinosaur taxa (known from the Late Triassic Period of Argentina), *Herrerasaurus* possesses an elongate skull and snout that is more rectangular in lateral view than the abovementioned wedge shape, with enlarged antorbital fenestrae with no lateral antorbital fossa (Sereno and Novas, 1993; Witmer, 1997) (fig. 4.2). The prominent circular orbits in herrerasaurids suggest it is a relatively weaker region of the skull compared to that in most derived large carnivorous theropods (except spinosaurids) but are still

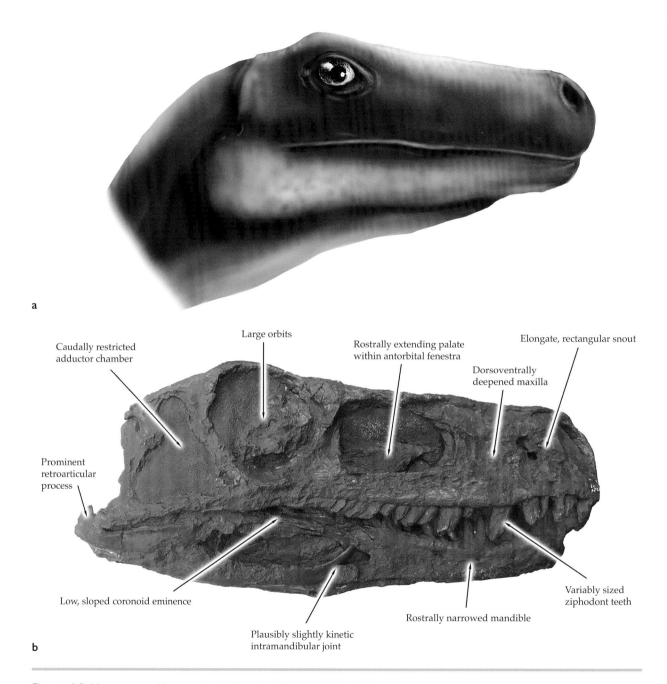

Figure 4.2. *Herrerasaurus* (a) reconstructed head and (b) skull in lateral view (length approximately 30 cm).

either comparable to or stronger than that of other smaller or similarly sized theropods like compsognathids, coelophysoids, and dromaeosaurids (Henderson, 2002). Herrerasaurids also possess subcircular supratemporal fenestrae and a caudally restricted postorbital region and adductor chamber with a quadrangular infratemporal fenestra at the caudal end of the skull (Sereno and Novas, 1993). The palate is triangular, with a transversely expanded caudal end and narrowed rostral extension with a dorsal m. pterygoideus dorsalis (mPTD) attachment that extends into the antorbital region (Witmer, 1997). The mandible is also elongate and rostrally narrow, with a low, sloped coronoid eminence; an elongate mandibular fenestra; and prominent, dorsally projecting retroarticular process.

Herrerasaurus shows variably sized ziphodont teeth along the oral margins, with 4 premaxillary teeth, about 18 maxillary teeth, and about 16 dentary teeth (Sereno and Novas, 1993). Maxillary teeth extend the entire length of the maxillary oral margin. Dental traits analyzed in a recent study describing the herrerasaurid *Gnathovorax*

concluded that the ecomorphological morphospace of herrerasaurids shows that they occupied an ecological role that statistically overlapped that of the later, post-Carnian medium-sized theropods (Pacheco et al., 2019). The maxilla in *Herrerasaurus* is both dorsoventrally deeper and labiolingually broad, supporting its strong dentition and creating a relative skull strength in this region comparable to that of much larger carnivorous theropods (Henderson and Weishampel, 2002). They also possess a plausibly kinetic sliding intramandibular joint, ideal for better manipulation and grasp of prey items.

Sakamoto (2010) found that *Herrerasaurus* jaws had among the highest mechanical advantage (MA) both mesially and distally among theropod dinosaurs (with comparable measures to the sauropodomorph *Plateosaurus*). Of note is the higher mesial MA values in abelisaurids and *Ceratosaurus* (Sakamoto, 2010), likely owing to their stouter snouts and jaws. Furthermore, Monfroy (2017) found that the bending strength would have been higher at the mesial dentition in *Herrerasaurus*, an adaptation ideal for varanid-like slashing bites (fig. 4.3).

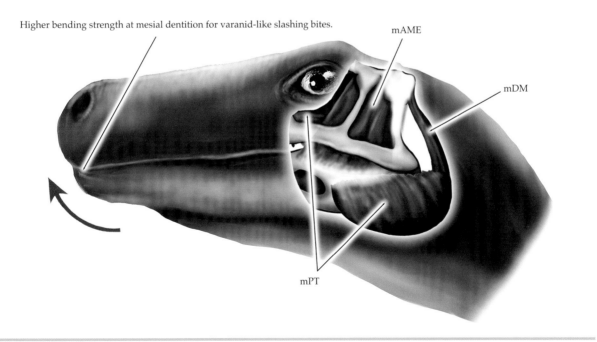

Higher bending strength at mesial dentition for varanid-like slashing bites.

mAME

mDM

mPT

Figure 4.3. *Herrerasaurus* head showing feeding function with jaw muscles. Red arrow shows orthal jaw action. Abbreviations: mAME, m. adductor mandibulae externus; mDM, m. depressor mandibulae; mPT, m. pterygoideus.

Because of their longer, more gracile jaws, however, Monfroy (2017) asserted that these bites would have been less powerful than that of dromaeosaurids. Healed puncture wounds in their teeth also point to the possibility of struggles in prey capture (Sereno and Novas, 1993). The skull length of *Herrerasaurus* is nearly equivalent with the length of the neck, suggesting that it was suited to capture larger prey items than the similarly sized coelophysoids with smaller heads (Holtz, 2003).

The forelimbs of early carnivorous dinosaurs generally exhibit high mobility of the wrist for prey capture and manipulation. The forelimbs of *Herrerasaurus*, for instance, are less than half the length of the hindlimbs and are highly specialized for grasping prey. Although there is more limited mobility at the elbow, a mostly fused carpus allows dorsoventral flexion and extension of the hand. The larger deltopectoral crest; the short but stout humerus, radius, and ulna; and a longer manus are all suggestive of powerful forelimb musculature (Sereno, 1993). The extralong penultimate phalanges in its manus, the visible metacarpal extensor muscle insertion sites, and large, recurved claws are ideal for grasping and raking motions for prey capture and manipulation (Sereno, 1993).

The small, 1-m-long, bipedal *Eoraptor* is another Late Triassic dinosaur from Argentina that possesses more of the basal "wedge-like" shaped skull, with a pointed snout, large, circular orbits, and an enlarged subrectangular antorbital fenestra (Sereno et al., 2013) with no clear maxillary sinus (Witmer, 1997). Its temporal region is rostrocaudally narrow with a vertical rectangular postorbital region and adductor chamber (Sereno et al., 2013). It has a small, slightly caudally expanded palate. Its mandible is long and narrow with a low to almost nonexistent coronoid eminence, small mandibular fenestrae, and small retroarticular process. The overall *Bauplan* of the skull is what one would expect a basal theropod to look like (Hendrickx et al., 2015b; Baron et al., 2017a), although *Eoraptor* has also been suggested by some to have been a basal sauropodomorph (Martinez et al., 2011; Sereno et al., 2013).

The tooth row in *Eoraptor* consists of 4 premaxillary teeth, 17 maxillary teeth, and at least 20 dentary teeth.

Their teeth are somewhat heterodont, with both ziphodont and folidont teeth along the tooth row with some constriction at the base of each tooth (Hendrickx et al., 2019). Generally, the tooth crowns in lateral view are gently arched along their mesial margins and nearly straight along their distal margins (with only a slight curvature). The denticulate margins are unique in that the denticles are slightly hooked toward the apex of the crown, indicating a gripping function (Hendrickx et al., 2019) or possibly even an herbivorous diet with an orthal pulping feeding mechanism (Sereno et al., 2013). The small size and overall gracile morphology of the skull of *Eoraptor* point to the assertion that it is relatively weak. Henderson (2002) attributed its larger, more circular orbits and small dentition as indicators of low bite force and cranial strength overall. Although there is no evidence of cranial kinesis, limited intramandibular flexion is plausible (Sereno et al., 2013).

Other early dinosaurs seem to have been experimenting with different feeding characteristics before the major clades came to fruition. *Daemonosaurus* is a Triassic genus of possible basal theropod with large orbits and uniquely large, procumbent front teeth and shorter rostrum not seen in other Triassic carnivorous dinosaurs (Nesbitt and Sues, 2021). *Liliensternus* was a gracile bipedal predator (roughly 5 m long) that may have fed upon some larger herbivores like the early sauropodomorphs (Paul, 1988). *Chilesaurus* is an example of a basal dinosaur with surprisingly herbivorous craniodental traits (including a beaklike mouth) with unknown origins, with different studies disagreeing about whether it is a basal tetanuran theropod (Novas et al., 2015; Müller et al., 2018) or a basal ornithischian (Baron and Barrett, 2017).

Tawa (from the Late Triassic of New Mexico) was less than 3 m long and is mostly recognized as an early-diverging theropod (Ezcurra and Brusatte, 2011; You et al., 2014; Baron et al., 2017a), although one phylogenetic analysis indicated that it might be a basal coelophysoid (Martinez et al., 2011). It has a long skull, large circular orbits, and a large antorbital fenestra with what looks to be a deeper maxilla than that of coelophysoids (described below). The postorbital region is caudally restricted and nearly vertical. The mandible is long and narrow, with a low to nearly nonexistent coronoid eminence, a visible mandibular fenestra, and a small retroarticular process reminiscent of coelophysoids (see below). Their teeth are ziphodont and of variable size (Nesbitt et al., 2009). In an extensive study that reconstructed forelimb musculature in *Tawa*, Burch (2014) showed that the muscles of the shoulder in *Tawa* (and other basal theropods) share more similarities with those seen in basal sauropodomorphs and basal ornithischians than to some later theropods (like dromaeosaurids). Powerful shoulder movements were likely based on enlarged m. supracoracoideus and m. deltoideus muscle bodies. Moreover, enlarged forearm muscles and intrinsic hand muscles also support the notion of extensive grasping and manipulation (Burch, 2014).

Coelophysoidea

Coelophysoid skulls are elongate and narrow in profile and narrow in dorsal view (e.g., *Coelophysis*, known from the Late Triassic of New Mexico, South Africa, and Zimbabwe) (fig. 4.4). They have bladelike teeth that run continuously down the jaws and a slight notch between the premaxilla and maxilla that accommodated the crown of a bladelike dentary tooth. The orbit is large and circular (although slightly smaller relative to those of *Herrerasaurus*; Henderson, 2002), as is the especially elongate and robust antorbital fenestra. The temporal region is small and caudally restricted, and the postorbital region and adductor chamber are also rostrocaudally narrow compared to the length of the skull. The palate is narrow and flat, with moderately sized pterygoid flanges. The mandible is also long and narrow, with a low, nearly nonexistent coronoid eminence, a visible mandibular fenestra, and prominent retroarticular process. Some *Coelophysis* specimens even show paired long hyoid bones (first ceratohyals with an expanded rostral end) beneath the jaw.

Coelophysis skulls show 4 premaxillary teeth, up to 26 maxillary teeth, and up to 27 dentary teeth (Colbert,

1989). Their teeth are mostly ziphodont (although the mesial-most teeth are unserrated) and are more uniform in size relative to more derived theropods (Colbert, 1989; Hendrickx et al., 2019). A recent histological analysis of *Coelophysis* teeth showed a tripartite periodontium, as seen in crocodilians and mammals, suggesting that this is a character that is plesiomorphic to dinosaurs as well. As mentioned above, a small subnarial toothless gap and slight notch is seen at the oral junction between the premaxilla and maxilla (a character that is more prominent in dilophosaurids), possibly used for helping grasp and manipulate prey items, as is also seen in crocodylomorphs (Holtz, 2003). Henderson and Weishampel (2002) showed that *Coelophysis* represented a more basal form in which the maxilla is narrower (not deflecting ventrally as in most other theropods) with weaker teeth at a greater quantity and that are roughly the same size.

Rayfield (2005a), performing FEA on a lateral view of its skull, showed that stress distributions throughout the skull are similar to *Allosaurus* and *Tyrannosaurus*, with dorsal compression and ventral tension. *Coelophysis*, like *Allosaurus*, exhibits the most shear and compression in the frontoparietal region of the skull (fig. 4.5) and much less stress on the nasal region (speaking to its relatively much more gracile morphology compared to that of *Tyrannosaurus*). Rayfield (2005a) suggested that *Coelophysis* shows signs of its own unique specializations among theropods—an assertion later confirmed by Sakamoto (2010), who showed that the skull morphology of coelophysoids occupies its own specialized functional morphospace in terms of mandibular lever arm mechanics. This specialization is especially interesting given that coelophysoids are located near the base of Theropoda. Owing to their long snouts, the mechanical advantages across the tooth row range from among the highest mechanical advantage values among theropods at the distal end of the tooth row (located near the attachment of m. adductor mandibulae externus, or mAME) to among the lowest mechanical advantage values at the mesial end of the tooth row.

Coelophysoids were small, gracile theropods (roughly 3 m in length) with long hindlimbs and tail, and an elon-

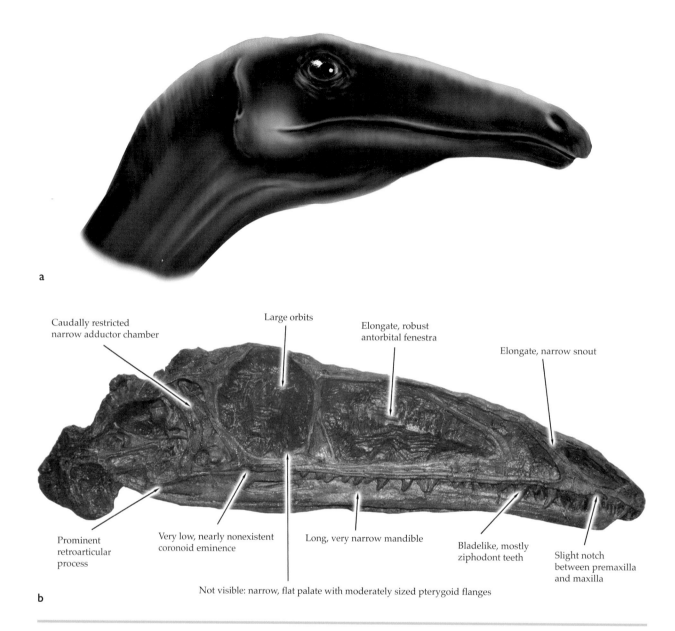

a

b

Caudally restricted
narrow adductor chamber

Large orbits

Elongate, robust
antorbital fenestra

Elongate, narrow snout

Prominent
retroarticular
process

Very low, nearly nonexistent
coronoid eminence

Long, very narrow mandible

Bladelike, mostly
ziphodont teeth

Slight notch
between premaxilla
and maxilla

Not visible: narrow, flat palate with moderately sized pterygoid flanges

Figure 4.4. *Coelophysis* (a) reconstructed head and (b) skull in lateral view
(length up to approximately 27 cm).

gated neck relative to its smaller skull, which suggests that they likely preyed on much smaller animals than other similarly sized theropods with larger heads (Holtz, 2003) (fig. 4.6). Their shorter forelimbs with large three-clawed hands were well suited for grasping for prey items (Colbert, 1989). Unlike most other theropod groups, coelophysoid forelimbs show more isometric growth with ontogeny than the usual negative allometry

(Palma Liberona et al., 2019), possibly indicating greater continued forelimb use in prey capture into adulthood. With a femur shorter than its tibia, *Coelophysis*, along with most other theropods, was clearly a cursor (Colbert, 1989; Persons and Currie, 2016). Computer modeling of the hindlimb musculature suggests that they stood with a more upright posture with extended hindlimbs for improved cursoriality (Carrano, 1999; Bishop

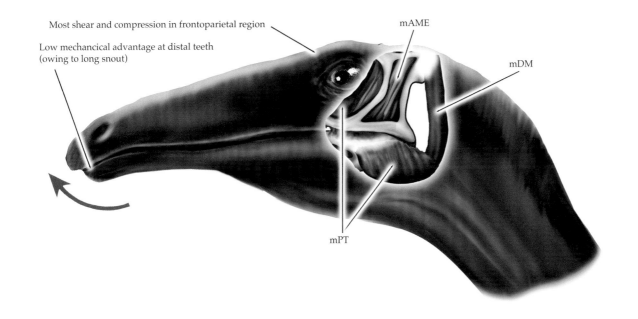

Figure 4.5. *Coelophysis* head showing feeding function with jaw muscles. Red arrow shows orthal jaw action. Abbreviations: mAME, m. adductor mandibulae externus; mDM, m. depressor mandibulae; mPT, m. pterygoideus.

Figure 4.6. Skeletons of *Coelophysis* adult with juvenile.

et al., 2021). Additionally, recent analysis of tail biomechanics in *Coelophysis* has suggested a great deal of lateral flexion of the tail in locomotion to regulate angular momentum (Bishop et al., 2021). Although cololites (preserved stomach contents) associated with *Coelophysis* were previously thought to have revealed cannibalistic behavior (Colbert, 1989), closer morphological analysis of these cololites showed that they were actually the remains of early crocodylomorphs—most likely *Hesperosuchus* (Nesbitt et al., 2006).

Dilophosauridae

Dilophosaurid skulls are also elongate, yet slightly more robust than that of coelophysoids, with elaborate cranial crests made of thin sheets of bone above the orbits (e.g., *Dilophosaurus* from the Early Jurassic of northern Arizona) (fig. 4.7). Their ziphodont teeth are variably robust and long and run down the jawline, with a much more prominent notch between the maxilla and down-hooked premaxilla that corresponds to a raised dentary tooth owing to a slight bump in that region. The skull of *Dilophosaurus* consists of about 4 premaxillary teeth, 12 maxillary teeth, and 17 dentary teeth (Welles, 1984). Their nares are elongate, and their orbits are more rostrocaudally compressed. Their antorbital fenestrae are relatively slightly smaller than in coelophysoids; however, they also have a small promaxillary fenestra just rostral to it. Their temporal region is caudally restricted, with the postorbital region and adductor chamber expanding as it extends rostroventrally, and their palates are broader compared to those of coelophysoids. The mandible is typically narrow as it extends rostrally (with an expanded symphyseal end) and is much more robust caudal to the dentition, with a deep postdentary region and gently sloped coronoid eminence. A small mandibular fenestra is visible, and the retroarticular is blunt yet bulky.

Dilophosaurus shows more prominent cranial features related to prey capture relative to more basal theropods. The constricted nature of the craniomandibular joint supports an orthal (up-and-down) jaw-closing mechanism. Welles (1984) suggested that *Dilophosaurus* might have had an additional, and very slight, propalinal (fore-aft) kinetic motion of the cranium while feeding; however, Holliday and Witmer (2008) argued against this in theropod dinosaurs as a whole owing to intracranial restrictions not allowing such movements to occur.

Therrien et al. (2005), using beam modeling methodology measuring the cross-sectional geometry of the mandible at different parts of the jaw, found that dorsoventral and labiolingual strength in *Dilophosaurus* jaws is focused more so at the more mesial section of the tooth row (at about the second alveolus) than at the middentary section. This is further supported by the symphyseal buttressing of the mandible to withstand more loading rostrally during prey capture and manipulation (also seen in felids and crocodilians with a powerful rostrally focused bite used to kill prey). A majority of the loads on the mandible were from the actual bite and less so from torsional and labiolingual loading patterns. These combined traits point to the assertion that *Dilophosaurus* shows mandibular adaptations suited for more mesially focused prey-slashing bites (Therrien et al., 2005), an idea also hypothesized by Paul (1988) (fig. 4.8). Therrien et al. (2005) suggested that dilophosaurids likely preyed on smaller animals, and these slashing bites were likely delivered after moving the prey to the concave part of the jaw opposed to the largest teeth of the maxilla, as is seen in crocodilians (Drongelen and Dullemeijer, 1982; Busbey, 1989).

Using lever arm mechanics analysis, Sakamoto (2010) also tested relative bite force of the mandible and therein binning *Dilophosaurus* as a coelophysoid, showed that it had among the highest distal relative bite forces (although slightly lower than in true coelophysoids) and among the lowest mesial relative bite forces (although slightly higher than in true coelophysoids as well as spinosaurids and basal coelourosaurs). These results show a clear focus on bite speed rather than mandibular mechanical advantage, although it is more mechanically advantageous than coelophysoid jaws as a whole.

Dilophosaurids were small- to medium-sized theropods that were generally larger than coelophysoids, with a skull that was itself larger and longer with respect to its

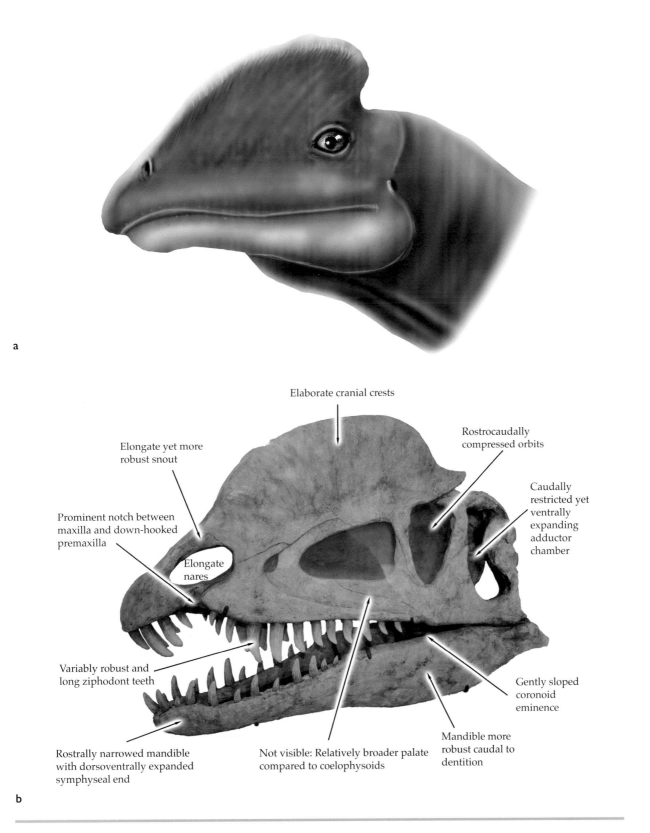

Figure 4.7. *Dilophosaurus* (a) reconstructed head and (b) skull in lateral view (length approximately 50 cm).

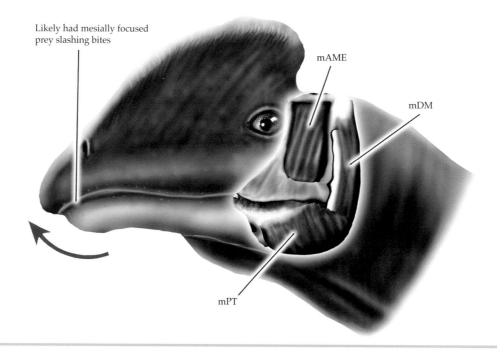

Likely had mesially focused
prey slashing bites

mAME

mDM

mPT

Figure 4.8. *Dilophosaurus* head showing feeding function with jaw muscles. Red arrow shows orthal jaw action. Abbreviations: mAME, m. adductor mandibulae externus; mDM, m. depressor mandibulae; mPT, m. pterygoideus.

neck length, implicating more strength in the neck for prey capture (Holtz, 2003; Senter and Juengst, 2016; Marsh and Rowe, 2018, 2020; Senter and Sullivan, 2019). In a morphological and range-of-motion study of the longer forelimbs in *Dilophosaurus*, Senter and Sullivan (2019) concluded that this genus was capable of a two-handed strong prehension of objects, a one-handed grip and hold on smaller objects, seizing prey just beneath its chest, clutching prey to its chest, hooking prey with its claws, and bringing objects to its mouth, among other movements. Although mandibular mechanics studies have suggested dilophosaurids ate smaller prey items (Therrien et al., 2005), other studies have noted they likely preyed on larger animals as well, as evidenced by bite marks in non-sauropod sauropodomorph *Sarahsaurus* with associated shed *Dilophosaurus* teeth (Marsh and Rowe, 2018, 2020). Their diet likely included a broad range of animals and possibly even fish, suggested because of their strong, longer arms with functionally tridactyl clawed hands and expanded rostral end of the jaw with interlocking teeth (Welles, 1984; Milner and Kirk-

land, 2007; Senter and Sullivan, 2019; Marsh and Rowe, 2020). With three individuals being found in the same locality, gregarious behavior has been hypothesized for *Dilophosaurus*, possibly indicating pack-hunting behavior (Rowe and Gauthier, 1990).

Ceratosauria

Averostra consists of two major theropod clades: Ceratosauria and Tetanurae. Averostrans evolved what is known as the "oreinorostral" skull shape condition, which basically means "hatchet head" owing to the deep and robust nature of the postorbital region of the skulls that becomes triangular rostrally (although they are relatively slenderer mediolaterally). This skull shape would have been ideal for quick prey attacks but were not strong enough to hold prey in the jaws (Busbey, 1995; Holtz, 2003; Rayfield, 2005a).

Ceratosauria includes ceratosaurids (e.g., *Ceratosaurus* from the Late Jurassic Morrison Formation of western

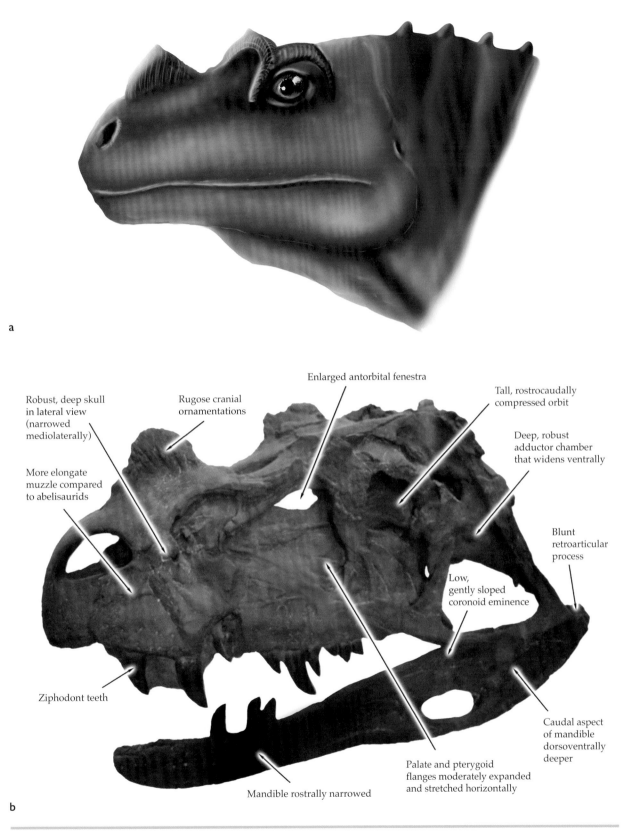

a

Enlarged antorbital fenestra

Robust, deep skull
in lateral view
(narrowed
mediolaterally)

Rugose cranial
ornamentations

Tall, rostrocaudally
compressed orbit

Deep, robust
adductor chamber
that widens ventrally

More elongate
muzzle compared
to abelisaurids

Blunt
retroarticular
process

Low,
gently sloped
coronoid eminence

Ziphodont teeth

Caudal aspect
of mandible
dorsoventrally
deeper

Mandible rostrally narrowed

Palate and pterygoid
flanges moderately expanded
and stretched horizontally

b

Figure 4.9. *Ceratosaurus* (a) reconstructed head and (b) skull in lateral view (length approximately 60 cm).

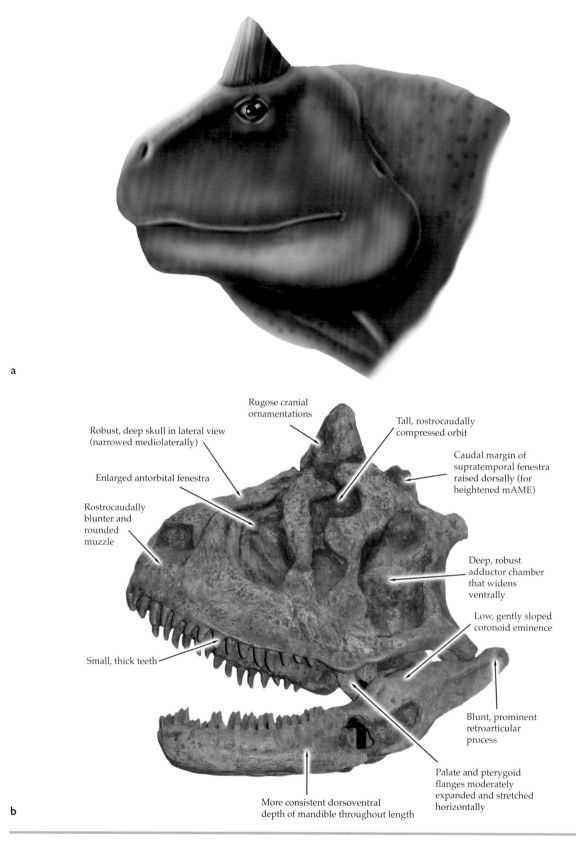

a

b

Rugose cranial ornamentations

Robust, deep skull in lateral view (narrowed mediolaterally)

Tall, rostrocaudally compressed orbit

Caudal margin of supratemporal fenestra raised dorsally (for heightened mAME)

Enlarged antorbital fenestra

Rostrocaudally blunter and rounded muzzle

Deep, robust adductor chamber that widens ventrally

Low, gently sloped coronoid eminence

Small, thick teeth

Blunt, prominent retroarticular process

More consistent dorsoventral depth of mandible throughout length

Palate and pterygoid flanges moderately expanded and stretched horizontally

Figure 4.10. *Carnotaurus* (a) reconstructed head and (b) skull in lateral view (length approximately 60 cm).

North America, and possibly elsewhere) (fig. 4.9); abelis-aurids (e.g., *Abelisaurus*, *Carnotaurus*, *Majungasaurus*, *Spectrovenator*, *Rugops*, and more, all from the Middle Jurassic on through the Cretaceous Period and ranging from Africa, including Madagascar, to the Indian subcontinent and South America); and noasaurids (e.g., *Masiakasaurus*, *Elaphrosaurus*, and *Limusaurus* from the Late Jurassic to Late Cretaceous of South America, Madagascar, and Asia). Both ceratosaurid and (especially) abelisaurid skulls are generally dorsoventrally robust and deep while mediolaterally narrow. Their skulls are ornamented along the skull roof with thick, rugose horned structures, an enlarged antorbital fenestra, and rostrocaudally compressed but tall orbit.

Abelisaurid crania (figs. 4.10 and 4.11) possess neurovascular channels within the nasal bones that have been linked to either thermal exchange, epidermal display structures, or possibly even integumentary sensory organs (Cerroni et al., 2020). As in other large-bodied theropods, Henderson (2002) found a decrease in orbital size from the more ancestral state (e.g., in coelophysoids), with a smaller orbit size in *Ceratosaurus* (oriented at a uniquely low angle with respect to the long axis of the skull) and, to a greater degree, the abelisaurids *Carnotaurus* and *Abelisaurus*. The orbits of these abelisaurid taxa show a rostrally oriented bony extension coming off of the postorbital bone that almost closes off the dorsal part of the orbit from the ventral part—a feature also seen in *Carcharodontosaurus* and *Tyrannosaurus* that is said to be an adaptation for stronger skulls as a whole (Henderson, 2002).

Ceratosaurid muzzles are generally more elongate

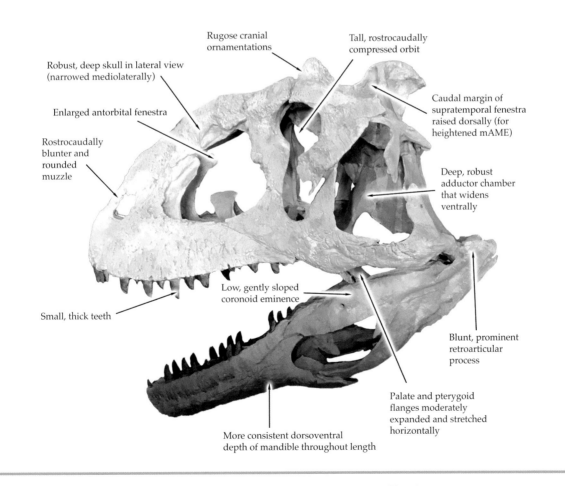

Figure 4.11. Skull of *Majungasaurus* in lateral view (length reaching approximately 60 to 70 cm).

and possess larger, strong ziphodont teeth (Hendrickx et al., 2019; Henderson and Weishampel, 2002). Conversely, abelisaurid muzzles are relatively rostrocaudally blunter and rounded, with smaller (but thick) teeth (fig. 4.12). Candeiro et al. (2017) found tooth wear in some abelisaurids showing tooth-on-tooth wear patterns in a scissorlike cutting action, typical of carnivorous theropods. This is further supported by findings of abelisaurid tooth replacement rates suggestive of increased wear, with a particularly high tooth replacement rate of 56 days in the abelisaurid *Majungasaurus* in contrast to other theropods such as the ceratosaurid *Ceratosaurus*, with tooth replacement rates of about 107 days (D'Emic et al., 2019). Noasaurid skulls are narrower in profile, with some (e.g., *Masiakasaurus*) showing procumbent premaxillary and symphyseal dentary teeth bizarrely hooking outward rostrally (with corresponding inflections of the bone). The teeth of the noasaurid *Limusaurus* are small in young individuals and are lost completely as the animal ages, likely indicating an ontogenetic shift from a more omnivorous to a more herbivorous lifestyle (Wang et al., 2016).

The temporal and postorbital regions (along with the adductor chamber) of ceratosaurids and abelisaurids are deep and robust, with the adductor chamber and infratemporal fenestra widening rostrocaudally as they extend

Figure 4.12. Comparison of teeth in (*left*) a ceratosaurid (*Ceratosaurus*) and (*right*) an abelisaurid (*Majungasaurus*).

straight ventrally. The caudal margin of the supratemporal fenestrae in the abelisaurid *Carnotaurus* is raised dorsally, which would correspond with a dorsally heightened origin of mAME. In noasaurids, the adductor chamber is vertically oriented and narrower. The palate and pterygoid flanges are moderately expanded and stretched horizontally, with visible striation indicating a substantial origin for m. pterygoideus ventralis (Snively and Russell, 2007a). Ceratosaur mandibles are generally robust caudally and narrowed rostrally along the tooth row with a symphysis that is dorsoventrally deeper than its rostrocaudal breadth (Therrien et al., 2005). In *Ceratosaurus*, the dorsoventral depth of the mandible narrows continuously as it extends rostrally, whereas in abelisaurid mandibles, this depth remains consistent (Therrien et al., 2005). As stated previously, some noasaurids (e.g., *Masiakasaurus*) have a rostrally deflected dentary symphysis with corresponding hooked teeth. The coronoid eminences of ceratosaurs are generally low and gently sloped (as in most theropods). Nonetheless, their mandibular fenestrae are large, and their retroarticular processes are blunt but prominent.

Sakamoto (2010) showed *Ceratosaurus* to have among the highest overall mechanical advantages both at the mesial and distal ends of the jaw compared to most other theropods, even abelisaurids, although these advantages are comparable to the megalosaurid *Eustreptospondylus* and the carcharodontosaurid *Carcharodontosaurus*. Abelisaurids were shown to have MA values closer to those of most allosauroids. Molnar (2013) similarly found a higher overall MA of the mandible (and relatively larger cranial adductor muscle sizes) in *Ceratosaurus* than tyrannosaurids and allosaurids, also suggesting a powerful bite. Brusatte et al. (2012) found ceratosaurs in general (mostly abelisaurids) to plot near most large-bodied carnivorous theropods in functional morphospace (with deeper skulls postorbitally and narrower orbits). An exception is *Limusaurus*, with its longer skull, shallower cheek region, and bigger orbits, presenting in a separate position in morphospace that supports the idea of this taxon as more herbivorous.

Ceratosaurus is a prime example of a more basal large-bodied carnivorous theropod, and as such, *Ceratosaurus* feeding studies often can tell us a lot about the evolution of theropod feeding ecology. Quantitative analysis by Therrien et al. (2005) showed that the *Ceratosaurus* was adapted for a more powerful mesial-to-middentary bite (showing more consistency in mandibular bite strength across the tooth row than in dilophosaurids), with a focus on capturing and biting down on and subduing prey with the rostral half of the muzzle (fig. 4.13). The middentary alveolar region is also where the largest teeth are seen in the mandible, so it makes sense that the bite would be of especially higher efficiency in that region. Therrien et al. (2005) found that the maximum bite force in *Ceratosaurus* was comparable to that of *Alligator mississipiensis* (the American alligator) and that the loads across the mandible were constrained and good for cutting meat. The mandibular strength at the rostral end of

the jaw of *Ceratosaurus* is comparable to that of *Varanus komodoensis* (the Komodo dragon), which, along with incisiform premaxillary teeth and an expanded symphysis, suggests that their prey was likely smaller bodied (relative to that of other larger carnivores) and that they didn't put up much of a fight while in the mouth. Mazzetta et al. (1998) suggested that *Ceratosaurus* had a stronger bite throughout the jaw because of a greater moment arm length of m. pterygoideus ventralis (therein called m. pterygoideus anterior). Therrien et al. (2005) rejected this assertion, however, showing that although *Ceratosaurus* constrains dorsoventral mandibular forces throughout the jaw, the actual bite strengths of abelisaurids are more powerful.

Snively and Russell (2007a) found that *Ceratosaurus* possessed rather enhanced muscle insertion sites for ventroflexors of the neck (i.e., muscles that pull the head

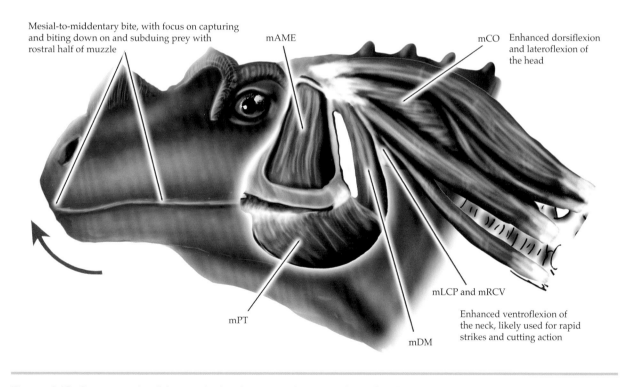

Figure 4.13. *Ceratosaurus* head showing feeding function with jaw muscles and neck muscles (redrawn from Snively and Russell, 2007a), with highlighted key neck muscles mentioned in text. Red arrow shows orthal jaw action. Abbreviations: mAME, m. adductor mandibulae externus; mCO, m. complexus; mDM, m. depressor mandibulae; mLCP, m. longissimus capitis profundus; mPT, m. pterygoideus; mRCV, m. rectus capitis ventralis.

downward), specifically, m. longissimus capitus profundus and m. rectus capitis ventralis (fig. 4.13). They also found larger insertion sites for m. complexus, which would have been instrumental in dorsiflexion and lateroflexion of the head. Because of this neck muscle morphology along with the relatively weaker jaw muscles compared to that of tyrannosaurids, Snively and Russell (2007a) concluded that *Ceratosaurus* was likely using rapid strikes and cutting actions to snap at and capture their prey (albeit not too forcefully) and were also raking their upper ziphodont teeth into the prey they captured. As ceratosaurs in general have shorter forelimbs and shorter digits in the hand, using their arms and hands in prey capture was not a particularly viable option (Holtz, 2003; Snively and Russell, 2007a; Carrano and Choiniere, 2016). They retained a lot of flexibility in their spines and tails, however, with supposedly weaker development of interspinous ligaments (Bakker and Bir, 2004), and the presence of a large cnemial crest on its tibia is suggestive of powerful knee extension and indicative of rapid acceleration (albeit still to a relatively slow pace) (Snively and Russell, 2007a).

Ceratosaurus likely fed on a broad range of animals that reached the size of smaller to medium sauropods as well as ornithischians. Fish and other aquatic animals have also been suggested as a possible food source owing to associations with shed teeth (Bakker and Bir, 2004). Perhaps most intriguing is a study by Drumheller et al. (2020) that noted a possibility of *Ceratosaurus* feeding on a theropod of comparable size, *Allosaurus*, owing to the presence of bite marks, although it is unclear whether these bite marks truly belonged to *Ceratosaurus*. If it indeed was *Ceratosaurus*, however, it would indicate the first known instance of a large theropod feeding on another theropod of equal size from the same environment (Drumheller et al., 2020).

Abelisaurid feeding mechanics have been studied in great detail as well (fig. 4.14). Therrien et al. (2005) found that the mandibular strength and function in the abelisaurids *Carnotaurus* and *Majungasaurus* (therein referred to as its prior name, *Majungatholus*) share even more similarities in their feeding apparatus with the Ko-

modo dragon, such as adaptations for fast slashing bites without grasping the prey. Monfroy (2017) calculated tooth bending strengths in *Carnotaurus* and came to a similar conclusion, with tooth strengths shown to be greater in the mesial half of the mandible. It is also possible that, like Komodo dragons, they wounded their prey with these bites and waited until the prey died before eating it (Therrien et al., 2005). They likely hunted larger prey and implemented a powerful bite with a maximum force up to twice as much as *Ceratosaurus* (contrary to Mazzetta et al., 1998). Bakker (1998) suggested they used their upper jaw like a club when they attacked, likely similar to a slashing bite proposed by Therrien et al. (2005). The cross-sectional strengths of abelisaurid mandibles are more even along the tooth row compared to that of *Ceratosaurus* and *Dilophosaurus* (Therrien et al., 2005), corroborating with the visibly uniform depth of the dentary along the tooth row.

Mazzetta et al. (1998) proposed that the bite of *Carnotaurus* was faster and, as a result, less powerful (compared to *Allosaurus*), a finding further found with finite element modeling of the mandibular stress patterns (Mazzetta et al., 2004) and overall three-dimensional skull (Mazzetta et al., 2009). This type of bite is more attributed to capturing small prey. But Mazzetta et al. (2009) suggested that the ability of the skull of *Carnotaurus* to withstand greater loads with struggling prey possibly means it could have preyed on larger animals, acknowledging that cranial stress patterns of simultaneous biting and tugging are not much different than that found by Rayfield (2004) in *Tyrannosaurus*. Their results also suggested that the skull was capable of withstanding a hatchet-like biting action (like that hypothesized for *Allosaurus*; Rayfield et al., 2001), with high-impact cranial blows to the head (Mazzetta et al., 2009) (fig. 4.14). Through FEA models, Mazzetta et al. (2009) found that when jaw muscles were activated, stress levels were diminished in many parts of the skull, including the skull roof, part of the braincase, the rostrum, the cheeks, and part of the pterygoid bones—an ideal adaptation for powerful prey capture.

Sampson and Witmer (2007), through osteological

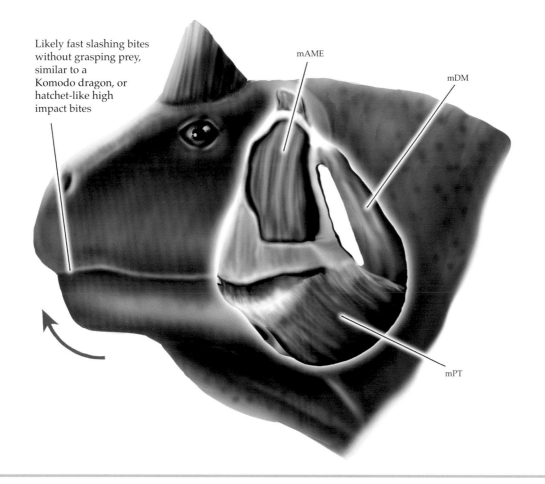

Likely fast slashing bites without grasping prey, similar to a Komodo dragon, or hatchet-like high impact bites

mAME

mDM

mPT

Figure 4.14. *Carnotaurus* head showing feeding function with jaw muscles. Red arrow shows orthal jaw action. Abbreviations: mAME, m. adductor mandibulae externus; mDM, m. depressor mandibulae; mPT, m. pterygoideus.

observations, also suggested that *Majungasaurus* had similar feeding strategies to that of *Carnotaurus* and indicated that *Majungasaurus* was a "headhunter," using criteria by Van Valkenburgh and Molnar (2002). Additionally, Mazzetta et al. (1998) hypothesized that the *Carnotaurus* skull was kinetic with a prokinetic hinge rostrodorsal to the orbits; however, as stated previously, skulls in non-avian theropods have since been deemed likely to be akinetic owing to constraints throughout the skull (Holliday and Witmer, 2008). A moderately kinetic intramandibular joint is also observed in abelisaurids like *Majungasaurus* (Sampson and Witmer, 2007). This feeding apparatus would have been ideal for killing large, slow prey like sauropods by biting and

holding on to their necks and suffocating them (Sampson and Witmer, 2007).

An interesting early abelisaurid known as *Spectrovenator* shows a suite of characters that stand out with respect to the rest of Abelisauridae (Zaher et al., 2020). *Spectrovenator* has a somewhat longer snout than most derived abelisaurids and a structurally different temporal region as well as mandible. Its temporal region created a small adductor origin at the supratemporal fenestra, and it lacked a dorsal projection of the parietal bones that would have supported more adductor muscle origin. The adductor chamber size was small overall as well. Additionally, the mandible shows a longer mandibular fenestra and apparently lacks a clear intramandibular joint, indicating a more generalized theropod feeding mechanism. These characters speak to the progression toward the derived, shorter-muzzled, more muscular abelisaurid feeding apparatus that was also likely influenced by increased body size (Zaher et al. 2020).

As indicated above, the abelisaurid skull was well suited for high stresses with fast strikes, and this is also indicative of powerful neck musculature for withstanding prey tugging while slicing and tearing (Sampson and Witmer, 2007; Mazzetta et al., 2009). As such, it is clear that the head was the only useful tool for prey acquisition because abelisaurid forelimbs are substantially reduced in length, with a shortened humerus and an even shorter radius and ulna that were no larger than their hands (Bonaparte et al., 1990; Holtz, 2003) (fig. 4.15). In a study reconstructing forelimb musculature, Burch (2017) found that *Majungasaurus* had reduced forelimb musculature overall, with much smaller attach-

Figure 4.15. Skeleton of *Carnotaurus*. Note its extremely short forelimbs relative to its elongate hindlimbs.

ment sites and lever arms of brachial musculature that crossed the shoulder joint, reducing their leverage and torque, and antebrachial musculature that was fused into combined sets of flexors and extensors. The muscles acting on the wrist and hands were maintained, however, indicating a small degree of function maintained in the forelimb.

There is variation in hindlimb length among abelisaurids. *Carnotaurus* is known to have longer hindlimbs and a uniquely massive femoral attachment site for m. caudofemoralis body that extends out to attach onto the tail—larger than in most other theropods. M. caudofemoralis retracts the femur caudally during running, meaning its massive size would have made *Carnotaurus* an exceptional cursor for sprinting in short bursts—ideal for hunting fast prey such as ornithopods (Persons and Currie, 2011a). The shorter limbs of *Majungasaurus* were likely not as strong as in *Carnotaurus* but would have been sufficient for ambushing larger titanosaurian prey, as indicated by tooth marks in their bones (Rogers et al., 2003; Sampson and Witmer, 2007). Bite marks in other *Majungasaurus* specimens have also suggested they might have engaged in cannibalistic behaviors (Rogers et al., 2003). At a broader scale, a calcium isotope study also supports the inference that abelisaurids fed on terrestrial animals like herbivorous dinosaurs (Hassler et al., 2018). Track sites have been described as indicating gregarious behavior among abelisaurids (Heredia et al., 2019), but whether this might indicate some degree of pack hunting is unknown.

Noasaurids were generally much smaller than their ceratosaurian relatives, with body plans that resembled those of earlier theropods. Still, some degree of variation is seen in their feeding strategies. The procumbent mesial (front) sharp teeth of *Masiakasaurus* suggest a unique feeding strategy not present in many theropods. Sampson et al. (2001) speculate that it might have had an insectivorous diet, as seen in some caenolestid marsupials that also have procumbent front teeth, or some other diet divergent from most other theropods.

In contrast, *Limusaurus* was an edentulous noasaurid with a keratinous beak in its adult form (although it still had teeth as a juvenile) (Wang et al., 2017) that has been deemed herbivorous (Xu et al., 2009; Zanno and Makovicky, 2011), with isotopic evidence and the presence of gastroliths furthering evidence for this assertion (Wang et al., 2017). Adults show larger and more numerous gastroliths than younger individuals, indicating a complex gut system for digesting plant material. The youngest juveniles show no signs of gastroliths at all. Isotopic evidence suggests that the toothed juveniles might have been omnivorous, showing an ontogenetic transition in feeding ecology (Wang et al., 2017). The forelimbs of *Limusaurus* were highly reduced as well (Carrano and Choiniere, 2016), likely meaning they had little to no use. Moreover, *Berthasaura*, a more recently discovered toothless noasaurid, has also been deemed either herbivorous or omnivorous owing to similar features, although it shows a toothless beak at an earlier growth stage compared to *Limusaurus* (de Souza et al., 2021).

CHAPTER 5

Early Tetanuran, Spinosaurid, and Allosauroid Theropods

Early Tetanurae and Non-Spinosaurid "Megalosauroidea"

For reference in understanding the relationships of theropods discussed in this chapter, see the phylogeny below (fig. 5.1). Tetanurae includes two major theropod groupings ("Megalosauroidea" and Avetheropoda) along with more basal members, such as *Monolophosaurus* and *Sinosaurus* (from the Early to Middle Jurassic of China) as well as *Cryolophosaurus* (from the Early Jurassic of Antarctica). For reference in understanding clade affiliation of non-allosauroid tetanurans discussed in this chapter, see the taxon list below (table 5.1).

Monolophosaurus possesses a long, more rectangular snout with horizontal narial openings and a relatively small, ventrally constricted, and more vertically oriented orbit compared to more basal theropods (Henderson, 2002); a deep cheek region (Brusatte et al., 2012); and a relatively small, rectangular, vertically oriented postorbital region and adductor chamber that is caudally restricted relative to the skull length (fig. 5.2). A longitudinal, dorsally projecting crest extends along the dorsal margin of the snout, likely used for display and not acting to strengthen the skull (Zhao and Currie, 1993; Henderson and Weishampel, 2002). They possess relatively small but strong ziphodont teeth along a horizontal maxilla with a low amount of ventral deflection (Henderson and Weishampel, 2002). The mandible is narrow with a low to nearly nonexistent coronoid eminence and only a slight deepening in the postdentary region.

Table 5.1.
Non-allosauroid tetanuran genera organized by clade

Major clade	Subclade	Genus
Early Tetanurae		
		Monolophosaurus
		Sinosaurus
		Cryolophosaurus
"Megalosauroidea" (non-spinosaurid)		
	Megalosauridae	
		Megalosaurus
		Afrovenator
		Debreuillosaurus
		Torvosaurus
		Eustreptospondylus
	Piatnitzkysauridae	
		Piatnitzkysaurus
		Marshosaurus
Spinosauridae		
	Spinosaurinae	
		Spinosaurus
		Irritator
	Baryonychinae	
		Barynoyx
		Suchomimus

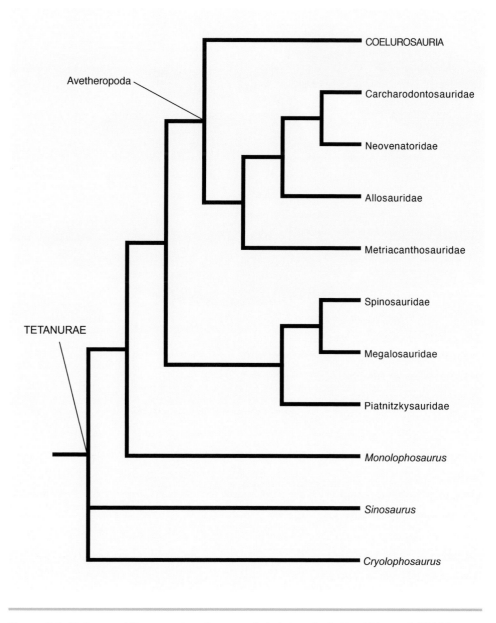

Figure 5.1. Phylogeny of Tetanurae, based on compiled phylogenies in Hendrickx et al. (2015a).

Mandibular bending strength in *Monolophosaurus* was relatively low, equivalent to other similar-sized non-tyrannosaurid theropods (Monfroy, 2017). Relative to most other theropods, *Monolophosaurus* presents with a relatively weak bite, with low mechanical advantage both mesially and distally (Sakamoto, 2010). Rayfield (2011) included *Monolophosaurus* in a comparative two-dimensional (2D) finite element analysis (FEA) modeling study of tetanurans and described intermediate feeding stress patterns and magnitudes in the skull (especially in the quadrate and postorbital region) compared with other tetanurans in the study, with most similarity to the spinosaurid *Suchomimus*. Direct evidence of predatory behavior in basal tetanurans can be seen

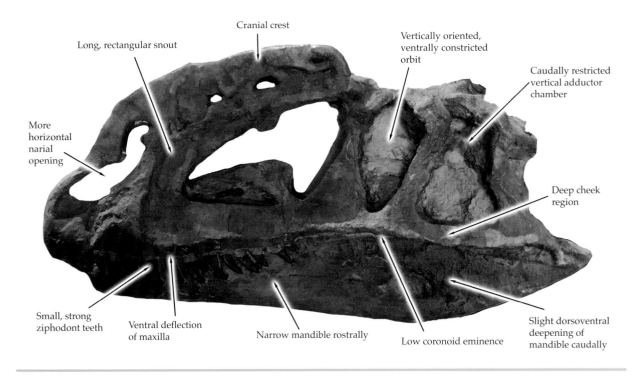

Figure 5.2. Skull of *Monolophosaurus* in lateral oblique view (length approximately 80 cm).

with bite puncture marks likely made by *Sinosaurus* that are seen in a specimen of the non-sauropod sauropodomorph *Lufengosaurus*, resulting in what is suggested as possible osteomyelitis after surviving the attack (Xing et al., 2018).

"Megalosauroidea" (the first of two groupings within Tetanurae; herein put in quotations because of recent phylogenetic analyses finding it a potentially paraphyletic group; Rauhut and Pol, 2019) includes megalosaurids (e.g., *Megalosaurus*, *Afrovenator*, *Debreuillosaurus*, *Torvosaurus*, and *Eustreptospondylus*) and piatnitzkysaurids (e.g., *Piatnitzkysaurus* and *Marshosaurus*)—together with worldwide distribution from the Middle to Late Jurassic—as well as spinosaurids (see below). Among megalosauroids, the megalosaurid skull is relatively more elongate and rostrocaudally narrower than that of ceratosaurids and abelisaurids, with a corresponding robust antorbital fenestra and rostrocaudally constricted orbit. The adductor chamber is subrectangular and oriented ventrally with a downward sloping of the jugal for greater depth. Their teeth are ziphodont (Hendrickx et al., 2015b) and variable in size. Megalosaurid mandibles show a very low to practically nonexistent coronoid eminence, a narrow and elongate dentary ramus, a clear mandibular fenestra, and a blunted yet robust retroarticular process. Sakamoto (2010) measured the adductor leverage of the skulls of *Debreuillosaurus* and *Eustreptospondylus* and found them both to have relatively high mechanical advantage, both distally and especially mesially. In Rayfield's (2011) tetanuran study, the skulls of the smaller-bodied *Dubreuillosaurus* and *Afrovenator* presented with low to intermediate feeding stress patterns and magnitudes in the skull, respectively (especially in the quadrate and postorbital region). *Dubreuillosaurus* showed especially low stresses within the quadrate relative to other tetanurans (Rayfield, 2011). Skull material for piatnitzkysaurids is more fragmentary and largely incomplete, but from the known material, it is fair to assert that their skulls were similar in morphology to those of megalosaurids.

Spinosauridae

Spinosaurids (e.g., *Spinosaurus*, *Baryonyx*, *Irritator*, and *Suchomimus*) ranged worldwide throughout the Cretaceous Period, particularly in Africa, Europe, South America, and Asia. They present with an extreme case of crocodile-like snout elongation (Holtz, 1998), extending its muzzle to nearly five times the rostrocaudal breadth of the subrectangular, vertically oriented postorbital adductor region (fig. 5.3). Their orbits are oriented rostroventrally, and their antorbital fenestra and naris are also rostrocaudally elongate. Although spinosaurids are of large body size, Henderson (2002) suggested that these cranial features contribute to the relatively weak build of the skull, but this assertion is challenged to a certain degree in later functional studies (see below).

Spinosaurids possess small, conical teeth (fig. 5.4) of slightly variable sizes that stretch all the way down the maxillae, premaxillae, and dentaries (the largest of which were positioned at the front of the jaws in a type of "pincerlike 'terminal rosette'" (Holtz, 1998). These conical teeth lack denticles altogether but instead indicate a high likelihood of small prey capture instead of hypercarnivorous feeding (Brink et al., 2015). Spinosaurids replaced their teeth every 59 to 68 days (Heckeberg and Rauhut, 2020)—faster than other known large theropod tooth replacement rates, such as tyrannosaurs (see D'Emic et al., 2019). The oral junction between the premaxilla and maxilla in spinosaurids also forms a prominent, dorsally arched notch with corresponding curvature of tooth placement, which allows room for the apically extended dentary teeth at that level (reminiscent of dilophosaurids and crocodiles). The ventral deflection of the maxilla just distal to that margin contributes to a greater ventral deflection of the maxilla, where the larger maxillary teeth are placed (Henderson and Weishampel, 2002).

The spinosaurid mandible, like that of megalosaurids, consists of a coronoid eminence that is very low to practically nonexistent, a narrow and elongate dentary ramus, a clear mandibular fenestra, and a blunted yet robust retroarticular process. The skull and overall skeleton of spinosaurids have been subject to many studies that elucidate its functional adaptations and paleoecological implications in possible semiaquatic feeding habits, similar to crocodilians and pike conger eels (Therrien et al., 2005; Rayfield et al., 2007; Sakamoto, 2010; Hendrickx et al., 2016; Vullo et al., 2016; Arden et al., 2019; Hone and Holtz, 2019). There is also evidence of some spinosaurids like *Baryonyx* feeding on smaller terrestrial animals, such as iguanodontians (Charig and Milner, 1997; Sues et al., 2002).

It is clear that spinosaurids were orthal feeders like most carnivorous theropods, with their jaw closing straight dorsoventrally. Hendrickx et al. (2016) additionally showed evidence of a mediolateral expansion of the ventral quadrate along with a similarly expanded mandibular articular surface. They suggest that this morphology along with a mobile mandibular symphysis facilitated a mediolateral expansion of the caudal part of the mandibular rami that would have helped open the throat for swallowing, as is seen in pelicans when they feed on fish (Hendrickz et al., 2016). Arden et al. (2019) argued that, in addition to some key postcranial features, the skull of spinosaurine spinosaurids (e.g., *Spinosaurus* and *Irritator*) is adapted for a more semiaquatic lifestyle (spending more of its time in the water) based on the retracted and laterally positioned narial openings and elevated orbits in spinosaurids as a whole, both allowing for breathing and seeing above the surface of water. Hone and Holtz (2019) countered these arguments in a prompt response. With regard to the narial opening, because baryonychine spinosaurids (e.g., *Suchomimus* and *Baryonyx*) have nares that are elevated slightly higher than the orbits, it goes against the argument that spinosaurines are more adapted for a semiaquatic lifestyle than baryonychines. Hone and Holtz (2019) suggested that the retracted lateral nares were used to position just above the water while waiting for passing fish during feeding specifically, as is seen in herons (which do not spend most of their time underwater). Furthermore, with regard to the raised orbits, Hone and Holtz (2019) argued that the orbits are not elevated enough to be held above the water without more of the head being elevated.

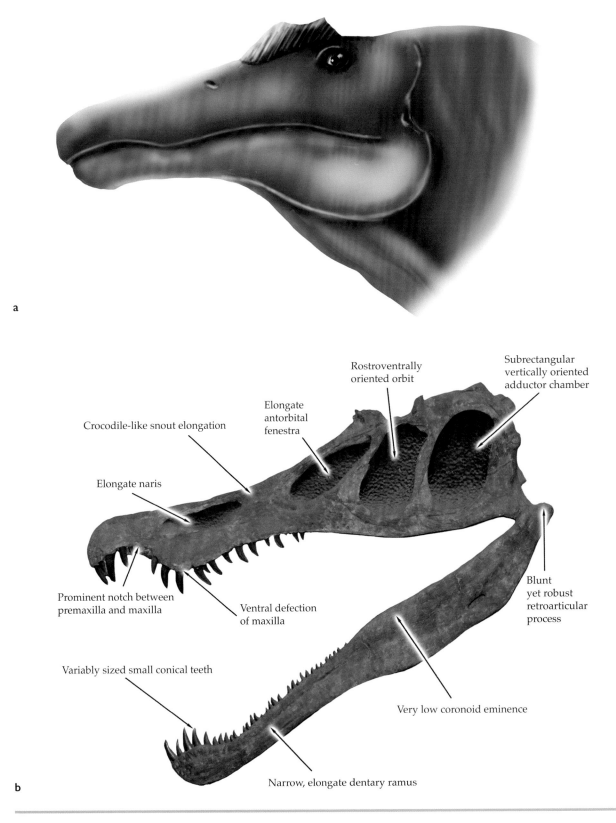

a

b

Crocodile-like snout elongation

Elongate
antorbital
fenestra

Rostroventrally
oriented orbit

Subrectangular
vertically oriented
adductor chamber

Elongate naris

Prominent notch between
premaxilla and maxilla

Ventral defection
of maxilla

Variably sized small conical teeth

Blunt
yet robust
retroarticular
process

Very low coronoid eminence

Narrow, elongate dentary ramus

Figure 5.3. Spinosaurids represented by (a) reconstructed head of *Spinosaurus* and
(b) skull of *Baryonyx* in lateral view (length approximately 90 cm).

Figure 5.4. *Spinosaurus* tooth.

Therrien et al. (2005) found the spinosaurid *Suchomimus* to have a bite with comparable strength to *Alligator mississipiensis* (the American alligator), as proposed for *Ceratosaurus* (see above). According to Therrien et al. (2005), *Suchomimus* exhibited the strongest dorsoventral and mediolateral mandibular bending strengths at the mesial-most landmark (the second alveolus), with a decrease in strength at the middentary mark. They attributed this strength to the rostrocaudal elongation and mediolateral broadening of the mandibular symphysis, providing extra bone to help withstand greater mesially focused forces in all directions. As such, Therrien et al. (2005) suggested that spinosaurids, like dilophosaurids, exhibited adaptations for capture and manipulation of small prey at the mesial end of the tooth row. Owing to a dorsoventral buttressing of the distal portion of the mandible in spinosaurids providing support specifically for the sagittal plane (and not transverse or torsional forces), however, Therrien et al. (2005) suggested a bite-and-hold predatory strategy in spinosaurids (possibly followed by shaking the prey item), contrary to the slashing bites presumed to have been used by dilopho-

saurids. Sakamoto (2010), in his theropod jaw leverage study, found spinosaurids to plot with relatively low mechanical advantage values at both mesial and distal tooth positions, suggesting that their feeding apparatus was focused more on speed of closure rather than overall bite strength and that this further supports the possibility that spinosaurids were piscivorous. Vullo et al. (2016) suggested that spinosaurids shared many convergent cranial adaptations for piscivorous prey capture with pike conger eels, including the mesial notch in the upper jaw supporting larger teeth as well as possible sensory reception in the snout.

Rayfield et al. (2007) performed an FEA study of spinosaurid snout biomechanics (specifically that of *Baryonyx*) compared to that of crocodilians and a generalized large theropod dinosaur to test spinosaurids as being crocodile mimics, as well as the effect of the reduced antorbital fenestrae and rudimentary secondary palate on their feeding mechanics. They concluded that *Baryonyx* was convergent on the extant gharial, a crocodilian with a rostrocaudally elongate and narrow snout (more so than with a generalized large theropod or an American alligator). The main difference they noted is that the snout in *Baryonyx* is mediolaterally compressed throughout its length (contrary to the dorsoventral compression of the gharial snout). Rayfield et al. (2007) found that both the reduction of antorbital fenestra and the addition of a secondary palate aid in strengthening animals with a tubular skull, as is seen in spinosaurids and gharials.

A few years later, using 2D FEA to compare spinosaurids skulls with other tetanurans, Rayfield (2011) found that although *Suchomimus* showed stress patterns and magnitudes comparable to most other tetanuran taxa in the study (with stresses focused around the quadrate), *Spinosaurus* experienced much greater stress across the skull overall (extending into the postorbital-jugal bar, lacrimal, and dorsal parietal region) (fig. 5.5). This suggests that *Spinosaurus* would have experienced bending forces at the snout, which is oddly not exhibited in *Suchomimus*. It is possible that this is an indicator that larger theropods like *Spinosaurus* are constrained in

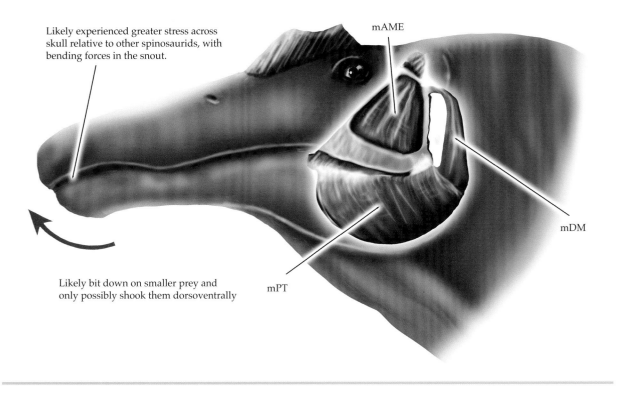

Likely experienced greater stress across skull relative to other spinosaurids, with bending forces in the snout.

mAME

mDM

Likely bit down on smaller prey and only possibly shook them dorsoventrally

mPT

Figure 5.5. *Spinosaurus* head showing feeding function with jaw muscles. Red arrow shows orthal jaw action. Abbreviations: mAME, m. adductor mandibulae externus; mDM, m. depressor mandibulae; mPT, m. pterygoideus.

what it could have eaten and are not as prone to resisting feeding stresses owing to skull lightening. Because this study is in two dimensions, however, Rayfield (2011) acknowledged that there might be a discrepancy in the scaling of adductor musculature because of the lack of the mediolateral dimension to provide more data.

Cuff and Rayfield (2013) measured cross-sectional properties (i.e., the second moment of area) of spinosaurid (and crocodilian) snouts using computed tomography data. They found that spinosaurid snouts have similar bending and torsion resistance when scaled to similar size (with both performing better than extant crocodilians). When true scale is accounted for, the larger, longer, and slenderer skull of *Spinosaurus* did not perform as well as *Baryonyx* in bending and torsion resistance as a whole. The snout of *Spinosaurus* was less resistant in dorsoventral bending than an alligator as well as less resistant to mediolateral bending and torsion than crocodilians as a whole—agreeing with the conclusion by

Rayfield (2011) that *Spinosaurus* overall had a weaker skull. The relatively more robust oreinorostral (triangular) snout of *Baryonyx* is found to have relatively high dorsoventral bending resistance, similar to what Therrien et al. (2005) found in the skull of *Suchomimus*. Contrarily, Therrien et al. (2005) found *Baryonyx* to show less resistance to mediolateral bending and torsion. As a whole, *Spinosaurus* had rostral bending resistance most similar to gharials, whereas the shorter but relatively more robust skull of *Baryonyx* did not, contrary to what Rayfield et al. (2007) concluded. In all, because of the focus on dorsoventral bending resistance on the spinosaurids rostrum, Cuff and Rayfield (2013) agreed with Therrien et al. (2005) that spinosaurids generally bit down on smaller prey and only possibly shook them dorsoventrally (fig. 5.5). They also agreed that spinosaurids were not strictly piscivorous and likely preyed on small terrestrial animals.

Further evidence of piscivory in spinosaurids is also

present outside of the feeding apparatus. Schade et al. (2020) show that the neuroanatomy and angled morphology of the semicircular canals in the spinosaurid *Irritator* indicate that it held its snout angled downward and that it was capable of rapid, controlled downward movements of the head for snatching prey like fish. Spinosaurid hands are especially well adapted for piscivory as well. The recurved, hooklike ungual claw in spinosaurid hands was exceptionally large and would have been ideal for puncturing fish while hunting (Charig and Milner, 1997; Sereno et al., 1998; Sues et al., 2002) (fig. 5.6). *Baryonyx* was possibly a facultative quadruped because of its long skull, long neck, and enlarged humerus (Charig and Milner, 1997). The same was possibly true for *Suchomimus*, but further analysis is likely needed in both cases.

The enigmatic, sail-backed spinosaurid *Spinosaurus* is a unique case in which various aspects of its anatomy have been described separately, but all of which have led to the belief that it may have lived a semiaquatic lifestyle. Ibrahim et al. (2014) found that *Spinosaurus* sported an elongate neck and trunk that shifted the center of mass cranially, a small pelvis, short hindlimbs with solid bones for buoyancy, an enlarged flexor muscle attachment site on the femur, and flat pedal claws that would have helped in foot-propelled locomotion. In a study modeling its buoyancy, Henderson (2018) challenged the idea of a semiaquatic lifestyle for *Spinosaurus*, stating that its center of mass was more caudally placed than previously thought and that it was likely a terrestrial shoreline animal. Subsequently, a recent study by Ibrahim et al. (2020) found that *Spinosaurus* actually possessed a massive, propulsive, flexible tail, with elongate neural spines and chevrons that would have allowed substantial lateral excursion through large bodies of water. Robotic testing of various tail shapes further supported that the tail of *Spinosaurus* was capable of more thrust and efficiency in swimming undulatory motions

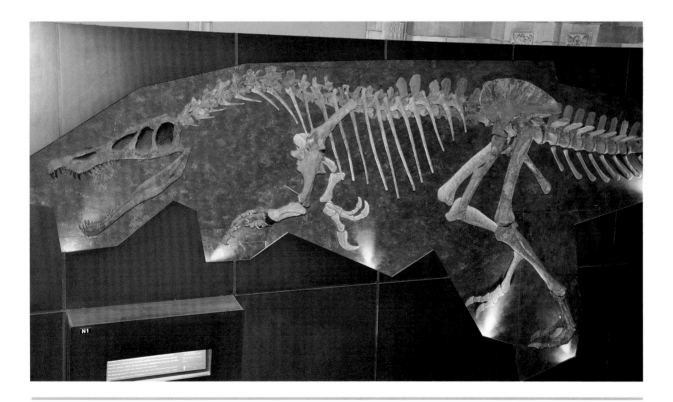

Figure 5.6. Skeleton of *Baryonyx*. Note hooklike ungual claws on hands.

that those of terrestrial dinosaurs, with similarity to living aquatic vertebrates with taller neural spines in the tail used in swimming (Ibrahim et al., 2020). An enlarged m. caudofemoralis as well as epaxial musculature extending down the tail would have acted to laterally pull and steer the elongate tail from side to side.

In a comprehensive study of *Spinosaurus* ecology, Hone and Holtz (2021) put together all known and argued aspects of *Spinosaurus* functional morphology and paleoecology to assess the validity of a specialized aquatic nature of the animal. They argued that although there are aquatic adaptations, the assertion that it is a "highly specialized aquatic predator" is not supported through a combination of multiple lines of evidence. For instance, they argued that the shape of the head, the orbits being on the side of the head, and the nostrils being retracted caudally are not indicative of a fully aquatic animal. The long head and apparently strong ventroflexing abilities of its neck would have allowed a dorsoventral slashing with its snout into the water. Furthermore, they argued the shape of its body would have produced too much variable drag in swimming. Instead of a fully or mostly aquatic lifestyle, they posited that the animal was likely a wader in the shorelines, waiting in shallow waters for fishes or other aquatic life to feed on (Hone and Holtz, 2021).

Still, a more recent study using bone density argues for subaqueous foraging behavior in *Spinosaurus* and even *Baryonyx* to a certain extent owing to increased bone density compared to *Suchomimus* (Fabbri et al., 2022). They make this argument by showing similarities in relative bone densities with living, somewhat more aquatic birds, reptiles, and mammals. With further studies continuing to challenge this assertion, however, and with additional arguments pertaining to anatomy, sampling, and statistics, the matter remains controversial from a morphological standpoint (Hone and Holtz, 2021; Myhrvold et al., 2022; Sereno et al., 2022). Isotopic evidence for both oxygen (Amiot et al., 2010) and calcium (Hassler et al., 2018) has provided further support for the assertion that spinosaurids foraged more so on aquatic life in some form. Amiot et al.

(2010) specifically found that the tooth enamel of spinosaurids was similar to other semiaquatic animals, but interestingly, they were of a different composition than those of other terrestrial theropods. With them found in coastal environments, Holtz (1998, 2003) has indicated that spinosaurids were better capable of accessing prey and traveling across different water sources than other large-bodied theropods and crocodyliforms, likening them to "heron-like stalkers of fish." The most direct evidence of piscivory is with known gut contents associated with *Baryonyx* that show that it preyed on fish, with scales found of the large gar *Lepisosteus* (Charig and Milner, 1997), although there is also some evidence of spinosaurids preying on terrestrial herbivorous dinosaurs, like bones of young *Iguanodon* individuals also associated with *Baryonyx* (Charig and Milner, 1997) and spinosaurid teeth (possibly *Irritator*) embedded in a pterosaur from northeastern Brazil (Buffetaut et al., 2004).

Allosauroidea

Avetheropods (the second tetanuran group) include carnosaurs and coelurosaurs. Carnosaurs are primarily made up of allosauroids, including metriacanthosaurids (e.g., *Sinraptor* and *Yangchuanosaurus* ranging from the Middle Jurassic to Early Cretaceous of Europe and Asia); allosaurids (e.g., *Allosaurus* [fig. 5.7] and *Saurophaganax* from the Late Jurassic Morrison Formation of North America); carcharodontosaurids (e.g., *Mapusaurus*, *Acrocanthosaurus*, *Giganotosaurus*, and *Carcharodontosaurus* ranging worldwide, particularly in Gondwana, from the Late Jurassic through the Late Cretaceous); neovenatorids (e.g., *Neovenator* and *Deltadromeus*); and the more basal *Asfaltovenator*. For reference in understanding clade affiliation of allosauroids discussed in this chapter, see the taxon list in table 5.2.

Allosauroid skulls are striking, whether you are looking at the smaller *Allosaurus*—the best-known allosauroid—or the carcharodontosaurids (fig. 5.8)—by far the largest allosauroids of all (and among the largest theropods as a whole). They possess tall, ovoid, and sometimes

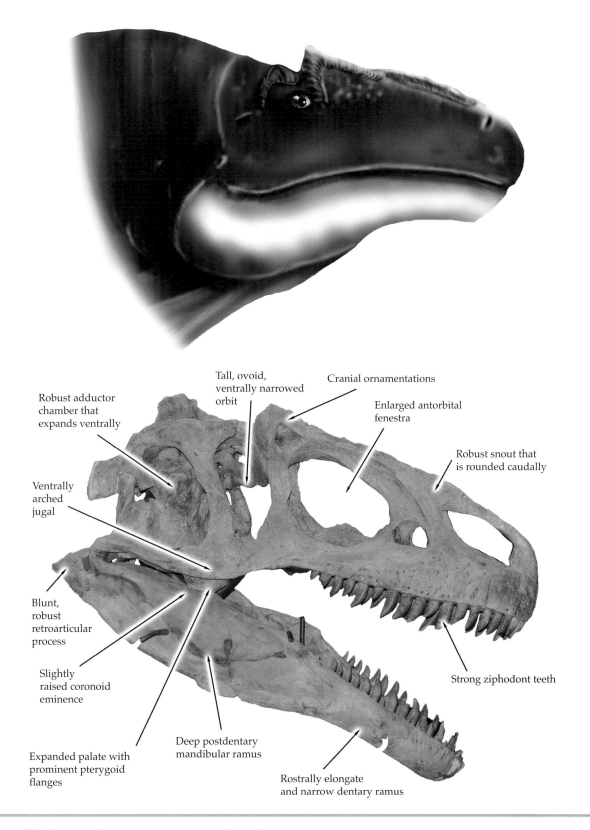

Figure 5.7. *Allosaurus* (a) reconstructed head and (b) skull in lateral view
(length reaching up to approximately 70 to 85 cm).

Table 5.2.
Allosauroid genera organized by clade

Major clade	Subclade	Genus
Allosauroidea		
	Metriacanthosauridae	*Asfaultovenator*
		Sinraptor
		Yangchuanosaurus
	Allosauridae	
		Allosaurus
		Saurophaganax
	Carcharodontosauridae	
		Mapusaurus
		Acrocanthosaurus
		Giganotosaurus
		Carcharodontosaurus
	Neovenatoridae	
		Neovenator
		Deltadromeus

ventrally narrowed orbits, some of which (e.g., *Carcharodontosaurus*, *Giganotosaurus*, and *Sinraptor*) show an additional rostrally projected process from the postorbital (midway down a rostrocaudally compressed orbit), plausibly indicative of greater strength in the cranium (Henderson, 2002). The cranial reconstructions of some allosauroids like *Allosaurus* and *Carcharodontosaurus* show that their binocular vision was restricted to a 20-degree overlap, which is comparable to that of crocodilians but does not match that of tyrannosauroids (Stevens, 2006). The snout is robust but rounded near the orbit with a greatly enlarged triangular-to-ovoid antorbital fenestra. Some, such as *Allosaurus*, possess rugose ornamentation along the skull roof. A small maxillary fenestra is also seen between the antorbital fenestra and the external naris (along with a miniscule subnarial foramen).

The ziphodont dentition (fig. 5.9) runs along the maxilla and dentary (anterior to the level of the orbit). Although many carnosaurs (e.g., *Allosaurus* and *Sinraptor*) have a low ventral deflection of the maxilla and smaller teeth, their teeth are still quite strong (Henderson and Weishampel, 2002). For instance, Hendrickx et

a

Figure 5.8. Carcharodontosaurids represented by (a) reconstructed head and (*facing page*) (b) skull of *Giganotosaurus* (length approximately 1.5 m) and (c) skull of *Acrocanthosaurus* (length approximately 1.2 m).

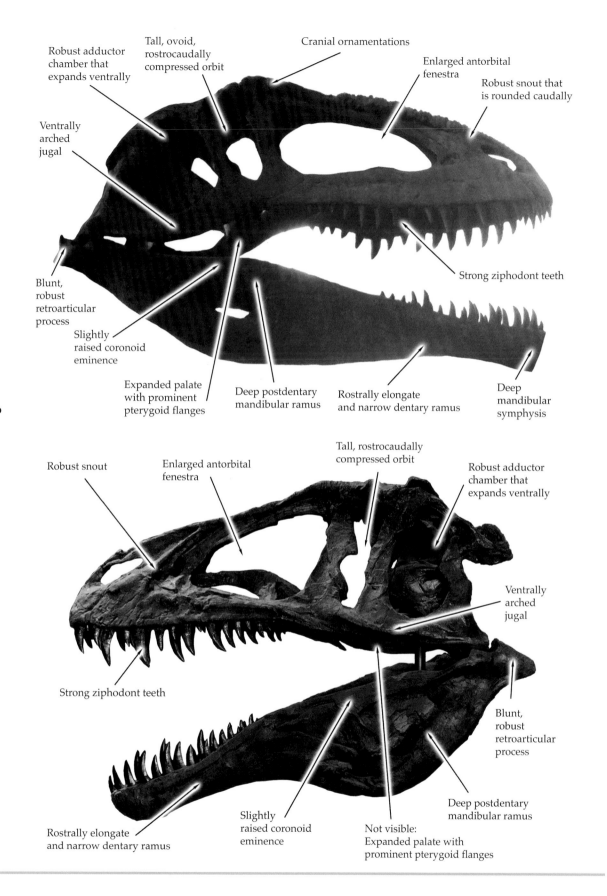

b

Robust adductor chamber that expands ventrally

Tall, ovoid, rostrocaudally compressed orbit

Cranial ornamentations

Enlarged antorbital fenestra

Robust snout that is rounded caudally

Ventrally arched jugal

Blunt, robust retroarticular process

Slightly raised coronoid eminence

Expanded palate with prominent pterygoid flanges

Deep postdentary mandibular ramus

Rostrally elongate and narrow dentary ramus

Deep mandibular symphysis

Strong ziphodont teeth

c

Robust snout

Enlarged antorbital fenestra

Tall, rostrocaudally compressed orbit

Robust adductor chamber that expands ventrally

Ventrally arched jugal

Strong ziphodont teeth

Blunt, robust retroarticular process

Deep postdentary mandibular ramus

Rostrally elongate and narrow dentary ramus

Slightly raised coronoid eminence

Not visible: Expanded palate with prominent pterygoid flanges

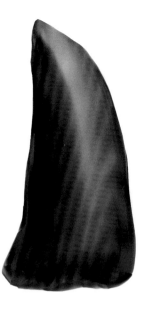

Figure 5.9. *Allosaurus* tooth.

al. (2020) found that the metriacanthosaurid *Sinraptor* was built for predation, with tooth crowns that would have been strong enough to withstand bone-to-bone contact to wound its prey with deep cuts. Although not quite reaching the rates of abelisaurids, *Allosaurus* had high tooth replacement rates relative to other theropods (but close to that of *Ceratosaurus*) at about one per 104 days (D'Emic et al., 2019).

Large neurovascular canals within the skull of the neovenatorid *Neovenator* have been suggested to house enlarged nerves for enhanced facial sensitivity (Barker et al., 2017), although vasculature would have also been housed in these canals. The postorbital adductor region is especially robust, widening rostrocaudally as it extends straight ventrally (as in ceratosaurids), with a jugal that has a ventrally arched body giving even more depth to the adductor chamber. The palate is rostrocaudally and transversely expanded with prominent pterygoid flanges. Allosauroid mandibles are also built strongly, with a slightly raised coronoid eminence seen in many forms as well as a deep postdentary mandibular ramus. Mandibular bending strength in *Allosaurus* was relatively low, equivalent to other similar-sized non-tyrannosaurid

theropods (Monfroy, 2017). The rostral extension of the dentary is elongate and narrowed with many bladed teeth of various sizes. In most larger individuals, the mandibular symphysis deepens ventrally, is deeper than it is long, and is also subvertically oriented (Therrien et al., 2005). The mandibular ramus is variable in size, and the retroarticular process is blunt yet robust. Morphometric analysis by Brusatte et al. (2012) found allosauroid crania also plotted near other large-bodied hypercarnivorous theropods owing to converging similarities in having caudally deeper skulls and narrow, keyhole orbits that add strength to the cranium when biting down.

At a broadscale evolutionary level, Sakamoto (2010) found that allosauroids (therein including metriacanthasaurids, allosaurids, and carcharodontosaurids) as a group generally possess highly efficient feeding systems (especially mesially), comparable to that of tyrannosaurs and ceratosaurs. Among allosauroids, metriacanthasaurids (e.g., *Sinraptor* and *Yanghuanosaurus*) plotted with the lowest mesial-to-distal mandibular mechanical advantage (although by a small margin), with allosaurids (e.g., *Allosaurus*) and especially carcharodontosaurids (e.g., *Acrocanthosaurus* and *Carcharodontosaurus*) plotting higher in the range of abelisaurids. Sakamoto (2010) expressed that his findings for *Allosaurus* contradicted that of Bakker (1998) and Rayfield et al. (2001; see below), who proposed rather weak jaw musculature for *Allosaurus* in particular, suggesting that neck musculature accounted for more of the force in an attack on large prey. Molnar (2013) also found that *Allosaurus* shows lower overall mandibular mechanical advantage values than tyrannosaurids and *Ceratosaurus*, signaling a weaker leverage for biting than these taxa. Furthermore, in a study investigating jaw adductor muscle recruitment and constraint in open-mouth gape, Lautenschlager (2015) found that *Allosaurus* was capable of creating and sustaining the widest gape compared to *Tyrannosaurus* and the herbivorous therizinosaur *Erlikosaurus* (which presented with the smallest gape among theropods by far), confirming the importance of wide gape in predatory behavior.

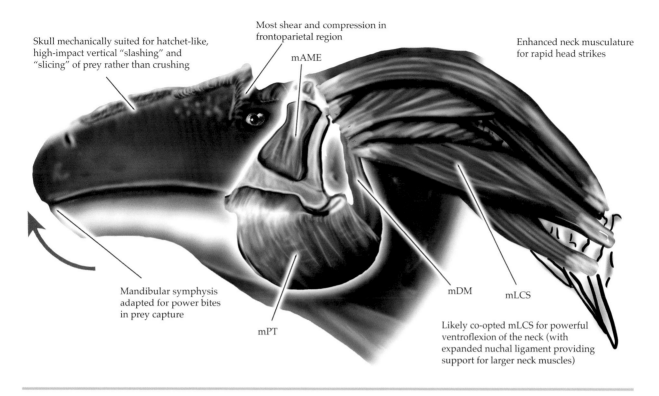

Skull mechanically suited for hatchet-like, high-impact vertical "slashing" and "slicing" of prey rather than crushing

Most shear and compression in frontoparietal region

mAME

Enhanced neck musculature for rapid head strikes

Mandibular symphysis adapted for power bites in prey capture

mPT

mDM   mLCS

Likely co-opted mLCS for powerful ventroflexion of the neck (with expanded nuchal ligament providing support for larger neck muscles)

Figure 5.10. *Allosaurus* head showing feeding function with jaw muscles and neck muscles (redrawn from Snively and Russell, 2007a), with highlighted key neck muscle mentioned in text. Red arrow shows orthal jaw action. Abbreviations: mAME, m. adductor mandibulae externus; mDM, m. depressor mandibulae; mLCS, m. longissimus capitis superficialis; mPT, m. pterygoideus.

Allosaurids (especially *Allosaurus*; fig. 5.10) have been key subjects in the advancement of using computer modeling techniques to study dinosaur feeding biomechanics. In the first study to perform FEA on a dinosaur skull, Rayfield et al. (2001) ran 3D FEA on the skull of *Allosaurus* to model the stresses and strains that likely passed through the head as it bit down on and subdued its prey. As expected, greater magnitudes of stress were seen at points along the skull where there was more bite force as well as force on the jaw joint. They found that *Allosaurus* likely had a substantially weaker muscle-driven bite than that of *Tyrannosaurus* and modern alligators, which suggests that *Allosaurus* was not able to crush bone with its bite. Rayfield et al. (2001) noted that the mediolaterally compressed and recurved teeth of *Allosaurus* means it is better suited for "slashing" and "slicing" at its prey rather than crushing. They also found that although the skull of

Allosaurus is lightly built and open, with many large fenestrae, it is actually strong and capable of withstanding high maximum forces along its tooth row (about 55,000 Newtons, N, of force) before tension could break its teeth or jaws, especially at the central teeth along the maxilla (Rayfield et al., 2001). They noted that should the teeth break off, the skull would still be protected from experiencing greater stresses.

Rayfield et al. (2001) went into great detail to show how the stresses and strains of biting would have traveled throughout the skull of *Allosaurus*. Although shear stresses were found to be relatively low, compressive and tensile stress were prevalent in their analysis. Specifically, the ventral aspect of the skull reflected tensile stress, while the dorsal aspect reflected more compressive forces. Because the skull is built for high-impact vertical forces on the teeth, it stands to reason that the stresses need to make their way upward in the skull. But the sides of the skull are made of strut-like bones (rather than a solid sheet) that form all of the visible fenestrae (Rayfield et al., 2001). After the initial contact with the maxilla, any of the compressive stresses that were not already absorbed by the maxilla would travel through these strut-like bones on the sides of the skull as well as the nasals in the rostral region. Ultimately, Rayfield et al. (2001) found that the dorsal-most compressive stresses were absorbed by its thickened skull roof, and the rest of the compressive forces would travel through "functional loops" around the bones surrounding the fenestrae, and this would have helped manage the stresses in an otherwise lightly built skull.

One of the main reasons Rayfield et al.'s (2001) paper was so revolutionary was because it was able to tell a quantitative story of the most likely predatory behavior of *Allosaurus* (or any dinosaur, for that matter) with nothing but the architecture of the skull. Their seemingly unusual results of a weak muscle-driven bite combined with a strong skull, along with their recurved, mediolaterally compressed sharp teeth and their powerful neck musculature, suggests that *Allosaurus* likely used a "hatchet"-like, high-speed strike of the skull onto its prey. Once its prey is in its mouth, it would have pull its head back while tearing off its flesh to swallow like a Komodo dragon (Rayfield et al., 2001). With a lighter skull and greater speed on impact of the upper jaw taking the place of higher bite force, Rayfield et al. (2001) suggested that *Allosaurus* likely went after smaller, agile prey animals like ornithopods more so than larger ones, although they note that should *Allosaurus* have chosen to capture larger prey, like stegosaurs or even sauropods, they would have chosen to ambush them before they could defend themselves. In reference to critique by Frazzetta and Kardong (2002) arguing the unlikelihood of predatory behavior in *Allosaurus*, Rayfield et al. (2002) promptly responded with counterarguments to their reasoning. For instance, in response to the critique that there are no living predators behaving in the manner they suggested, Rayfield et al. (2002) noted that there is no way to compare it to living predators because there aren't any to the

scale of *Allosaurus*, and it is not appropriate to compare it to living small carnivorous reptiles. Additionally, contrary to what Frazzetta and Kardong (2002) argued, Rayfield et al. (2002) clearly showed that the teeth of *Allosaurus* are variable in shape (suggesting predatory action) and that the cranium is not, in fact, kinetic at the basal joint.

Subsequently, Rayfield (2005a) compared the lateral view FEA results of *Allosaurus* skulls with those of *Coelophysis* and *Tyrannosaurus* and showed the similarities in stress distribution throughout all skulls—specifically, compressive forces dorsally and tensive forces ventrally. When compared to the more robust skull of *Tyrannosaurus*, *Allosaurus* (as in *Coelophysis*) shows the most shear and compression in the frontoparietal region of the skull (which is thickened and has tight interdigitations for extra strength) and much less in the gracile nasal region, which shows an often patent internasal suture (Rayfield, 2005a). With the addition of sutures in her 3D FEA analysis, Rayfield (2005b) further found that cranial sutures in *Allosaurus* effectively accommodated biting stress and strain patterns. Regions that were specifically tested were the quadrate-squamosal joint, the palate, the nasofrontal contact, the nasal-nasal contact, the quadratojugal-jugal contact, joints around the postorbital bones, and the frontoparietal skull roof bones. What is most interesting is how individual joint stresses, strains, and movements combine to create a complex functional framework that, when working as one unit, create the optimal biting capabilities for the taxon (Rayfield, 2005b).

In their comprehensive study of theropod feeding mechanics, Therrien et al. (2005) included the mandibular force profiles of a variety of allosauroids, both allosaurids like *Allosaurus* and what was therein deemed "*Antrodemus*" (possibly another specimen of *Allosaurus* itself) as well as carcharodontosaurids like *Acrocanthosaurus* and *Giganotosaurus*. Among allosaurids, they found that the mandible of *Allosaurus* is nearly as strong dorsoventrally at the mandibular symphysis as it is at the middentary region, suggesting adaptations at the rostral extent of their jaws suitable for powerful bites in prey capture. Labiolingually, the second alveolus is shown to be stronger than at the middentary region. Conversely and interestingly, their results for "*Antrodemus*" show a more linear decrease in mandibular strength toward the rostral extent of the dentary, which is likely a manifestation of its smaller symphysis. This finding would suggest a possible slashing bite behavior like that seen in Komodo dragons, and more similar to the findings of Rayfield et al. (2001) for *Allosaurus*. Whether this is solely a sign of variability in *Allosaurus* itself or if "*Antrodemus*" is indeed its own genus is still unclear. Therrien et al. (2005) did, however, note ontogenetic differences in some *Allosaurus* mandibular properties, like bending rigidity, which implies differences in overall feeding style from juvenile to adult (with juveniles delivering slashing bites). Allosaurid bite forces are suggested to have been comparable to that of American alligators, as suggested for the ceratosaurid *Ceratosaurus* and the spinosaurid *Suchomimus* (see above).

Therrien et al. (2005) noted that the mandibles of the carcharodontosaurids *Acrocanthosaurus* and *Gigano-*

tosaurus (fig. 5.11) were of similar force profiles and are stronger both dorsoventrally and labiolingually at the second alveolus than at the middentary region (as in *Allosaurus*)—a manifestation of their dorsoventrally deeper symphysis. This suggests a feeding behavior similar to what they have suggested for *Allosaurus*, with powerful "slicing" bites in prey capture with loads focused more toward the rostral extent of the jaw. Carcharodontosaurids, especially the larger *Giganotosaurus*, are said to have high bite forces but are still outdone by tyrannosauroids in terms of magnitude and bone-crushing capabilities. Additionally, an FEA study of a *Giganotosaurus* tooth by Mazzetta et al. (2004) found that its teeth were able to withstand up to 10 kN of pulling forces exerted by its prey.

Rayfield (2011), in her FEA study of tetanuran skulls, found that the larger carcharodontosaurids, namely *Acrocanthosaurus* and *Carcharodontosaurus*, both exhibited relatively higher stresses in the quadrate that progress toward the squamosal and caudal region of the skull roof,

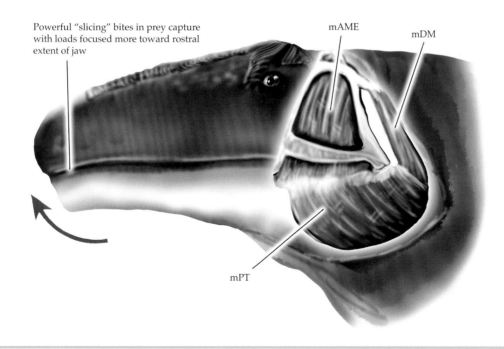

Powerful "slicing" bites in prey capture with loads focused more toward rostral extent of jaw

mAME

mDM

mPT

Figure 5.11. *Giganotosaurus* head showing feeding function with jaw muscles. Red arrow shows orthal jaw action. Abbreviations: mAME, m. adductor mandibulae externus; mDM, m. depressor mandibulae; mPT, m. pterygoideus.

with general stresses being surpassed in magnitude only by the larger *Spinosaurus*. This lends evidence to their assertion that larger-bodied theropods, in general, show an allometric relationship between the length of the skull and the stresses experienced by the skull, with higher stresses seen throughout the skull in these larger-bodied tetanurans than smaller- to medium-sized taxa (Rayfield, 2011).

With so much going on in terms of prey capture, it is obvious that the head is not the only entity in need of functional adaptation. In predators like allosauroids, the neck is equally, if not more so, integral to every motion involved in prey capture, and perhaps no predatory non-avian dinosaur (outside of *Tyrannosaurus*) has had its neck function studied by paleontologists like *Allosaurus* itself. Bakker (1998) highlighted the importance of the neck muscles in *Allosaurus* for increasing leverage of flexion of the neck to account for its allegedly comparatively small jaw musculature. He noted that the paroccipital processes are lowered relative to the occipital condyle, which would have co-opted some neck muscles in ventral flexion of the neck. Snively and Russell (2007a), in reconstructing muscles in the neck in large theropods, argued that muscle attachment sites confirm that *Allosaurus* likely co-opted m. longissimus capitis superficialis for more powerful ventroflexion of the neck, in contrast to how *Ceratosaurus* mainly used other neck muscles (fig. 5.10). This coincides with assertions of a transversely expanded nuchal ligament in *Allosaurus* that made way for larger neck muscles (Paul, 1988; Snively and Russell, 2007a). Snively and Russell (2007a) argued that *Allosaurus* used rapid head strikes in prey capture, as in *Ceratosaurus*, and that this enhancement of neck musculature was key to their success as top predators.

Using multibody dynamics analysis of the head and neck, Snively et al. (2013) were able to model the feeding motions of *Allosaurus* to gain a quantitative perspective of neck mechanics in feeding. They indeed found that the ventrally displaced m. longissimus capitis superficialis created nearly twice the ventroflexive acceleration in *Allosaurus* than of its typical placement in other theropods like *Tyrannosaurus*. This corroborates Rayfield et al.'s (2001) study asserting that bite force was augmented by ventroflexing its upper jaws. They additionally infer a raptor-like head-retracting de-fleshing mechanism is more likely than lateral shaking of the prey. As for neck use in larger carcharodontosaurids, a study by Henderson and Nicholls (2015) 3D modeled and analyzed the bodies of two *Carcharodontosaurus* adults and found that although its head would have been powerful, the neck muscles of one individual could not have lifted a medium-sized sauropod (*Limaysaurus*) easily, which is why they suggest it they may have paired up to help each other lift a sauropod carcass.

Allosauroids are known for having forelimbs that were well built for grasping, with large, well-built hands sporting enhanced, "trenchant," recurved claws (largest on the first digit) that would have been ideal for holding on to (grasping and clutching) or tearing into its prey (fig. 5.12). This is especially notable in *Allosaurus* (Gilmore, 1915; Carpenter, 2002; Holtz, 2003). A range-of-motion study on the carcharodontosaurid *Acrocanthosaurus* found that it had limited shoulder, elbow, and wrist mobility, but more strength and dexterity within the digits (especially in hyperextension; Senter and Robins, 2005), all of which suggested that it likely couldn't have used its forelimbs as well in initial prey capture but would still have effectively impaled its prey beneath its chest with its shorter forelimbs (Senter and Robins, 2005; Guinard, 2020).

Allosauroids were bipeds with powerfully built legs and tails (fig. 5.13) but were likely not fast runners, according to a variety of functional morphological studies (Blanco and Mazzetta, 2001; Hutchinson, 2004; Sellers and Manning, 2007; Bates et al., 2012a). Indeed, hindlimb muscles such as the major hip extensor m. caudofemoralis show strong evidence for negative allometry in larger allosauroids compared to medium-sized ones (Bates et al., 2012a). It has been suggested that allosauroids were "grounded runners" (rather than aerial bounding runners), which would have provided more stability in running at a reasonable speed in pursuit of prey, even though its energy expenditure was higher, as well as possibly helping improve head stability and visual acuity (Bishop et al., 2017).

Figure 5.12. Skeleton of *Allosaurus*. Note the recurved, "trenchant" claws on hands.

Figure 5.13. Skeleton of *Acrocanthosaurus*.

Allosauroids, especially the larger carcharodontosaurids, are calculated to have exhibited lower agility (and higher rotational inertia) relative to comparatively sized tyrannosaurids (Snively et al., 2019) and were generally less efficient runners with respect to body size and limb proportions (Dececchi et al., 2020a). This stands to reason, however, because for large allosauroids like *Giganoto-saurus*, higher rotational inertia would have been more important for holding on to large, slow-moving prey like titanosaurian sauropods and their young, with which they coexisted (Snively et al., 2019; Dececchi et al., 2020a). Although the smaller *Allosaurus* would have likely been slightly more agile than *Giganotosaurus*, it also would have been pursuing slower prey like sauropods and stegosaurs (rather than somewhat faster prey like hadrosaurs and ceratopsians that coexisted with tyrannosaurids).

More direct evidence of predatory behavior also exists from calcium isotopes that suggest that carcharodontosaurids fed on terrestrial animals, likely herbivorous dinosaurs (Hassler et al., 2018). Feeding bite marks in various dinosaurs, including striations, punctures, scores, furrows, and pits, have also suggested predatory feeding behavior in *Allosaurus* or possibly even *Saurophaganax*. These feeding traces were seen in a broad range of taxa, including sauropods, the ankylosaur *Mymoorapelta*, and even other theropods, possibly indicating carnivore-on-carnivore feeding or maybe even cannibalistic behavior (Drumheller et al., 2020). Bite marks on a mamenchisaurid sauropod also confirm feeding behavior by a metriacanthosaurid (Augustin et al., 2020).

CHAPTER 6

Early Coelurosaurian and Tyrannosauroid Theropods

Early Coelurosauria and Compsognathidae

For reference in understanding relationships and clade affiliations of theropods discussed in this chapter, see the phylogeny (fig. 6.1) and taxon list (table 6.1) below.

Coelurosauria is the second group within Avetheropoda (in addition to Carnosauria; see chap. 5). It is the main clade that leads to the evolution of famous dinosaurs like *Tyrannosaurus*, *Velociraptor*, and of course all birds, and as such, many characters throughout this grouping are informative of the evolution of bird anatomy, such as the structure and distribution of cranial (and even mandibular) pneumaticity and pneumatic sinuses that are prevalent in coelurosaurs in general (Witmer, 1997; Tahara and Larsson, 2011; Gold et al., 2013; Aranciaga Rolando et al., 2020). Small, bipedal early coelurosaurs like *Bicenteneria* (Aranciaga Rolando et al., 2020), *Juravenator* (Bell and Hendrickx, 2020), and *Ornitholestes* show anatomical similarities to the first major grouping of coelurosaurs, Compsognathidae. Compsognathids (e.g., *Compsognathus* [fig. 6.2], *Sinosauropteryx*, *Sinocalliopteryx*, and *Scipionyx* ranging at least from the Late Jurassic to Early Cretaceous of Asia, Europe, and South America) make up the first of three major clades within Coelurosauria (the other two being Tyrannosauroidea and Maniraptoriformes).

Compsognathids and their early coelurosaur relatives are some of the smallest non-avian dinosaurs to have ever lived, which means their heads are also among the smallest. The skull is narrow and lightly built with thin bone and small ziphodont teeth. Their orbit is large and horizontally ovoid (indicative of a weaker skull; Henderson, 2002) and possesses large scleral rings. The antorbital fenestra is also large (with other smaller maxillary fenestrae), and their external naris is ovoid. The temporal region and supratemporal fenestrae are narrow, and the postorbital adductor region is vertically oriented and subrectangular (with a thin infratemporal arch). The mandible of compsognathids is narrow with a nearly nonexistent coronoid eminence and an elongate, narrow retroarticular process. Sakamoto's (2010) analysis of mandibular mechanical advantages in *Compsognathus* and *Ornitholestes* plot among relatively weak mechanical advantage values both distally and mesially. Likewise, Brusatte et al. (2012) found compsognathids plot among the weaker skulled theropods, especially because of its longer skull, large orbits, and shallow cheek region. It is not surprising that compsognathids would have weak bites given their tiny stature, but it is true even in relative shape of the skull regardless of size.

Compsognathids and other early coelurosaurs were small theropods (roughly 1 m long) with a typical bipedal early theropod body plan, short forelimbs with three-clawed hands, and longer, slender hindlimbs and tail (Ostrom, 1976). The early coelurosaur *Juravenator* has been described as potentially showing integumentary sense organs in their tails, which might hold clues to their paleoecology and the possibility that they were nocturnal hunters (like crocodilians; Bell and Hendrickx, 2020). The first digit has a large ungual claw on a stocky metacarpal with an elongate second digit and shorter

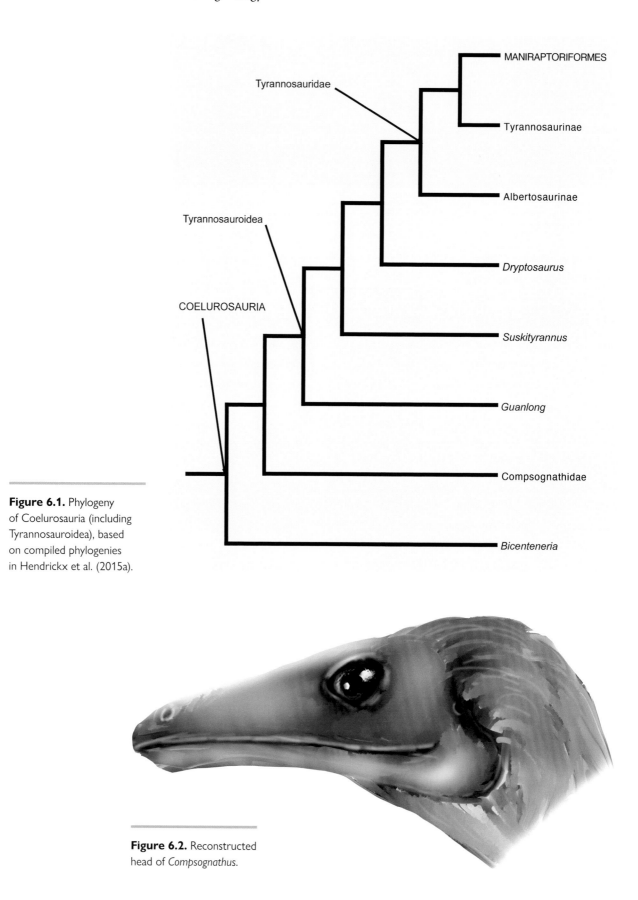

Figure 6.1. Phylogeny of Coelurosauria (including Tyrannosauroidea), based on compiled phylogenies in Hendrickx et al. (2015a).

Figure 6.2. Reconstructed head of *Compsognathus*.

Table 6.1.
Early coelurosaurian and tyrannosauroid genera organized by clade

Major clade	Subclade	Genus
Early Coelurosauria		
		Bicenteneria
		Juravenator
		Ornitholestes
Compsognathidae		
		Compsognathus
		Sinosauropteryx
		Sinocalliopteryx
		Scipionyx
Tyrannosauroidea		
		Guanlong
		Proceratosaurus
		Suskityrannus
		Moros
		Dryptosaurus
	Albertosaurinae	
		Albertosaurus
		Gorgosaurus
	Tyrannosaurinae	
		Alioramus
		Qianzhousaurus
		Lythronax
		Tarbosaurus
		Teratophoneus
		Daspletosaurus
		Tyrannosaurus

third digit (Gishlick and Gauthier, 2007) possibly for acquisition of very small prey, such as insects and lizards, as shown by associated lizard gut contents in *Compsognathus* (Ostrom, 1976) and *Sinosauropteryx* (Smithwick et al., 2017). Another study also found abdominal contents of the large compsognathid *Sinocalliopteryx* show-

ing that it fed on animals that were slightly larger than itself, such as the dromaeosaurid *Sinornithosaurus*, an unnamed ornithischian, and the confusciusornithid bird *Confuciusornis* (Xing et al., 2012b). Owing to a lack of arboreal characteristics, Xing et al. (2012b) inferred that *Sinocalliopteryx* was a stealth predator, capable of capturing flying prey. Furthermore, Dal Sasso and Maganuco (2011) found intact soft tissue with the morphology of abdominal organs in *Scipionyx*, showing a generally short and deep gut with a concomitant high absorption rate.

Tyrannosauroidea

Tyrannosauroids (the second major coelurosaurian clade) included such taxa as the basal crested *Guanlong* as well as *Proceratosaurus*, *Suskityrannus*, *Moros*, *Dryptosaurus*, and many tyrannosaurids. Tyrannosaurids are divisible into albertosaurines (*Albertosaurus* [fig. 6.3], *Gorgosaurus*) and tyrannosaurines (*Alioramus*, *Qianzhousaurus*, *Lythronax*, *Tarbosaurus*, *Teratophoneus*, *Daspletosaurus*, and, of course, the infamous *Tyrannosaurus*), ranging from the Middle Jurassic to the Late Cretaceous, with worldwide distribution and particularly prevalent in North America and Asia. It is possible that Tyrannosauroidea also included megaraptorids (e.g., *Megaraptor*), although this is still debated.

The tyrannosauroid skull varies from being elongate and gracile (e.g., *Alioramus* and *Qianzhousaurus*) to some of the largest and most robust skulls of any theropod (e.g., *Tyrannosaurus*; fig. 6.4). Pneumatic sinuses are variously prevalent in tyrannosauroid skulls, including within the ectopterygoid, quadrate, palate, jugal, and nasal recesses, and are particularly extensive in *Alioramus*, including within its snout (Gold et al., 2013). The orbit is variably shaped (Henderson, 2002). In *Albertosaurus* and *Daspletosaurus*, for instance, the orbit is only slightly rostrocaudally compressed but still quite ovoid (indicative of a relatively weaker skull), whereas *Tyrannosaurus* has a much more rostrocaudally compressed orbit and a rostrally oriented bony protrusion that extends from the postorbital bone that nearly closes communication be-

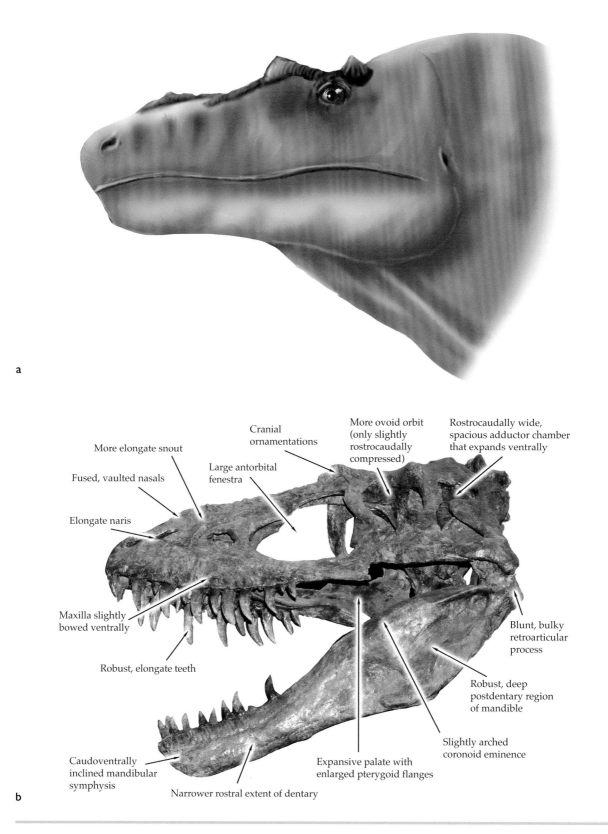

a

b

Cranial
ornamentations

More ovoid orbit
(only slightly
rostrocaudally
compressed)

Rostrocaudally wide,
spacious adductor chamber
that expands ventrally

More elongate snout

Large antorbital
fenestra

Fused, vaulted nasals

Elongate naris

Maxilla slightly
bowed ventrally

Robust, elongate teeth

Blunt, bulky
retroarticular
process

Robust, deep
postdentary region
of mandible

Slightly arched
coronoid eminence

Caudoventrally
inclined mandibular
symphysis

Expansive palate with
enlarged pterygoid flanges

Narrower rostral extent of dentary

Figure 6.3. *Albertosaurus* (a) reconstructed head and (b) skull in lateral view
(length approximately 1 m).

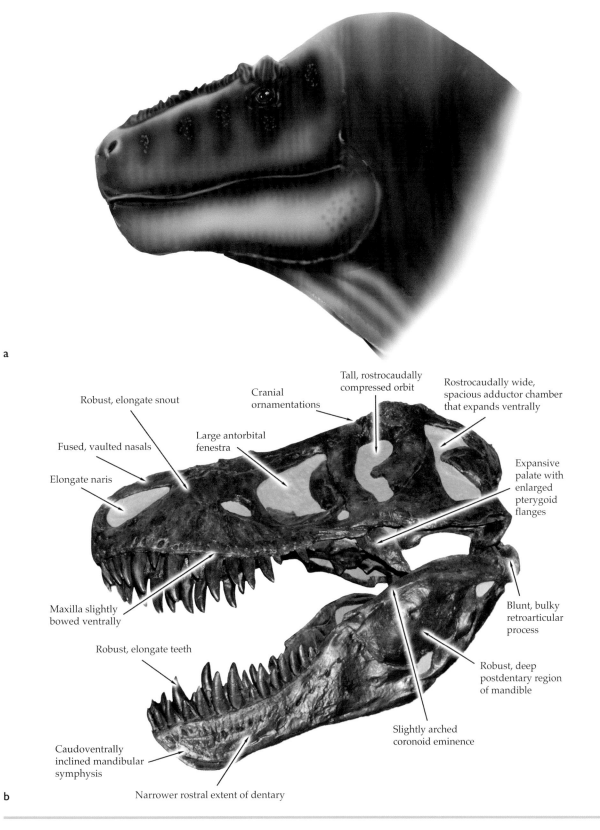

a

b

Robust, elongate snout

Cranial
ornamentations

Tall, rostrocaudally
compressed orbit

Rostrocaudally wide,
spacious adductor chamber
that expands ventrally

Fused, vaulted nasals

Large antorbital
fenestra

Elongate naris

Expansive
palate with
enlarged
pterygoid
flanges

Maxilla slightly
bowed ventrally

Blunt, bulky
retroarticular
process

Robust, elongate teeth

Robust, deep
postdentary region
of mandible

Slightly arched
coronoid eminence

Caudoventrally
inclined mandibular
symphysis

Narrower rostral extent of dentary

Figure 6.4. *Tyrannosaurus* (a) reconstructed head and (b) skull in lateral view
(length reaching up to approximately 1.5 m).

tween the dorsal and ventral parts of the orbits, interpreted as an adaptation for stronger skulls during bite force (Henderson, 2002). It is clear from the skulls of such tyrannosauroids as *Tyrannosaurus* and *Daspletosaurus* that many of them possessed effective binocular vision, with 45- to 60-degree overlap of visual fields, similar to that of living birds of prey (Stevens, 2006). Their snout is generally elongate, ornamented with small rugosities, and deep at its caudal extent, with a large, ovoid-to-trapezoidal antorbital fenestra, a relatively large promaxillary fenestra, an elongate naris, and a rostrocaudally narrow (yet dorsoventrally tall) and rostrally facing orbit. The infratemporal fenestra is unique in that there is a process extending rostrally within it from the rostrolateral margin of the quadrate.

The rostral extent of the snout is broad and rounded, with possible hypersensitivity at the oral margin (as suggested by Carr et al., 2017), although this assertion has been challenged (Bouabdellah et al., 2022). The jawline possesses many thickened, robust, deep-rooted, and elongate teeth with rounded cross-sections—most notable by far in *Tyrannosaurus* (fig. 6.5)—with some teeth up to 0.3 m in length. These strong, robust teeth can be described as being banana-shaped, variable in size and girth along the jawline (i.e., heterodont), and are well suited for biting orthally, crushing bone, holding on to prey, and pulling away in what is known as a "puncture-pull" mechanism (Farlow et al., 1991; Abler, 1992, 1999, 2001; Erickson et al., 1996; Henderson and Weishampel, 2002; Holtz, 2003; Rayfield, 2004; Smith, 2005; Snively et al., 2006; Reichel, 2010a; Brink et al., 2015; Monfroy, 2017; Torices et al., 2018; Hendrickx et al., 2019). This mechanism was especially shown with dental microwear analysis of *Gorgosaurus* (Torices et al., 2018). The premaxillary teeth were much smaller, incisiform, and D-shaped in cross-section (possibly for scraping meat; Holtz, 2003), while teeth running down the jawline were larger and variable in size, with the largest maxillary teeth positioned at the most ventrally bowed part of the he maxilla for extra bony support (Henderson and Weishampel, 2002), with high bending strength in the teeth at the middentary section (Monfroy, 2017).

This is a possible indicator that tyrannosaurids used their more distal teeth to crush bone (Therrien et al., 2005, 2021; Monfroy, 2017).

The small serrations along tyrannosaur teeth have thick, rounded-to-flattened enamel caps that are consistent in size and shape, are tightly packed, and show deep interdental folds that allow the teeth to cut much more like a smooth knife blade rather than a serrated blade (Farlow et al., 1991; Abler, 1992, 2013; Brink et al., 2015). This type of denticulation allows for high serration density and strengthens the cutting edge, much like similar morphology seen in gorgonopsian synapsids (Whitney et al., 2020). Wear facets are parallel, elliptical, and positioned in a way that suggests tooth-tooth contact between maxillary teeth lingually and dentary teeth labially (Schubert and Ungar, 2005). Tyrannosaurids had a position-alternating tooth replacement pattern between odd and even teeth (Hanai and Tsuihiji, 2019; Sattler and Schwarz, 2021), which would have been helpful in

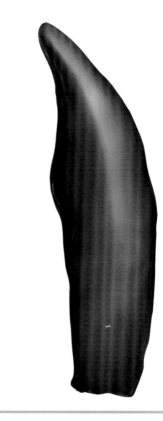

Figure 6.5. *Tyrannosaurus* tooth.

maintaining force pressures in biting (e.g., Gignac and Erickson, 2017).

Tyrannosauroid nasals are also notably fused and arched ("vaulted"), an adaptation that has been shown to withstand forces of a stronger bite and lateral movements with a more massive head (Snively et al., 2006). The temporal region is transversely expanded and rostrocaudally broad (with corresponding large supratemporal fenestrae). The postorbital adductor region is also rostrocaudally widened (as is the infratemporal fenestra) and widens even further as it extends straight ventrally. This adductor chamber morphology generally points to a much larger temporal muscle complex (as noted by Molnar, 2013). The jugal curves ventrally to deepen the adductor chamber even more (with some rugosity seen on its external surface). The tyrannosauroid palate is expansive rostrocaudally and mediolaterally with enlarged pterygoid flanges. The rostral palatal process of the maxilla helps form a unique secondary palate as well, which may have given extra support for higher bite forces (Molnar, 1998; Holtz, 2003), as suggested by Busbey (1995) in crocodilians with a similar secondary palate.

The tyrannosauroid mandible is robust in the postdentary region (with a coronoid eminence slightly arched more so than in other theropods) and a narrower rostral extent of the dentary. The mandibular symphysis is caudoventrally inclined, which is unique to tyrannosaurids compared to other theropods, and varies between being subequal in length and depth to longer than it is deep—dimensions that would have given it substantial torsional strength in biting (Therrien et al., 2005). Also of note are the reduced mandibular fenestra and the relatively bulky and rugose (yet rostrocaudally blunt) retroarticular process. Although an intramandibular joint exists, the supradentary and coronoid bones are fused and cross the intramandibular joint, restricting any kind of movement at this junction (Hurum and Currie, 2000).

A noteworthy aspect of tyrannosaurid cranial anatomy is the enhanced sensory system. For instance, cranial endocasts have shown that tyrannosaurids have hypertrophied olfactory bulbs in the brain, making their sense of smell incredibly important (Witmer and Ridgely,

2009). The elongate cochlea of the ear and tympanic pneumaticity show that they were capable of hearing low-frequency sound waves. The orientation of endosseous labyrinth within the ear indicates that tyrannosaurid alert postures likely involved holding their heads a bit lower than the horizontal to varying degrees. Overall, these sensory adaptations are consistent with predatory function; coordinated, quick movements of the eyes and head; low-frequency hearing capacity; and a keen sense of smell (Witmer and Ridgely, 2009).

Cranial musculature in tyrannosauroids has been reconstructed in numerous studies for more than a century (Gregory and Adams, 1915; Adams, 1919; Gregory, 1920; Molnar, 1973, 2008, 2013; Witmer, 1997; Rayfield, 2004, 2005a; Holliday, 2009; Sakamoto, 2010; Lautenschlager, 2015; Gignac and Erickson, 2017; Bates and Falkingham, 2012, 2018a, 2018b; Carr, 2020; Therrien et al., 2021; for a detailed description of cranial musculature in tyrannosauroids and all other dinosaur groups, see chap. 3). These muscle reconstructions as well as a combination of the abovementioned cranial characters are together essential in assessing cranial feeding function, and no other group of dinosaurs has been subject to studies of cranial biomechanics more than tyrannosauroids, especially *Tyrannosaurus*, which had massive cranial muscles, both those contributing to the m. adductor mandibulae externus (mAME) complex and the m. pterygoideus (mPT) complex (Molnar, 2008, 2013; Holliday, 2009; Bates and Falkingham, 2012, 2018b; Gignac and Erickson, 2017).

Sakamoto (2010) showed that in comparison with other theropods, tyrannosauroids exhibit shallower slopes in mechanical advantage across the tooth row (meaning it has a more consistent mechanical advantage along the jaw), with the highest efficiencies being comparable with (but lower than) allosauroids and ceratosaurs—with the caveat that size and muscular force are not factors in this study. The consistency along the jaw suggests an ability to keep a greater bite force anywhere along the jawline (Sakamoto, 2010). Molnar (2013) noted that most of the adductor musculature of *Tyrannosaurus* (mAME and mPT), in addition to being larger overall in cross-sectional area, had greater leverage than that of *Daspletosaurus* and that both of these tyrannosaurid taxa exceed both *Allosaurus* and *Ceratosaurus*. Conversely, m. depressor mandibulae (mDM) has less leverage in tyrannosaurids than in *Allosaurus* (Molnar, 2013), likely an indicator of the larger gape seen in *Allosaurus*, as found by Lautenschlager (2015) in his study modeling muscular constraint and recruitment in gape. Additionally, Brusatte et al.'s (2012) morphometric analysis of theropods found tyrannosauroid crania plot similarly to other large-bodied hypercarnivorous theropods because of converging similarities in having caudally deeper skulls and narrow, keyhole orbits that add strength to the cranium when biting down. A later geometric mor-

phometric study by Schaeffer et al. (2020) that examined theropod mandibles noted that although the more basal tyrannosauroids *Proceratosaurus*, *Guanlong*, and *Dilong* overlap with some maniraptoriforms in morphospace, tyrannosauroids in general plot in a more restricted morphospace of their own (signaling lower disparity than maniraptoriforms).

The largest and most famous tyrannosauroid of all, *Tyrannosaurus*, has fascinated dinosaur paleontologists for decades, especially in terms of its feeding function and behavior (fig. 6.6). Many studies speculated as to just how powerful of a feeding apparatus it had (mainly because its head was likely the primary organ used in prey capture (Holtz, 2003), but until the mid-1990s,

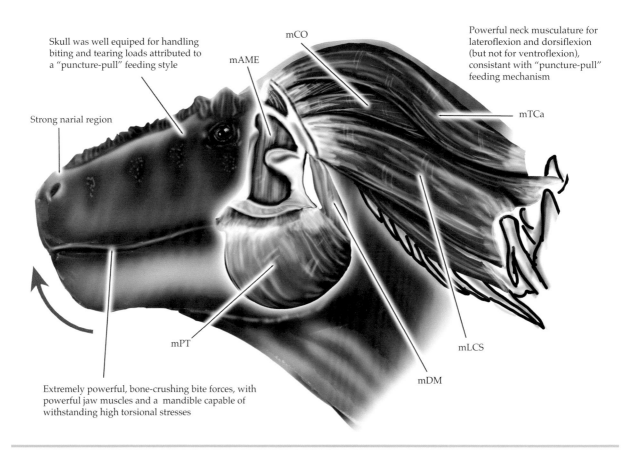

Figure 6.6. *Tyrannosaurus* head showing feeding function with jaw muscles and neck muscles (redrawn from Snively and Russell, 2007a), with highlighted key neck muscles mentioned in text. Red arrow shows orthal jaw action. Abbreviations: mAME, m. adductor mandibulae externus; mCO, m. complexus; mDM, m. depressor mandibulae; mLCS, m. longissimus capitis superficialis; mPT, m. pterygoideus; mTCa, m. transversospinalis capitis.

these assertions had mainly been based on qualitative anatomical studies (e.g., Molnar, 1991). The first study to quantitatively investigate bite force in *Tyrannosaurus* was that of Erickson et al. (1996). With the help of a *Triceratops* pelvis with *Tyrannosaurus* bite marks, they were able to predict bite forces by simulating the same kind of bite (with replica teeth) on a bovine pelvis, which they found to have similar bony microstructure as *Triceratops*. Their results showed a predicted bite force ranging from 6,410 to 13,400 N of force at different tooth positions, comparable to that of American alligators. These force estimates suggest *Tyrannosaurus* possessed strong teeth able to withstand extremely high forces. Erickson et al. (1996) argued that, although not direct proof, these findings were at least supportive of the idea that tyrannosauroids were predators, as argued by Farlow et al. (1991) and Abler (1992), instead of scavenging behavior, as argued by Lambe (1917), Barsbold (1983), and Horner et al. (2011). Meers (2002), in comparing extant predators to the body mass of their prey, predicted a force of about 7,600 to 9,800 N at a single tooth position, with a full powerful bilateral bite calculated as having a maximum of between 183,000 and 235,000 N of force when considering the entire jaw apparatus—putting it in range to prey on large herbivores like adult *Triceratops*. Therrien et al.'s (2005) study extrapolated Erickson et al.'s (1996) study based on Meers's (2002) calculation and tentatively estimated a maximum bite force of between roughly 153,600 and an astounding 321,600 N of force (an even higher estimate than Meers, 2002).

In later years, these bite force estimates have changed drastically with more advanced modeling methodologies (e.g., multibody dynamics analysis, or MDA), but even with this technology, disagreements and vast differences in *Tyrannosaurus* bite force estimates still exist. (e.g., Bates and Falkingham, 2012, 2018a, 2018b; Gignac and Erickson, 2017). Using three-dimensional (3D) MDA, Bates and Falkingham (2012, 2018a) digitally reconstructed cranial musculature on a *Tyrannosaurus* skull and simulated its maximum biting capacity, using estimates for physiological cross-sectional area, fiber

lengths, maximum contraction velocity, force unit per area, and pennation of the muscles they modeled. Their results showed a range between 33,123 and 53,735 N of force at a single tooth position (Bates and Falkingham, 2018a)—a correction of their earlier study's value (Bates and Falkingham, 2012)—a much higher value than previously predicted and measuring up to 10 times that of an American alligator.

In an effort to further investigate the bone-crushing abilities in *Tyrannosaurus*, Gignac and Erickson (2017) used *Tyrannosaurus* bite puncture marks, observations of the skull and teeth (and tooth crowns), reconstructed cranial musculature (based on archosaurian traits and their estimations of muscle sizes, and also assuming parallel muscle fiber orientations) to estimate bite forces, and deduced tooth pressures on bone to assess method of bone pulverization. Their results showed maximum bite force estimations ranging from 8,526 to 32,522 N of force. They also found that the penetration of teeth in bone was performed at between 718 and 2,974 megapascals (mPa), which would have effectively created cracks in the bone being eaten and would crack it open efficiently. Their analysis of dental and palatal anatomy showed how they were set up in specific configuration of three- and four-point loading to help localize bite and fracture of bone. Additionally, a more recent analysis of bite marks on an *Edmontosaurus* suggests a bite force of up to 5,641 N in a smaller, more juvenile *Tyrannosaurus* (Peterson et al., 2021).

Because of the discrepancies in bite force estimates, Bates and Falkingham (2018b) noted that there are differences in how muscle architecture is reconstructed in various studies (of *Tyrannosaurus* and extinct animals in general), such as modeling parallel versus pennate muscle fibers as well as accurately modeling muscle fiber lengths. The paucity of data on carnivorous animals speaks especially to the need for further testing of more sophisticated models in the future to gain an accurate set of bite force values.

The structural performance of the tyrannosauroid skull has also been well studied over the decades. Molnar (1998), with space frame analysis (a methodology normally used for mechanical engineering), presented the first quantitative biomechanical analysis of the architecture of the cranium and mandible of *Tyrannosaurus*, with a simplified interpretation of the skull in a geometric framework of cantilever beams. He showed how the robust skull of *Tyrannosaurus* was much more structurally sound compared to the narrower skull of *Allosaurus* and that it was well built for, and stable in, withstanding vertically oriented forces while biting down. Additionally, he showed how fenestrae and sinuses of the skull were placed in areas where the stresses were not as concentrated and that various anatomical structures were modified to account for tension, compression, torsion, and generally higher stressed areas. Although the simplified framework did not represent the full extent of anatomical nuance in skull anatomy, Molnar (1998) still provided a thorough basis for *Tyrannosaurus* feeding biomechanics studies that followed.

Rayfield (2004) performed the first FEA of a *Tyrannosaurus* skull. With a two-dimensional (2D) profile model, her model showed that the skull of *Tyrannosaurus* was overall equally well equipped for handling biting and tearing loads attributed with a "puncture-pull" feeding style (i.e., biting down forcefully and then pulling away the meat). Specifically, Rayfield (2004) tested how bidirectional feeding loads (stresses and strains) were handled across skull elements, if seemingly mobile sutures resist or dissipate forces, and what effect these sutures have on the skull as a whole. Results showed patterns of stresses and loads that coincide with morphological architecture. For instance, the fused, enhanced nasal bones withstand shear and compressive forces (as also later seen by Snively et al., 2006) and localize forces in that region. Additionally, the lacrimals withstand a variety of stresses, and the maxilla-jugal suture dissipates, forces, withstands, and reduces tension as a shock-absorbing mechanism (although this in turn makes the skull functionally weaker; Rayfield, 2004).

Subsequently, in her study comparing the FEA model of *Tyrannosaurus* with those of *Coelophysis* and *Allosaurus*, Rayfield (2005a) again emphasized the importance of the nasal bones in *Tyrannosaurus* withstanding and

localizing shear and compressive forces in the snout region rather than those forces being experienced in the frontoparietal region, as seen in *Allosaurus* and *Coelophysis*. Rayfield (2005a) also noted how similar the skulls are in other stress patterns shown in the FEA models, such as ventral tension and dorsal compression. Since there is no neutral bending region, however, Rayfield (2004, 2005a) argued that it is not reasonable to think of these skulls as cantilever beams when assessing function.

The potential for cranial kinesis in *Tyrannosaurus* has been a subject of scrutiny. Molnar (1991) interpreted the otic joint as saddle-shaped and suggested possible mobility of the quadrate in a streptostylic manner. Larsson (2008), with osteological observation, described a looser articulation of palatal bones with surrounding skull bones as being suggestive of a fore-aft ("propalinal") palatal kinesis (through a horizontal plane) playing another role in feeding mechanics in *Tyrannosaurus*. This would have eased strain on the bones that would have also been linked by ligaments when biting down on prey. Concurrently, however, Holliday and Witmer (2008) suggested that *Tyrannosaurus* (and most other non-maniraptoran theropods) were likely functionally akinetic in terms of the feeding mechanism as a whole, even if there might have been slight movement at different joints. More than a decade later, Cost et al. (2020) tested this further with finite element models, comparing it to the kinetic skull of a parrot and a gecko. They integrated articular tissues in joints, protractor muscle loads, and postural changes of bones and found that the nature of the anatomy of *Tyrannosaurus* renders it functionally akinetic when compared to the parrot and gecko. Even though a fore-aft motion of the palate is more plausible than a mediolateral motion, Cost et al. (2020) noted that constraints around the otic joint and the higher loads it would have had to endure make it all the more unlikely to be functionally kinetic.

In assessing mandibular adaptations, Therrien et al. (2005) found that tyrannosaurids showed adaptations capable of withstanding high torsional stresses dorsoventrally and labiolingually across the jaw, even mesially (at the rostral end of the jaw; third alveolus), which would have been essential in capturing prey and puncturing through bone. Although results for *Albertosaurus* (fig. 6.7) showed maximum bite forces comparable to that of abelisaurids, *Tyrannosaurus* and *Daspletosaurus* showed the highest estimates of bite force, proving consistent with bone-crushing capabilities suggested previously. The mandibular symphysis of tyrannosauroids was resistant to bending and torsional loads that, according to Therrien et al. (2005), made it possible for them to crush bone with their distal (i.e., back) teeth more forcefully, like a Nile monitor lizard crushing mollusks. But the high bite force values at the third alveolus and middentary region also suggest prey capture adaptations seen in modern crocodilians. A combination of all of these data shows just how well adapted tyrannosauroids (and especially

Tyrannosaurus) were for biting forcefully and maintaining a hold on struggling prey. Furthermore, in studying growth series of albertosaurines and tyrannosaurines, Therrien et al. (2005) found no major differences throughout growth in mandibular properties (except for their smaller size, meaning smaller bite forces), suggesting that even younger tyrannosaurids were able to handle prey capture on their own (albeit with smaller prey). Similar results were seen years later in a study by Therrien et al. (2021), which further tested the mandibular force profiles of both albertosaurines and tyrannosaurines. Again, they found that the mandibular symphysis was ontogentically consistently stronger in resisting bending forces than the middentary region in albertosaurines, making it important in prey capture and holding all throughout

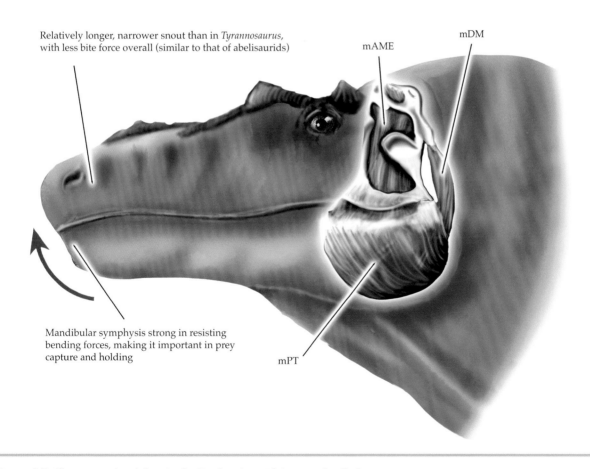

Relatively longer, narrower snout than in *Tyrannosaurus*, with less bite force overall (similar to that of abelisaurids)

mAME

mDM

Mandibular symphysis strong in resisting bending forces, making it important in prey capture and holding

mPT

Figure 6.7. *Albertosaurus* head showing feeding function with jaw muscles. Red arrow shows orthal jaw action. Abbreviations: mAME, m. adductor mandibulae externus; mDM, m. depressor mandibulae; mPT, m. pterygoideus.

life (with more adult forms preying on larger animals). Still, *Tyrannosaurus* performed better in resisting torsional stresses, making it more likely to have bone-crushing capabilities than albertosaurines. A shift from more ziphodont to thicker conical teeth in adult forms coincides with an exponential increase in bite force with increased body size (Therrien et al., 2021).

To test differences between albertosaurines and tyrannosaurines in a different light, Reichel (2010a) performed FEA on different teeth of various root lengths in both *Albertosaurus* and *Tyrannosaurus*. She found similar patterns of shear stresses in both mesiodistal and labiolingual axes of the teeth in the upper jaws of *Albertosaurus* and *Tyrannosaurus*—specifically, mid-maxillary teeth were better suited for labiolingual stresses, and premaxillary and distal maxillary teeth were more capable of withstanding mesiodistal and apicobasal stresses. The difference between the two taxa, however, was in the dentary teeth, where *Albertosaurus* showed similar stress distribution patterns as in the upper jaw, whereas *Tyrannosaurus* showed higher shear along the labiolingual axis of mid-dentary teeth (meaning the opposite of the upper jaw), suggesting a shift in tooth function and based on position along the jaw line. This speaks to the differences in jaw proportions and how important heterodonty is in understanding the function of the jaw as a result. The mesial-most dentary teeth in *Albertosaurus* are aligned more with the premaxillary teeth, while the mesial-most dentary teeth of *Tyrannosaurus* are aligned more with the larger maxillary teeth (meaning it had an overbite, and the mesial dentary teeth took over in function in prey capture).

To further test mandibular function of tyrannosaurines, Rowe and Snively (2021) performed 3D FEA on three tyrannosaurines of various sizes as well as 2D FEA on other ontogenetic stages of *Tyrannosaurus* and other theropods for comparison. Generally, younger tyrannosaurs showed lower stresses owing to smaller muscle forces with smaller body size and lower muscle forces. Relatedly, younger tyrannosaurines performed comparably to adult forms of other, smaller tyrannosaurine taxa, such as *Tarbosaurus* and *Daspletosaurus*. With relatively longer snouts, however, if larger muscles are implemented, the stress and

strain values would increase significantly because of their smaller size. Interestingly, Rowe and Snively (2021) found that because there is so much stress on the rostral extent of the jaw (as previously determined), the bending stresses in that region are diminished by the opposing compressive forces created on the angular bone by the tension of m. pterygoideus ventralis (mPTV) wrapping ventrolaterally to it. Additionally, ligaments help absorb stresses and strain in the caudal aspect of the dentary. All these features together help to deduce how the ontogenetic increase in size is accounted for in terms of adaptations of the mandible for feeding on larger prey items.

In a larger-scale evolutionary mandibular FEA study, Ma et al. (2022) also found similar ontogenetic trends in reduced feeding stresses from juvenile to adult, both in *Tyrannosaurus* and *Tarbosaurus*. Furthermore, in analyzing craniomandibular function in a wider range of tyrannosauroids using FEA, Johnson-Ransom et al. (2021) found that tyrannosauroids (as a whole, including non-tyrannosaurids) generally show lower cranial stress values than other theropods, therein further showing the importance of the unique evolutionary adaptations the entire group had undergone to produce the powerful bite force seen in the largest tyrannosaurid of all, *Tyrannosaurus*.

As in most large predatory theropods, the function of the neck is critical in our understanding of tyrannosauroid feeding function. Certainly, with the "puncture-pull" mechanism proposed by numerous authors (see above), the neck would have needed sufficient power to handle the forces involved in tearing off large amounts of flesh upon each bite, and described craniocervical muscle attachment sites suggest this to have been true (Snively and Russell, 2007a, 2007b, 2007c; Tsuihiji, 2010; Samman, 2013; Snively et al., 2014).

Snively and Russell (2007a, 2007b, 2007c), in describing preserved muscular attachment sites to reconstruct tyrannosaurid and other theropod neck musculature and in analyzing their functional morphology, showed tyrannosaurid necks were powerful in lateroflexion and dorsiflexion but not well adapted for ventroflexion. This is consistent with a "puncture-pull" feeding style rather than one of rapid head strikes, as in *Ceratosaurus* and *Allosaurus*. Snively and Russell (2007a) showed that the tyrannosaurines *Tyrannosaurus* and *Daspletosaurus* were more efficient in lateroflexion of the neck than albertosaurines because of a larger moment arm of m. longissimus capitis superficialis and m. complexus. Additionally, the m. complexus moment arm for neck dorsiflexion was also greater in tyrannosaurines, especially in *Tyrannosaurus*, which shows much larger origin sites (and in turn larger muscle cross-sectional areas) for neck dorsiflexors as a whole (including m. transversospinalis capitis and m. transversospinalis cervicis). These traits are notable because a distinguishing trait of tyrannosaurines is that they have evolved relatively shorter necks compared to albertosaurines (Holtz, 2003), and the need for neck dorsiflexion and lateroflexion for subduing potentially larger prey was likely a factor in their development (Snively and Russell, 2007a, 2007b, 2007c; Snively et al., 2014).

Furthermore, modeling of the skull and cervical bones of *Tyrannosaurus* has suggested that its shorter neck allows for less flexibility than other theropods, although regionalization allows greater mobility near the cranial end of the neck than at the caudal end, a condition similar to that seen in birds (Samman, 2013). Snively et al. (2014) performed electromyographic studies of m. complexus in chickens to test its lateroflexive and dorsiflexive abilities and confirmed not only its use in these actions, but also that m. complexus helped sustain neck stabilization by applying long-axis roll function of the neck while the contralateral ventroflexors were at work. Raptorial birds of prey were also filmed to show the involvement of their neck in attacking prey and tearing of flesh. These modes of lateroflexive and dorsiflexive motion as well as distinct neck stabilization cranially further support the assertion that tyrannosaurids were efficient "puncture-pull" predators.

Tyrannosauroids were bipedal (fig. 6.8) and ranged from small, agile predators to some of the largest predatory carnivorous dinosaurs to have lived, with *Tyrannosaurus* reaching up to more than 12 m (about 40 feet) long. They had powerful, muscular hindlimbs, a large, horizontally held tail for counterbalance, and (especially in the case of all tyrannosaurids) greatly reduced fore-

Figure 6.8. Skeleton of *Albertosaurus* (hunting the ankylosaurid *Euoplocephalus*).

limbs that in the tyrannosaurids possessed only two digits. It is safe to assume that the reduction in forelimb size was made up for by the fact that the head of *Tyrannosaurus* was so large and powerful that it did not need to use its forelimbs for prey capture (Holtz, 2003; Padian, 2022). Despite their small size, however, Carpenter and Smith (2001) showed that the bony properties and muscle scars on *Tyrannosaurus* forelimb muscles suggest they had strong elbow flexor muscles (especially m. biceps)—able to lift up to 180 kg—and were mechanically advantageous. These anatomical adaptations might mean they used their forelimbs to latch and hold on to struggling prey while they were biting down on them (Carpenter and Smith, 2001; Holtz, 2003; Krauss and Robinson, 2013). Megaraptorids are known for their enlarged sickle-shaped hand claws, which they likely used in prey capture but would be a deviation from other tyrannosauroids if they are indeed part of this clade.

Countless studies have used limb proportions, muscular reconstruction, and biomechanical analyses to examine the stance and locomotor abilities of tyrannosauroids, particularly *Tyrannosaurus* (Newman, 1970; Alexander, 1985; Holtz, 1994; Farlow et al., 1995, 2000; Carrano, 1999; Hutchinson and Garcia, 2002; Hutchinson et al., 2005, 2011; Persons and Currie, 2011b, 2014, 2016; Sellers et al., 2017; Snively et al., 2019; Dececchi et al., 2020a). Although a faster-running *Tyrannosaurus* would have been a sight to see, most studies have pointed toward it having a slower running speed—more of a "fast-walking" pace (Alexander, 1985; Farlow et al., 1995, 2000; Hutchinson and Garcia, 2002; Hutchinson et al., 2005, 2011; Dececchi et al., 2020a).

Hutchinson and Garcia (2002) performed an analysis in which they modeled the musculoskeletal system of *Tyrannosaurus* to estimate the minimum mass of extensor muscles required for it to run at high speeds, since the extensor group of the hindlimb is essentially the counteracting support system of animals that run at fast speeds. With a modeling method validated for modern alligators and chickens, they estimated that *Tyrannosaurus* would have needed extremely large extensor muscle mass in order to run quickly—much larger than would

be expected for an animal its size. These estimates would hold true with most large dinosaurs, however, and would not preclude the idea that *Tyrannosaurus* hunted other larger, slower dinosaurs, such as *Triceratops* (Hutchinson and Garcia, 2002). In a later study using computer modeling to assess muscular moment arms and functional morphology of bones, joints, and muscles, Hutchinson et al. (2005) found first that the mechanical advantages of hindlimb flexor and extensor musculature were generally higher at upright postures and second that their muscular moment arms would not have grown allometrically to the extent needed to attain high speeds, also suggestive of slower running speeds for *Tyrannosaurus* and larger animals in general (Hutchinson et al., 2005).

Hutchinson et al. (2011) furthermore took their work a step further and modeled the ontogeny of *Tyrannosaurus* growth and locomotor abilities, showing that the torso grew larger and longer while the limbs became relatively shorter into adulthood, although the elongate metatarsals helped increase limb length (Holtz, 1995; Persons and Currie, 2016; see below). With this increase, they infer that although it likely had large, strong hip and thigh muscles, there is a relative shortening of the hindlimb extensor muscles with a cranially shifting center of mass, including m. caudofemoralis longus, which extends out to attach to the tail (and is responsible for extreme extension of the hindlimb; Hutchinson et al., 2011). But some anatomical studies have argued for a much larger m. caudofemoralis muscle mass (Persons and Currie, 2011b) and suggested *Tyrannosaurus* might have been built for sprinting but with low endurance for when they chase after prey like hadrosaurs (Persons and Currie, 2014). With the use of MDA as well as an analysis of skeletal stresses, Sellers et al. (2017) created models of *Tyrannosaurus* running to simulate what types of loads would have been applied to its hindlimbs in a running gait and found that these loads would have been much too high, also supporting a slower gait.

Elongated metatarsals create the arctometatarsalian foot structure of tyrannosauroids (also seen in some other coelurosaurs like ornithomimosaurs, oviraptorosaurs,

Figure 6.9. Skeleton of *Tyrannosaurus*.

and troodontids) (Holtz, 1994; Snively and Russell, 2003; Snively et al., 2004; Persons and Currie, 2016), and this structure would have allowed their ankles to act as shock absorbers and stabilizers for cursorial "fast-walking" abilities relatively more adequately than other predators of equal size (Holtz, 1994, 2003; Farlow et al., 1995, 2000; Hutchinson and Garcia, 2002; Snively and Russell, 2003). The larger legs of an adult *Tyrannosaurus* (fig. 6.9) were really better suited for reducing energy expenditures while foraging (Dececchi et al., 2020a). In contrast, smaller, more basal tyrannosauroids such as *Moros* (Zanno et al., 2019) and *Suskityrannus* (Nesbitt et al., 2019) and even juvenile tyrannosaurids had a *Bauplan* more suited for higher speeds. Interestingly, Snively et al. (2019) showed that tyrannosaurids had twice the agility of allosauroids, being able to plant their feet and turn rapidly to catch relatively smaller, more agile prey items, like juvenile ornithischians and prey with en-

hanced defense mechanisms, such as ceratopsians and hadrosaurs.

Direct evidence of tyrannosauroid diets comes from many sources. Carbon isotopic evidence suggests tyrannosaurids were the apex predators of their ecosystem (Cullen et al., 2020; Owocki et al., 2020). Coprolite evidence has shown traces of fragmented bone from various ornithischians, such as ornithopods (especially hadrosaurs), pachycephalosaurs, and ceratopsians (Chin et al., 1998, 2003; Varricchio, 2001), some of which are fragments of bone that lack secondary osteons histologically, indicating predation on juvenile individuals (e.g., Chin et al., 1998). Partially undigested bone fragments and muscle fibers in tyrannosaurid coprolites indicate the possibility that, at least in some cases, food passed through the gut relatively quickly, and in turn, the gut was not able to digest everything in a timely manner (Chin et al., 1998, 2003). This rapid transit of digestion may be due to instances of gorging, as is frequent in many living carnivores, and may fluctuate with the amount of prey available (Chin et al., 2003).

Evidence of tyrannosaurid bite marks and tooth scores is well known in other herbivorous dinosaurs (mostly, but not always, on postcrania), including in hadrosaurs and ceratopsians (Fiorillo, 1991; Erickson and Olson, 1996; Erickson et al., 1996; Carpenter, 1998; Jacobsen, 1998; Farlow and Holtz, 2002; Wegweiser et al., 2004; Fowler and Sullivan, 2006; Happ, 2008; Hone and Watabe, 2010; Fowler et al., 2012; DePalma et al., 2013; Dalman and Lucas, 2018; Peterson and Daus, 2019), with some of these studies speculating on various styles of feeding, with "punctures," "drag marks," and "bite-and-drag marks." Healed bite marks on some individuals give further support for the assertion that tyrannosaurids were active predators (e.g., DePalma et al., 2013; Dalman and Lucas, 2018). Evidence of bite marks from a late-stage juvenile tyrannosauroid shows that they were already preying on the larger-bodied herbivores that adults fed on, although they did not have the bone-crushing bite force of an adult at that stage (Peterson and Daus, 2019). Interestingly, although it is rarer, evidence of facial biting and possible cannibalism has been detected on other tyrannosaurid specimens as well (Jacobsen, 1998; Bell and Currie, 2009; Peterson et al., 2009; Longrich et al., 2010; McLain et al., 2018; Dalman and Lucas, 2021).

As has been made clear by the extent of biomechanical and direct evidence shown above, the likelihood that *Tyrannosaurus* was a predator rather than a strict scavenger is high, but if it were to have happened upon free carrion, it likely would not have passed up the opportunity (as indicated by a possible instance of scavenging described by Hone and Watabe, 2010). But although some studies have suggested the possibility of a primarily scavenger lifestyle of tyrannosaurids based on morphology (Lambe, 1917; Barsbold, 1983) and abundance (Horner et al., 2011), numerous studies have compiled data from previous studies as well as their own anatomical and paleoecological observations to convincingly argue for a mainly predatory lifestyle (see studies discussed above as well as Lingham-Soliar, 1998; Holtz, 2008; Witmer and Ridgely, 2009; Carbone et al., 2011; Carpenter, 2013; Krauss and Robinson, 2013).

Smaller, non-tyrannosaurid tyrannosauroids would have likely had more competition with similar-sized predators. However, a large amount of the dietary partitioning and structuring of dinosaur communities between tyrannosaurids in any ecosystem was most likely intraspecific, as indicated by the fact that any particular tyrannosaurid taxon represents the main apex predator of its environment toward the end of the Late Cretaceous (Holtz, 2021). Schroeder et al. (2021), in analyzing various dinosaur community structures, showed that the relatively weaker-skulled yet more agile and fast-running juveniles of a particular tyrannosaurid taxon most likely filled in the mesocarnivore (i.e., "medium-sized") niche, taking down smaller, faster, and likely more juvenile prey (i.e., thescelosaurines, pachycephalosaurs, ornithomimosaurs, and juveniles of larger ornithischians). Meanwhile, older tyrannosaurid individuals with bone-crushing bite forces were better able to take down larger adult ceratopsids, ankylosaurs, and hadrosaurs, therein setting up the structure and otherwise limited predatory diversity (Schroeder et al., 2021). Numerous adaptive functional morphological changes throughout ontogeny clearly made way for tyrannosaurids to assimilate to different niches throughout their respective ecosystems (Holtz, 2021).

CHAPTER 7

Maniraptoriform Theropods

Introducing Maniraptoriformes

The final major coelurosaurian clade is Maniraptoriformes, which includes ornithomimosaurs and maniraptorans (e.g., therizinosaurs, oviraptorosaurs, alverezsaurids, scansoriopterygids, dromaeosaurs, troodontids, and, of course, birds and their avialan relatives). These animals are especially known for their increased relative brain size as well as small skull and tooth size (with increased number of teeth in many), a morphological deviation from the typical theropod ziphodont tooth morphotype. Many maniraptoriforms, especially those that are toothless, likely sported a keratinous rhamphotheca on the snout for stress dissipation in feeding (Lautenschlager et al., 2013). Three maniraptoriform subclades (ornithomimosaurs, therizinosaurs, and oviraptorosaurs) are well supported for being herbivorous, with many sporting special adaptations for reducing mandibular stress, such as a downcurved mandibular symphysis and deepened postdentary region (Ma et al., 2022). The highly derived skulls and body plans of avialans (which includes all birds) are also discussed at the end of this chapter. For reference in understanding relationships of theropods discussed in this chapter, see the phylogeny below (fig. 7.1).

Ornithomimosauria

The first of the herbivorous theropod groups is Ornithomimosauria (Zanno and Makovicky, 2011). Ornithomimosaur skulls are small relative to their body size,

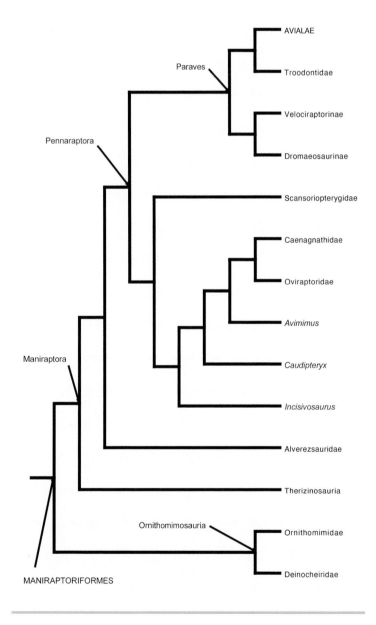

Figure 7.1. Phylogeny of Maniraptoriformes, based on Cau (2018).

with a slender profile and, in most cases, a narrow and toothless snout covered by a rhamphotheca (which has been found preserved; Norell et al., 2001; Barrett, 2005) and with pneumaticity throughout (Tahara and Larsson, 2011). They are divided into deinocheirids (e.g., *Deinocheirus* and *Garudimimus*) and ornithomimids (e.g., *Gallimimus*, *Struthiomimus*, *Sinornithomimus*, and *Ornithomimus*), ranging from North America to Europe and Asia throughout the Cretaceous. (For

Table 7.1.
Ornithomimosaurian, therizinosaurian, and alverezsaurid genera organized by clade

Major clade	Subclade	Genus
Ornithomimosauria		
		Harpymimus
	Deinocheiridae	
		Deinocheirus
		Garudimimus
	Ornithomimidae	
		Struthiomimus
		Ornithomimus
		Gallimimus
		Sinornithomimus
Therizinosauria		
		Erlikosaurus
		Alxasaurus
		Beipiaosaurus
		Jianchangosaurus
		Falcarius
		Segnosaurus
		Nothronychus
		Erliansaurus
		Nanshiungosaurus
		Therizinosaurus
Alverezsauridae		
		Mononykus
		Shuvuuia
		Patagonykus
		Qiupanykus

reference in understanding clade affiliation of ornithomosaurs, therizinosaurs, and alverezsaurids, see the taxon list in table 7.1).

The ornithomimosaur skull (fig. 7.2) possesses a large, rostrocaudally ovoid orbit (likely indicative of relatively lower stresses in feeding) (Henderson, 2002). The temporal region and supratemporal fenestra are narrow and restricted caudally in the skull and are smaller compared to most other theropod groups (and smallest among herbivorous theropod groups) with a short supratemporal bar. With the exception of the more basal ornithomimid *Sinornithomimus* (with a subrectangular and more vertical postorbital region), the postorbital region and adductor chamber of most ornithomimid and deinocheirid ornithomimosaurs are rotated at a lower angle relative to the horizontal skull plane (prominently seen in *Gallimimus*, especially). Most ornithomimosaurs show a straight, continuous maxilla-jugal suture and consistent rostral margin of the adductor chamber, except for *Deinocheirus*, which possesses a slightly flaring jugal like that of most ornithischians. The ornithomimosaur palate is slightly raised at an angle and visible laterally, with a moderately sized pterygoid flange. The quadrate is mostly oriented vertically except in the ornithomimids *Ornithomimus* and *Gallimimus*, with a lower quadrate angle relative to the horizontal plane of the skull. Most ornithomimosaur paroccipital processes are curved ventrally, with the exception of *Garudimimus* and *Sinornithomimus*, with a more horizontally oriented paroccipital process. The mandible in ornithomimosaurs possesses a gently sloped and low to nearly nonexistent coronoid eminence. Deinocheirids show a rostral expansion of the rostral ridge of the coronoid (rostral to the surangular). The mandibular fenestra is small and displaced ventrally. The retroarticular process is reduced (Barrett, 2005) and even curves dorsally in *Struthiomimus* and *Ornithomimus*.

Ornithomimosaurs were orthal feeders and mainly herbivorous (Barrett, 2005; Cuff and Rayfield, 2015) (fig. 7.3). Although some have suggested that their skulls were kinetic (Russell, 1972), this was likely not the case owing to multiple constraining factors (Holliday and Witmer, 2008). Muscular reconstructions show they had relatively

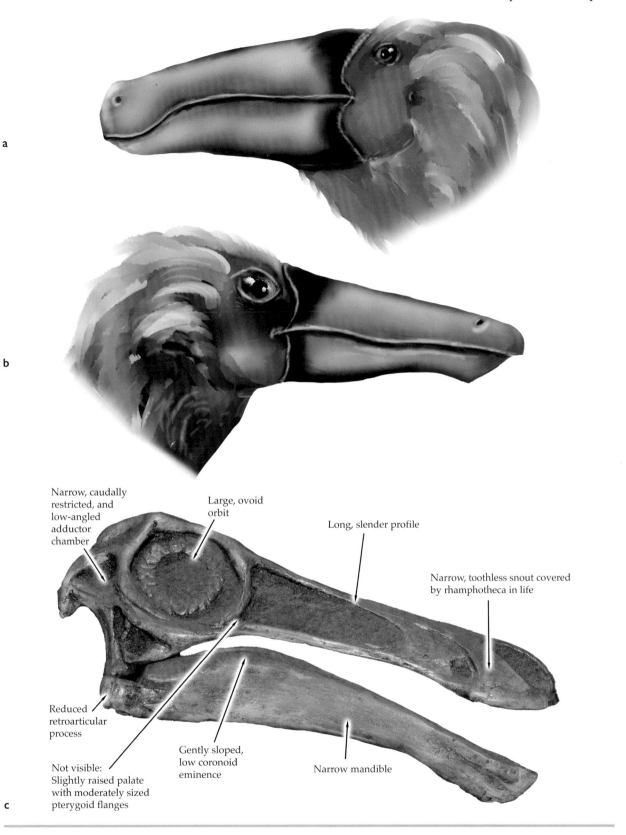

Narrow, caudally restricted, and low-angled adductor chamber

Large, ovoid orbit

Long, slender profile

Narrow, toothless snout covered by rhamphotheca in life

Reduced retroarticular process

Not visible: Slightly raised palate with moderately sized pterygoid flanges

Gently sloped, low coronoid eminence

Narrow mandible

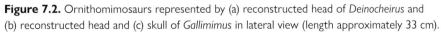

Figure 7.2. Ornithomimosaurs represented by (a) reconstructed head of *Deinocheirus* and (b) reconstructed head and (c) skull of *Gallimimus* in lateral view (length approximately 33 cm).

smaller temporal musculature (m. adductor mandibulae externus, or mAME) with low mechanical advantage compared to other theropods (indicated by both the small temporal origin and low coronoid process; Barrett, 2005; Sakamoto, 2010; Cuff and Rayfield, 2015; Nabavizadeh, 2020b). Variation in adductor chamber morphology determines the general architecture of mAME as it extends from the temporal region to the mandible. Few ornithomimosaurs, such as the early-diverging ornithomimid *Sinornithomimus*, possess a more vertically oriented adductor chamber, but most others have either a triangular or a more rostroventrally directed mAME body with a caudally displaced and abbreviated temporal origin and a rostroventrally expanded insertion along the coronoid region (Nabavizadeh, 2020b).

Deinocheirids, such as *Garudimimus* and especially *Deinocheirus* (fig. 7.4), have even more room for a greater rostroventral expansion of the most superficial division of mAME (m. adductor mandibulae externus superficia-

lis, or mAMES, with the aid of the abovementioned slightly flaring jugal flange) that inserts laterally along a rostrally lengthened surangular (Nabavizadeh, 2020b). If present, this rostrally expanded muscle in deinocheirids would have added more strength and greater mechanical advantage to taxa that have otherwise been deemed to have relatively weaker bite forces owing to their elongate rostra (Cuff and Rayfield, 2015). The rostroventrally oriented, tunnel-like adductor chamber in the ornithomimids *Gallimimus* and *Ornithomimius* coincides with a low-angled quadrate that pushes the jaw joint rostrally (relative to the position of the temporal region) and creates a low-angled mAME complex, with the jaw joint properly braced against the quadrate while lifting the jaw upward in orthal, isognathous closure (Barrett, 2005; Nabavizadeh, 2020b).

The size of m. pterygoideus (mPT) in ornithomimosaurs was relatively reduced, and the raised palate would have directed mPT toward its insertion along the ven-

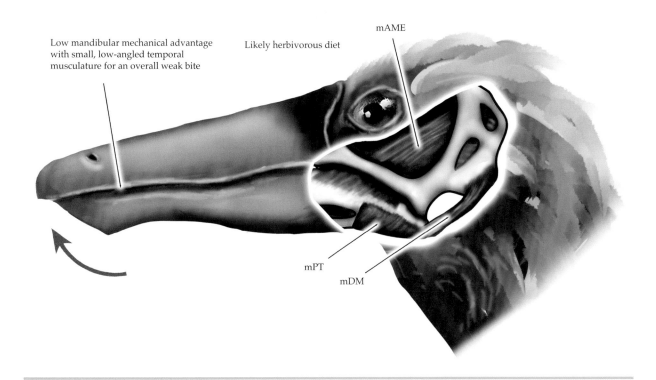

Figure 7.3. *Gallimimus* head showing feeding function with jaw muscles. Red arrow shows orthal jaw action. Abbreviations: mAME, m. adductor mandibulae externus; mDM, m. depressor mandibulae; mPT, m. pterygoideus.

tral margin of the postdentary region and the small retroarticular process. The rostrally displaced jaw joint in the ornithomimids *Gallimimus* and *Ornithomimus* are consequently situated directly beneath the palatal origin of mPT, which would have created a more vertical muscle vector and maybe higher force in jaw closure than even mAME (Nabavizadeh, 2020b), although Cuff and Rayfield (2015) still show a lower bite force in *Ornithomimus* overall (see below). The jaw opener m. depressor mandibulae (mDM) originated from (in most cases) more downturned paroccipital processes that directed the muscle belly to the retroarticular process insertion site.

Given the fact they have such slender, elongate skulls, no teeth, and small muscle mass, it stands to reason that the mechanical advantage of ornithomimosaur jaws is generally low throughout the jawline relative to most other theropods (Sakamoto, 2010). As one of the few herbivorous theropod groups, *Ornithomimus* has a skull that is morphologically distinct from carnivorous theropods in morphospace, although not to the extent of oviraptorosaurs (Brusatte et al., 2012; Schaeffer et al., 2020). Interestingly, Button and Zanno (2020) found some adaptational overlap in functional morphospace across ornithomimosaurs, some therizinosaurs (e.g., *Segnosaurus*), caenagnathid oviraptorosaurs, and diplodocoid and titanosaurian sauropods, although, clearly, the vast difference in body sizes between these animals shows that there is much more to herbivorous adaptations than just cranial characters.

Cuff and Rayfield (2015) reconstructed cranial muculature and performed three-dimensional (3D) biomechanical analyses of mechanical advantage and bite force in three ornithomimosaur skulls (post-retrodeformation); namely, those of the deinocheirid *Garudimimus* and the ornithomimids *Struthiomimus* and *Ornithomimus*. They confirmed that, overall, ornithomimosaurs had lower bite forces, with both ornithomimid taxa showing higher

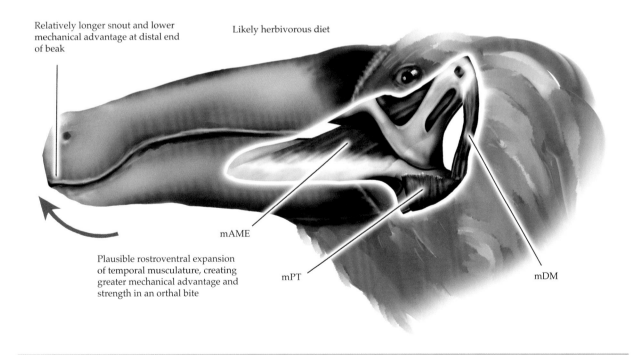

Relatively longer snout and lower mechanical advantage at distal end of beak

Likely herbivorous diet

Plausible rostroventral expansion of temporal musculature, creating greater mechanical advantage and strength in an orthal bite

mAME

mPT

mDM

Figure 7.4. *Deinocheirus* head showing feeding function with jaw muscles. Red arrow shows orthal jaw action. Abbreviations: mAME, m. adductor mandibulae externus; mDM, m. depressor mandibulae; mPT, m. pterygoideus.

muscular mechanical advantage than in the deinocheirid, likely because deinocheirids (including the large *Deinocheirus*) generally have relatively longer snouts compared to the smaller size of the adductor chamber (Cuff and Rayfield, 2015). Ornithomimosaur bite forces are also the lowest among herbivorous theropods, especially compared to therizinosaurs. *Struthiomimus* showed by far the highest bite force among the three taxa (at 57.6 Newtons, N, at beak tip; 75.2 N at mid-beak), with *Ornithomimus* (at 22 N at beak tip; 28.6 N at mid-beak) performing only slightly better than *Garudimimus* (at 19 N at beak tip; 23.9 N at mid-beak). The lower bite force in the latter two likely indicated a shift to eating soft vegetation. Although there is variation in terms of degree of muscle use, both temporal (mAME) and palatal (mPT) muscle complexes were useful in orthal feeding mechanics (Cuff and Rayfield, 2015).

Ornithomimosaurs were feathered, bipedal, mainly cursorial animals with long necks, long forelimbs (especially *Deinocheirus*; fig. 7.5) and hindlimbs (with fused three-toed arctometatarsalian feet; Holtz, 1994), and tails held horizontally. Although some were smaller, others showed body masses of more than 100 kg, with *Deinocheirus* going far beyond that and exceeding 6 tons (well over 5,000 kg) (Zanno and Makovicky, 2013). Muscular reconstruction of the pectoral girdle and forelimb in the ornithomimid *Struthiomimus* infers the potential for protracting, retracting, and adducting the humerus efficiently (Nicholls and Russell, 1985). Additionally, with an offset first digit in extension and a somewhat coalesced second and third digits that were strong in flexion, it is likely that most ornithomimosaurs used their claws as a hooking and clamping mechanism (instead of ground-raking or grasping) to acquire small branches (somewhat like a tree sloth) or fern fronds and cycads while using its long neck to reach it to eat (Nicholls and Russell, 1985; Barrett, 2005), although this might have developed from a somewhat more "raptorial" grasping hand in earlier-diverging ornithomimosaurs, such as *Sinornithomimus* and *Harpymimus* (Kobayashi and Lü, 2003; Barrett, 2005). Interestingly, ornithomimosaurs are one of the few theropod clades that show

more of an isometric trend in forelimb growth through ontogeny rather than the normally seen negative allometry in most theropods (Palma Liberona et al., 2019).

Ornithomimid ornithomimosaurs (fig. 7.6) were highly cursorial animals, capable of high speeds relative to many other theropods (Russell, 1972; Paul, 1998; Carrano, 1999; Dececchi et al., 2020a; Rhodes et al., 2021). As in tyrannosaurids, the fusion of metatarsals to create arctometatarsalian feet would have helped in loading and shock absorption in running, confirming further the cursorial function of their long hindlimbs (Holtz, 1994;

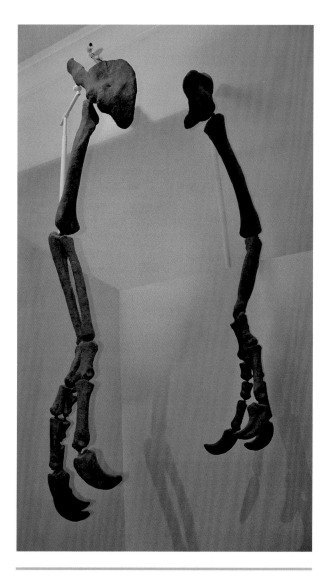

Figure 7.5. Large forelimb bones of *Deinocheirus*.

Figure 7.6. Skeleton of *Ornithomimus.*

Snively and Russell, 2003; Snively et al., 2004). Although ornithomimosaurs in general clearly had strong pelvic musculature, including powerful hip extensors for running (Russell, 1972; Rhodes et al., 2021), limb proportions in deinocheirids, such as *Garudimimus* and the gigantic *Deinocheirus*, show that they were not adapted for fast running like ornithomimids were (which is especially understandable in something with as large of a size as *Deinocheirus*) (Lee et al., 2014).

In addition to many anatomical indicators of herbivory in ornithomimosaurs (see above; see also Zanno and Makovicky, 2011), another strong indicator that ornithomimosaurs were most likely herbivorous is the presence of gastroliths that have been found associated with the abdominal regions of some such as *Sinornithomimus* and *Deinocheirus*, indicating a gastric mill, or gizzard-like structure, that would have assisted in crushing up plant material after eating it with a weaker, toothless beak (Kobayashi et al., 1999; Kobayashi and Lü,

2003; Barrett, 2005; Lee et al., 2014). Even though ornithomimosaurs have been characterized as eating softer plant material (like leaves and fruits) and possibly even eggs, insects, or small soft-bodied animals if given the chance (Russell, 1972), their force-resistant skulls with a cranial bracing mechanism (described above) likely helped them eat high-resistance plant material, especially by shearing the plants with its rhamphotheca (Barrett, 2005), which may have been taken over in processing vegetation to some extent in derived ornithomimids that do not show evidence of a gastric mill (Zanno and Makovicky, 2011). Furthermore, although some have stated that ornithomimosaurs have been found in mesic environments and may have strained food through their rhamphothecae (Norell et al., 2001), some of these environments were more semiarid and, with the abundances of ornithomimosaurs found in a given place, suspension feeding would not have been the most viable trophic option (Barrett, 2005). This is not to say, however, that some ornithomimosaurs didn't also eat the occasional fish, as has been indicated with the presence of fish remains within gastroliths associated with *Deinocheirus*, possibly characterizing it as more of a mega-omnivore (Lee et al., 2014).

Therizinosauria

Therizinosaurs (e.g., *Erlikosaurus* [fig. 7.7], *Beipiaosaurus*, *Segnosaurus*, *Falcarius*, *Alxasaurus*, and *Jianchangosaurus*, ranging from North America to China and Mongolia throughout the Cretaceous) possess small heads compared to their massive bodies, with elongate rectangular snouts and many leaf-shaped teeth with large denticles along a long, narrow jawline. Histological study reveals that the tooth crown volume relative to body size in therizinosaurs is more comparable to other dinosaurs with a similar diet than to other, more carnivorous dinosaurs (Button et al., 2017a). Differences in tooth morphology among therizinosaurs, however, are indicative of various feeding strategies, such as the teeth of *Segnosaurus* having more complex denticles, carinae, and facet morphologies that would have been useful in shredding plant material more so than other therizinosaurs with more simplistic tooth designs (Zanno et al., 2016).

A keratinous rhamphotheca likely enveloped the rostral tip of the snout. Biomechanical modeling using finite element analysis (FEA) in the therizinosaur *Erlikosaurus* has suggested that such a beak would help dissipate forces when biting down, allowing the animal to have a more powerful bite than just its skull alone could withstand (Lautenschlager et al., 2013). Their temporal regions are expanded transversely, but with a shorter supratemporal bar in lateral view. Their postorbital region and adductor chambers are more trapezoidal (expanding ventrally), which mimics that of most carnivorous theropods (e.g., tyrannosaurids and allosauroids). Their palate lies flat within the skull with moderately sized pterygoid flanges. The mandible variably shows either a straighter or downturned mandibular symphysis, gently sloped and low coronoid eminences, a ventrally displaced and small mandibular fenestra, and reduced retroarticular process (Zanno et al., 2016).

Therizinosaurs fed orthally and isognathously, with craniomandibular characteristics and musculature that are broadly suggestive of an herbivorous lifestyle (Zanno et al., 2009; Zanno and Makovicky, 2011; Lautenschlager, 2013, 2015, 2017; Lautenschlager et al., 2013, 2016) (fig. 7.8). The architecture of their temporal regions indicates transversely broadened temporal (mAME) musculature that fans out rostroventrally through the adductor chamber to insert along a rostrocaudally elongate yet low and rounded to subtriangular coronoid eminence of the mandible, with a more rostral attachment that would have added mechanical advantage and small, ventrally displaced mandibular fenestra, possibly giving room for more laterally attached mAMES muscle fibers (Lautenschlager, 2013; Nabavizadeh, 2020b). Although palatal (mPT) musculature added strength in feeding, it is likely that it was only second to enhanced temporal musculature in terms of its impact on bite force, given the small- to medium-sized lateral m. pterygoideus ventralis (mPTV) insertion site along the angular and reduced retroarticular process (e.g., Lautenschlager, 2013). The mDM muscle bodies were oriented rostroventrally from horizontal paroccipital processes toward a reduced retroarticular process, acting to open the mouth (Smith, 2015). Lautenschlager (2015) found, through multibody dynamics analysis (MDA) of the skull of *Erlikosaurus*, that constraints created by all jaw muscles working together made it difficult to open its mouth as wide as carnivorous theropods like *Tyrannosaurus* and *Allosaurus*, which makes sense given its herbivorous lifestyle, where a large gape would have otherwise been of relatively little use.

Mechanical advantage of the mandible across the jawline in *Erlikosaurus* shows midrange values across the jawline that are comparable to the closely related oviraptorosaurs and, compared to other theropods, indicate a possible shift in jaw function as a whole in maniraptorans (Sakamoto, 2010). In terms of skull morphology and functional traits, the therizinosaur skulls plot only slightly outside of other carnivorous theropods

(Brusatte et al., 2012) but rather broadly across functional morphospace among herbivorous dinosaurs, indicating variation in functional characteristics and proportions of the skull and a broad range of ecologies (Button and Zanno, 2020).

Lautenschlager (2013) created a 3D reconstruction of cranial musculature in *Erlikosaurus* to investigate function and bite force in the therizinosaur skull. Volume, cross-sectional area, length, and angle of the muscles were all re-created based on the geometry of the

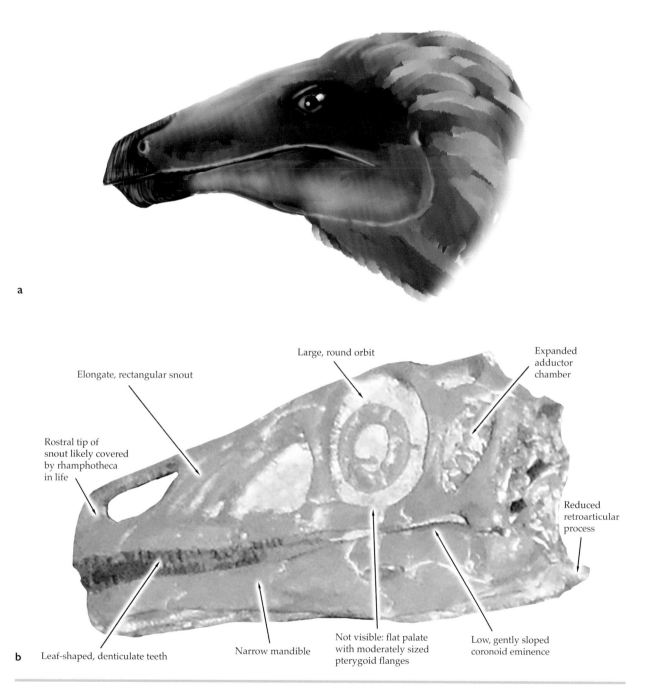

Figure 7.7. Therizinosaurs represented by (a) reconstructed head of *Erlikosaurus* and (b) reconstructed therizinosaur skull in lateral view (length reaching up to approximately 25 cm).

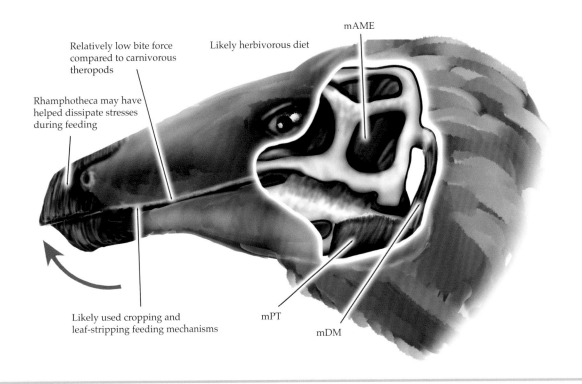

Relatively low bite force compared to carnivorous theropods

Rhamphotheca may have helped dissipate stresses during feeding

Likely herbivorous diet

mAME

Likely used cropping and leaf-stripping feeding mechanisms

mPT

mDM

Figure 7.8. *Erlikosaurus* head showing feeding function with jaw muscles. Red arrow shows orthal jaw action. Abbreviations: mAME, m. adductor mandibulae externus; mDM, m. depressor mandibulae; mPT, m. pterygoideus.

skull and adductor chamber to then calculate bite performance. Bite force across the jawline ranged from 43–65 N at the tip of the snout to 59–88 N at the first tooth and 90–134 N at the most distal (back) tooth, which in all are relatively low bite forces compared to many carnivorous theropod dinosaurs. This led Lautenschlager (2013) to deduce that these animals used cropping and leaf-stripping feeding mechanisms rather than a more sophisticated oral processing system.

Furthermore, Lautenschlager et al. (2013) performed FEA on the *Erlikosaurus* skull while incorporating various rhamphothecal beak morphotypes (small, large, none, etc.) on the rostral tip of the snout to test the stress dissipating capacity of such a structure and how the rest of the skull reacts to such force distributions. They clearly show that a rhamphotheca makes the rostral extent of the skull more stable (i.e., less bendable and less likely to be displaced) and helps dissipate stresses and strains throughout the snout, particularly while bit-

ing down. Mandibular stresses were focused more in the caudal, postdentary region, which makes sense given the attachment of jaw musculature there. Additionally, neck musculature was likely enhanced to help in feeding, and the stresses they impose on the skull require the tensive counteraction of the jaw muscles to help brace the entire feeding apparatus (Lautenschlager et al., 2013). In comparing skull FEA and MDA performance of *Erlikosaurus* with herbivorous dinosaurs of similarly shaped skulls (e.g., the non-sauropod sauropodomorph *Plateosaurus* and the ornithischian *Stegosaurus*), Lautenschlager et al. (2016) showed that stresses with beak use were highest in *Erlikosaurus*. Its bite force is in the same range as *Plateosaurus*, however, with higher bite force seen in *Stegosaurus*.

Examining across different therizinosaur taxa, Lautenschlager (2017) compared mandibular performance with FEA in different feeding scenarios, such as clipping, upward pull, downward pull, and lateral pull. He found that various morphological differences throughout the mandible were responsible for fluctuations in feeding strategy, which in turn may be indicative of functional niche partitioning. Overall, the expanded postdentary region in therizinosaurs like *Erlikosaurus* and *Beipiaosaurus* allowed them to experience the lowest amounts of stress and strain throughout the mandible in all feeding strategies. Because of their more elongate dentaries, however (and a consequently longer distance between the symphysis and the jaw joint), these two taxa also present with the lowest overall relative bite forces. Still, they could have been rather flexible in their feeding strategies given these traits. A recurved, down-turned mandibular symphysis in some taxa (*Erlikosaurus* and *Segnosaurus*) were found to be effective stress-mitigating characteristics where it was lacking in other taxa with elongate, straight dentaries and symphyses, such as *Falcarius*, *Jianchangosaurus*, and *Alxasaurus* (Lautenschlager, 2017).

Lautenschlager (2017) also found that the highest relative bite forces were seen in *Falcarius* and *Alxasaurus* (even though there was more stress and strain in their jaws), likely (in part) because their tooth row extends back closer to the jaw joint, which would result in increased mechanical advantage. It is possible that *Falcarius* shows higher mechanical advantage in the jaw because it may have shown signs of a transition to an omnivorous lifestyle from carnivorous ancestors. In turn, the lower mechanical advantages of other, more derived therizinosaurs might suggest a shift to a more gut-based plant processing system rather than sophisticated oral processing, like in many large herbivorous ornithischians and mammals (Lautenschlager, 2017).

The long necks of therizinosaurs have been suggested to have partially compensated for their lower bite force (Lautenschlager et al., 2013). Smith (2015) reconstructed neck musculature in the therizinosaurs *Falcarius* and *Nothronychus*, and showed that both taxa possessed muscular anatomy that favored an increase in ventroflexion of the neck with a ventral displacement of ventroflexors (e.g., m. rectus capitis anterior and m. longissimus capitis profundus), creating a longer lever arm for downward mo-

tions. Where the two taxa differed is that, although well developed in both taxa, lateroflexors (e.g., m. longissimus capitis superficialis, m. intercostalis capitis, and m. splenius capitis, and part of m. complexus) and dorsiflexors (m. transversospinalis capitis and some of m. complexus) were more emphasized in *Falcarius* than in *Nothronychus*. Still, lateroflexion was highly depended upon in both taxa. Since therizinosaur skulls have a flat occipital condyle and the vertebrae have shallower articular facets (lacking a more substantial ball-and-socket joint between them), their necks were generally less mobile and moved more like an ostrich neck. The amount of movement would have been dependent on the lengths of each individual vertebrae and the sum of how much movement was allowed between them (Smith, 2015).

Smith (2021a) reconstructed musculature in the forelimbs of *Nothronychus* and found that it retained mostly plesiomorphic traits of the forelimb, suggesting a focus on the hindlimb being more modified. He suggests that the humerus may have been able to be elevated to some extent, but not past the level of the vertebral column, and that some level of supination and pronation may have been plausible. Attachments for m. latissimus dorsi do support strong retraction of the humerus, which would have been ideal for pulling back on foliage. Humeral range of motion was reduced as well as protraction and abduction (owing to a small deltopectoral crest). Interestingly, Smith (2021a) also shows the possibility of a triosseal canal having been developed in therizinosaurs based on the shape of the furcula (perhaps for allowing a more birdlike m. supracoracoideus tendon to pass through for humeral elevation and rotation).

The most striking feature in therizinosaur forelimbs, however, is their claws. Lautenschlager (2014) performed FEA on a variety of claw morphologies across different therizinosaur taxa. He showed that some therizinosaurs (e.g., *Alxasaurus* and *Erliansaurus*) had more generalized, short, compact claw morphology that were built to withstand forces in a variety of actions such as scratch-digging or piercing, while others had more enlarged, elongated, flattened claws, such as *Falcarius*, *Beipaiosaurus*, and especially *Therizinosaurus*—with the longest claw of any ter-

Figure 7.9. Skeleton of *Nothronychus* standing upright showing large torso and elongate claws on hands.

gastralia circumference, that would have allowed for plenty of space for an enormous gut for plant fermentation (Zanno et al., 2009; Rhodes et al., 2021; Smith, 2021b). Pelvic and lower limb muscular reconstruction in therizinosaurs like *Falcarius* and *Nothronychus* indicates a general reduction in cursoriality, showing convergence with ornithischians and birds with an opisthopubic pelvis that altered muscle lengths and showed reduced hip extensors, like m. caudifemoralis longus and brevis (coinciding with a loss of a fourth trochanter in maniraptorans), and reduced tibial flexors, among others (Rhodes et al., 2021; Smith, 2021b). With the reduction in tail length, these muscle reconstructions would shift the center of mass cranially in therizinosaurs, requiring more lower limb support in standing bipedally (Smith, 2021b).

Alverezsauridae

Alverezsaurids (e.g., *Mononykus* [fig. 7.10], *Shuvuuia*, *Patagonykus*, and *Qiupanykus* from the Late Cretaceous of Asia and South America) possess small and narrow skulls with a beak-like muzzle and incredibly tiny teeth (thought to have aided in a possibly insectivorous diet in most cases). Their temporal regions are small and caudally restricted with a vertical quadrangular postorbital region and adductor chamber, very thin, rodlike jugals, and flat palate. Their mandibles are thin as well, with a low coronoid eminence and thin rostral extent. Alverezsaurids, like all coelurosaurs, are small bipedal cursorial animals with long gracile limbs, but the most notable feature in alverezsaurids is their incredibly reduced forelimbs with fused metacarpals and only a single functional digit with an enlarged claw. The pectoral girdle shows a long deltopectoral crest and long olecranon process of the ulna, both indicating strong muscle attachment for humeral depression and forearm extension (Perle et al., 1993; Senter and Robins, 2005). It has been suggested that these specialized forelimbs and claws could be an indication for burrow-digging function, although the long hindlimbs would suggest otherwise (Perle et al., 1993).

restrial animals—that were adapted for a hook-and-pull function during browsing on trees because they would have otherwise been prone to high stresses and strains in other actions, thereby further supporting a herbivorous lifestyle (Lautenschlager, 2014).

In some cases, Therizinosaurs represented some of the largest coelurosaurian herbivores to have existed (fig. 7.9). The largest body sizes have been calculated at more than 6,000 kg, including those of *Therizinosaurus* and *Nanshiungosaurus* (Zanno and Makovicky, 2013). This, combined with their long necks and enlarged, hooking claws, helped them reach plants at great heights. Another major indication for an herbivorous lifestyle is the enlarged flared hips (or synsacrum), in addition to trunk rib and

Figure 7.10. Reconstructed head of *Mononykus*.

Senter (2005) studied the orientation and range of motion in the forelimb of the alverezsaurid *Mononykus* and found that their humeri sprawled laterally and their forearms were likely held subvertically, with ventrally oriented palms and restricted near-parasagittal flexion and extension within the hand—a deviation from the typical theropod hand (held medially against the chest and with transverse flexion and extension within the hand). Senter (2005) concluded that *Mononykus* was likely using scratch-digging and hook-and-pull movements to open mounds of termite nests, much like anteaters and pangolins of today. He also suggested that *Shuvuuia* and *Patagonykus* had similar forelimb function. Furthermore, Lü et al. (2018) described the alverezsaurid *Qiupanykus* and suggested they were egg-eaters because the oviraptorosaur eggshells associated with it seem to have been pierced by the enlarged claw. Meso et al. (2021) suggested that tail anatomy implicates the capacity to change rotational inertia and that this, combined with likely digging forelimbs, may suggest a niche similar to aardvarks, anteaters, and pangolins. Whether or not alverezsaurids were insect-eaters, egg-eaters, or a combination of the two is still a mystery, but each present with positives and negatives, and it is possible that a combination of the two would have been the ideal adaptive scenario.

Oviraptorosauria

Oviraptorosaurs—divided mostly into oviraptorids (e.g., *Citipati*, *Oviraptor*, and *Khaan*; fig. 7.11) and caenagnathids (e.g., *Anzu*, *Chirostenotes*, *Caenagnathus*, and *Gigantoraptor*) in addition to early-diverging taxa (e.g., *Incisivosaurus*, *Caudipteryx*, and *Avimimus*)—ranged mainly from North America to Asia throughout the Cretaceous. (For reference in understanding clade affiliation of oviraptorosaurs, scansoriopterygids, dromaeosaurids, and troodontids, see the taxon list in table 7.2.)

Oviraptorosaurs have rostrocaudally shorter and more box-shaped skulls, exhibiting cranial crests above the orbits that are of various shapes and sizes reminiscent of cassowary casques and possess a number of ridges, fossae, and fenestrae (e.g., Barsbold, 1977; Smith, 1992; Funston et al., 2018). Most oviraptorosaurs (with the exception of some early-diverging oviraptorosaurs such as *Incisivosaurus* and *Caudipteryx*) are toothless and possess a rounded, rhamphotheca-encased beak-like muzzle, which was effective in a strong nipping bite (Funston et al., 2018). Most caenagnathids possess lateral occlusal grooves and ridges on the oral margin and mandibular symphysis (Longrich et al., 2013; Ma et al., 2017) that fuses early on in life (Funston et al., 2020).

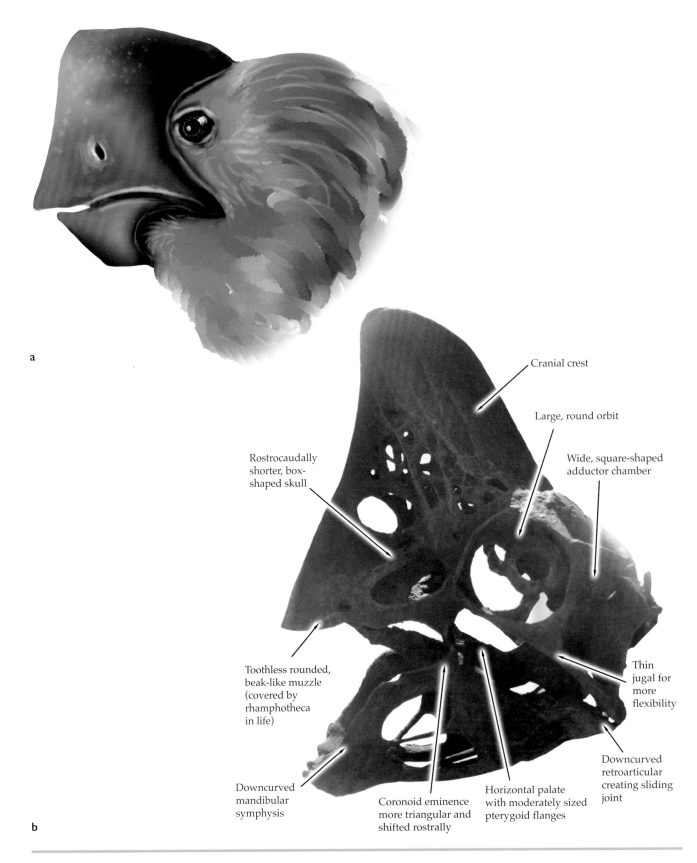

Cranial crest

Large, round orbit

Wide, square-shaped adductor chamber

Rostrocaudally shorter, box-shaped skull

Toothless rounded, beak-like muzzle (covered by rhamphotheca in life)

Thin jugal for more flexibility

Downcurved mandibular symphysis

Coronoid eminence more triangular and shifted rostrally

Horizontal palate with moderately sized pterygoid flanges

Downcurved retroarticular creating sliding joint

a

b

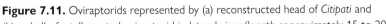

Figure 7.11. Oviraptorids represented by (a) reconstructed head of *Citipati* and (b) a skull of a tall-crested oviraptorid in lateral view (length approximately 15 to 20 cm).

Table 7.2.
Oviraptorosaurian, scansoriopterygid, dromaeosaurid, and troodontid genera organized by clade

Major clade	Subclade	Genus
Oviraptorosauria		
		Caudipteryx
		Avimimus
		Incisivosaurus
	Oviraptoridae	
		Oviraptor
		Citipati
		Khaan
	Caenagnathidae	
		Caenagnathus
		Gigantoraptor
		Chirostenotes
		Anzu
Scansoriopterygidae		
		Yi
		Epidexipteryx
		Scansoriopteryx
		Ambopteryx
Dromaeosauridae		
		Halzskaraptor
		Microraptor
		Bambiraptor
	Unenlagiinae	
		Buitreraptor
		Austroraptor
		Unenlagia
	Dromaeosaurinae	
		Dromaeosaurus
		Utahraptor
		Deinonychus (?)
	Velociraptorinae	
		Velociraptor
		Saurornitholestes
		Deinonychus (?)
Troodontidae		
		Troodon
		Jinfengopteryx
		Mei
		Saurornithoides

Although these grooved structures around the oral margin have been described as remnants of alveoli as a result of ontogenetic tooth reduction patterns (Wang et al., 2018), a histological ontogenetic study by Funston et al. (2020) on caenagnathid dentaries found that these structures are a result of secondary remodeling of bone due to food processing.

Oviraptorosaur temporal regions and supratemporal fenestrae are mostly widely expanded both rostrocaudally and transversely (with the exception of more basal oviraptorosaurs with transversely narrower temporal regions, e.g., *Incisivosaurus*, *Caudipteryx*, and *Avimius*). The postorbital region and adductor chambers are vertical and subrectangular to squared laterally. Their jugals are thin and rodlike, which potentially adds extra flexibility, as seen in birds. The palate is largely horizontal with moderately sized pterygoid flanges. Oviraptorosaur mandibles show variation in coronoid eminence morphology (Barsbold, 1977; Smith, 1992; Ma et al., 2017). Basal oviraptorosaurs as well as caenagnathids (fig. 7.12) show the typical low, gently sloped, or almost nonexistent coronoid eminence of a theropod, whereas oviraptorid coronoid eminences are shifted rostrally, project dorsally, and are triangular with a gentle slope downward at the caudal margin. The beak-like mandibular symphysis is curved downward in oviraptorids, whereas it is more horizontal in other oviraptorosaurs. The mandibular fenestra is large and, in oviraptorids, shifted rostrally. Although shorter and more horizontal in basal oviraptorosaurs, the retroarticular process in the more derived oviraptorids and caenagnathids is uniquely downcurved, creating a sliding jaw joint, much like the condition seen in dicynodont therapsids (Cracraft, 1971; Funston and Currie, 2014; Ma et al., 2017).

Temporal morphology in many oviraptorosaurs shows large sites of mAME muscle origin relative to overall skull size, indicating large muscle bodies both in rostrocaudal and transverse breadths, with the exception of the early-diverging toothed oviraptorosaurs *Incisivosaurus* and *Caudipteryx*, in which the transverse breadth is relatively narrower. The vertically oriented subrectangular adductor chambers contained the ventrally extend-

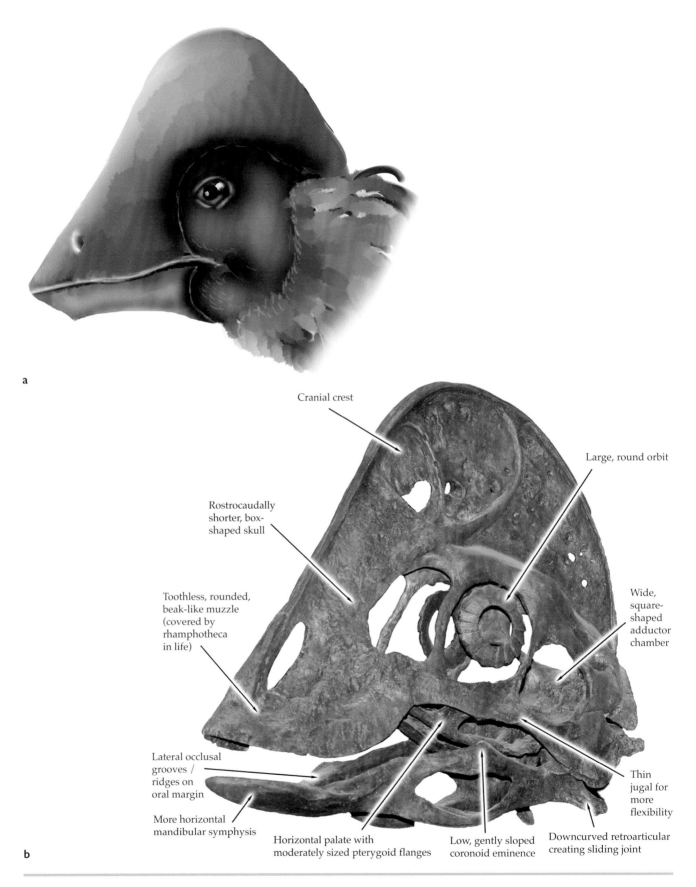

a

b

Cranial crest

Large, round orbit

Rostrocaudally
shorter, box-
shaped skull

Toothless, rounded,
beak-like muzzle
(covered by
rhamphotheca
in life)

Wide,
square-
shaped
adductor
chamber

Lateral occlusal
grooves /
ridges on
oral margin

Thin
jugal for
more
flexibility

More horizontal
mandibular symphysis

Horizontal palate with
moderately sized pterygoid flanges

Low, gently sloped
coronoid eminence

Downcurved retroarticular
creating sliding joint

Figure 7.12. *Anzu* (a) reconstructed head and (b) skull in lateral view (length approximately 35 to 40 cm).

ing mAME muscle bodies to envelop the coronoid eminence of the mandible. The rostrally displaced, taller triangular coronoid eminence of oviraptorids gives the temporal jaw muscles a more rostral insertion site that would create higher mechanical advantage compared to caenagnathids and early-diverging oviraptorosaurs (with low coronoid eminences) and, in turn, likely more crushing, orthopalinal biting motions (Barsbold, 1977; Smith, 1992; Ma et al., 2017; Nabavizadeh, 2020b).

Fossae on the rostrolateral surface of the mandible in the caenagnathids *Gigantoraptor* and *Microvenator* and also some oviraptorids have been reconstructed as insertions for a novel, potentially apomorphic muscular division of mAME (somewhat like m. adductor mandibular externus ventralis, or mAMEV, seen in parrots) (Ma et al., 2017), but they may alternatively also be insertion sites for a more rostrally extending mAMES, as seen in ornithischians, to assist in the palinal feeding component (Nabavizadeh, 2020b). A more rostral insertion site of mAME in these oviraptorosaurs may have given it an extra amount of mechanical advantage and muscle power to help in a palinal shearing motion to process vegetation (or, in the possible case of some caenagnathids, some meat as well along its grooved triturating oral margin) (Funston and Currie, 2014; Ma et al., 2017, 2020a; Nabavizadeh, 2020b). The assertion of oviraptorosaurs using palinal feeding motions is greatly supported morphologically—especially with the unique structure of the jaw joint, which consists of a down-curved retroarticular process and a sliding articular facet that would have allowed the jaw to glide back and forth against the quadrate, much like what is also seen in the jaw joint of dicynodont therapsids (Cracraft, 1971; Ma et al., 2017). Additionally, the flattened palatal morphology would have given a large platform for prominent palatal mPT musculature, m. pterygoideus dorsalis (mPTD) dorsally and mPTV ventrally, that would have enveloped the retroarticular process and ventral angular margin, also assisting in added force to the bite of oviraptorosaurs.

The mechanical advantage of oviraptorosaur jaws is midrange at the distal end of the jaw but among the strongest at the tip of the bill compared to other theropods (Sakamoto, 2010), likely a result of their short muzzles. The oviraptorid *Citipati* performed the highest of all the tested oviraptorosaurs (mostly consisting of oviraptorids) (Sakamoto, 2010), probably because of the taller triangular coronoid process (Ma et al., 2017) (fig. 7.13). These results mark a shift in theropod skull plan and diets from the normally seen theropod morphology, and their geometric morphometric analysis has further enlightened this aspect of their cranial and mandibular evolution, with oviraptorosaurs deviating from the typical skull plan (Brusatte et al., 2012; Schaeffer et al., 2020). The highly variable nature of the mandibles, however, has also led to show that oviraptorosaurs cover a broad range of morphotypes (Schaeffer et al., 2020). Among herbivorous dinosaurs, Button and Zanno (2020) found some overlap in functional morphospace between oviraptorids, ankylosaurs, and non-titanosaur macronarian sauropods, showing more robust skull features and between caenagnathids, ornithomimosaurs, diplodocids, and titanosaurs with weaker skull features, although there is no real evidence that much of this is due to any convergence between groups.

It has previously been suggested that oviraptorosaur skulls may have had a kinetic component wherein the rostrum and rostral dentary would flex rostroventrally against the rest of the skull, with the assistance of a powerful, mechanically efficient bite force based on early muscle reconstruction (Barsbold, 1977). Although this particular hypothesis has not been further tested, Holliday and Witmer (2008), in discussing normally kinetic intracranial joints, pointed out that although oviraptorosaurs likely had synovial basal and otic joints as well as protractor musculature, the kinetic linkages are not substantial enough to allow any functional cranial kinesis. Smith (1992) reconstructed cranial musculature and tested cranial lever arms in *Oviraptor* and suggested that its bite was most similar to Ostrom's (1966) results for ceratopsian jaw mechanics (and less like carnivorous dinosaurs like *Allosaurus*), with the heightened coronoid eminence in *Oviraptor* acting in a similar fashion to the coronoid process in ceratopsids in increasing mechani-

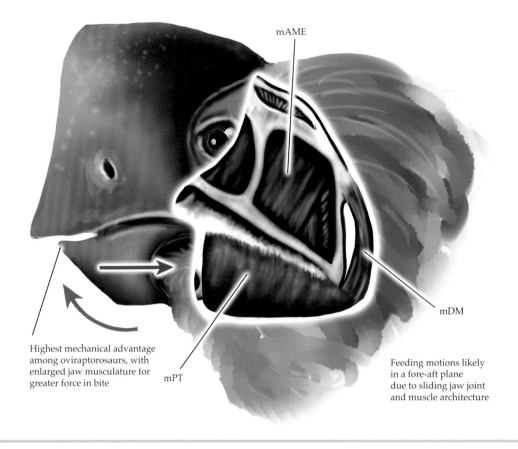

mAME

mDM

Highest mechanical advantage among oviraptorosaurs, with enlarged jaw musculature for greater force in bite

mPT

Feeding motions likely in a fore-aft plane due to sliding jaw joint and muscle architecture

Figure 7.13. *Citipati* head showing feeding function with jaw muscles. Red arrows show orthal and palinal jaw action. Abbreviations: mAME, m. adductor mandibulae externus; mDM, m. depressor mandibulae; mPT, m. pterygoideus.

cal advantage. Of course, these broad-scope comparisons by both Barsbold (1977) and Smith (1992) were not nearly enough to get a good sense of oviraptorosaur feeding ecology, but they gave earlier indications that these animals were perhaps herbivorous.

Funston and Currie (2014) performed FEA on the mandible of the caenagnathid *Chirostenotes* and found that the grooved lingual ridges of the mandible are effective in stress dissipation, which they interpret as meaning they would have been effective food processing structures. From this, they deduced that *Chirostenotes* (and possibly most caenagnathids) likely used its lower jaw in a shearing action through palinal motion and that it likely sheared vegetation and possibly even the meat of small prey, asserting a potentially more omnivorous lifestyle. The upturned beak in *Chirostenotes* and

other caenagnathids is suggested as another feature that is especially effective in shearing meat (Funston and Currie, 2014).

Ma et al. (2017), through anatomical comparisons and description of muscle attachments, performed a more in-depth investigation of oviraptorosaur mandibles in light of a large specimen of the mandible of the caenagnathid *Gigantoraptor*. They confirm that oviraptorosaurs were capable of mandibular feeding motions in a propalinal plane owing to the arched gliding morphology of the jaw joint (see above). Ma et al. (2017) also described fundamental differences between the mandibles of oviraptorids and caenagnathids. Oviraptorids are more built for powerful, crushing bites because they generally have deeper mandibles with a taller coronoid eminence and larger medial mandibular fossa for extra adductor muscle attachment and a downcurved mandibular symphysis, possibly built to withstand stresses of tougher foods. Conversely, caenagnathids generally have weaker jaws with hardly any kind of coronoid eminence and straighter (or sometimes upturned) symphysis, which would be most ideal for plant-shearing (and possible even meat-shearing) purposes against their triturating oral margin (Funston and Currie, 2014) (fig. 7.14). *Gigantoraptor*, although very large, is a more early-diverging caenagnathid, still possessing a relatively deeper mandible and lacking a triturating shelf like oviraptorids, possibly indicating it is a transitional form to a less mechanically advantageous feeding system in later caenagnathids (for eating softer foods) (Ma et al., 2017, 2020b).

In a study using two-dimensional (2D) geometric morphometrics, Ma et al. (2020b) were able to analyze the

diversity of skull shapes in oviraptorosaurs to see if there were any clues into their ecological roles. They found that oviraptorosaurs had wide-ranging variability of the snout, again with the downturned snouts of later oviraptorosaurs likely being an adaptation for more powerful bites. Additionally, Ma et al. (2020b) showed that the shape of the mandible and the beak as a whole are indeed correlated with phylogeny (but the cranium is not). Also, the mandible and lower beak are particularly correlated with function. These findings add more confidence in the above-mentioned studies showing that mandibular morphology has great functional importance (e.g., Ma et al., 2017). The oviraptorids and caenagnathids are distinctly separated in morphospace, both showing difference of function as well as giving insight into niche partitioning within their ecosystems. Furthermore, oviraptorids that coexisted in the

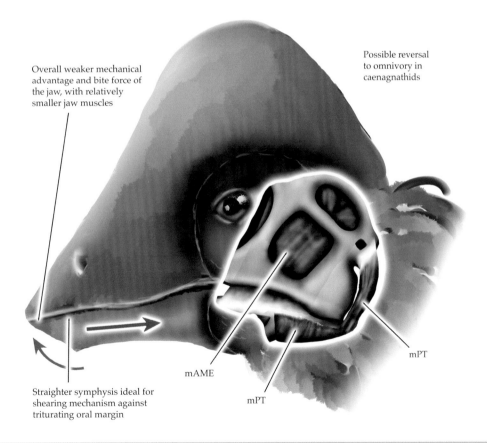

Overall weaker mechanical advantage and bite force of the jaw, with relatively smaller jaw muscles

Possible reversal to omnivory in caenagnathids

Straighter symphysis ideal for shearing mechanism against triturating oral margin

mAME

mPT

mPT

Figure 7.14. *Anzu* head showing feeding function with jaw muscles. Red arrows show orthal and palinal jaw action. Abbreviations: mAME, m. adductor mandibulae externus; mDM, m. depressor mandibulae; mPT, m. pterygoideus.

same space also show variability in mandibular functional morphology, while caenagnathid taxa that coexisted with each other varied in beak and body sizes, showing how members within both groups were able to fit into their own ecological roles in their own ways.

Ma et al. (2020a), in comparing oviraptorosaur cranial function to that of scansioriopterygids, showed how oviraptorosaurs are generally more adapted for herbivory given the suite of craniomandibular characters mentioned above, especially their comparatively much higher mechanical advantage of the jaw systems. They also point out how, again, caenagnathids have a lower coronoid region and dentary height and that they have a slightly lower jaw joint relative to the oral margin, possibly indicative of a reversal to omnivory or carnivory in caenagnathids (as previously proposed by Funston and Currie, 2014, and Ma et al., 2017). In oviraptorids, however, the jaw joint is displaced further ventrally, ensuring

an ornithischian-like simultaneous occlusion of the jawline and higher mandibular mechanical advantage (Ma et al., 2020a).

Oviraptorosaurs ranged from small- to medium-sized (over 200 kg) (fig. 7.15) to much larger, as in the case of *Gigantoraptor*, at more than 3,000 kg (Zanno and Makovicky, 2013). They were feathered, bipedal animals with long necks; elongate, grasping hands with small claws; and gracile hindlimbs of variable cursoriality (Funston et al., 2018; Dececchi et al., 2020a). They likely used their long necks to direct their heads to help acquire vegetation at various heights. Their forelimbs show uniquely isometric trends in growth rather than the more typical negative allometry in theropods, likely meaning they used their arms for acquiring food more often (Palma Liberona et al., 2019). Funston et al. (2018) and Funston (2020) suggested that while oviraptorids were less built for running, caenagnathids

Figure 7.15. Skeleton of *Anzu*.

(and avimimids) were typically better runners (and better graspers with their elongate fingers). Some caenagnathids, like *Citipes*, show enhanced cursorial adaptations such as a fused arctometatarsalian foot (for shock absorption; Holtz, 1994; Snively et al., 2004) and enlarged pelvic musculature (Rhodes et al., 2020). Cursoriality could correlate with either the need to outrun predators or to catch prey, so if caenagnathids were indeed omnivorous in some cases, these postcranial features would have been of significant help (Funston, 2020). Rhodes et al. (2020, 2021), in reconstructing pelvic musculature, found that there was a larger attachment area for hip flexors (with an enlarged preacetabular blade of the ilium) and less so for hip extensors in *Chirostenotes*, which led them to propose more of a wading locomotion rather than cursoriality, but they simultaneously noted that more variation exists among caenagnathids than previously expected.

Given the above, it is clear that oviraptorosaur feeding ecology has been a great mystery to paleontologists for many decades. The very name *Oviraptor* means "egg thief," speaking to the initial thought that oviraptorosaurs were egg-eaters because a specimen was found alongside what were thought to be *Protoceratops* eggs (Osborn, 1924), but they were later found to be eggs of *Oviraptor* itself (Norell et al., 1995). Although Barsbold (1977) suggested a durophagous (mollusk-eating) diet because of its powerful feeding mechanism, Smith (1992), Jansen (2008), and Longrich et al. (2010) have rejected this idea because of the structure and toothless nature of the jaw. An herbivorous lifestyle has been the most recent consensus for all oviraptorids. For caenagnathids, some have been deemed herbivorous (e.g., *Anzu* and *Caenagnathus*), while others possibly show signs of omnivory (as described above) (Smith, 1990, 1992; Lamanna et al., 2014; Ma et al., 2017, 2020a, 2020b, 2022; Funston et al., 2018; Funston, 2020). Gastroliths have been found associated with *Caudipteryx*, indicating likely plant-crushing within the gut (Ji et al., 1998), but further direct evidence is needed to fill in these gaps in our knowledge of oviraptorosaur feeding ecology as a whole.

Scansoriopterygidae

Scansoriopterygids (e.g., *Yi*, *Epidexipteryx*, *Scansoriopteryx*, and *Ambopteryx* from the Middle to Late Jurassic of China) possess small heads that are somewhat triangular in profile and bear resemblance to the closely related early-diverging oviraptorosaurs (Ma et al., 2020a). They have large orbits and moderately sized antorbital fenestrae and nares. Their temporal region ranges from rostrocaudally shorter (e.g., *Yi*) (fig. 7.16) to broader (e.g., *Epidexipteryx*), showing variation in the breadth of a subvertically oriented mAME muscle complex within the adductor chamber, which passes medial to narrow jugal bones. The mandible is straight and narrow throughout its length (with no visible coronoid eminence), with a downward curved rostral symphysis, long mandibular fenestra, and short retroarticular process. They have small unserrated teeth that are mesially positioned both on the upper and lower jaw, with the dentary teeth being elongate and highly procumbent (unlike any oviraptorosaurs).

Although they speculate that it was mainly herbivorous based on the likely functionality of their mesially positioned teeth, Ma et al. (2020a) found that scansoriopterygids had relatively low mandibular mAME mechanical advantage in jaw-closing and jaw-opening (compared to the herbivorous oviraptorids), particularly driven by the lack of a coronoid eminence, and generally smaller temporal musculature because of the shorter height of the skull and smaller infratemporal fenestra. The smaller jaw-opening mechanical advantage may also indicate faster jaw-opening than oviraptorosaurs, which might have been helpful in hunting small insects or very small vertebrates. But the articular is ventrally offset relative to the tooth row in *Yi* (and less so in *Epidexipteryx*), which is also suggestive of a trend toward omnivory or herbivory (Ma et al., 2020a). Wang et al. (2019) recently described gut contents in *Ambopteryx* consisting of at least some bone as well as gastroliths, which would seem to support the notion of a more omnivorous lifestyle.

Figure 7.16. Reconstructed head of *Yi*.

The most notably bizarre postcranial feature in scansoriopterygids is its highly elongate third digit, which we now know held a winglike membrane (i.e., patagium), similar to that in bats or pterosaurs, which has been preserved in at least *Yi* and *Ambopteryx* (Xu et al., 2015; Dececchi et al., 2020b). Dececchi et al. (2020b), upon further examining these membranes with laser-stimulated fluorescence, indicated that these animals were most likely arboreal and that their wings were not built for long-distance gliding or powered arm-flapping flight. Instead, they were likely short-distance gliders at best, and they represent a short-lived failed attempt in the evolution of flight.

Dromaeosauridae

Dromaeosaurids include *Halzskaraptor*, unenlagiines (*Buitreraptor*, *Austroraptor*, *Unenlagia*), *Microraptor*, *Sinornithosaurus*, eudromaeosaurs such as *Bambiraptor*, dromaeosaurines (*Dromaeosaurus*, *Utahraptor*, possibly *Deinonychus*; fig. 7.17), and velocirapatorines (*Saurornitholestes*, *Velociraptor*; fig. 7.18). They had worldwide distribution throughout the Cretaceous and possessed skulls of various lengths and robusticity. For example,

velociraptorines (e.g., *Velociraptor*) have a relatively more elongate and narrower muzzle, whereas dromaeosaurines (e.g., *Dromaeosaurus*) tend to possess a more robust and bulkier skull. The orbits are mostly large and round (likely indicting less skull strength in biting; Henderson, 2002), and the antorbital fenestrae are large in most cases. They all had many small, bladelike teeth along a narrow jawline, in many cases with small denticles along their distal edges of their teeth that helped hold on to struggling prey (Torices et al., 2018). But there were some exceptions, such as the conical, unserrated teeth of some unenlagiines (Gianechini et al., 2020). Their temporal regions are caudally restricted within the skull with a circular supratemporal fenestra. The postorbital region and adductor chamber are subrectangular, vertically oriented, and narrow relative to the total length of the skull, especially in velociraptorines. Their palates are horizontally oriented with variably sized pterygoid flanges. Their mandibular symphyses range from equal length and breadth to a bit taller than long (Therrien et al., 2005), and their mandibles possess low, gently sloped to almost nonexistent coronoid eminences, a large mandibular fenestra, and reduced retroarticular process.

Perhaps unsurprisingly, dromaeosaurid jaw systems

generally fall in morphospace with other carnivorous theropods (Schaeffer et al., 2020). Comparing them to other theropods, Sakamoto (2010) found that the mechanical advantages of dromaeosaurid jaws are generally

on the lower end of the spectrum across the tooth row—likely a result of their elongate snouts and their need to bite down quickly. *Dromaeosaurus* plotted higher in mechanical advantage than the other dromaeosaurids tested

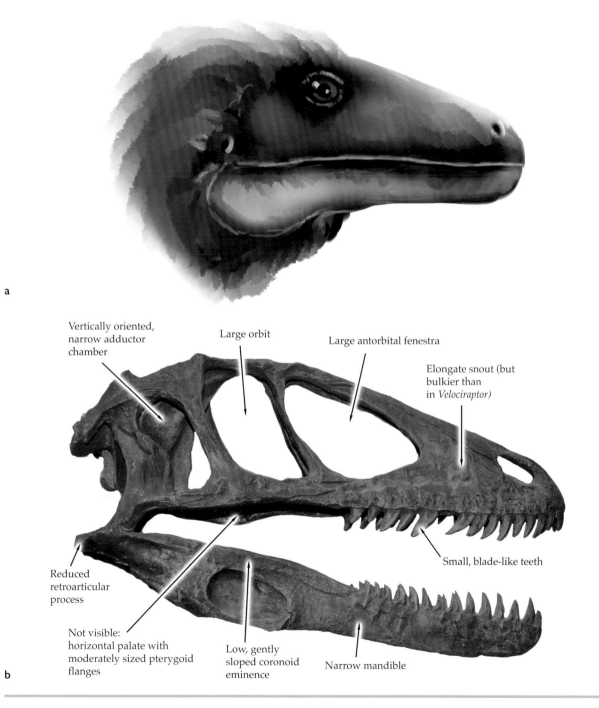

a

Vertically oriented, narrow adductor chamber

Large orbit

Large antorbital fenestra

Elongate snout (but bulkier than in *Velociraptor)*

Small, blade-like teeth

Reduced retroarticular process

Not visible: horizontal palate with moderately sized pterygoid flanges

Low, gently sloped coronoid eminence

Narrow mandible

b

Figure 7.17. *Deinonychus* (a) reconstructed head and (b) skull in lateral view (length approximately 33 cm).

(Sakamoto, 2010), which is also reinforced with its stouter, deeper skull and larger teeth comparatively (Barsbold, 1983), possibly allowing it to bite through the bone of small prey (Therrien et al., 2005; Monfroy, 2017). The maximum bending strength of teeth in *Dromaeosaurus* is located near the mesial (front) end of the jawline, further improving its ability to produce effective slashing bites (Therrien et al., 2005; Monfroy, 2017).

More in-depth biomechanical analyses by Therrien et al. (2005) shows further insights to these trends. They suggested that dromaeosaurids as a whole had the strongest bending strengths at around the second alve-

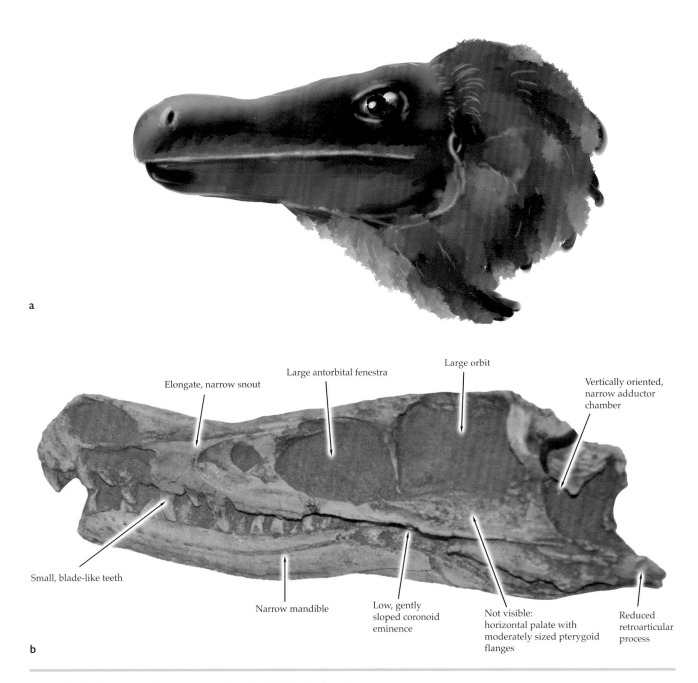

Figure 7.18. *Velociraptor* (a) reconstructed head and (b) skull in lateral view (length approximately 23 cm).

olus, likely used in fast slashing bites like Komodo dragons. *Dromaeosaurus*, again, showed the strongest values in dorsoventral and labiolingual force values (followed next by *Deinonychus* [fig. 7.19] and with the elongate- and slender-snouted *Velociraptor* and *Saurornitholestes* plotting on the weaker end of the spectrum). Ultimately, there is a clear functional distinction between the more robust-skulled dromaeosaurines and the smaller, narrower-skulled velociraptorines. Furthermore, Therrien et al. (2005) noted that, as pointed out by Currie et al. (1990), the higher bite force of dromaeosaurines would have been additionally useful in crushing small bones with the short, broad denticles of their teeth, whereas the longer, hooked denticles of velociraptorine (e.g., *Velociraptor*) teeth would have been better suited for slicing flesh in a direction parallel to the bone's surface.

Additionally, the fact that the bite force of dromaeosaurids—particularly dromaeosaurines—was most fo-cused at the rostral end of the snout has led to the assumption that they used fast slashing bites to kill instead of being able to hold on to struggling prey with their heads quite as much (Therrien et al., 2005). Still, even though *Dromaeosaurus* shows signs of not being able to handle struggling prey with its head, the structure of the teeth may have made up for it. In using dental microwear analysis and FEA to compare tooth denticle function of the dromaeosaurids *Dromaeosaurus* and *Saurornitho-lestes*, Torices et al. (2018) showed that the broadened shape of the small denticles of dromaeosaurids was indeed reasonably well suited for handling struggling prey, at least to some extent compared to the weaker-toothed troodontids. Their results indicated that the teeth of *Dromaeosaurus*, in particular, were not as prone to failing while biting down, even at nonoptimal bite angles, which is good for side-to-side jerking motions of struggling prey, which they point out is also seen in Komodo dragons (Torices et al., 2018).

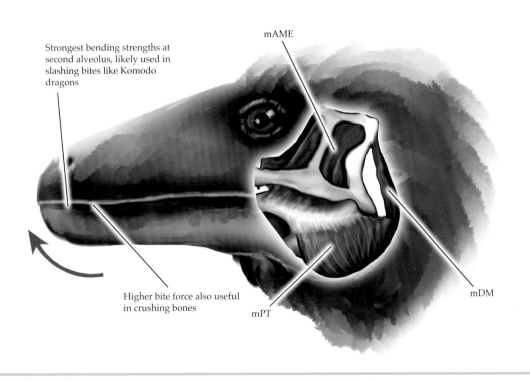

Strongest bending strengths at second alveolus, likely used in slashing bites like Komodo dragons

mAME

Higher bite force also useful in crushing bones

mPT

mDM

Figure 7.19. *Deinonychus* head showing feeding function with jaw muscles. Red arrow shows orthal jaw action. Abbreviations: mAME, m. adductor mandibulae externus; mDM, m. depressor mandibulae; mPT, m. pterygoideus.

Regardless of whether *Deinonychus* is more closely related to dromaeosaurines (DePalma et al., 2015) or to velociraptorines (Currie and Evans, 2020), the fact that *Dromaeosaurus* shows a stronger bite force than *Deinonychus* is interesting in itself, as *Deinonychus* is a larger animal. Still, *Deinonychus* was an adept predator in many ways, including a bone-crushing bite of its own. Gignac et al. (2010) used deep bite puncture marks found in the bones of the ornithopod dinosaur *Tenontosaurus* (presumably made by *Deinonychus*) to simulate and estimate the bite force of *Deinonychus*. A nickel-alloy scale replica of a *Deinonychus* tooth was used to puncture a bovine long bone to simulate its bite. Their results showed an estimated range of between 4,100 and 8,200 N to generate the puncture marks—which is similar to comparatively sized crocodilians and certainly enough to bite through bone (Gignac et al., 2010).

Dromaeosaurids are feathered and strictly bipedal animals. They had S-shaped necks that likely helped with precise attacks of smaller prey items, and they had long forelimbs with plenty of mobility in their joints and hands that were well adapted for grasping and apprehending prey (Ostrom, 1969; Gishlick, 2001; Carpenter, 2002; Senter, 2006). Senter (2006) performed a manual manipulation range-of-motion study using the forelimbs of *Deinonychus* and *Bambiraptor*. Much of the forelimb function was found to be similar between the two taxa, with reasonable retraction and elevation of the humerus, strong elbow flexion, weaker elbow extension, and a lack of active pronation and supination, meaning their palms faced medially against their torso. Both were likely able to use their arms for two-handed prehension with flexed wrists as well as clutching, swinging and hooking, and keeping balance. Additionally, *Bambiraptor* may have had more opposability in its fingers than most other theropods, making it capable of one-handed prehension, although large feathers on the hands might have made this and any other general use of the hands in dromaeosaurids more difficult (Senter, 2006).

Interestingly, studies of cursorial limb characteristics and limb proportions suggested that many eudromaeosaurs may not have been able to run quite as fast as previ-

ously thought, and therefore not as dependent on pursuit predation as is normally suspected, as their limbs do not seem to be modified as such (Carrano, 1999; Persons and Currie, 2012, 2016; Rhodes et al., 2021). Persons and Currie (2012, 2016) and Rhodes et al. (2021) indicated that dromaeosaurids also have relatively reduced hip extensor musculature (such as m. caudofemoralis and other related muscles) compared to other theropods, leading to less efficient hip extension (albeit showing more flexibility, which might be a remnant of flight capabilities before secondary flightlessness in many dromaeosaurs; Persons and Currie, 2012). As Dececchi et al. (2020a) pointed out, however, it is important to understand that the expectations for certain types of cursoriality may be a bit more nuanced, with likely variability of cursorial pursuit strategies possibly coming into play. For instance, Gianechini et al. (2020) show how unenlagiine dromaeosaurs (such as the slender-snouted *Buitrerator*, *Austroraptor*, and *Unenlagia*) seem to be an exception to the above, as they possessed long tibiae and long, near-arctometatarsalian feet, which would have provided much greater cursorial ability, whereas eudromaeosaurs had relatively shorter tibiae and wider metatarsals. Although both subclades shared elongation of phalanges for grasping, eudromaeosaur feet do have wider phalanges for gripping larger prey, while unenlagiines likely pursued smaller prey items (Gianechini et al., 2020). The small microraptorines (e.g., *Microraptor*) also show a unique adaptation in having two sets of wings with elongated feathers (two wings on the forelimbs and two on the hindlimbs), possibly for gliding purposes (Longrich, 2006). A unique lateral pubic tubercle in microraptorines was initially thought of as muscular attachment, but further analysis by Rhodes and Currie (2020) has indicated this structure is homologous with pubogastralial ligaments that anchor the caudal gastralia to the hip. Surrounding musculature was affected by the position of this tubercle, however, including m. puboischiofemoralis externus, which protracts and laterally rotates the femur—a possible key to gliding control (Rhodes and Currie, 2020).

Perhaps the best-known predatory tool in dromaeosaurids is the hypertrophied "sickle claw" sported on their second toe (fig. 7.20). This mobile toe claw has been of great interest, particularly when considering the functional restrictions of generally weaker skulls and slower pursuit capabilities in many dromaeosaurid taxa. Ostrom (1969) argued that the greatly curved, laterally compressed, trenchant killer claw in the dromaeosaurid *Deinonychus* had the right amount of curvature to produce the optimal force in maximum penetration, and that flexor tubercles present underneath the second digit were especially indicative of strong flexion of that digit, with strong extension needed as well to retract the claw. The same is present in other dromaeosaurids like *Velociraptor* (Norell and Makovicky, 1999). Carpenter (1998) suggested that instead of slashing at prey, dromaeosaurids may have used their sickle claws to pierce through major blood vessels in the throat, like those in the carotid sheath or around the trachea. But Therrien et al. (2005) point out that this may not have been habitual given the need for great precision for such an act. To test its function mechanically, Manning et al. (2006) created a robotic model of a *Deinonychus* sickle claw and found that it was likely able to use this claw to pierce and grip the flesh of its prey instead of slashing and using it to disembowel. They also posit a climbing function for it because the arc angle of the claw is comparable to that of some extant birds. Therefore Manning et al. (2006) suggest it might have leapt onto its prey and applied a foothold while gripping and piercing with its feet and holding on to it with its grasping hands.

To test these assertions further, Manning et al. (2009) performed FEA on a *Velociraptor* (fig. 7.21) sickle claw and confirmed that it was resistant to forces that acted in one longitudinal plane but was less resistant to forces acting tangentially. This adds further support to the notion that the tip of the claw was used in puncturing and gripping. It also allowed *Velociraptor* to be capable of tree-climbing (therefore giving it a scansorial lifestyle), with the widened proximal end of the toe claw lessening stresses onto the digit itself. Fowler et al. (2011) rejected the assertion of tree-climbing and instead compared the enlarged second digit to that of hawks and eagles and

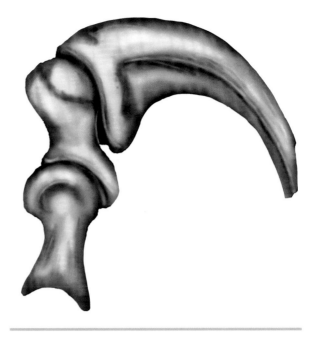

Figure 7.20. *Deinonychus* toe "sickle" claw.

how they use their claws to maintain grip on their prey. Fowler et al. (2011) show how the dromaeosaurid foot is built for grasping functions, and they discussed the likelihood of using their feathered forelimbs while feeding by flapping them as they have a hold on their prey to keep stability during the initial immobilization. Using 3D musculoskeletal modeling techniques applied to the tendons surrounding the sickle claw, Bishop (2019) found that, given the right criteria, the highest forces seen in *Deinonychus*'s sickle claw are made with the animal in a crouching position of the hindlimbs (with the knees and ankles flexed) and with larger flexor musculature overall. Also, the tip of the claw would not have been able to apply much force, so the force transmission was more dependent on the more proximal part of the foot, also supporting the notion of a grasping function of the second digit for struggling prey (Bishop, 2019).

A few notable discoveries have provided evidence of diet and predation in dromaeosaurids. One of the most famous discoveries is that of the "fighting dinosaurs"—a *Velociraptor* seemingly locked in battle with the early ceratopsian *Protoceratops* (Kielan-Jaworowska and Bars-

Figure 7.21. Skeleton of *Velociraptor* jumping and displaying recurved claws.

bold, 1972; Unwin et al., 1995; Carpenter, 2000). Broadly, this can be interpreted as direct evidence of predator–prey interaction, with the *Velociraptor* with its foot and sickle claw pushed up on the *Protoceratops*'s torso while the *Protoceratops*'s jaws are clamped down on the *Velociraptor*'s forelimb. It is difficult to know for certain, however, how this fight really started, and it is clear that *Protoceratops* had a handle on the situation before they both met with their sudden demise. More recent evidence of tooth marks on bone and associated shed teeth has shown that *Velociraptor* indeed at least fed upon the carcass of *Protoceratops*, so it is safe to assume that *Protoceratops* was a common meal (Hone et al., 2010). Additionally, neuroanatomical studies implicate that *Velociraptor* would have been capable of detecting a wide range of sound frequencies and was likely an agile predator that also scavenged upon available carcasses when given the chance (King et al., 2020).

Another example of possible evidence of predator–prey interaction is the discovery of a specimen of the medium-sized ornithopod *Tenontosaurus* that was found with shed teeth and bones of at least three to maybe even six or more *Deinonychus* associated with it. This has led to speculation of possible pack-hunting behavior in *Deinonychus*, with a number of individuals working together to take down a larger herbivore (with some dying in the process; Maxwell and Ostrom, 1995) (fig. 7.22). Additionally, preserved footprints of deinonychosaurs traveling together have led to speculation of gregarious behavior (Li et al., 2007). It is hard to know for certain, however, if this is truly evidence of pack hunting or even scavenging or something else. Roach and Brinkman (2007) considered pack-hunting behavior unlikely and instead hypothesized that these animals were more solitary and likely more agonistic, mobbing, and possibly even cannibalistic. Isotopic data have also suggested that *Deinonychus* ate a range of different sizes of prey throughout life, with juveniles eating much smaller prey than adults, which itself is not consistent with pack-hunting behavior (Frederickson et al., 2020).

Dietary variation is also broadly more variable than previously assumed. Analysis of the snout in eudromae-

Figure 7.22. Skeletons of *Deinonychus* feeding on young *Tenontosaurus*.

osaurs has suggested that Asian taxa possessed rostro-caudally longer snouts while North American eudromaeosaurs possess shorter snouts, which might indicate variation in prey and feeding mechanics (Powers et al., 2020). Isotopic evidence has suggested variation in the environment of predation as well, with some feeding in specifically floodplain settings and others showing a broader range (Frederickson et al., 2018). Furthermore, evidence of gut contents from microraptorine *Microraptor* suggests it was an opportunist that fed on a wide variety of small vertebrates, like fish, mammals, reptiles, and birds (O'Connor et al., 2011; Xing et al., 2013; O'Connor and Zhou, 2020). Although some have suggested a venomous bite in some dromaeosaurs like *Sinornithosaurus* based on alleged grooved teeth and possible venom pits in the gumline (Gong et al., 2010), others have deemed the evidence presented to be weak and unlikely (Gianechini et al., 2011). Perhaps one of

the more unique *Bauplans* among dromaeosaurids is that of halszkaraptorines (e.g., *Halszkaraptor*), an early-diverging dromaeosaurid group that has been deemed amphibious, likened to a present-day merganser, with a long narrow snout with tiny teeth, hyper-elongated neck, and inferred forelimb-propelled swimming (Cau et al., 2017; Cau, 2020). Although it has been claimed to show signs of omnivory or herbivory (Brownstein, 2019), these arguments have been rebutted in favor of a more carnivorous semiaquatic lifestyle (Cau, 2020).

Troodontidae

Troodontids (e.g., *Troodon, Jinfengopteryx, Saurornithoides*, and *Mei*, known from the Late Jurassic to Late Cretaceous of North America and Asia) possess narrow snouts that are built lightly and possess many small

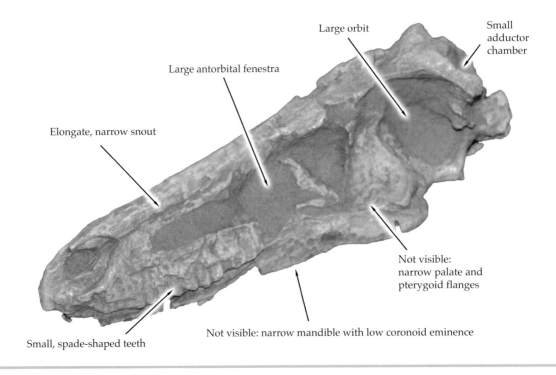

Large orbit

Small adductor chamber

Large antorbital fenestra

Elongate, narrow snout

Not visible: narrow palate and pterygoid flanges

Not visible: narrow mandible with low coronoid eminence

Small, spade-shaped teeth

Figure 7.23. Skull of *Saurornithoides* (troodontid) in lateral view (length approximately 10 to 11 cm).

spade-shaped teeth (fig. 7.23). Their orbits are especially large and face more rostrally, and their antorbital fenestrae are also large. As in dromaeosaurids, troodontid temporal regions are caudally restricted. The postorbital region and adductor chamber are rostrocaudally narrow in relation to the total length of the skull, and their palates and pterygoid flanges are more narrowed. Troodontid mandibles possess low to nearly nonexistent coronoid eminences, a moderately sized mandibular fenestra, and reduced retroarticular processes.

In *Troodon*, the bending strength of the mandible is relatively higher at the mesial end of the tooth row than the rest of the mandible, implicating a more mesially focused bite. Troodontid teeth can be classified as more modified versions of ziphodont teeth that have transitioned to folivorous morphology with asymmetrical hooked denticles (Brink et al., 2015; Torices et al., 2018; Hendrickx et al., 2019). Holtz et al. (1998), using morphometric measurements, suggested troodontids were most likely omnivorous because the denticles along the

mesial and distal margins of the dentition are fewer, larger, and wider, traits more associated with herbivorous feeding habits. Holtz et al. (1998) argued against full herbivory because of certain anatomical features suited for carnivory, such as a loose mandibular symphysis and intramandibular joint in the skull (as well as other postcranial features). Fiorillo (2008) suggested that troodontids likely ate soft food items because the wear facets on their teeth show fine wear patterns and lack the pitting seen with a hard food diet. Later, using phylogenetic methods, Zanno and Makovicky (2011) found a mixture of traits among troodontids that showed inheritance of both signs of herbivory and carnivory across taxa. Brink et al. (2015), looking at dental microstructure, showed that troodontid teeth lacked deep interdental folds seen in carnivorous theropods. Furthermore, Torices et al. (2018), using FEA methodologies, found that troodontid teeth exhibited high stress patterns at nonoptimal cutting angles, suggesting that hyper-carnivory was not an option because the stresses of struggling prey would have broken the teeth easily. For this reason, Torices et al. (2018) also suggested that troodontids preferred softer foods.

Troodontids had somewhat reduced forelimbs with clawed hands and gracile lower limbs and feet with traits of cursoriality as a means of strengthening a more predatory ecology (Carrano, 1999; Fowler et al., 2011; Dececchi et al., 2020a). A highly mobile first digit claw and a shallowly recurved sickle claw on the retractable second digit may have been used in manipulating small prey items, but different foot bone proportions point to a divergence in predatory strategy from the more powerful feet of dromaeosaurids (Fowler et al., 2011). Still, direct evidence of omnivory certainly may be present. The possible troodontid *Anchiornis* (which alternatively may be an early-diverging avialan) has been discovered with associated gastric pellets (like those regurgitated by raptorial birds today) that contain lizard bones and fish scales, indicating a transition to an avian-style digestive system (Zheng et al., 2018). Furthermore, a recent description of gastric pellets plausibly attributed to *Troodon* contained small mammal skeletons (Freimuth et al., 2021). Conversely, oval-shaped structures have been found preserved in the gut contents of *Jinfengopteryx* that might be either seeds or nuts (or possibly even developing follicles or eggs) (Ji et al., 2005).

Mesozoic Avialae and Feeding in Living Birds

Birds are among the most fascinating vertebrate groups living today, and the fact that they are dinosaurs themselves makes them incredibly important in our understanding of non-avian dinosaur biology as a whole. Birds (or, really, the avialan ancestors of birds) were present during the Late Jurassic, as indicated by the presence of the famous toothed "first bird" *Archaeopteryx*. But most of our knowledge of Mesozoic bird life comes from the Early Cretaceous of China (Jehol Biota), with a plethora of specimens of all kinds of birds preserving gut contents of many taxa (that show signs of

all types of diet ranging from carnivory to omnivory to herbivory of various sorts).

Although all of today's birds (Aves) belong to the avian group Euornithes (more specifically, Ornithurae), most Cretaceous birds belonged to the now-extinct Enantiornithes, which had teeth and fingered wings. Enantiornithes is the primary group of avialans that interacted with non-avian dinosaurs until their demise. Both major clades of birds are known to have developed powered flight capabilities (although multiple lineages have lost that capability). A suite of cranial features, including adaptations for multiple types of complex cranial kinesis and endless forms of toothless bill shape, are found throughout all birds and continue to be present in the more than 10,000 species of bird that live today.

The famous *Archaeopteryx* (fig. 7.24), well known to many tentatively as the "first bird," was one of the earliest diverging avialans (Xu et al., 2011). This Late Jurassic bird ancestor from Germany was no more than half a meter long, and although superficially it had many traits showing a clear transition to the birdlike body plan we know of today (most striking of all being its birdlike feathers and large wings for powered flight), it still possessed numerous features that show its maniraptoran theropod dinosaur ancestry. The skull of *Archaeopteryx* looks more similar to that of troodontids and other closely related maniraptorans, including a still-complete infratemporal bar (Rauhut et al., 2018; O'Connor, 2019). *Archaeopteryx* also possessed stout, unserrated teeth with recurved apices and oval cross-sections that were evenly spaced, running along the premaxillary-maxillary and dentary oral margins (O'Connor, 2019). Other traits reminiscent of maniraptoran ancestry include three-fingered claws; a long, feathered tail; and maniraptoran-like toe claws (capable of withstanding stresses in perching; Birn-Jeffery and Rayfield, 2009). *Archaeopteryx* may have had an omnivorous or possibly insectivorous diet, using its teeth to snatch tiny insects; however, it is difficult to tell its exact diet owing to a lack of direct evidence (i.e., no known gastric mill or stomach contents) (Ostrom, 1976; Elzanowski and Chiappe, 2002; Wellnhofer, 2008; O'Connor, 2019). It is also

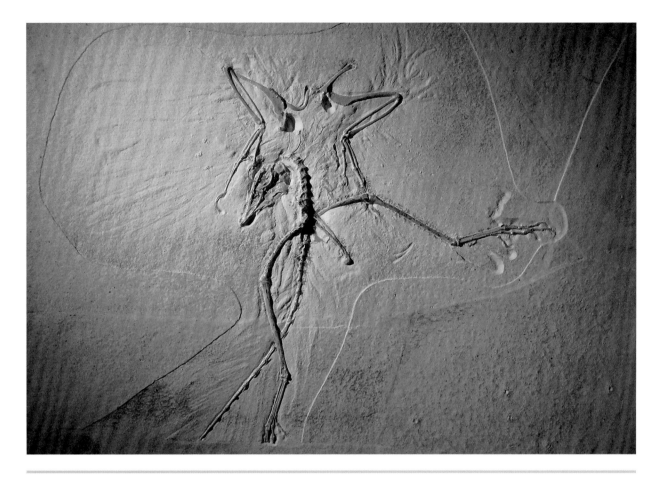

Figure 7.24. Early avialan *Archaeopteryx*.

possible that there was some dietary variation within the genus owing to morphological disparity in skull shape (Rauhut et al., 2018; Miller and Pittman, 2021).

Another long-tailed, early-diverging avialan is *Jeholornis* from the Cretaceous Jehol Biota. Its skull looks similar to *Archaeopteryx* but with a somewhat shorter snout and reduced dentition (in many cases, no dentition in the upper jaw), and ventrally curved, expanded mandibular symphysis (O'Connor, 2019; Zhou and Zhang, 2002). Its characteristics suggest it was herbivorous and granivorous, with associated stomach contents showing a diet of seeds with the assistance of small gastroliths ingested for digestion (O'Connor et al., 2018; O'Connor, 2019; O'Connor and Zhou, 2020).

Confuciusornis and its close relatives were early-diverging pygostylians (containing a "pygostyle" or short-ened set of fused caudal vertebrae—a grouping that includes all living birds). *Confuciusornis* was edentulous and presumably had a blunt, recurved keratinous rhamphotheca (or beak). Studies of mandibular mechanical advantage (Navalón, 2014) and FEA (Miller and Pittman, 2021) have suggested its beak would have been suited mostly for an herbivorous or granivorous lifestyle (Miller and Pittman, 2021), but there is no direct evidence of this. *Confuciusornis* beaks have been shown to have a burst in growth at a younger age and continued growing steadily, opening it up to dietary niche shifts throughout its life (Marugán-Lobón and Chiappe, 2022). Some potentially associated disarticulated fish bones may indicate some piscivory, but this association is not for certain either (O'Connor, 2019).

Sapeornis, another early-diverging pygostylian, shows

a blunt, deep skull and bill with reduced dentition. Some specimens have been found preserving small gastroliths as well as seeds in a crop (an outgrowth of the esophagus) for storage, as seen in living birds (Zheng et al., 2011; O'Connor, 2019). These findings are in agreement with mandibular mechanical advantage results showing *Sapeornis* possessed a skull mechanically suited for herbivory or granivory (Navalón, 2014; Miller and Pittman, 2021). Furthermore, its metacarpals were partially fused (similar to later birds), meaning it likely used its feet for grasping (Pittman and Xu, 2020; Miller and Pittman, 2021).

Enantiornithes was the most diverse group of Cretaceous avialans, and yet, surprisingly, little direct evidence of diet is known from this group. Most were arboreal, with wider ecological diversity seen in the Late Cretaceous (O'Connor, 2009, 2019). They had diverse snout and overall cranial morphologies (Zhang et al., 2004; Morschhauser et al., 2009; O'Connor and Zhou, 2013; O'Connor, 2019; Miller and Pittman, 2021). Although few later taxa were edentulous with expanded and fused beaks (e.g., *Gobipteryx*), possibly suggesting some herbivory, most enantiornithines had diverse, and sometimes heterodont, dentition across the upper and lower jaws (O'Connor, 2019; Miller and Pittman., 2021).

Enantiornithine teeth most frequently resemble the recurved teeth of *Archaeopteryx*, plausibly indicating manipulation of food items (O'Connor and Zhou, 2013; O'Connor, 2019). Strengthened enamel in some enantiornithines like *Sulcavis* and *Monoenantiornis* (along with robust cranial features) may suggest an element of durophagy in their diet as well, puncturing the exoskeletons of insects (O'Connor, 2019; Miller and Pittman, 2021). Other enantiornithies like longipterids (e.g., *Longipteryx* and *Rapaxavis*) show hyper-carnivorous or piscivorous qualities, such as uniquely large, recurved, bladelike teeth in *Longipteryx* (O'Connor et al., 2011; O'Connor, 2019) and a low mechanical advantage of the jaw system of *Rapaxavis* (relating to increased bill length, leading to increased speed of jaw closure) (Navalón, 2014). The elongate bills of *Rapaxavis* and *Longirostravis* may also suggest probing behavior for finding small invertebrates (Hou et al., 2004; Morschhauser et al., 2009; Miller and Pittman, 2021). This is further supported by a quantitative study showing most longipterids to have been either generalists or to have fed on invertebrates with possible raptorial behavior (Miller et al., 2022). Pengornithids like *Pengornis* also showed hyper-carnivorous features, but in the form of more numerous, low-crowned teeth (Zhou et al., 2008; O'Connor, 2019) and intermediate mechanical advantage of the jaw system (Navalón, 2014).

Limited known palatal material from enantiornithines makes it difficult to assess whether cranial kinesis was present in this group. *Chiappeavis* shows a medio-laterally extensive palate with characteristics that suggest a more akinetic skull. A possible vomer of *Gobipteryx*, however, shows a distinct morphology suggestive of limited cranial kinesis, although more material is needed for further assessment (Hu et al., 2019; O'Connor, 2019). Many enantiornithines also show foot and body proportions like avian raptors, further suggesting possible carnivorous habits (Zhou and Zhang, 2005; Wang et al., 2010, 2014; O'Connor, 2019). No gastroliths are known in enantiornithines, as they appear to not have had crops, leading to the assumption that they may have fed on softer food items (O'Connor and Zhou, 2020; Miller and Pittman, 2021).

Euronithes (a vast majority of which is grouped as Ornithuromorpha) is the group leading to today's birds, and although there are still differences in cranial and manus morphology, many of the skeletal characteristics of Ornithurae modern birds are present in the earliest ornithuromorphs, including those of the flight apparatus (O'Connor et al., 2011). Many skull bones remain unfused, and the rostra often do not show the proportions of those of modern birds, but it is possible that most did not possess a free postorbital or infratemporal bar (O'Connor, 2019). Many early ornithuromorphs show tiny teeth that are slightly recurved and variable in number (sometimes being reduced), although the tip of the snout is generally toothless (Zhou and Martin, 2011; O'Connor and Zhou, 2013; O'Connor, 2019). Of course, many later ornithurines (including all of today's birds) show complete edentulism. Many toothed ornithuromorphs show a unique avian version of a predentary bone at the rostral tip of the lower jaw between the dentaries (similar to that seen in ornithischians; see chaps. 10 through 13), which likely played a role in intramandibular kinesis (Zhou and Martin, 2011; Bailleul et al., 2019).

Most known Mesozoic birds were either terrestrial or arboreal (Mayr, 2017; Serrano and Chiappe, 2017; Cobb and Sellers, 2020), but two well-known Mesozoic marine ornithurines were *Ichthyornis* and the diving

Hesperornis (fig. 7.25), both of which had elongate beaks with a predentary and many sharp teeth, and which likely had a primarily piscivorous diet (Rees and Lindgren, 2005; Hinić-Frlog and Motani, 2010; Field et al., 2018; Miller and Pittman, 2021). Furthermore, the skull of *Ichthyornis* may have been kinetic, further aiding in oral manipulation, but otherwise kinesis may have mainly been more generally present in the bird skull, starting with later neognaths (Zusi, 1984; Bout and Zweers, 2001; Bhullar et al., 2016; Field et al., 2018;

Figure 7.25. Skeleton of *Hesperornis*.

Hu et al., 2019). Mandibular leverage and cololite evidence from *Yanornis* (as well as the presence of many small teeth) point to a piscivorous diet as well (Zhou et al., 2004; Navalón, 2014; Zheng et al., 2014). *Asteriornis*, another ornithuromorph, shows more modern skull traits, including a lack of teeth and a ventrally recurved bill that is possibly indicative of shoreline feeding (Field et al., 2020). *Hongshanornis* may have been granivorous or herbivorous, as suggested by mandibular leverage as well as gastrolith evidence (Chiappe et al., 2014; Navalón, 2014; Miller and Pittman, 2021). The vast range of evolutionary experimentation of the avian feeding apparatus across the Mesozoic and after led to the incredible diversity of feeding ecology in birds as we know it today.

The evolution and fusion of cranial and mandibular bones as well as highly modified joint structures of the palate and other regions of the skull (Bock, 1964; Zusi, 1984; Zweers, 1991; Bailleul et al., 2017; Olsen, 2019; Cost et al., 2020) led to the kinetic nature of the bird cranium along with a toothless beak surrounded by a rhamphotheca. Zusi (1984) showed that the mechanisms of bird cranial kinesis can largely be grouped into three categories. The first is prokinesis (fig. 7.26), the most widespread mechanism among birds, where the upper beak protracts and retracts as one unit at the craniofacial hinge. The second grouping, amphikinesis, involves this hinge as well, but the beak has a longer narial opening, making the beak more flexible and allowing the upper jaw the be pushed upward more toward the distal end when protracted. The last grouping, rhynchokinesis, takes a larger variety of forms, with multiple points of flexion of the beak, depending on the taxon. It can be seen in birds such as shorebirds, cranes, and hummingbirds (Zusi, 1984). Variation in beak shape also plays a large role in feeding ecology (e.g., in different duck species; Olsen, 2017).

The modification of the forelimb into a flight apparatus is also one of the key influencers of ecological expansion. Flight opens the door to many more food options, and in turn, evolutionary innovation for a variety of diets is more likely. The pectoralis muscles (modified for downward flapping motion) and supracoracoideus muscles (modified for upward flapping motion) attach along a keel on the sternum that projects forward, with different sizes determining how big the flight muscles might be. The gut system in birds is also highly modified and contains a crop (a pouch off the esophagus for storage and softening of food), a proventriculus (involved in chemical breakdown of foods), the gizzard (used in rotation and crushing of foods, sometimes with stones that have been swallowed), and of course the intestinal tract. The highly modified nature of the gut system of birds has given them the efficiency needed for digestion,

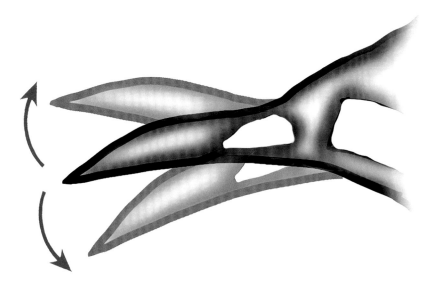

Figure 7.26. Depiction of prokinesis in a bird beak, with red arrows showing movement of beak against craniofacial hinge (redrawn from Zusi, 1984).

Figure 7.27. Passerine bird (Bay-Headed Tanager).

Figure 7.28. Ostrich head.

with such active lifestyles that include such high-energy activities like flight.

We will not be covering the immense diversity of feeding in birds, with dietary ecologies ranging from granivory (e.g., geese and grouse), frugivory (e.g., tanagers; fig. 7.27), folivory (e.g., hoatzins), nectarivory (e.g., hummingbirds), and herbivory in general (e.g., ostriches [fig. 7.28] and swans) to omnivory (e.g., pheasants and quails), insectivory (e.g., thrushes, woodpeckers, swallows, and cuckoos), piscivory (e.g., penguins, herons, darters, loons, pelicans, and storks), crustacivory (e.g., rails and flamingos), detritivory (e.g., vultures), and general faunivory (e.g., birds of prey such as hawks, falcons, eagles, and owls, as well as shrikes). But the enormous span of ecological roles seen across all of today's birds comes with an astounding array of unique and amazing adaptations that have allowed these dinosaurs to continue to thrive to this day.

CHAPTER 8

Early Sauropodomorphs and the Rise of Sauropods

Introducing Sauropodomorphs

Sauropodomorphs were iconic dinosaurs in most cases, thanks to their immense body size, long necks (with relatively miniscule heads), columnar limbs, and long, muscular tails. Sauropodomorph evolution has been subject to a great deal of study because of their impressive taxonomic and phenotypic diversity. Specifically, although sauropodomorph heads are small compared to the size of their bodies (relative to other herbivores; Christiansen, 1999), the diversity of sauropodomorph skulls integrates many combinations of traits in various subclades (Button et al., 2017b). Early sauropodomorphs, or "prosauropods" like *Plateosaurus* and *Massospondylus*, are themselves a highly diverse paraphyletic "grouping" of animals phylogenetically leading up to the derived sauropods, including such forms as *Brachiosaurus*, *Camarasaurus*, and *Diplodocus*, among many others. "Prosauropods" are generally distinguishable because they are less robust compared to sauropods, and across sauropodomorphs, there are generally three skull morphotypes, two of which are represented by the derived sauropods (Young and Larvan, 2010; Button et al., 2014, 2017b; MacLaren et al., 2017). All three of these morphotypes are represented in multiple clades independently, even though there is strong phylogenetic signal within each individual clade, especially with regard to body mass evolution (Button et al., 2017b).

Early Sauropodomorpha ("Prosauropoda")

For reference in understanding relationships and clade affiliation of early sauropodomorphs and early-diverging sauropods, see the phylogeny (fig. 8.1) and taxon list (table 8.1) below.

"Prosauropods" experimented with different trait morphologies, leading up to derived sauropod dinosaurs. "Prosauropods" included a morphologically diverse grouping of small- to medium-sized early sauropodomorphs, including *Buriolestes*, *Panphagia*, *Pantydraco*, *Efraasia*, *Leyesaurus*, *Coloradisaurus*, *Thecodontosaurus*, *Yunnanosaurus*, *Bagualosaurus*, riojasaurids (e.g., *Riojasaurus*), unaysaurids (e.g., *Macrocollum*), massospondylids (e.g., *Massospondylus* [fig. 8.2] and *Lufengosaurus*), anchisaurids (e.g., *Anchisaurus*, *Irisosaurus*), and, the largest and most famous of all, plateosaurids (e.g., *Plateosaurus*). They ranged largely from the Late Triassic to Early Jurassic (with worldwide distribution) before the generally larger sauropod dinosaurs took over.

"Prosauropod" skulls were reduced in size relative to body size (a trend taken to the extreme in sauropods), and they possessed prominent orbits, antorbital fenestrae, and nares. "Prosauropod" cheek teeth are mostly variably leaf-shaped with denticles on the edges (absent in *Yunnanosaurus* and *Irisosaurus* teeth; de Fabrègues et al., 2020), sometimes with a few conical front teeth for

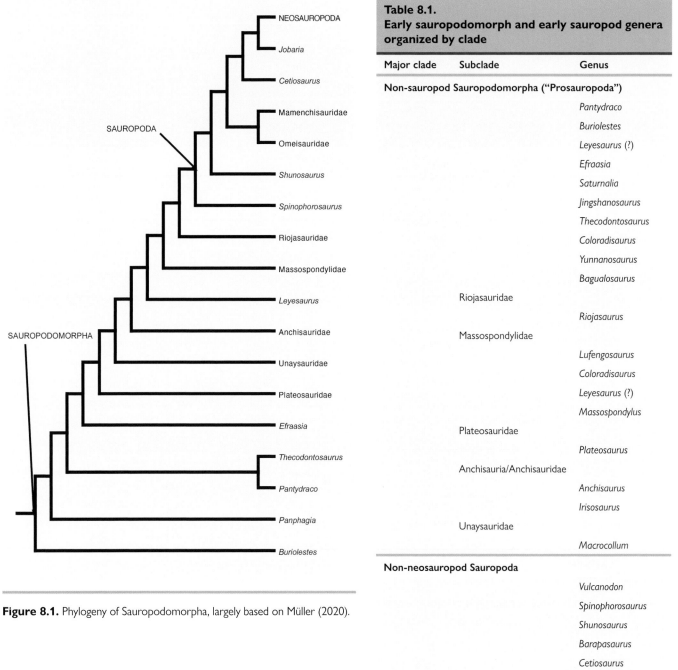

Figure 8.1. Phylogeny of Sauropodomorpha, largely based on Müller (2020).

Table 8.1.		
Early sauropodomorph and early sauropod genera organized by clade		
Major clade	**Subclade**	**Genus**
Non-sauropod Sauropodomorpha ("Prosauropoda")		
		Pantydraco
		Buriolestes
		Leyesaurus (?)
		Efraasia
		Saturnalia
		Jingshanosaurus
		Thecodontosaurus
		Coloradisaurus
		Yunnanosaurus
		Bagualosaurus
	Riojasauridae	
		Riojasaurus
	Massospondylidae	
		Lufengosaurus
		Coloradisaurus
		Leyesaurus (?)
		Massospondylus
	Plateosauridae	
		Plateosaurus
	Anchisauria/Anchisauridae	
		Anchisaurus
		Irisosaurus
	Unaysauridae	
		Macrocollum
Non-neosauropod Sauropoda		
		Vulcanodon
		Spinophorosaurus
		Shunosaurus
		Barapasaurus
		Cetiosaurus
		Jobaria
	Mamenchisauridae	
		Mamenchisaurus
	Omeisauridae	
		Omeisaurus
	Turiasauria	
		Mierasaurus
		Turiasaurus

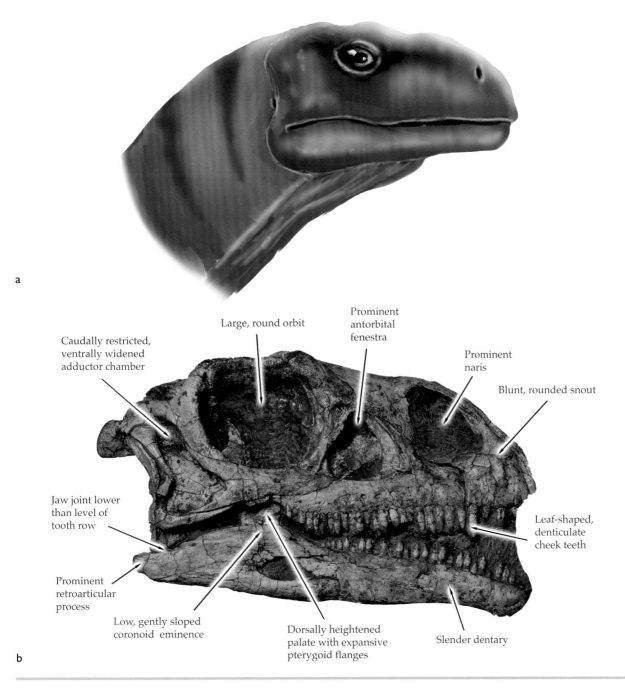

Figure 8.2. *Massospondylus* (a) reconstructed head and (b) skull in lateral view (length approximately 21 cm).

plucking vegetation. This dentition is likely indicative of a transition from omnivory to herbivory (Barrett, 2000; Barrett and Upchurch, 2007; see below), with the exception of a few such as *Buriolestes*, an early sauropodomorph with recurved, bladelike dentition suited for a more carnivorous diet (Cabreira et al., 2016; Müller and Garcia, 2019, 2020a). Interestingly, although teeth are replaced one by one with root absorption through ontogeny from hatchling to adulthood, histology and computed tomography (CT) images of embryonic *Lufengo-*

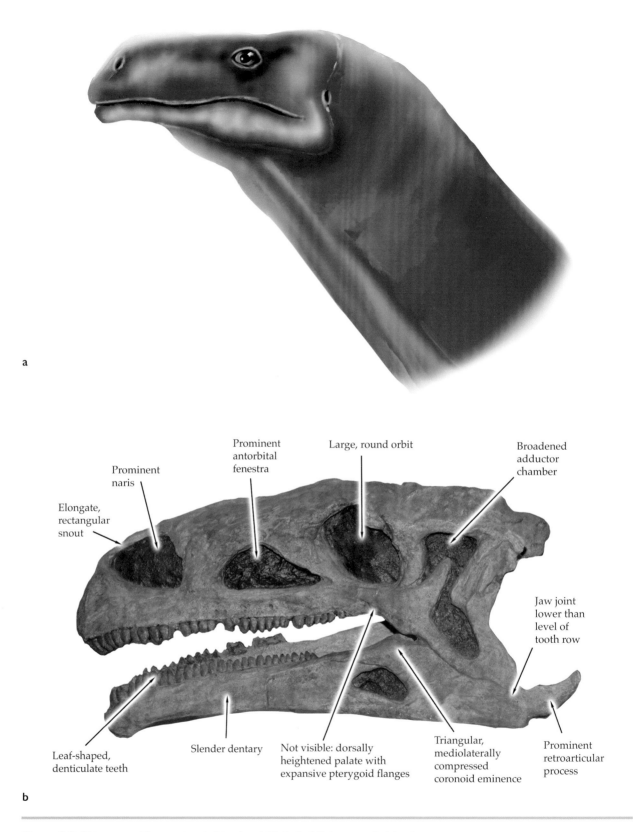

Figure 8.3. *Plateosaurus* (a) reconstructed head and (b) skull of *Plateosaurus* in lateral view (length up to approximately 36 cm).

saurus show that they initially developed multiple cycles of teeth within the same alveolus. This morphology is reminiscent of the assumed derived condition of pencil-like teeth in diplodocoids and titanosaurs, and suggests these sauropod groups' teeth evolved through paedomorphosis (Reisz et al. 2020), although their skulls as a whole tell a different story (Fabbri et al., 2021; see below).

Many "prosauropods" possessed elongate triangular to rectangular skulls (e.g., *Plateosaurus*; fig. 8.3), and others presented blunt skulls (e.g., *Massopondylus*) with a jaw joint lowered relative to the level of the tooth row. Snout widths and heights show great variability across "prosauropods," sometimes suggesting ranges across the omnivory-to-herbivory spectrum between major groups (e.g., plateosaurids vs. massospondylids; Barrett and Upchurch, 2007) to even similar taxa, showing signs of niche partitioning with slight cranial variations (e.g., between *Massospondylus* and *Ngwevu*; Chapelle et al., 2019a). Gow et al. (1990) suggested there was potential for mobility between the premaxilla and maxilla in *Massospondylus*. But Barrett and Upchurch (2007) noted that the junction between the premaxilla and nasal would not have allowed room for kinetic mobility and instead suggested the premaxilla-maxilla junction might have acted as more of a shock absorber. Button et al. (2016) also noted that rotation at this juncture would also heighten bending stresses. Few "prosauropods," including *Massospondylus* (Crompton and Attridge, 1986) and *Riojasaurus* (Bonaparte and Pumares, 1995), may have possessed a keratinous rhamphotheca (beak) at the front end of the snout in life. Barrett and Upchurch (2007) noted that although there is possible evidence of the presence of a rhamphotheca in *Riojasaurus* owing to slight rugosity of the rostral tip of the premaxilla, no such evidence exists that this is the case in *Massospondylus*.

Temporal regions of "prosauropods" are highly variable. Some "prosauropods" (e.g., *Buriolestes*) possessed a transversely narrow yet rostrocaudally broad temporal region and supratemporal fenestra. Other "prosauropods" (e.g., *Riojasaurus*, *Lufengosaurus*, and plateosaurids, like *Plateosaurus*) possessed both a rostrocaudally

and transversely broadened temporal region. Anchisaurids (e.g., *Anchisaurus*), massospondylids (e.g., *Massospondylus*), and *Coloradisaurus* temporal regions are rostrocaudally short and caudally restricted yet transversely widened. The smallest temporal regions among "prosauropods" are seen in *Pantydraco*, *Efraasia*, and *Leyesaurus*, and are all caudally restricted and transversely narrow. Postorbital regions, infratemporal fenestrae, and adductor chambers in "prosauropods" are subrectangular in most cases, except for in *Leyesaurus* and *Anchisaurus*, where the infratemporal fenestrae are lower-angled, triangular, and positioned more ventral to the orbit instead of solely caudal to it. "Prosauropod" palates are more dorsally heightened and vertical (visible laterally) and in most cases (except for *Pantydraco*, *Buriolestes*, and *Riojasaurus*) are rostrocaudally shorter and positioned more caudally within the skull. In plateosaurids and massospondylids especially, the pterygoid flanges are prominent and moderately expansive. Most "prosauropods" possessed a vertically oriented quadrate, except for anchisaurids, which had a quadrate situated at a lower angle relative to the horizontal plane. "Prosauropod" mandibles variably possessed low, gently sloped coronoid eminences (e.g., *Buriolestes*, *Efraasia*, *Leyesaurus*, *Anchisaurus*, *Massospondylus*) to more triangular, mediolaterally compressed coronoid eminences (e.g., *Riojasaurus*, *Lufengosaurus*, and *Plateosaurus*) heightened relative to the level of the tooth row. Mandibular fenestrae are prominently visible in "prosauropods," as are the retroarticular processes (which is especially long in *Plateosaurus*). Hyobranchial elements have only been found in a select few taxa (e.g., *Anchisaurus* and *Massospondylus*; Hill et al., 2015; Li et al., 2018), so information is still scarce regarding possible tongue morphology.

Galton (1984, 1985a, 1985b, 1986) performed more comprehensive studies of *Plateosaurus* and other "prosauropod" skull and cranial muscle anatomy as well as feeding mechanisms as a whole. Although some "prosauropods," especially *Massospondylus* (fig. 8.4), were previously suggested to have had a more carnivorous lifestyle based on the morphology of the masticatory apparatus resembling that of a carnivore (Cooper, 1981),

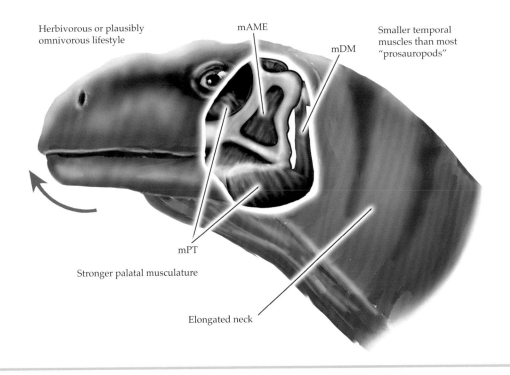

Herbivorous or plausibly
omnivorous lifestyle

mAME

mDM

Smaller temporal
muscles than most
"prosauropods"

mPT

Stronger palatal musculature

Elongated neck

Figure 8.4. *Massospondylus* head showing feeding function with jaw muscles.
Red arrow shows orthal jaw action. Abbreviations: mAME, m. adductor mandibulae
externus; mDM, m. depressor mandibulae; mPT, m. pterygoideus.

Galton (1984, 1985b) rejected this notion, stating that "prosauropods" were herbivores. The dentition of "prosauropods" like *Plateosaurus* have been likened to that of iguanas, consisting mostly of leaf-shaped teeth (as noted above) with sharp, denticulate serrations around the crown ridge at a 45-degree angle to the cutting edge, suggesting an isognathous orthal pulping feeding mechanism (Galton, 1985b; Crompton and Attridge, 1986; Norman et al., 1991; Barrett, 2000; Barrett and Upchurch, 2007; Button et al., 2017b). In addition to tooth shape, Galton (1984, 1985b) suggested that "prosauropods" were herbivorous based on the offset craniomandibular articulation ventral to the tooth row that created more mechanical advantage for increasing force evenly distributed across the dentition for grinding vegetation. Galton (1985a, 1985b) also suggested "prosauropods" were herbivorous based on their more precise dental occlusion, an inset tooth row with a labial dentary ridge (which he suggests might indicate the presence of a fleshy "cheek"-like structure), the presence of gastric mills, and

the fact that they had an exceptionally long neck with a proportionately small head inefficient for a carnivorous lifestyle. Some "prosauropods" (e.g., *Aardonyx*) lack the lateral dentary ridge and inset tooth row, possibly indicating the loss of a fleshy cheek for increased gape in bulk browsing (Upchurch et al., 2007; Yates et al., 2010).

Through comprehensive analyses across a broad range of sauropodomorphs, Barrett (2000), Barrett and Upchurch (2007), and Button et al. (2017b) have all shown that in many "prosauropod" genera, many characters are indicative of herbivory, such as the offset jaw articulation, coronoid eminence, precision of dental occlusion, higher mechanical advantage of jaw adductor musculature, and others are quite variable. In some cases, the aforementioned traits are not as well developed as in others, suggesting that "prosauropods" represented a spectrum of omnivorous (i.e., *Anchisaurus* and *Massospondylus*) and herbivorous (i.e., *Plateosaurus*) taxa that ultimately led to the strictly herbivorous nature of the derived sauropod dinosaurs (Barrett, 2000; Barrett and Upchurch, 2007;

Button et al., 2017b; Button and Zanno, 2020). A few exceptions (e.g., *Jingshanosaurus*) are similar to more carnivorous dinosaurs, such as homodont recurved teeth and a low mechanical advantage of temporal adductor muscles, indicating faster jaw closure (Button et al., 2017b). *Yunnanosaurus* also differed from other "prosauropods" in having more homodont, minimally overlapping, narrow teeth with no denticles, some of which are similar to (but separately evolved from) what is seen in the derived sauropods, indicating its own specializations toward strict herbivory (Button et al., 2017b). Furthermore, Button and Zanno (2020) found that, among biomechanical trait morphospaces of herbivorous dinosaurs as a whole, "prosauropods" plot in the same general morphospace as therizinosaurs, some ornithomimosaurs, heterodontosaurids, and thyreophorans.

In shape analyses using micro-CT scans of two *Anchisaurus* individuals (one adult and one juvenile), Fab-

bri et al. (2021) found that the sauropod-like reoriented adductor chamber (stretching ventral to the orbit) and tilted braincase developed later in ontogeny, contrary to previous assertions that sauropods show paedomorphic cranial morphology (e.g., Reisz et al., 2020). Analyses of "prosauropod" and sauropod skull shapes show that sauropod-like traits in adult "prosauropods" are already present in the juveniles of sauropod taxa, meaning they had more room to experiment with various skull types at adulthood (Fabbri et al., 2021).

The anatomy of cranial musculature is also surprisingly diverse across "prosauropod" taxa. The basic attachment sites are mostly like that typical of dinosaurs in general (i.e., Galton, 1985a; Holliday, 2009; Button et al., 2016). Following Galton's (1985a) reconstruction of *Plateosaurus* (fig. 8.5) cranial musculature, Fairman (1999) reconstructed cranial musculature in the "prosauropods" *Plateosaurus* and *Anchisaurus*. Specifically,

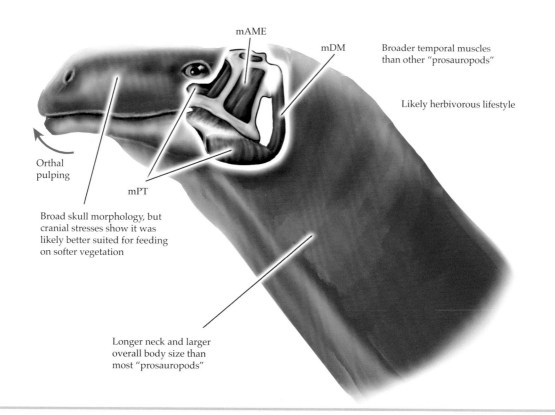

mAME

mDM

Broader temporal muscles
than other "prosauropods"

Likely herbivorous lifestyle

Orthal
pulping

mPT

Broad skull morphology, but
cranial stresses show it was
likely better suited for feeding
on softer vegetation

Longer neck and larger
overall body size than
most "prosauropods"

Figure 8.5. *Plateosaurus* head and neck showing feeding function with jaw muscles. Red arrow shows orthal jaw action. Abbreviations: mAME, m. adductor mandibulae externus; mDM, m. depressor mandibulae; mPT, m. pterygoideus.

Fairman (1999) performed a comparative study with dissection of iguanian lizard jaw musculature, comparing jaw muscle orientation in iguanians and "prosauropod" skull structure. She showed some key differences between *Plateosaurus* and *Anchisaurus* in adductor chamber morphology: *Plateosaurus* resembled the herbivorous *Iguana*, while *Anchisaurus* resembled the omnivorous *Sceloporus*. These analogies with living animals are primarily based on similar traits of the adductor chamber and mandibular attachments, which are informative of muscle architecture. As noted, most if not all "prosauropods" primarily used a strictly orthal pulping feeding mechanism, with isognathous elevation of the mandible to occlusion (meaning both sides of the jaw occluding at the same time) (Galton, 1985b; Crompton and Attridge, 1986; Norman et al., 1991; Barrett and Upchurch, 2007; Button et al., 2017b).

There are different ways muscle architecture can change with skull shape while remaining able to create the same jaw motions, however. In comparing cranial muscle anatomy in a wide variety of herbivorous dinosaurs, Nabavizadeh (2020b) qualitatively explored the extent of diversity in "prosauropod" cranial musculature based on cranial anatomy, following muscle attachment descriptions by Galton (1985a) and Button et al. (2016). The rostrocaudally expanded temporal region and more vertically oriented adductor chambers in a majority of "prosauropods" implicate a rostrally expanded and vertical temporal muscle complex (m. adductor mandibulae externus, or mAME). Many possess low, rounded to subtriangular coronoid eminences on the lower jaw; however, some, including *Riojasaurus*, *Lufengosaurus*, and especially *Plateosaurus*, present with a transversely broadened temporal region with larger surface area for greater muscle origin—a raised, triangular coronoid eminence that is ventrally expanded. This morphology indicates that the mAME complex insertion site was higher and greater in surface area, creating a greater mechanical advantage of the adductors and a larger, more fan-shaped muscle insertion site (see also Button et al., 2016). Together, these traits create a vertical muscle vector with optimal mechanical advantage and greater

strength in orthal feeding. Furthermore, the palatal muscle attachment sizes (mainly of m. pterygoideus, or mPT) of these "prosauropods" are generally smaller (e.g., *Riojasaurus*) to moderately expansive (e.g., *Lufengosaurus* and *Plateosaurus*), indicating that they only aided the larger temporal muscles in jaw closure strength (Button et al., 2016; Nabavizadeh, 2020b).

Although *Anchisaurus* and *Massospondylus* are also orthal feeders (e.g., Galton, 1985b; Button et al., 2016), muscular anatomy of their feeding system looks a bit different (Fairman, 1999; Nabavizadeh, 2020b). Both of these taxa present with caudally displaced and rostrocaudally short temporal regions and an adductor chamber and quadrate that are rostroventrally oriented at a lower angle, indicating relatively much smaller mAME muscle origin and narrower, low-angled insertion of mAME complex than most other "prosauropods." The low-angled quadrate also places the jaw joint and corresponding mPT insertion site much closer to its more dorsal palatal origin, creating a more vertical mPT muscle vector and assisting in greater strength and mechanical advantage in jaw closure (as is seen in diplodocoids; Button et al., 2014 [see below]) (Nabavizadeh, 2020b).

Recent advances in computer modeling technology have greatly improved our understanding of early sauropodomorphs, particularly *Plateosaurus*. Using finite element analysis (FEA), Button et al. (2016) compared stresses and strains in feeding performance of the skull of *Plateosaurus* with that of the macronarian sauropod *Camarasaurus* and found that *Plateosaurus* was generally better suited for softer foods with its comparably longer, less robust snout. Specifically, they find that because the mandible of *Plateosaurus* has a longer tooth row and relatively smaller adductor muscles, the overall mechanical advantage and bite force of the jaw are lessened as a result of increased biting speed—an adaptation they suggest signals *Plateosaurus* and other "prosauropods" as facultative bipeds, as proposed by previous authors (noted above).

Lautenschlager et al. (2016), using FEA as well as multibody dynamics analysis (MDA), compared the performance of the skull of *Plateosaurus* with that of the therizinosaur *Erlikosaurus* and the derived stegosaur *Stegosaurus* to test whether unrelated taxa with similar cranial morphotypes (i.e., elongate, triangular snout) were generally equal in the way stresses travel throughout the skull during feeding. Lautenschlager et al. (2016) found that stresses in *Plateosaurus* were generally lower and comparable to that of *Erlikosaurus*, although they also note that *Plateosaurus* overall has a more robust skull morphology, which they interpret as a generalist broadly feeding on softer vegetation.

In order for "prosauropods" to procure vegetation, most of them took advantage of the development of an elongated neck and, many times, larger body size and powerful limbs (Galton, 1985b; Barrett and Upchurch, 2007). Galton (1985b) pointed out that, like giraffes, "prosauropods" likely could have extended their long necks vertically (at least higher than horizontal) to reach higher levels of vegetation—an adaptation that was taken to extremes in the derived sauropods. Having a proportionately small skull also would have made it easier to hold its head up with such a long neck. Most "prosauropods" have roughly 10 amphicoelous cervical vertebrae, creating a neck nearly the length of their torso in some cases (Barrett and Upchurch, 2007). Cooper (1981) suggested that the restrictive morphology of the pre- and post-zygapophyses in the cervical vertebrate would have limited the ability for *Massospondylus* to flex its neck laterally. Upon closer examination, however, Barrett and Upchurch (2007) explained that these restrictions were specifically at the base of the neck among "prosauropods" and that the cranial end of the neck was actually more flexible than previous realized, with the ventral cervical ribs of each vertebrae overlapping that of each successive vertebra in these movements. Given this type of neck flexibility, Barrett and Upchurch (2007) consider it likely that "prosauropods" were able to feed in both the horizontal and vertical planes with ease. Although there is no direct evidence of pneumaticity in "prosauropod" necks or torso, it stands to reason that they likely had pneumatic bone and birdlike air sacs because of its presence in early saurischians as well as sauropods (Wedel, 2007). Additionally, three-dimensional (3D) modeling

has shown that *Plateosaurus* likely held a more horizontal posture of the neck in locomotion (Mallison, 2010a) and had the ability for a fair amount of dorsoventral and lateral mobility, with no torsion (Mallison, 2010b).

During feeding, "prosauropods" likely used their strong hindlimbs and stout tail to make a tripodal rearing posture for reaching up to reach tall vegetation (Galton, 1985b), with the tallest "prosauropods" like *Plateosaurus* reaching heights of more than 3 m (Bakker, 1978; Weishampel and Norman, 1989; Barrett and Upchurch, 2007; Mallison, 2010a, 2010b). Their hands were capable of grasping at vegetation, with an enlarged first pollex (thumb) claw, but this digit was not opposable (Galton, 1971; Cooper, 1981; Barrett and Upchurch, 2007; Mallison, 2011a, 2010b; Reiss and Mallison, 2014). This thumb claw could have also been co-opted for use in defense from predators or even intraspecific combat (Galton, 1985b; Bonnan and Senter, 2007). If *Massospondylus* had a somewhat facultatively omnivorous lifestyle, this claw could have also been used in dismembering carrion (Cooper, 1981; Barrett and Upchurch, 2007).

Because of the large variety of locomotor features seen in "prosauropods," there has been much debate about whether these animals were bipedal, facultatively bipedal, or quadrupedal (e.g., Coombs, 1978; Christian et al., 1996; Langer, 2003; Barrett and Upchurch, 2007; Bonnan and Senter, 2007; Mallison, 2010a, 2010b, 2011a; Reiss and Mallison, 2014) (fig. 8.6). Most "prosauropods" possess forelimbs roughly half the length (or slightly more) of their hindlimbs, which is usually a clear giveaway that they are bipedal to a certain degree; however, variations in trunk length (and other factors) also play a role in determining stance (Barrett and Upchurch, 2007). For instance, although Galton (1971) suggested that the second and third digits were capable of load-bearing in quadrupedality, Cooper (1981) believed it would have been too much weight for two digits to bear. Christian and Preuschoft (1996) suggested that *Plateosaurus* must have been a habitual quadruped due to sustained bending moments along its axial skeleton. Years later, however, Bonnan and Senter (2007) performed a range-of-motion study by manipulating forelimb ele-

Figure 8.6. Skeleton of *Lufengosaurus*.

ments of *Plateosaurus* and *Massospondylus* (as well as some extant crocodilians and birds) and compared their results to the related sauropods and theropods. They discovered only limited humeral flexion and abduction as well as an articulation of the radius and ulna that oriented the palms medially, keeping them semi-supinated. This morphology would have prevented any pronation of the hands for quadrupedal stance, therefore restricting these animals to bipedality. Bonnan and Senter (2007) also note that tracks sometimes associated with "prosauropods" likely were not "prosauropods" at all, and so would not provide evidence of quadrupedality. Reiss and Mallison (2014) also showed the range of motion of the manus in *Plateosaurus* does not support quadrupedality. Three-dimensional modeling has been used to calculate the center of mass in *Plateosaurus*, supporting a bipedal stance, again with a horizontal neck in slow and fast movement (Mallison, 2010a, 2011a) as well as structured hindlimb motion and a laterally flexible tail (Mallison, 2010b). Furthermore, 3D musculoskeletal modeling of *Plateosaurus* has shown that its hindlimbs would have been capable of high-velocity bipedal locomotion, with a more proximally located fourth trochanter de-

creasing the moment arm leverage of the major hip extensor (m. caudofemoralis longus) and with knee flexors and extensors also showing decreased moment arm leverage (Klinkhamer et al., 2018). More recent limb muscle reconstruction in *Thecodontosaurus* similarly supports it as an agile biped, with decreased moment arms in its hip extensors and flexors as well (Ballell et al., 2022).

These results combined with the other traits generally support a bipedal stance, although variation among "prosauropods" still makes it possible for some to have been facultative bipeds (or maybe even quadrupedal), including the possible transitional form *Aardonyx* (Yates et al., 2010). Some, such as *Riojasaurus*, show longer forelimbs, suggesting quadrupedality, while others show ontogenetic transitions in stance from a more quadrupedal stance as a hatchling to bipedality in adulthood, such as *Mussaurus* (Reisz et al., 2005; Barrett and Upchurch, 2007; Chapelle et al., 2019b; Otero et al., 2019). According to Otero et al. (2019), the development of the neck and tail influenced a shift of the center of mass in *Mussaurus* caudally from the mid-torso toward the pelvis early on, favoring a more bipedal stance later in life. Further quantitative analysis of ontogenetic limb robusticity

Figure 8.7. Skeleton of *Plateosaurus*.

agrees with this postural transition in *Mussaurus* while also inferring that the same is not the case in *Massospondylus* (Chapelle et al., 2019b).

"Prosauropods" represented some of the largest terrestrial animals of the Late Triassic and Early Jurassic (before the larger sauropods took over), with the largest, *Plateosaurus* (fig. 8.7), weighing somewhere between 630 and 1073 kg (Seebacher, 2001; Gunga et al., 2007; Sander et al., 2011). Accordingly, they would have had large guts for effective fermentation of plant material, which likely consisted of ferns and conifers, among others. Gastric mills likely aided in the breakdown of plant material, as indicated by the presence of rounded gastroliths associated with the abdominal region of some "prosauropod" taxa, like *Massospondylus* (Raath, 1974;

Galton, 1985b; Weishampel and Norman, 1989; Barrett and Upchurch, 2007).

The fact that gastric mills are known in "prosauropods" lends further support for herbivory among taxa. The extent of which part of the omnivory-to-herbivory spectrum a particular taxon can be placed, however, is really dependent on the overall combination of traits mentioned above. Early origins of "prosauropods" show craniodental and neuroanatomical traits, suggesting predatory lifestyles (e.g., *Buriolestes*; Cabreira et al., 2016; Müller et al., 2020). Neuroanatomical evidence is especially useful because specific parts of the brain consistent with predatory lifestyles can be vastly informative, such as an elongated olfactory track (for a better sense of smell), smaller pituitary relative to the derived sauropods, and well-developed flocculus in the cerebellum for vestibular and visual coordination, suggesting an agile predatory behavior in *Buriolestes* (Müller et al., 2020). Similar neuroanatomical adaptations for at least some behavior (and, additionally, bipedalism) were also described for *Thescelosaurus*, which together with some herbivorous craniodental adaptations hint at a possibly omnivorous lifestyle (Ballell et al., 2020). Craniodental characters do sometimes distinguish some more omnivorous forms from later, more herbivorous forms (e.g., *Saturnalia*; Bronzati et al., 2019). Likewise, there are taxa that show clear transitions to herbivory (e.g., *Macrocollum*; Müller et al., 2018). Still, it is most likely the case that the largest "prosauropods"—the plateosaurids—were adapted to herbivory (Galton, 1985b; Barrett and Upchurch, 2007), and it is upon these characteristics that the later sauropods were built, becoming impressively gigantic megaherbivores.

General Sauropod Anatomy and Biology

Sauropod dinosaurs dominated terrestrial landscapes as the largest terrestrial herbivores to have ever lived, which makes it all the more comical that their skulls are incredibly small compared to the size of their massive necks, bodies, limbs, and tails. More basal eusauropods include *Shunosaurus*, mamenchisaurids (e.g., *Mamenchisaurus*), omeisaurids (e.g., *Omeisaurus*), and *Jobaria*. The main division of neosauropods is between Diplodocoidea (e.g., diplodocids, dicraeosaurids, and rebbachisaurids) and Macronaria (e.g., camarasaurids, brachiosaurids, euhelopodids, and titanosaurs; see below). Button et al. (2014), quantitatively comparing craniodental characters across 35 taxa, found that there is a distinct dichotomy in sauropod skull morphotypes, with one more robust morphotype they deemed "broad-crowned" sauropods, like camarasaurids, euhelopodids, and non-neosauropod eusauropods, and a more gracile morphotype they deemed "narrow-crowned" sauropods, like all diplodocoids and titanosaurs. Brachiosaurids are a special case, as they are

plotted somewhere in the middle of the two, implicating likely similarities with both. As described below, diplodocoids and titanosaurs tend to possess dorsoventrally shorter, rostrocaudally more elongate skulls, while non-titanosaur macronarians, mamenchisaurids, omiesaurids, and other eusauropods tend to have a somewhat shorter but broader snout, with a heightened dorsal margin of the skull. Nevertheless, the small size of the sauropod head is a sign it played only a small role in mechanical digestion of vegetation compared to ornithischian and especially mammalian herbivores (Christiansen, 1999).

The general form of the sauropod head resulted from a unique combination of regional reorientations. The temporal region in all sauropods is caudally displaced and rostrocaudally short. The narial opening in all of these forms is retracted caudally, giving the initial illusion that the external nose itself was retracted or that possibly the animals even had a proboscis, or "trunk," a position that has since been convincingly rejected with soft tissue data from extant reptiles (Witmer, 2001) as well as paleoneurological work on *Diplodocus* (Knoll et al., 2006). These retracted nares, however, altered the shape of the head in general (e.g., Dollo, 1884; Haas, 1963; Salgado and Calvo, 1992; Barrett and Upchurch, 1994; Upchurch and Barrett, 2000). Their snout was also variable in width, with the diplodocoids and titanosaurs typically possessing narrower snouts with thin, pencil-like teeth compared to the more common robust, spatulate-toothed sauropods (e.g., non-titanosaur macronarians and non-neosauropod sauropods). Sauropod teeth are often used to assess the extent of resource partitioning among sauropods in different periods and locations based on their morphology, enamel wrinkling, and wear patterns (e.g., Fiorillo, 1998; Holwerda et al., 2015, 2018; Becerra et al., 2017; Mocho et al., 2017; McHugh, 2018). In general, the tooth-bearing bones are more robust, and the tooth rows are arched and variably U-shaped (e.g., macronarians) to rectangular (e.g., diplodocoids). In many sauropods, the snout is prominently well rostral to the enlarged external nares. An antorbital fenestra is present, although reduced in size, and a smaller preantorbital fenestra is also present in many cases. The orbit is enlarged but pinched ventrally, with the infratemporal fenestra and adductor chamber extending ventral to it in lateral view. The maxilla-jugal suture in sauropods is continuous laterally (and the medially connected ectopterygoid creates the rostroventral margin of the adductor chamber). The sauropod palate is generally vertically heightened and visible laterally, yet rostrocaudally shorter in breadth and caudally restricted within the skull, with moderately expanded pterygoid flanges. As stated in chapter 9, sauropod-like skull traits (a tilted braincase and reoriented adductor chamber) are already present in juvenile sauropods (whereas in some "prosauropods" like *Anchisaurus*, these traits did not develop until later in life; Fabbri et al., 2021).

In the mandible, sauropod coronoid eminence ranges from low to triangular (see below), a mandibular fenestra is mostly present (although highly reduced in many cases), and the retroarticular process is generally prominent. The quadrate-mandibular jaw joint is approximately in line with the maxillary tooth row in many Middle Jurassic sauropods such as cetiosaurids and some Late Jurassic sauropods such as diplodocids; however, the jaw joint migrated ventral to the maxillary tooth row in other Late Jurassic sauropods, such as camarasaurids and brachiosaurids, representing an efficient cropping mechanism similar to prosauropods (Galton, 1986). MacLaren et al. (2017) found that mandibular shape disparity in sauropod mandibles continued to grow throughout the Mesozoic, as in other herbivorous dinosaurs, and hit its highest amount in the Late Jurassic before it decreased again in the Cretaceous. Only a select few sauropod taxa have hyobranchial elements associated with them (e.g., *Europasaurus* and *Omeisaurus*; Hill et al., 2015), meaning there is insufficient evidence about the morphology of sauropod tongues.

Sauropods are primarily orthal feeders, with slight variations between each kind (e.g., Calvo, 1994a; Upchurch and Barrett, 2000; Barrett and Upchurch, 2005; Whitlock, 2011, 2017; Button et al., 2014, 2016). With microwear data, Whitlock (2011) found evidence of minimal yet diverse jaw movements during feeding in a wide variety of sauropod dinosaurs—mechanisms that can to a large extent be explained by cranial anatomy. It stands to reason that sauropods were constantly eating, so tooth replacement rates were quite high in order to keep up with constant wear (e.g., every 35 days in *Diplodocus* and every 62 days in *Camarasaurus*; D'Emic et al., 2013). Interestingly, sauropod tooth replacement rate is correlated with tooth complexity rather than with diet as in ornithischians and mammals, which has been suggested as an adaptation for large body sizes among other things (Melstrom et al., 2021). Wiersma and Sander (2017) suggested that soft tissue impressions found in *Camarasaurus* might indicate that gingiva partially covered the teeth. They also suggest that *Camarasaurus* might have also had a keratinous rhamphotheca to also help protect and hold in the teeth (as described in a few "prosauropods") because of grooves and foramina along the rostral oral margin. But this remains uncertain, with another explanation for the foramina being the possible presence of a "lip-like" structure, as in modern lizards (Morhardt, 2009).

Sauropod cranial musculature has been reconstructed with information about attachment sites and general inferred morphology by numerous authors, including Janensch (1935), Haas (1963), Zhang (1988), Barrett and Upchurch (1994), Holliday (2009), Button et al. (2014, 2016), and Nabavizadeh (2020b). These references are the basis for cranial muscle anatomy described throughout the remainder of this chapter (also see chap. 3 for a general description of cranial muscle anatomy).

With such relatively small heads, the hyper-elongated necks of sauropods are what truly give them their unique identity. A long neck allowed sauropods to eat large amounts of vegetation in all directions, from side to side to the tops of tall trees. Among different sauropod groups, the anatomy and functionality of their necks were highly variable, and understanding the differences between them can tell us a lot about their evolutionary ecology. The functional mechanics of the sauropod neck is probably the most controversial concept in sauropod anatomy. Numerous studies have investigated cervical musculoskeletal anatomy and mechanics in sauropods, assessing cervical articulation; the effect of soft tissue structures such as intervertebral joints, ligaments, and musculature; and even energetic costs and cardiovascular consequences of holding a long neck up higher or lower. These anatomical and physiological details help paleobiologists determine which types of sauropods normally held their heads up high in stance and locomotion, whether certain sauropods could even lift their necks up high during feeding alone, or whether they normally held their necks more horizontally and fed side to side at lower vegetation levels from a wide horizontal range (Holland, 1910; Janensch, 1929, 1950; Bakker, 1971; Coombs, 1975; Alexander, 1985; Martin, 1987; Christian and Heinrich, 1998; Martin et al., 1998; Paul, 1988, 2017; Stevens and Parrish, 1999, 2005a, 2005b; Seymour and Lillywhite, 2000; Upchurch and Barrett, 2000; Christian, 2002, 2010; Wedel et al., 2002; Tsuihiji, 2004; Berman and Rothschild, 2005; Christian and Dzemski, 2007, 2011; Dzemski and Christian, 2007; Schwarz et al., 2007a, 2007b; Schwarz-Wings and Frey, 2008; Schwarz-Wings, 2009; Seymour, 2009; Taylor et al., 2009; Klein et al., 2012; Christian et al., 2013; Cobley et al., 2013; Henderson, 2013; Preuschoft and Klein, 2013; Stevens, 2013; Taylor and Wedel, 2013a, 2013b; Taylor, 2014; Hughes et al., 2016; Woodruff, 2017; Vidal et al., 2020a, 2020b). Taylor and Wedel (2013a) indicated that the interconnected nature of adaptations is responsible for sauropod hyper-elongate necks, including large body size, quadrupedal stance, muscular attachments, and distinctive cervical morphologies. Furthermore, pneumaticity and the structure of pneumatic diverticula (air sacs) of the cervical (and general postcranial) skeleton have also been highly studied, with interest of how the cervical skeleton related to the evolution of the neck and its mechanics, as well as of how it related to a possible avian-style respiratory system (Wedel, 2003a, 2003b, 2005, 2006, 2009; Schwarz et al., 2007a; Schwarz-Wings and Frey, 2008; Schwarz-Wings, 2009).

Because sauropods were graviportal quadrupeds that reached gargantuan sizes, their appendicular and axial musculoskeletal and joint anatomy were also enormously complex, they had to support such heavy weight and hold postures that allowed them to hold their necks up, stand upright, and locomote (Alexander, 1985; Carrano, 2005; Henderson, 2006; Schwarz et al., 2007b; Otero, 2010, 2018; Hohn, 2011; Mallison, 2011a; Ibiricu et al., 2013; 2018; Sellers et al., 2013; Tsai and Holliday, 2015; Ullmann et al., 2017; Tsai et al., 2018; 2020; Klinkhamer et al., 2018; 2019; Diaz et al., 2020; Otero et al., 2020; Vidal et al., 2020c; Voegele et al., 2020, 2021).

Additionally, and perhaps unsurprisingly, sauropod gigantism, metabolic capabilities, and overall body plans have been of great interest because of the incomprehensible nature of an animal growing to such enormous sizes (e.g., Colbert, 1993; Sander et al., 2011; Bates et al., 2016). Despite early studies speculating a possible aquatic or amphibious feeding behavior in sauropods, the overall form and functional implications of sauropod anatomy indicate a terrestrial feeding lifestyle (e.g., Coombs, 1975). With such large gut systems, sauropods would have been able to digest massive amounts of vegetation at a time, mostly including higher-energy conifers and, likely to a lesser extent, ferns and cycads, with varying food intake across taxa owing to varying environments, physiologies, and nutritional value of different plants (Hummel et al., 2008; Gee, 2011; Gill et al., 2018). Carbon isotope research has also suggested that there was possible resource partitioning between various sauropods owing to differences in height (e.g., conifers versus ferns and horsetails; Tütken, 2011). Sauropod coprolites have been found to contain microflora showing signs that sauropods indeed ate taller conifers, cycads, and other gymnosperms as well as some angiosperms (possibly even grasses of the Late Cretaceous) (e.g., Matley, 1939; Mohabey, 2005; Prasad et al., 2005; Sonkusare et al., 2017).

The presence of a gastric mill (gizzard) has been speculated in at least some sauropods, with the discovery of polished stones associated with some specimens (Calvo, 1994b; Christiansen, 1996), although Calvo (1994b) acknowledges that it would not have been the primary mode of digestion. Accordingly, sauropod gastroliths are rarely found (Calvo, 1994b; Christiansen, 1996; Wings, 2007; Wings and Sander, 2007). Wings and Sander (2007) revealed that these stones did not show textural characteristics of an avian-style gastric mill and that the amount of gastroliths found with sauropods (far less than 0.1% of body mass) are an order of magnitude less than that of ostriches and other herbivorous birds; thereby they rejected the likelihood of gastric mill presence altogether. Still, Malone et al. (2021) identified certain gastroliths—that have likely already gone through the digestive tracks of sauropods—as possibly indicative of migration behavior.

With enormous needs for digestion in such a massive animal with enormous trunk regions, and the unlikelihood of a gastric mill that might have assisted in it, sauropod digestive tracts would have been gargantuan, and gut retention time of plant material would have been incredibly long (possibly 6 to 12 days), especially since oral processing in sauropods is so minimal compared to ornithischians (Franz et al., 2009). Long fermentation periods

also create an environment for proper nutrient absorption (Hummel and Clauss, 2011).

Early-Diverging Sauropoda

Early-diverging (non-neosauropod) sauropods include many forms, such as *Vulcanodon*, *Spinophorosaurus*, *Shunosaurus* (fig. 8.8), *Barapasaurus*, mamenchisaurids (e.g., *Mamenchisaurus*), omeisaurids (e.g., *Omeisaurus*), *Cetiosaurus*, *Jobaria*, turiasaurs (e.g., *Mierasaurus* and *Turiasaurus*) and others. They ranged from the Early Jurassic on through the Early Cretaceous mostly (with neosauropods mainly taking over throughout that time) with worldwide distribution.

These sauropods generally present with deep, robust, and blunt muzzles that are variably broadened (e.g., omeisaurids) to more narrowed (e.g., mamenchisaurids; fig. 8.9), indicating variation in feeding selectivity, and U-shaped in dorsal view. They possess large ex-

Figure 8.8. Reconstructed head of *Shunosaurus*.

ternal nares and worn spatulate teeth with a slender crown showing some mesiodistal expansion (Upchurch and Barrett, 2000). The temporal region in early-diverging sauropods are generally transversely broadened yet caudally restricted relative to the length of the skull. The postorbital adductor region and infratemporal fenestra widens rostrally as it extends rostroventrally beneath the orbit. The palate is set at an inclined plane. The mandibular coronoid eminence of these sauropods is generally raised and triangular in a mediolaterally compressed sheet of bone. There is variation in the level of the jaw joint, with *Shunosaurus* showing a jaw joint level with the tooth row, while forms like *Mamenchisaurus* and *Omeisaurus* have ventrally offset jaw joints for added mechanical advantage (Upchurch and Barrett, 2000).

As non-neosauropod sauropods are primarily considered relatively "broad-crowned" sauropods (confirmed with morphometric analysis by Button et al., 2014) that

were primarily orthal feeders with slight shearing action (Christiansen, 2000; Upchurch and Barrett, 2000), it is fair to suggest that feeding strategy and cranial performance is, as a whole, more comparable to that of the macronarian *Camarasaurus* than that of diplodocoids (Button et al., 2014, 2017b; see below). *Vulcanodon* likely did not use as much of a shearing action (if any at all) as other early-diverging sauropods. *Shunosaurus* shows signs of possible terminal wear facets (Zhang, 1988; Upchurch and Barrett, 2000), and wear patterns in *Omeisaurus* and *Patagosaurus* show heavy wear, resulting in "shoulders" on the crown margins (with both mesial and distal wear facets in the cetiosaurid *Patagosaurus*) (Upchurch and Barrett, 2000). The inferred triangular mAME complex with a near-vertical caudal border and a rostroventral expansion would have expanded onto the lateral surface of a raised, triangular coronoid eminence with palatal musculature also in-

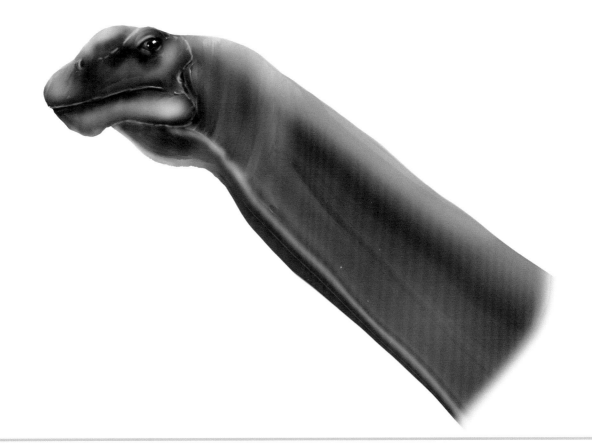

Figure 8.9. Reconstructed head of *Mamenchisaurus*.

clined (Upchurch and Barrett, 2000; Button et al., 2014; Nabavizadeh, 2020b).

The long necks and overall posture of sauropods are best discussed together because of the cohesive functional relationship between these systems. The earliest-diverging sauropods, such as *Antetonitrus* and *Vulcanodon*, are generally small and quadrupedal (albeit with bipedal-like limb proportions that are transitional from earlier sauropodomorphs) and likely fed on very low plant material (Upchurch and Barrett, 2000; Yates and Kitching, 2003). Although *Antetonitrus* has characteristics of the foot seen in later sauropods, it still retained the grasping hand morphology seen in "prosauropods" (Yates and Kitching, 2003). *Spinophorosaurus* was substantially bigger than both *Antetonitrus* and *Vulcanodon*, and with more material to work from, this characteristic is much more informative of the early evolution of neck position and overall posture of sauropods (Vidal et al., 2020a, 2020b).

Vidal et al. (2020a) analyzed the osteologically neutral pose (ONP, meaning the pose as determined by the articular anatomy of just the cervical bones without soft tissue) of the neck of *Spinophorosaurus* as well as giraffes (with ontogenetic series sampling of both). They observed the anatomy and tested the range of motion of the neck with virtual bones and determined that *Spinophorosaurus* likely had large relative cervical intervertebral spaces earlier in ontogeny that decreased as the animal got older. But adult forms still showed the highest range of motion because of elongation of the individual vertebrae as well as elongation of the neck as a whole. *Spinophorosaurus* showed a greater range of motion of the neck than large "prosauropods" like *Plateosaurus*, meaning a wider range of postures were possible in the earliest sauropods than in their predecessors and even some later sauropods. The results of this study, along with craniodental features, suggest that the earliest sauropods engaged in high browsing with a wide range of neck motion in the vertical direction. Although the flexibility between vertebrae was less than that of giraffes, the overall ranges of motion of the neck were similar because, with 12 cervical vertebrae, *Spinophorosaurus* has nearly twice the number of vertebrae as a giraffe. Additionally, because of its apparent inability to ventroflex its neck to bend down very far, *Spinophorosaurus* might have had to splay their legs out like a giraffe to bend down and drink. The long cervical ribs of *Spinophorosaurus*, as in many other sauropods, overlapped one another and likely had an impact on neck flexibility in some way (more on this discussed in chap. 9) (Vidal et al., 2020a).

Spinophorosaurus possesses another key anatomical feature that is relevant to neck and body posture of not only itself, but also all sauropods that evolved thereafter. Vidal et al. (2020b) described a unique arrangement of post-cervical vertebrae in which the dorsal vertebrae are oriented craniodorsally owing to the wedged nature of the sacral and caudal dorsal vertebrae, creating an upward slope of the vertebral column prior to reaching the neck that, with its elongated scapulae and humeri, creates a tall cranial extent of the torso and, by default, pushes the cervical vertebrae to be oriented higher as well. Modification of the first dorsal vertebrae and zygopophyseal articular morphology also help create a more vertical neck posture and feeding range. This craniodorsal sloping of the vertebral column seems to be consistent across most sauropods, with more than 10 degrees of wedging in the sacrum. For some later, lower-browsing sauropods, other modifications of the skeleton (such as a rostroventral curvature of more cranial dorsal vertebrae) would have been needed to make up for the wedged sacrum (Vidal et al., 2020b).

Shunosaurus, another early non-neosauropod, was a smaller sauropod and also possessed 12 to 13 rather short cervical vertebrae, creating a relatively short neck compared to other sauropods. The cervical ribs were also relatively short, and mobile zygopophyseal articular morphology allowed its neck to be flexible in vertical and especially lateral movement. Short forelimbs and small body size of *Shunosaurus* suggest it was a low-level browser (Upchurch and Barrett, 2000). Other closely related sauropods, such as *Barapasaurus*, grew larger (to about 15 m in length) and may have been able eat higher foliage, although *Shunosaurus* might have still eaten tougher vegetation because it shows signs of tooth wear, as stated above (Upchurch and Barrett, 2000).

Mamenchisaurids and omeisaurids are both characterized by their incredibly elongated necks that made up greater than 40 percent of the entire length of the animal (Upchurch and Barrett, 2000; Christian et al., 2013). This extreme elongation of the neck—even for normal sauropod standards—came as a result of lengthening individual cervical vertebrae as well as an increase in their number of from 12 or 13 (which was the norm) to around 17 cervical vertebrae. They also possessed cervical ribs that reached up to 3 m in length, passing up to

Figure 8.10. Skeleton of *Mamenchisaurus* showing cervical vertebrae (with cervical ribs) in a hyper-elongate neck.

three vertebral segments each, which caused overlapping cervical ribs to form bundles of up to four thin processes that jutted caudally at each vertebral level (Martin et al., 1998; Upchurch and Barrett, 2000).

Martin et al. (1998), in detailing various modes of sauropod neck stability, described both *Mamenchisaurus* (fig. 8.10) and *Omeisaurus* as having "ventrally braced" necks, which are inherently stable but inflexible relative to dorsally braced necks (see below). Biologically, this means that the overlapping cervical rib bundles (acting as incompressible struts) served to stabilize the neck as a ventral bracing mechanism with extra reinforcement, only transmitting compressive loads while simultaneously limiting the amount of mobility in the neck because of their overlap. The cervical ribs would have been bound together by surrounding connective tissues and bundles of small, short-fibered muscle. Additionally, the neural spines in mamenchisaurid and omeisaurid vertebrae are relatively low, and the transverse processes are relatively short compared to other sauropods, and the centra were also low and elongate (as mentioned above). Martin et al. (1998) noted that a smaller mass helped in the increase in length and segment number in the neck of mamenchisaurids and omeisaurids and that they likely held their heads lower like a beam (as they suggested for most other sauropods).

Later, Klein et al. (2012), through histological study of the cervical ribs of *Mamenchisaurus* (as well as *Giraffatitan*—therein still called *Brachiosaurus*—and what is likely *Diplodocus*), found that periosteal bone tissue was absent and that longitudinal fibers were the dominant microstructure. This essentially means that cervical ribs are ossified tendons, which are inherently more flexible, and that it was tensile, not compressive, forces that likely acted along the neck, which would have shifted neck musculature back to the trunk and lightened the neck. This structural composition contradicts the ventral bracing hypothesis posed by Martin et al. (1998), which implicates compressive forces created by the cervical ribs. The cervical ribs would have likely acted as a cervical muscle attachment and suggests that m. scalenus (and possibly m. longissimus colli ventralis) were indeed present along the ventral aspect of the neck, helping in ventroflexion (Klein et al., 2012; Preuschoft and Klein, 2013).

Additionally, Preuschoft and Klein (2013) analyzed torsional or twisting abilities in sauropods, including mamenchisaurids, and found that m. longus colli ventralis and mm. scaleni likely counteracted torsional moments when contracted unilaterally, to keep intervertebral joints stable. When contracted bilaterally, these muscles would have kept the neck balanced when extended upright (which requires less energy). Whatever forces were produced by the cervical muscles would have transmitted dorsally from the transverse processes to the spinous processes by deep epaxial muscle fibers (e.g., m. multifidus cervicis and m. splenius capitus). Preuschoft and Klein (2013) also noted that the center of mass of the animal would have shifted depending on how it held its neck: A more horizontal neck would have shifted the center of mass cranially, whereas a more vertical neck would have moved it caudally through the base of the neck. The latter would mean cervical muscles needed to exert tensile force for intervertebral stability, which again saves more energy. This agrees with the assertion made by Upchurch and Barrett (2000), who speculated that *Mamenchisaurus* were high-browsing specialists, reaching treetop heights of up to 10–11 m. This is also supported by the length of the forelimb in *Omeisaurus*, measuring up to 85% of the length of the hindlimb (Upchurch and Barrett, 2000), which would add to the base height of the neck at least.

Biomechanical testing of *Mamenchisaurus* neck function by Christian et al. (2013) has added further insight into the mechanical loads implemented in its neck. They calculated levels of stress on the intervertebral cartilages and suggested that *Mamenchisaurus* likely held its head more horizontally relative to other sauropods (both macronarian and diplodocoid; see below) that have previously been testing through similar methods. They indicate that its neck was likely held nearly straight and nearly horizontal, browsing on vegetation at low to medium heights, but that it could have lifted its head more for alert posture or even at rest. They also specify functions of various neck regions, with the cranial end more mobile for quick head movements, the midsection more capable of ventroflexion or even slight dorsiflexion to medium heights, and caudal cervical region used in lateral movements of the entire neck or more extreme dorsiflexion. Christian et al. (2013) pointed out that *Mamenchisaurus* may have held its head inclined up to 30 or 40 degrees at rest and slightly lower (but still inclined) in locomotion. The stabilized lateral motions to sweep a wide transverse range may have been necessary, especially in places where vegetation density was lower. In general, the extremely long neck of *Mamenchisaurus* would have been advantageous for multiple reasons, including resource partitioning as well as selectivity (Christian et al., 2013).

Cetiosaurids possess the more standard 12 to 13 cervical vertebrae, with each individual cervical of relatively short length, equaling a shorter neck overall. Interestingly, their cervical ribs were long and slender and likely overlapped more proximal cervical ribs. Cetiosaurid cervicals also possessed an elongate rostral process, especially in distal cervicals. This cervical rib morphology would have helped stabilize the neck in a vertical direction (likely restricting lateral motion of the neck). This along with their long forelimbs, which were 84% the length of their hindlimbs, has been cited as support for higher lever browsing up to 5–6 m (Upchurch and Barrett, 2000). Previously, however, Martin (1987) assessed articulations of the neck and estimated intervertebral cartilage thickness, noting that the cervicals of *Cetiosaurus* may not have allowed the neck to be lifted higher than the shoulders (no more than 3.5 m high with a 4.5-m lateral swinging range). Martin et al. (1998) described *Cetiosaurus* as both dorsally and ventrally braced, with tall neural spines that indicated the presence of nuchal and other cervical ligaments, while the ventral aspect was braced by cervical ribs, which again argues for a more horizontal neck posture. Stevens and Parrish (2005a, 2005b) echoed this more horizontal pose with the aid of 3D digital reconstructions of numerous sauropods, including *Cetiosaurus*, suggesting that the cervical series were straight extensions of the dorsal vertebral columns and that therefore it held its head at or below the height of the shoulder in neutral pose. Regardless, both arguments are strongly dependent on what structures are examined and the extent of soft tissue covering (Cobley et al., 2013). Turiasaurs are large animals, but generally less is known of their feeding habits than other sauropods. Yet it is likely that they show similar adaptations to cetiosaurids and mamenchisaurids, with which they share similar "broad-crowned" cranial morphospace (Button et al., 2014). Furthermore, although *Atlasaurus* may or may not be a cetiosaurid, it presents its own special case as having an unusually short neck relative to other sauropods while simultaneously possessing long forelimbs, which would have made up for height lost with its shorter neck (Monbaron et al., 1999). Lastly, the giant non-neosauropod sauropod *Jobaria* shows a humerus-to-femur ratio comparable to that of elephants, which has led to the suggestion that that its hindlimb-directed weight distribution would have allowed it to rear up on its hindlimbs, possibly to reach higher vegetation.

CHAPTER 9

Neosauropods

Diplodocoidea

For reference in understanding relationships of neosauropod taxa, see the phylogeny below (fig. 9.1).

As noted in chapter 8, neosauropods are divided into two main groups: Diplodocoidea and Macronaria. Diplodocoidea includes diplodocids (e.g., *Diplodocus* [fig. 9.2] and *Apatosaurus, Barosaurus, Tornieria*), dicraeosaurids (e.g., *Dicraeosaurus, Suuwassea,* and *Amargasaurus* [fig. 9.3a]), and rebbachisaurids (e.g., *Rebbachi-*

saurus and *Nigersaurus* [fig. 9.3b]). They ranged from the Middle Jurassic through the Late Cretaceous, with worldwide distribution. (For reference in understanding clade affiliation of diplodocoids, see the taxon list in table 9.1.)

These sauropods possess gracile skulls that are dorsoventrally short and rostrocaudally elongate in lateral view with reduced nares that are retracted caudally between the eyes. They also have a broad muzzle and rostrally U-shaped or squared-off mandible and typically a

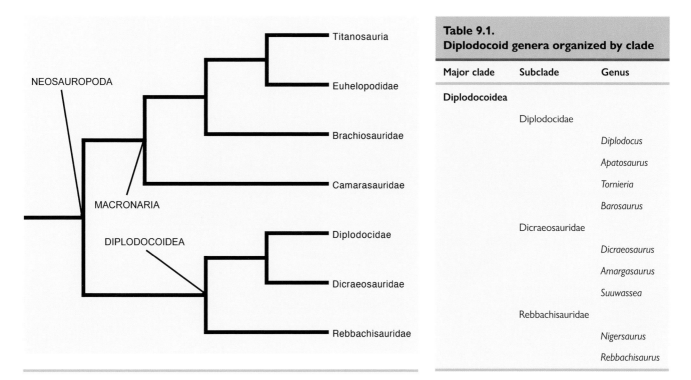

Figure 9.1. Phylogeny of Neosauropoda, largely based on Upchurch et al. (2004).

Table 9.1.
Diplodocoid genera organized by clade

Major clade	Subclade	Genus
Diplodocoidea		
	Diplodocidae	
		Diplodocus
		Apatosaurus
		Tornieria
		Barosaurus
	Dicraeosauridae	
		Dicraeosaurus
		Amargasaurus
		Suuwassea
	Rebbachisauridae	
		Nigersaurus
		Rebbachisaurus

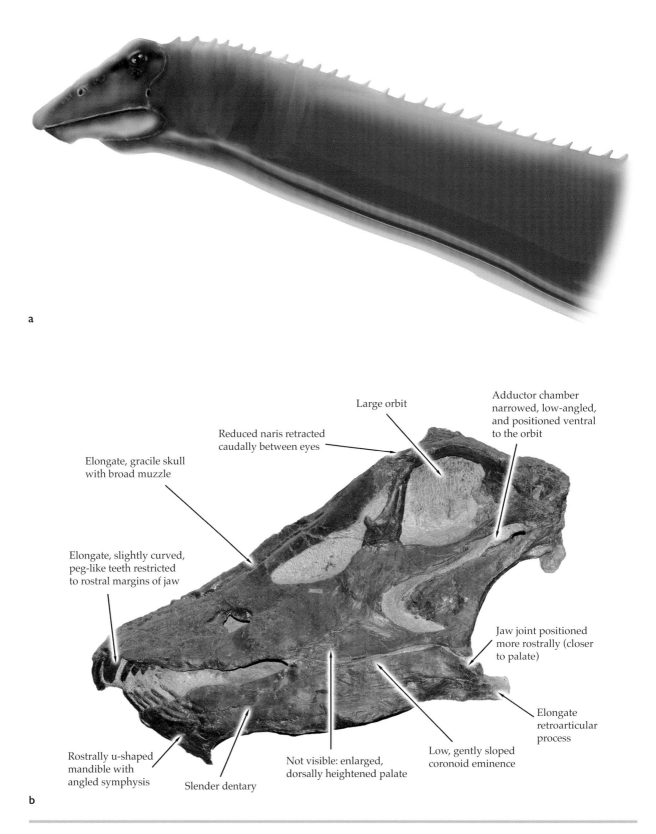

a

b

Large orbit

Adductor chamber narrowed, low-angled, and positioned ventral to the orbit

Reduced naris retracted caudally between eyes

Elongate, gracile skull with broad muzzle

Elongate, slightly curved, peg-like teeth restricted to rostral margins of jaw

Jaw joint positioned more rostrally (closer to palate)

Elongate retroarticular process

Rostrally u-shaped mandible with angled symphysis

Slender dentary

Not visible: enlarged, dorsally heightened palate

Low, gently sloped coronoid eminence

Figure 9.2. *Diplodocus* (a) reconstructed head and (b) skull in lateral view (length approximately 52 cm).

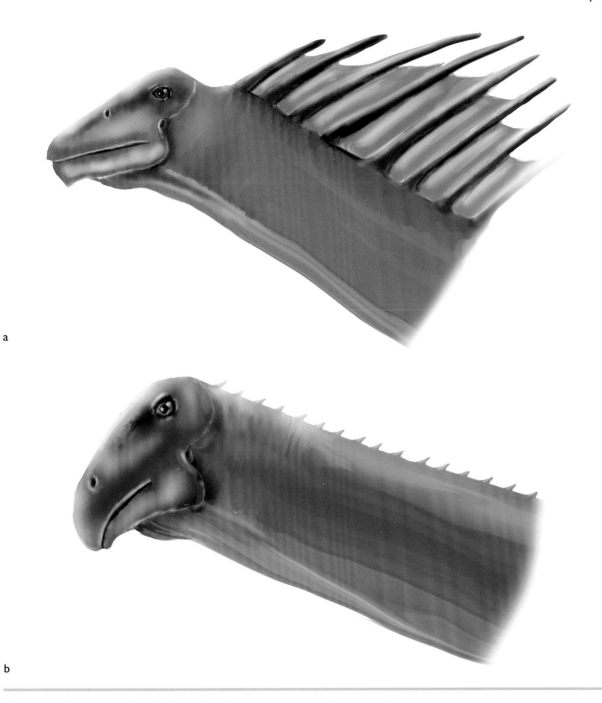

a

b

Figure 9.3. Reconstructed heads of diplodocoids (a) *Amargasaurus* (a dicraeosaurid) and (b) *Nigersaurus* (a rebbachisaurid).

caudodorsally oriented mandibular symphysis, especially in diplodocids.

Most diplodocoids have elongate, slightly curved, peg-like teeth (fig. 9.4) restricted to the rostral margins of the upper and lower jaw. These teeth develop with multiple tooth cycles in a common alveolar crypt, a quality that has been posited as a paedomorphic condition seen in some embryonic "prosauropods" (Reisz et al., 2020). The lightly built thin-boned skull of the rebbachisaurid *Nigersaurus* takes transverse broadening of

Figure 9.4. *Diplodocus* tooth.

the snout to an extreme, with a unique dental battery of more than 100 packed teeth, creating a shearing surface for low-level feeding (Sereno and Wilson, 2005; Sereno et al., 2007). The basipterygoid processes are narrow, elongate, and oriented more rostrally in diplodocoids compared to other sauropods.

The temporal region in diplodocoids is generally transversely smaller and caudally restricted relative to the length of the skull. The postorbital adductor chamber and infratemporal fenestra are positioned ventral to the orbit and are narrowed and rotated at a low angle relative to the horizontal plane. The mandibular coronoid eminence is low and gently sloped to a slight elevation in diplodocids and (likely) dicraeosaurids, but more dorsally heightened in a mediolaterally compressed triangular sheet in rebbachisaurids. Also of note in diplodocoids is a slightly caudodorsally oriented mandibular symphysis, a rostrocaudally expanded articular surface of the jaw joint, and an elongate retroarticular process in diplodocoids (all features especially apparent in *Diplodocus*).

Calvo (1994a) analyzed tooth structures in four major sauropod clades (including diplodocids, like *Diplodocus*), finding differences in tooth function in each. He noted the tooth-tooth contact and slightly inclined jaw joint in the jaws of *Diplodocus*, which would have allowed a jaw action in the propalinal (fore-aft) plane. He also suggested that slight horizontal "slicing" was likely, as shown by the precise tooth-tooth occlusion. Barrett and Upchurch (1994) offered further explanation as to why propaliny (or in this case just palinal motion, i.e., strictly caudal jaw motion during occlusion) was advantageous for *Diplodocus*, because it permitted wider gape for stripping foliage from high branches. Transverse movement was also possible, but it was not supported by tooth wear studies. The expanded mandibular symphysis in diplodocids likely added strength to the dentulous portion of the jaw, as their teeth are restricted to the rostral portion of the jaw (Barrett and Upchurch, 1994; Upchurch and Barrett, 2000). Furthermore, dicraeosaurids were suggested to have had "precision-shear" tooth occlusion, snipping through plant tissues (Upchurch and Barrett, 2000). Christiansen (2000) largely corroborated these studies with dental wear patterns and noted that *Diplodocus* and *Dicraeosaurus* show raking movements. Whitlock (2017) largely rejected the idea that *Diplodocus* chewed "propalinally" (especially proally—i.e., rostrally oriented jaw movement while the teeth were in occlusion) on the basis of the morphology of the skull bones. He did note that if palinal motion in particular was possible, it would have only been minimal owing to the subvertical orientation of m. depressor mandibulae (mDM) just caudal to the jaw joint (Barrett and Upchurch, 1994; Christiansen, 2000; Whitlock, 2017). Many of these combined traits put diplodocoids in a similar morphospace to titanosaurian macronarians, indicating similarities in niche space, although there are still differences (Button et al. 2017b; Button and Zanno, 2020).

Dental microwear studies are especially helpful in elucidating divergent diets across taxa. Fiorillo (1998), with detailed dental microwear, found that *Diplodocus* generally ate softer foods than *Camarasaurus*, although he found some overlap in diet between adult *Diplodocus* and juvenile *Camarasaurus*. Whitlock (2011), with dental microwear studies as well as morphometric analyses of diplodocoid snout shape, found that diplodocoids with differently shaped snouts had diverse mechanisms as well. Squarer-snouted diplodocoids, such as the diplodocids *Apatosaurus* and *Diplodocus* and the rebbachisaurids *Rebbachisaurus* and *Nigersaurus*, exhibited pits and fine scratches, suggesting that they were ground-height generalist browsers (with a somewhat "grazer"-like feeding strategy) on softer herbaceous plants. Alternatively, rounder-snouted diplodocoids, such as the dicraeosaurids *Dicraeosaurus* and *Suuwassea* and the diplodocid *Tornieria*, exhibited gouges and more prominent scratches, suggesting that they were more mid-height selective browsers on brittle, woody foliage. This further exemplifies the diversity of feeding mechanisms and preferences of closely related sauropod dinosaurs. Furthermore, Woodruff et al. (2018) studied the craniodental anatomy of a juvenile diplodocid skull and found unique traits indicative of ontogenetic change in feeding preference and associated dental morphology. They found that, instead of the adult condition with a broader snout and only narrow, peg-like teeth confined to the mesial margin of the oral cavity, juvenile diplodocids exhibit a shorter, narrower snout and a transition from narrow, peg-like teeth mesially to wider, more spatulate teeth with *Camarasaurus*-like wear facets distally (with a relatively longer tooth row compared to adult forms). Woodruff et al. (2018) hypothesized that these features give further evidence of a transitional ontogenetic niche-partitioning, from browsing feeding behavior in juveniles to a "grazer"-like feeding behavior in adults, as was suggested by Whitlock et al. (2010), but with less complete dental material.

For over half a century, diplodocoid cranial muscle anatomy research has been overwhelmingly represented by studies of the skull of *Diplodocus* (Haas, 1963; Barrett and Upchurch,1994; Holliday, 2009; Young et al., 2012; Button et al., 2014) (fig. 9.5), although osteological correlates and other anatomical traits relating to cranial musculature in all other diplodocoids (diplodocids, dicraeosaurids, and rebbachisaurids alike) have

since been deemed generally consistent (Nabavizadeh, 2020b). Diplodocoid cranial morphology suggests that the m. adductor mandibulae externus (mAME) complex would have had a caudally restricted and rostrocaudally and transversely abbreviated temporal origin. The mAME would have stretched rostroventrally through a tunnel-like adductor chamber at a low angle with respect to the horizontal plane of the skull, to insert onto the coronoid eminence. Contrary to the low coronoid eminence in diplodocids, the raised, more triangular coronoid eminence of rebbachisaurids like *Nigersaurus* would have lengthened the moment arm and therefore created a greater mechanical advantage of mAME. The orientation of the temporal adductor muscle group in diplodocoids, with a respectively low-angled quadrate at its caudal border, positions the jaw joint more rostrally, placing it near and just below the m. pterygoideus (mPT) palatal origin. This creates a hinged bracing mechanism during isognathous elevation of the lower jaw. This orientation also helps redirect the muscle vector (and, in turn, substantially increase the mechanical advantage) of the enlarged mPT body more vertically for a more powerful orthal closure, making the palatal muscles the primary muscle group acting in jaw closure (Button et al., 2014; Nabavizadeh, 2020b).

Young et al. (2012) performed finite element analysis (FEA) on a *Diplodocus* skull and found that it is not necessarily designed for high bite forces. Bark-stripping—suggested by Holland and Peterson (1924) and Bakker (1986)—was rejected owing to the presence of too much stress in the skull with this jaw movement. Branch-stripping and precision biting were supported, however, owing to the presence of less stress and strain induced in the skull. The peak stresses were seen in the premaxillary-maxillary lateral plates, dissipating and inflicting stresses throughout the jaw while feeding (Young et al., 2012). These lateral plates, found in most sauropods, are ridges found on the labial side of the premaxillary, maxillary, and dentary dentition, protecting the basal third of each tooth labially, most prominent in the rostral region and decreasing in height toward the caudal end of the jaw (Upchurch and Barrett, 2000).

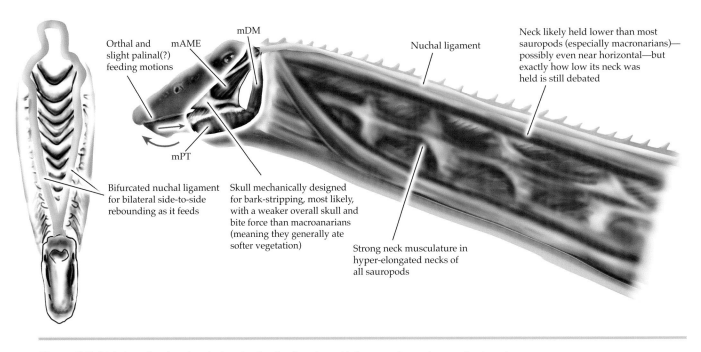

Figure 9.5. *Diplodocus* head and neck showing feeding function with jaw muscles and generalized neck musculature. A portrayal of the bifurcated nuchal ligament is also seen on the left in craniodorsal view (redrawn from Woodruff 2017). Red arrows show orthal and slight palinal jaw actions. Abbreviations: mAME, m. adductor mandibulae externus; mDM, m. depressor mandibulae; mPT, m. pterygoideus.

In an FEA study that compared the skull of *Diplodocus* with the more robust *Camarasaurus* skull, Button et al. (2014) clearly demonstrated that *Diplodocus* has a much weaker overall skull morphology. Regardless, Button et al. (2014) confirmed that the skull of *Diplodocus* was better equipped to withstand palatal feeding forces with its enlarged mPT palatal origin (which corresponds with its enlarged mandibular insertion site). Because overall cranial musculature in *Diplodocus* showed relatively low mechanical advantage and bite force, Button et al. (2014) showed that craniocervical musculature was suited to compensate with caudally directed motions of the head and neck to pull on and strip tree branches with its mesially constricted peg-like dentition. One mechanism that might have created additional cranial support during feeding came in the form of intracranial joint kinesis. Tschopp et al. (2018) reported the presence of multiple overlapping intracranial joints that may have allowed for minimal rostral sliding motions of the snout acting as a "shock absorber," allevi-

ating stresses while biting down. But the amount of possible movement within these joints has not yet been studied, and previous studies have generally concluded that sauropod skulls were likely akinetic (Haas, 1963; Holliday and Witmer, 2008).

Diplodocoid necks range from relatively shorter, with 11 to 13 short cervical vertebrae (e.g., rebbachisaurids and dicraeosaurids) to very long, with 13 to 16 elongate cervical vertebrae (e.g., diplodocids). Diplodocid and dicraeosaurid neural spines present a unique bifurcated morphology, taken to the extreme in dicraeosaurids with tall bifurcated neural spines (like *Amargasaurus*). Contrarily, their cervical ribs are relatively short, possibly allowing more flexibility (Upchurch and Barrett, 2000; Preuschoft and Klein, 2013). Wedel et al. (2002) showed that the cervical vertebrae of *Apatosaurus* possess most of the osteological correlates seen for cervical muscle attachments in birds, including dorsiflexors, ventroflexors, lateroflexors, and more intrinsic muscles of the neck.

For more than a century, numerous neck posture reconstructions have been proposed for diplodocoids, especially diplodocids (fig. 9.5). These reconstructions range from an upright to a horizontal neck posture. Holland (1910) reconstructed *Diplodocus* with an upright and curved S-shaped neck posture (like a swan) along with incorrectly sprawled limbs. Bakker (1971) also noted the likelihood of treetop browsing in diplodocids like *Barosaurus*. Other earlier studies suggested more restrictions on the neck that would not have allowed it to raise vertically, including the inclusion of a possible elastic nuchal ligament that stretched down the length of the neck of diplodocids and dicraeosaurids (fig. 9.6) along the dorsal margin of the bifurcated neural spines. This apparatus would have created a "suspension bridge"–type of body form, where the torso creates an anchor with firmly planted limbs while the neck acts as a cantilever arm with a cable-like structure (nuchal ligament) that stretches across (e.g., Janensch, 1929; Alexander, 1985). Alexander (1985) additionally speculated that the short, deep epaxial cervical muscles between the neural spines

in *Diplodocus* would not have been able to lift the entire neck up from a horizontal to a more vertical position. The potential presence of a nuchal ligament with taller cervical neural spines and moderate-length cervical ribs led Martin et al. (1998) to characterize diplodocoid necks as "dorsally-braced," in which a nuchal ligament would brace the neck like a cable, enhancing mobility at the expense of stability. They characterized *Diplodocus* and *Barosaurus* as a combination of both "dorsally-braced" and "ventrally-braced"; however, their cervical ribs do not exceed one vertebral segment.

The presence of a nuchal ligament has since been consistently supported by those studying sauropod necks. Tsuiji (2004) reconstructed more ligamentous structures in addition to a dorsal nuchal ligament, including interspinous and interlaminar ligaments along the bifurcated neural spines of diplodoicoids, also adding tensile strength to a more horizontal neck posture. Later, Schwarz et al. (2007a) proposed that supraspinous ligaments may have been split up with pneumatic diverticula throughout, which might have also had an effect on neck function.

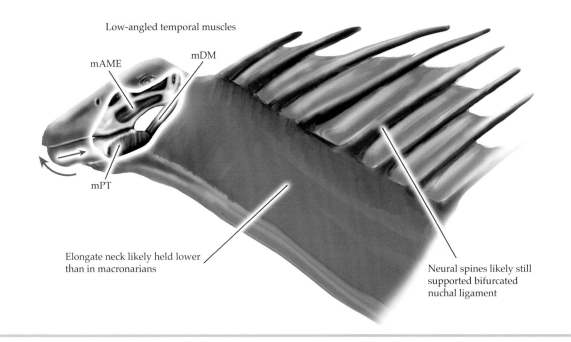

Figure 9.6. *Amargasaurus* head and neck showing feeding function with jaw muscles. Red arrow shows orthal and possible slight palinal jaw action. Abbreviations: mAME, m. adductor mandibulae externus; mDM, m. depressor mandibulae; mPT, m. pterygoideus.

This hypothesis was further analyzed anatomically and biomechanically, showing the importance of stress reduction patterns through pneumaticity in cervicals (pleurocoels) and also describing the effect of finely segmented muscles on cervical morphology (Schwarz-Wings and Frey, 2008; Schwarz-Wings, 2009). Furthermore, Woodruff (2017) proposed a split nuchal ligament morphology that would have created bilateral tensile forces in lateral motions (see below).

Stevens and Parrish (1999) performed the first three-dimensional (3D) digital reconstructions of diplodocid neck posture and feeding envelope (i.e., the maximum vertical and horizontal movements of which their necks would have been capable). With reconstructions of both *Diplodocus* and *Apatosaurus*, they found that the cervical vertebrae were best articulated as a straight continuation of the direction of the dorsal vertebrae, which creates a more horizontal neck posture that gently curved downward with the head angled down as well. This would have meant their neutral posture was effectively at ground level. Stevens and Parrish (1999) noted that although *Diplodocus* and *Apatosaurus* were able to ventroflex their necks easily, dorsal and lateral flexion was not as easy, with *Apatosaurus* capable of slightly more dorsal and lateral curvature than *Diplodocus*. These results implicated a ground-feeding or low-browsing ecology in diplodocids, with *Apatosaurus* reaching 6-m heights and *Diplodocus* only reaching 4-m heights. A nuchal ligament would have also helped keep the neck in this position while taught (Stevens and Parrish, 1999), as noted above.

In a rebuttal to Stevens and Parrish (1999), Upchurch (2000) expressed that although diplodocids were indeed likely low browsers, he questioned the extent of validity of their model, indicating a few anatomical nuances that conflict with their model. The first is that even though their model was fixed at the base of the neck for a reference point, the cranial-most three or four dorsal vertebrae in diplodocids show ball-and-socket joints as well as broad zygapophyseal articular morphology (features also seen in the caudal-most cervical vertebrae), meaning the trunk would have aided in dorsiflexion of the head and neck with magnified movement down the

length of the neck to the head. Stevens and Parrish (2000) responded that although it was likely mobile, further analysis of the constraints of the pectoral girdle is needed to test how much movement could have occurred in the cranial trunk region. In addition, Upchurch (2000) pointed out that the caudal-most cervicals of their *Apatosaurus* model were based on reconstructed bones, to which Stevens and Parrish (2000) responded with confidence in their reconstruction based on published descriptions as well as trends seen in the cervical skeleton of *Diplodocus*.

Upchurch and Barrett (2000) agreed that diplodocoids were lower-level feeders (based on tooth wear, their shorter forelimbs relative to hindlimbs, and other features), though with a large amount of lateral motion of the neck for greater food intake and selectivity. They did not consider a high browser or low browser to be necessarily mutually exclusive, however, and found that greater flexibility could still be possible for selectivity and proposed branch-stripping. They note that if diplodocoids could have had a vertical motion of the neck, they could potentially reach 10- to 12-m heights (or even greater if they could have reared up on their hindlimbs). The shorter necks of the smaller dicraeosaurids likely would not have been able to lift their heads as high, especially with the taller cervical neural spines impeding such a motion, so they were likely specialized low browsers (Upchurch and Barrett, 2000). An especially extreme case of this is the incredibly shortened neck of the dicraeosaurid *Brachytrachelopan*. The shorter-necked rebbachisaurids like *Nigersaurus* (Sereno et al., 2007) would have also kept their head low to the ground. Interestingly, rebbachisaurids have been shown to have enlarged olfactory regions of the brain relative to other sauropods, suggesting a possible increased reliance on smell for finding vegetation (Paulina-Carabajal and Calvo, 2021).

Further support followed of a lower-height neck posture in diplodocoids (although to a lesser extent). In accordance with their study mentioned above, Stevens and Parrish (2005a, 2005b), compared 3D models of the diplodocids they used previously (*Diplodocus* and *Apatosaurus*) with other sauropods, including the dicraeosaurid *Dicraeosaurus* as well as a few macronarians (see below). Again, they showed that all sauropods (diplodocoids and otherwise) have cervical vertebrae that were cantilevered far forward as straight extensions of their cranial dorsal vertebrae, with none of them showing signs of osteological correlates for more erect posture, similar to many birds or even giraffes. They postulated that diplodocoids indeed showed downcurved necks and skulls indicative of lower-level food intake (Stevens and Parrish, 2005a, 2005b). Using x-rays of vertebral centra, Berman and Rothschild (2005) added further support for diplodocoid horizontal neck posture (specifically analyzing *Diplodocus*, *Barosaurus*, and *Haplocanthosaurus*), indicating that external morphology and internal distribution of compact and increased cancellous bone (and pleurocoels) of their "gracile-type" centra (longer than wide) are supportive of horizontal neck posture because of lower capacity to withstand compressive forces.

Christian (2002) calculated stresses on intervertebral discs in *Diplodocus* and *Dicraeosaurus*, showing that the optimal stress patterns indicate a lower- to medium-height neck posture, though not completely horizontal and still with the potential to lift its neck higher if necessary in some cases. Comparing with other sauropods using similar methods, Christian and Dzemski (2011) found that macronarians held their heads higher than diplodocids. Additionally, they proposed that *Diplodocus* could have bent its neck backward in a "camel-like" fashion, whereas *Apatosaurus* likely created more of an S-curve with its neck, ending with a more horizontal head (Christian and Dzemski, 2011). Taylor and Wedel (2013b) calculated intervertebral cartilage thickness in sauropods of around 5% to 10% of centrum length, based on an intermediate width between birds and other amniotes, and noted that the addition of these cartilages in calculations of flexibility creates longer, more flexible necks that would be able to elevate the heads higher than if osteological articulation alone were used. Taylor (2014) tested flexibility in the necks of diplodocids with a wider range of intervertebral cartilage thicknesses (4.5%, 10%, and 18% centrum length). Specifically, for *Apatosaurus*, he calculated an extension of neutral pose ranges between 5.5 and 21.3 degrees, and for *Diplodocus* he calculated ranges between 8.4 and 33.3 degrees, highlighting how the neck's capacity to extend and flex could be wider-ranging than previously thought, similar to many living animals with longer necks. Additionally, Taylor et al. (2009) used extant animal neck movements observed in the wild as a model for sauropods in general to suggest that these animals likely all extended their necks up to an extent with their heads flexed downward, noting that the "osteological neutral pose" is somewhat near the halfway point of actual neck posture in both extant animals and sauropods. Taylor et al. (2009) also noted that even though semicircular canal data suggested horizontal head posture, considerable variation in actual head posture exists in animals today.

The surrounding soft tissues like muscle and ligaments play an important role in flexibility as well, providing a reliable idea of how much flexibility exists in a sauropod neck (Cobley et al., 2013). Taking reference from an ostrich neck dissection, Dzemski and Christian (2007) hypothesized that *Diplodocus* had a regionalized neck vertebral column that switched between dorsiflexive and vetroflexive functions. In their model, dorsiflexion was possible cranially but decreased as the neck extended caudally until it hit its minimum around the eighth cervical vertebrae. It is at this point that ventroflexion would have been at its highest. From the ninth cervical on, flexibility in dorsiflexion increases at the expense of ventroflexion. The base of the neck again allows more ventroflexion at the expense of dorsiflexion. With these characteristics, Dzemski and Christian (2007) predicted that *Diplodocus* likely held its head low in feeding, with lateral movement possible as well and with an elastic nuchal ligament helping store energy. It was, however, likely still able to lift its head up in alert poses or if it tried to reach higher food. Also, as in non-neosauropods, torsion and twisting of the head and neck as well as neck balance would have been possible with coordinated activation of mm. longus colli ventralis and m. scalenus tendons, especially considering the shorter cervical ribs that allowed more room for flexibility in diplodocoids (Preuschoft and Klein, 2013).

Therefore the general consensus for those arguing higher neck posture in diplodocoids is that it was still lower than macronarian neck posture but higher than horizontal—at about mid-height with flexibility based on requirements (e.g., Upchurch and Barrett, 2000; Christian, 2002; Dzemski and Christian, 2007; Taylor et al., 2009; Christian and Dzemski, 2011; Preuschoft and Klein, 2013; Taylor and Wedel, 2013b; Taylor, 2014; Paul, 2017). Given various other issues, however—including the abovementioned modeling results (e.g., Stevens and Parrish, 1999; 2005a, 2005b); concerns raised regarding blood circulation to the head that required an extremely large, unthinkably thick-walled heart (Seymour and Lillywhite, 2000; Seymour, 2009); constraints in the neck, especially those induced by muscles and ligaments; as well as variations in extant animal neck-moving behavior with differences in soft tissue constraints in articulating joints not seen in the fossilized bone (e.g., Stevens, 2013)—there is also continued support for a straighter, more horizontal neck posture in line with the dorsal vertebrae in diplodocoids (e.g.,

Stevens and Parrish, 1999, 2005a, 2005b; Berman and Rothschild, 2005; Stevens, 2013; Woodruff, 2017).

As noted above, one of the biggest constraints would have been the ligamentous attachments between the cervical vertebrae and especially the presence of a nuchal ligament. Recently, Woodruff (2017) proposed a unique soft tissue morphology in which the bifurcated neural spines of diplodocids and dicraeosaurids would have served as anchoring points for a split nuchal ligament, with the trough between the bifurcation filled with interspinal ligaments (instead of muscles of pneumatic diverticula proposed beforehand). With morphological and histological data as well as knowledge of how nuchal ligaments work in extant animal in terms of elastic energy storage, Woodruff (2017) argued that this novel split ligament in these diplodoids, and *Diplodocus* in particular, would have helped increase efficiency in elastic rebound in lateral motion of the neck sweeping side to side in a horizontal feeding plane. While laterally flexing muscles contract to pull the neck to one side, the elastic structure of the nuchal ligament branch on the

Figure 9.7. *Diplodocus* skeleton.

opposite side will be tensed and store energy, resulting in elastic rebounding to pull the neck back to center efficiently. This mechanism is well designed for horizontal sweeping of low vegetation such as ferns (Woodruff, 2017). Furthermore, Cerda et al. (2022) used histological data to show that there was likely strong interspinous ligaments connecting across all consecutive tall cervical hemispinous processes in *Amargasaurus* and that this likely created a sail with mechanical forces, creating greater compression strains.

The limbs of diplodocoids are generally more gracile and shorter in stature relative to trunk length (and with shorter strides) compared to macronarians, with their forelimbs shorter than their hindlimbs and a shorter manus (Carrano, 2005) (fig. 9.7). Schwarz et al. (2007b) reconstructed the pectoral girdle with a scapulocoracoid at a 60- to 65-degree inclination that would orient the serratus musculature into a sling (in a crocodilian-like reconstruction) that supported the large body. Should a more avian model be reconstructed, the m. serratus complex would be more horizontal and shorter (given the architecture of attachment sites), and the shoulder girdle would be more restricted in movement (Schwarz et al., 2007b; Schwarz-Wings, 2009; Hohn, 2011). Henderson (2006) suggested a more caudally positioned center of mass in *Diplodocus* and calculated better stability with a narrow-gauge trackway, although this still would have varied among diplodocoids of various sizes, such as the heavier *Apatosaurus* with a more cranial center of mass (although still less so than macronarians).

Klinkhamer et al. (2019) demonstrated with 3D musculoskeletal modeling of the forelimb that the more caudally positioned center of mass in narrow-gauged diplodocids like *Apatosaurus* proved relatively less reliant on forelimb action in locomotion, preferring more shoulder abduction and less adduction than in macronarians. In the hindlimb, Klinkhamer et al. (2018) found greater leverage in knee flexion and extension as well as relatively high mediolateral rotation of the hip, adding support for a more focused use of the hindlimbs. This result also meant that *Diplodocus* and other diplodocoids were especially capable of tripodal rearing

ability (i.e., rearing up on hindlimbs and using their long tail as support) with less exertion than other sauropods (fig. 9.8), as modeled by Mallison (2011b). This stance would have allowed *Diplodocus* to reach even higher treetops than one would assume with their relatively more horizontal neck posture. The characteristically hyper-elongate, whiplike tail of diplodocoids would likely have been horizontally used as a counterbalance in locomotion, and although it is possible this kind of tail might have been used for physical self-defense or even for making loud cracking noises, it is hard to know whether that is indeed the case (Myhrvold and Currie, 1997).

Figure 9.8. Skeleton of *Barosaurus* rearing up on hind limbs.

Macronaria

Macronaria (the other neosauropod group) includes camarasaurids (e.g., *Camarasaurus*), brachiosaurids (e.g., *Brachiosaurus, Giraffatitan, Europasaurus,* and *Abydosaurus*), euhelopodids (e.g., *Euhelopus*; fig. 9.9), and titanosaurs (e.g., *Sarmientosaurus, Nemegtosaurus, Tapuiasaurus, Bonitasaura, Quaesitosaurus, Malawisaurus, Opisthocoelicaudia, Patagotitan, Dreadnoughtus, Diamantinasaurus, Savannasaurus,* and *Rapetosaurus*). They ranged from the Middle Jurassic to the Late Cretaceous, with worldwide distribution. (For reference in understanding clade affiliation of macronarians, see the taxon list in table 9.2.)

Non-titanosaur macronarians generally possess rostrocaudally shorter yet dorsoventrally deeper and transversely wider skulls. The snout is U-shaped and widened, and the external nares are enlarged and placed within a dorsally arched skull roof created by a thin, midline, dorsally arching bone that created a hump above the head (with a clearly demarcated fossa surrounding it). Camarasaurid skulls (fig. 9.10) are more moderately sloped along the narial arch, whereas brachiosaurids show the most extreme arching. Euhelopodid skulls are more gradually sloped compared to camarasaurids and brachiosaurids (fig. 9.11). Non-titanosaur macronarians also possess large orbits that are pushed upward by the triangular infratemporal fenestra, fitting caudoventral to it. Their teeth are robust and spatulate and run along both sides of the entire U-shaped margin of the jaw line, good for resisting higher forces in biting (Janensch, 1935).

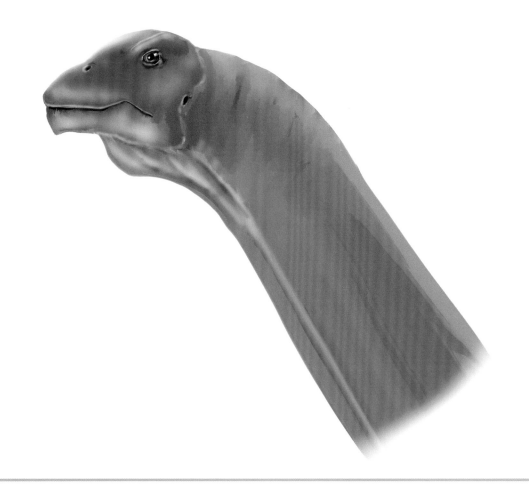

Figure 9.9. Reconstructed head of *Euhelopus.*

Titanosaur skulls (fig. 9.12) are unusually similar to those of diplodocoids (Button et al., 2017b; Button and Zanno, 2020) in that they are gracile in lateral view, dorsoventrally short, and rostrocaudally elongate. As embryos, their skulls show unique heterochronic change in patterns of ossification and synarthrosis compared to massospondylid "prosauropods," with a delayed closure of the skull roof, possibly for the development of the dorsal dural sinus (Kundrát et al., 2020). Their teeth are broader, yet still mostly cylindrical, than those of diplodocoids and extend a greater margin of the jawline, both rostrally and laterally. As in diplodocids, titanosaurs have multiple teeth in a common alveolar crypt as in some embryonic "prosauropods" (Reisz et al., 2020).

The temporal region in all macronarians is transversely broadened (to variable extents) yet caudally restricted within the skull. In non-titanosaur macronarians, as in basal eusauropods, the postorbital adductor region and infratemporal fenestra widen rostrally as they extend rostroventrally beneath the orbit. In titanosaurs (as in diplodocoids), the postorbital adductor chamber and infratemporal fenestra are much narrower and rotated at a low angle relative to the horizontal plane, while still positioned ventral to the orbit (Button et al., 2017b). Unique to titanosaurs is a dorsally arched maxilla-jugal junction that allows more room for larger muscle attachment along the mandible (Nabavizadeh, 2020b). The mandibular coronoid eminence of all macronarians is generally raised and triangular in a mediolaterally compressed sheet of bone, especially in *Camarasaurus*, in which it is raised quite high (Galton, 1986). Apesteguía (2004) noted a heightened ridge near the coronoid eminence (distal to the mesially restricted dentition) in the titanosaur *Bonitasaura*, inferring that it might have been covered by a blade made from something like a keratinous rhamphotheca; however, Button et al. (2017b) noted (1) that this was unlikely a result of its similarity with other titanosaurs, which exhibited no neurovascular foramina in the coronoid region necessary to have a rhamphotheca, and (2) that a blade in this part of the jaw would have been of minimal to no use in feeding anyway.

As in diplodocoids, macronarian tooth morphology (fig. 9.13) and wear have presented useful data on dietary preference. Calvo (1994a) noted that the interlocking, spoon-shaped dentition in camarasaurids formed a cutting edge for cropping and slight "chewing" of vegetation. Dental wear consists of scratches that are parallel to the labiolingual angle, leading Calvo (1994a) to suggest that the jaws likely had a "propalinal" action as well as a transverse chewing movement of the mandibles during feeding, as suggested by White (1958). Calvo (1994a) also highlighted the noninterlocking cone-chisel-like dentition in brachiosaurids, with the crown wider than the root. He noted that brachiosaurid dental wear shows high angled wear, indicating an orthal, isognathous, high-angle shearing and slicing in a "cut-and-crop" mechanism. Furthermore, Calvo (1994a) and Salgado et al. (1997) indicated that the diplodocid-like long,

Table 9.2.
Macronarian genera organized by clade

Major clade	Subclade	Genus
Macronaria		
	Camarasauridae	
		Camarasaurus
	Brachiosauridae	
		Brachiosaurus
		Giraffatitan
		Europasaurus
		Abydosaurus
	Euhelopodidae	
		Euhelopus
	Titanosauria	
		Sarmientosaurus
		Nemegtosaurus
		Malawisaurus
		Rapetosaurus
		Tapuiasaurus
		Bonitasaura
		Quaesitosaurus
		Opisthocoelicaudia
		Patagotitan
		Diamantinasaurus
		Savannasaurus

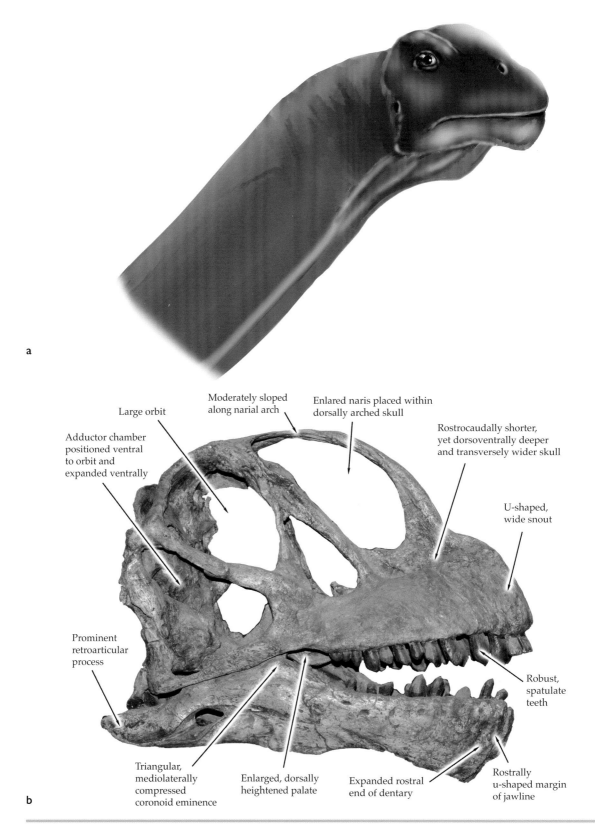

Figure 9.10. *Camarasaurus* (a) reconstructed head and (b) skull in lateral view (length approximately 57 cm).

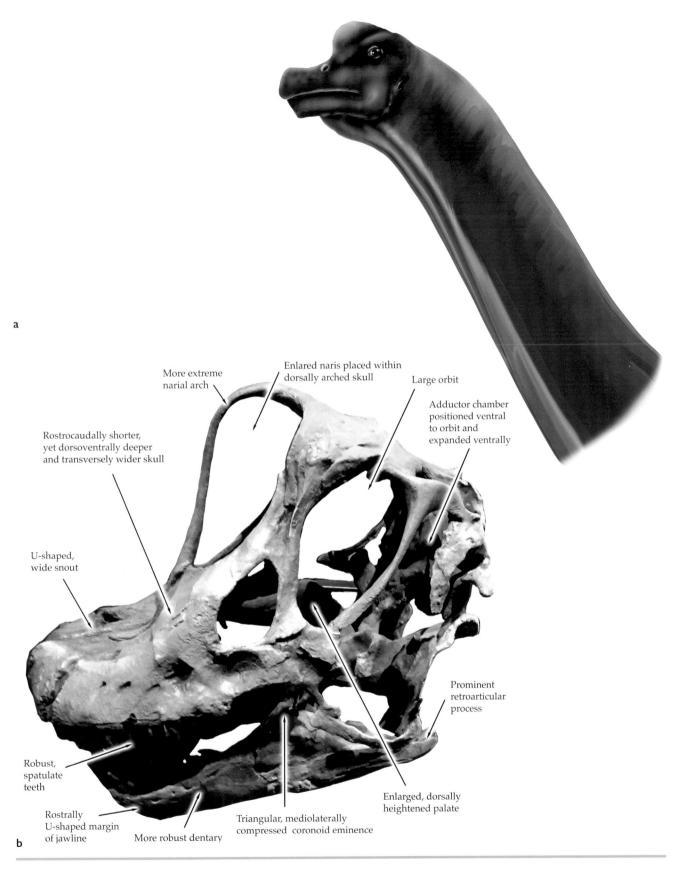

a

More extreme
narial arch

Enlared naris placed within
dorsally arched skull

Large orbit

Adductor chamber
positioned ventral
to orbit and
expanded ventrally

Rostrocaudally shorter,
yet dorsoventrally deeper
and transversely wider skull

U-shaped,
wide snout

Robust,
spatulate
teeth

Rostrally
U-shaped margin
of jawline

More robust dentary

Triangular, mediolaterally
compressed coronoid eminence

Enlarged, dorsally
heightened palate

Prominent
retroarticular
process

b

Figure 9.11. Brachiosaurids represented by (a) reconstructed head of *Brachiosaurus*
and (b) skull of *Giraffatitan* in lateral view (length approximately 70 to 90 cm).

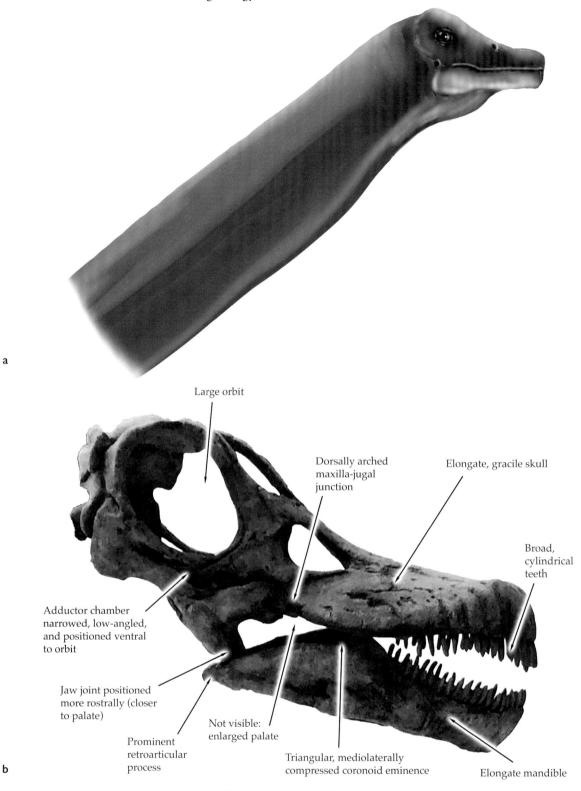

a

b

Large orbit

Dorsally arched
maxilla-jugal
junction

Elongate, gracile skull

Broad,
cylindrical
teeth

Adductor chamber
narrowed, low-angled,
and positioned ventral
to orbit

Jaw joint positioned
more rostrally (closer
to palate)

Prominent
retroarticular
process

Not visible:
enlarged palate

Triangular, mediolaterally
compressed coronoid eminence

Elongate mandible

Figure 9.12. Titanosaurs represented by (a) reconstructed head of (based largely on *Sarmientosaurus*) and (b) reconstructed titanosaur skull (based largely on *Sarmientosaurus*) in lateral view. The *Sarmientosaurus* skull is approximately 43 cm; other larger titanosaurs would have a considerably larger skull, such as that reconstructed here.

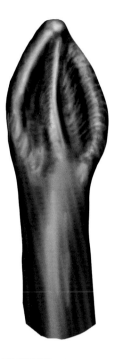

Figure 9.13. *Camarasaurus* tooth.

thin, chisel-like dentition in titanosaurs, with straight axes ending in a slightly lingually bent apex, showed high and sharp tooth wear angles in line with the tooth axis, also indicating an orthal, isognathous jaw action used to shear vegetation in a "cut-and-crop" mechanism. This is notably different from the mechanism seen in diplodocids, given their overall cranial similarities.

Christiansen (2000) agreed that the robust skulls and teeth of brachiosaurids and camarasaurids are well built for withstanding higher stresses in primarily orthal, isognathous feeding, although he notes the possibilities of raking and propalinal motions in both groups. Fiorillo (1998) showed how the spatulate teeth of *Camarasaurus* were generally more resistant to tougher foods than the thinner teeth of *Diplodocus* (see above), indicative of coarser browsing. Whitlock (2011), with microwear data, agreed that *Camarasaurus* (as well as *Brachiosaurus*) browsed on coarser plant material (from middle to higher canopy levels), presenting with a more rounded snout (rather than squared off), increasing selectivity.

Upchurch and Barrett (2000), in a comprehensive analysis of evolutionary trends in sauropod feeding mechanisms (based on skull morphology, tooth morphology and wear, joints, and jaw muscles), maintained that euhelopodids and camarasaurids showed signs of orthal sheer with some orthal processing during feeding, although they noted that camarasaurid feeding was more complex (fig. 9.14). Alternatively, they found that brachiosaurids (fig. 9.15) and titanosaurs showed signs of precision shear bites. *Sarmientosaurus*, a more basal titanosaur, presents with a skull transitional from the brachiosaurid-like condition to the more derived titanosaurian morphology. In *Sarmientosaurus*, there is some occlusion, contrary to what is seen in other titanosaurs (Martínez et al., 2016). Still, although titanosaur skulls in general are weaker than in other macronarians, they still show a diverse range of adaptations in feeding selectivity across different taxa, with some, like *Nemegtosaurus*, resembling other selective-feeding macronarians and others look like more generalist-feeding diplodocoids (Button et al., 2017b).

Macronarian cranial musculature has been reconstructed or more broadly described in a few studies in the past century (e.g., Janensch, 1935; Upchurch and Barrett, 2000; Holliday, 2009; Button et al., 2014, 2016; Nabavizadeh, 2020b). In non-titanosaur macronarians, the morphology of the adductor chamber indicates a triangular mAME complex, with a more vertical quadrate creating the caudal boundary of the muscle fan and with the rostral extent widening rostroventrally fanning to attach on the triangular coronoid eminence. In titanosaurs, much like diplodocoids, the adductor chamber is more tunnel-like, with a caudally displaced and abbreviated temporal region, a low-angled quadrate relative to the horizontal plane of the skull, and a thin infratemporal fenestra. This orientation of the adductor chamber implicates a respectively low-angled mAME complex that attaches to the coronoid eminence. The coronoid eminence is raised and more triangular than in diplodocids, possibly creating room for more rostral fanning as well as a greater mechanical advantage of this muscle group. Additionally, in *Sarmientosaurus* (fig. 9.16), the

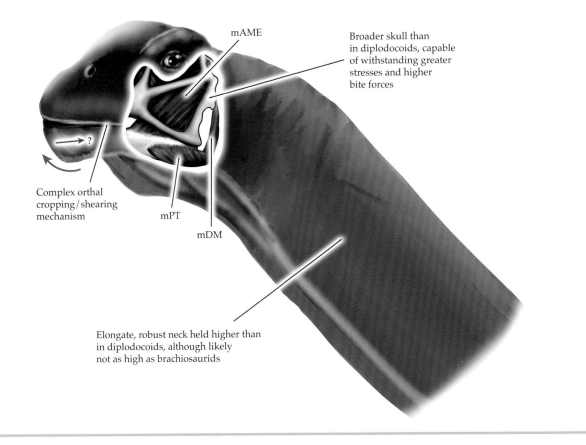

mAME

Broader skull than
in diplodocoids, capable
of withstanding greater
stresses and higher
bite forces

Complex orthal
cropping/shearing
mechanism

mPT

mDM

Elongate, robust neck held higher than
in diplodocoids, although likely
not as high as brachiosaurids

Figure 9.14. *Camarasaurus* head and neck showing feeding function with jaw muscles.
Red arrow shows orthal cropping and/or shearing jaw action. Abbreviations: mAME,
m. adductor mandibulae externus; mDM, m. depressor mandibulae; mPT, m. pterygoideus.

maxilla-jugal junction is curved dorsally, which might have increased the amount of room for an even larger fan of mAMES to extend even more rostrally for more muscle power and mechanical advantage (Nabavizadeh, 2020b). A braced hinge is created for jaw closure because, also similar to diplodocoids, the adductor chamber orientation positions the jaw joint (and associated mPT insertion sites) more rostrally, closer to the palatal origin of mPT just rostrodorsal to it. This orientation makes the muscle vector of mPT more vertical and thus more mechanically advantageous for orthal, isognathous closure of the jaw.

In their study of stresses and strains as well as bite forces during feeding in both *Camarasaurus* and *Diplodocus*, Button et al. (2014) found that *Camarasaurus* had almost four times the bite force as the more gracile-skulled *Diplodocus*, corroborating with Whitlock's (2011) microwear data showing a coarser diet in *Camarasaurus* with its larger, more spatulate dentition. This is in part thanks to its robust skull and larger adductor musculature acting on the jaw. Although overall stresses were less than in *Diplodocus*, maximum stresses in the skull of *Camarasaurus* were primarily in the quadrate region, with the pterygoids, teeth, and some thin bars of bone in the skull also showing high-stress patterns. Additionally, Button et al. (2016) found that the robust skull of *Camarasaurus* was much better equipped for higher bite forces during feeding than the relatively more gracile-skulled, longer-snouted "prosauropod" *Plateosaurus*.

Macronarian necks range from relatively shorter relative to body size (e.g., camarasaurids and some smaller titanosaurs like *Saltasaurus*)to much longer necks, with

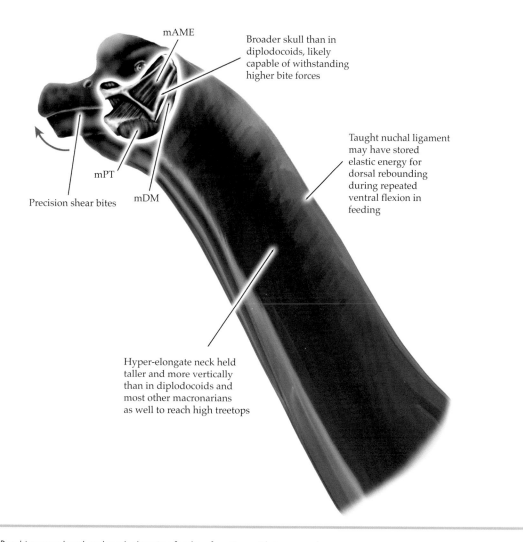

mAME

Broader skull than in
diplodocoids, likely
capable of withstanding
higher bite forces

Taught nuchal ligament
may have stored
elastic energy for
dorsal rebounding
during repeated
ventral flexion in
feeding

mPT

Precision shear bites mDM

Hyper-elongate neck held
taller and more vertically
than in diplodocoids and
most other macronarians
as well to reach high treetops

Figure 9.15. *Brachiosaurus* head and neck showing feeding function with jaw muscles. Red arrow shows orthal jaw action. Abbreviations: mAME, m. adductor mandibulae externus; mDM, m. depressor mandibulae; mPT, m. pterygoideus.

12 or 13 large, elongated individual vertebrae in brachiosaurids and large titanosauriforms and even upward of 17 vertebrae in the extremely long-necked euhelopodids (Upchurch and Barrett, 2000). Camarasaurids (fig. 9.17) show slight bifurcations in their cervicals, reminiscent of many diplodocoids. Macronarian necks also typically show long, caudally overlapping cervical ribs that bear a prominent rostral projection, possibly restricting movement to a higher degree (Upchurch and Barrett, 2000). Although some macronarians possess forelimbs that were slightly shorter than the hindlimbs (e.g., titanosaurs and camarasaurids, with forelimbs

reaching nearly 80% of hindlimb length), many in fact bear the opposite arrangement, with much longer forelimbs relative to hindlimbs (e.g., brachiosaurids and euhelopodids). With this forelimb morphology in addition to their extremely large body size, these animals could reach their necks up to incredible heights (upward of 12 m) for tree-browsing (Upchurch and Barrett, 2000), no matter if they held their heads more horizontally or more vertically.

The main feature that sets macronarians apart from diplodocoids is the broad consensus that macronarians were generally able to hold their heads up higher than

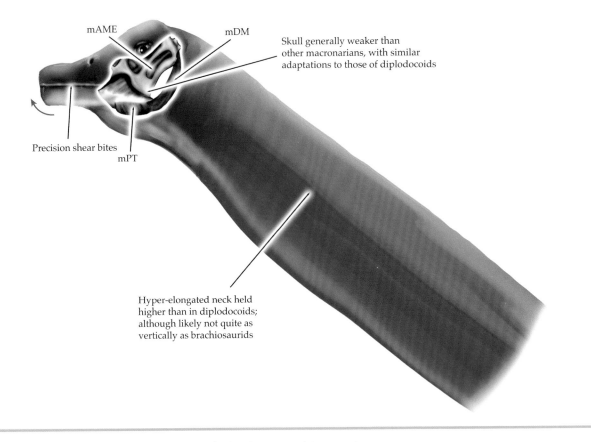

mAME

mDM

Skull generally weaker than
other macronarians, with similar
adaptations to those of diplodocoids

Precision shear bites

mPT

Hyper-elongated neck held
higher than in diplodocoids;
although likely not quite as
vertically as brachiosaurids

Figure 9.16. Titanosaur head and neck showing feeding function with jaw muscles.
Red arrow shows orthal jaw action. Abbreviations: mAME, m. adductor mandibulae
externus; mDM, m. depressor mandibulae; mPT, m. pterygoideus.

diplodocoids. In doing so, they were able to eat coarser
woody vegetation on taller coniferous trees, with the
somewhat shorter-necked (but still bulky) camarasaurids and smaller titanosauriforms eating foliage at mid-
height and brachiosaurids, euhelopodids, and larger
titanosauriforms eating higher vegetation. Like for diplodocoids, there have been many studies seeking to understand neck posture in macronarians and what it
might mean for the evolution of sauropods. Janensch
(1950), Bakker (1971), and Paul (1988) all reconstructed
the brachiosaurids *Brachiosaurus* and *Giraffatitan* with
a completely upright, somewhat swanlike neck posture,
making it the most recognizable pose for brachiosaurids
for decades (fig. 9.18). But some studies in more recent
years have cast doubt on the feasibility of this erect posture, arguing for a possible horizontal neck posture in

brachiosaurids and macronarians as a whole. For instance, as in other sauropods, a nuchal ligament would
have imposed restrictions on the mobility of the neck.
Tsuihiji (2004) reconstructed a cable-like nuchal ligament in *Camarasaurus* that attaches to the dorsal vertebrae and arches above the neural spines of the cervicals
giving off several branches, along with additional ligamentous structures reconstructed within the trough of
the unique neural spine bifurcation of *Camarasaurus*.
This bifurcation might have been an indication of a split
nuchal ligament that would have acted in lateral rebounding of the neck in *Camarasaurus*, as proposed by
Woodruff (2017) in diplodocoids (see above). Woodruff (2017) also proposed that a taut nuchal ligament in
Brachiosaurus would have wrapped over the hump of
the dorsal vertebrae and extended over the cervical ver-

Figure 9.17. Skeleton of *Camarasaurus*.

tebrae, noting that this structure would have stored elastic energy for dorsal rebounding of the neck during repeated ventral flexion in feeding. Martin et al. (1998) tentatively described macronarians as either "ventrally-braced" (e.g., *Camarasaurus* and *Brachiosaurus*) or even possibly a combination both "ventrally-braced" and "dorsally-braced" (*Brachiosaurus* and *Opisthocoelocaudia*). They proposed that the long cervical ribs in these macronarians served as a ventral bracing mechanism, limiting mobility of the neck, and only transmitted compressive loads (as they also proposed for mamenchisaurids and omeisaurids). The moderate neural spines and transverse processes would have allowed ample attachment for the nuchal ligament and other ligaments as well as the surrounding musculature noted above, adding more stability all around the neck. Klein et al.'s (2012) histological study of cervical ribs, which included those of *Giraffatitan*, found that these structures were transformed into ossified tendons, and were thereby likely more flexible and tensive than Martin et al. (1998) pro-

posed, and that they likely supported muscular ventroflexors as well (Klein et al., 2012; Preuschoft and Klein, 2013). Preuschoft and Klein (2013) also noted that the torsional or twisting abilities of the necks of brachiosaurids were controlled by m. longus colli ventralis and m. scalenus tendons in coordinated movements, as proposed in mamenchisaurids. With the tensile forces of more mobile cervical ribs, a vertical neck posture would move the center of mass caudally through the base of the neck and would mean that cervical muscles exerted tensile forces for stability (Preuschoft and Klein, 2013).

Still, the arguments for a horizontal neck posture in macronarians are similar to those made for diplodocoids and other sauropods (see above). Stevens and Parrish (2005a, 2005b), in comparing 3D digital reconstructions of many sauropods, argued that *Brachiosaurus*, *Camarasaurus*, and *Euhelopus* would have articulated the cervical vertebrae horizontally as a straight continuation of the cranial dorsal vertebrae, with slight arching, arguing that there are no osteological correlates suggest-

Figure 9.18. Skeleton of *Giraffatitan*.

ing a more vertical pose. With this horizontal neck posture, the largest taxon they tested, *Brachiosaurus*, would have held its head at about 6 m high. Berman and Rothschild (2005), with x-rays of vertebral centra, found slight variability in external morphology and internal distribution of compact and cancellous bone (and pleurocoels) among the macronarians in their study. The vertebrae of *Camarasaurus* and a titanosaur used in their study are categorized as having "robust-type" centra (wider than long), which supports a more vertical

neck posture with a higher capacity to withstand compressive forces. Notably, however, *Brachiosaurus* is characterized as having more "gracile-type" (longer than wide) cervical centra, and the internal structure shows more bending moments than compressive forces that would travel along the neck, suggestive of a more horizontal neck posture.

Stevens (2013) reconstructed a lower neck posture as a continuation of the dorsal vertebrae in brachiosaurids like *Giraffatitan*, noting that the osteological correlates needed for a more upright, swanlike cervical articulation and posture (i.e., extent of opisthocoely and zygopophyseal structure and spacing) are not present. Stevens (2013) showed that the articulation of cervical vertebrae actually arches slightly ventrally (the reverse from that seen in giraffes) and that therefore a lower reconstruction is plausible, although still arching higher than other horizontal reconstructions, meaning the neck is raised to more of a mid-height range between horizontal and vertical.

As sauropod neck anatomy is complex, integration of all soft tissues is important in trying to understand its function, although it is difficult to know the extent of soft tissues surrounding the structures and how much it would constrain movement (Cobley et al., 2013). Taylor et al. (2009) argued for more flexibility in the neck of *Brachiosaurus*, referencing the behavior of most modern long-necked animals that showed a wide range of flexion and extension of the neck. As noted above, Taylor and Wedel (2013b) estimated that the intervertebral cartilages would have measured roughly 5% to 10% of centrum length, adding to potential neck length and flexibility in more vertical posture. Potential intervertebral cartilage size, flexibility, and stress dissipation have been analyzed in macronarians, just as in other sauropods (see above).

Christian and Heinrich (1998), Christian (2002), and Christian and Dzemski (2007, 2011) calculated how the distribution of compressive forces would have acted on the intervertebral discs and joints along the neck of *Brachiosaurus* and concluded that its cervicals could have supported the weight to hold its neck more vertically in a slight S-shaped curve, although more likely not completely vertical (Christian and Dzemski, 2011). *Brachiosaurus* would have been able to flex and extend its neck in the vertical plane, browsing at great heights, with higher neural spines at mid-length of the neck, increasing lever arm of epaxial musculature. This increased leverage possibly implies more flexion in the front half of the neck and a potential dorsal flexion at the base of the neck (although the bones at the base of the neck be-

tween the cervical series and dorsal series are not as well preserved) (Christian and Dzemski, 2007, 2011). Christian (2010) and Christian and Dzemski (2011) also tested intervertebral force distributions of *Euhelopus* and noted that it, too, would have elevated its extremely long neck and that although its neck was more at an incline, it ultimately could not have reached as high as *Brachiosaurus* (with both between 45 and 60 degrees above horizontal). Paul (2017) estimated further that there would have been a 10-degree extension in the caudal cervical vertebrae, noting that tall brachiosaurids like *Giraffatitan* could have possibly reached up to 14.5-m heights. In response to Seymour's (2009) argument that using a vertical neck is too costly from an energetic standpoint for any sauropod, Christian (2010) argued that, despite the fact that they needed a high metabolism, high browsing in sauropods like euhelopodids and brachiosaurids would have actually been ultimately more energetically cost-effective. When resources are far apart spatially (or if there is a general shortage of food), being able to browse at high levels with a vertical neck allows for more food intake without having to walk long distances to find more vegetation (Christian, 2010).

Many large titanosaurs (fig. 9.19) likely had the ability to stretch their necks rather high, but there is still some uncertainty about their feeding height, with variability across taxa as far as determining whether they were lower-level browsers or mid- to higher-level browsers. *Sarmientosaurus* shows elongate cervical vertebrae, high pneumatization of the cervical vertebrae, and a downward orientation of the skull roughly based on semicircular canal orientation—features that are consistent with a lower-level browser—but they could have reached up to mid-height ranges as well (Martínez et al., 2016). Still, its craniodental features show more similarities to higher-level browsers like brachiosaurids (see above) (Poropat et al., 2021), so this might indicate a transition in feeding height. Variation among more elongate cervicals to somewhat shorter cervicals can also be seen in *Patagotitan* (Carballido et al., 2017) and *Diamantinasaurus* (Poropat et al., 2021), respectively, po-

Figure 9.19. Skeleton of *Futalognkosaurus*, a titanosaur.

tentially also indicating variabilities between low- to mid-height browsing. *Savannasaurus* shows more elongate cervical ribs, also indicating restrictions of the neck seen in higher-level browsers (Poropat et al., 2020, 2021). If a large titanosaur like *Diamantinasaurus* was at least a mid-height browser, it would have ensured less competition with other lower-level browsers, such as rebbachisaurid diplodocoids (Poropat et al., 2021). Paul (2017) also noted that smaller titanosaurs, like *Bonitasaura*, would have had a lower reach, but that possibly rearing up would have allowed them to feed higher.

As noted above, some macronarians show nearly equal lengths of forelimbs and hindlimbs (e.g., camarasaurids and titanosaurs), but others (e.g., brachiosaurids and euhelopodis) generally have longer forelimbs than hindlimbs (Upchurch and Barrett, 2000; Carrano, 2005). Brachiosaurid, camarasaurid, and euhelopodid limbs are typically gracile and are long relative to trunk length, and with the addition of longer metacarpals, they indicate longer stride lengths (Carrano, 2005). These forms also show a reduction in pelvic girdle size relative to the pectoral girdle, which shows a tendency to depend more

on the forelimb in locomotor action and stability. Henderson (2006) suggested a more central location of the center of mass (and more cranial than in diplodocoids) in *Brachiosaurus* and other macronarians like *Camarasaurus* and *Opisthocoelocaudia*, providing more stability in a wider gait. Although these brachiosaurids and camarasaurids show some "wide-gauge" adaptations of the limbs, titanosaurs show these types of characteristics more strongly. Titanosaurs possess a laterally canted femur, a wide sacrum, and a more oblong femoral circumference, which are clear indicators of "wide-gauge" locomotion, as seen in trackways (Wilson and Carrano, 1999; Carrano, 2005). They also show a modified shoulder joint and a wider range of flexion and extension of the elbow, with a long olecranon process of the ulna for powerful extensor musculature. A small, laterally flared ilium pulled the origin of m. iliofemoralis and m. iliotibialis laterally as well to pull the femur medially for support in movement with a more mobile forelimb (Carrano, 2005; Klinkhamer et al., 2018; Voegele et al., 2021).

Numerous studies have reconstructed forelimb, hindlimb, and tail musculature and function in macronarians—most notably in the gargantuan titanosaurs but also in brachiosaurids and others—speaking to the vast interest in how such large animals could have supported their immense, deeply barreled body size and shape in locomotion (Schwarz et al., 2007b; Schwarz-Wings, 2009; Ibiricu et al., 2013, 2018; Sellers et al., 2013; Ullmann et al., 2017; Vidal and Diaz, 2017; Klinkhamer et al., 2018, 2019; Otero, 2018; Vidal et al., 2020c; Voegele et al., 2020, 2021). These studies inform about not only the function of these regions, but also how the skeleton was structured in life.

For instance, as in diplodocids (see above), Schwarz et al. (2007b) reconstructed the pectoral girdle with a scapulocoracoid at a 60- to 65-degree inclination in *Camarasaurus* and a 55- to 65-degree inclination in *Opisthocoelocaudia*, both implying a postural decline from the dorsal series to the sacrum in these macronarians. Serratus musculature would have been oriented ventrally with this scapular reconstruction, forming a sling to support such massive bodies in locomotion along with strong epaxial musculature (Schwarz et al., 2007b; Schwarz-Wings, 2009). With computer modeling simulation, Sellers et al. (2013) demonstrated that when sauropod body sizes were taken to the extreme—in this case, that of *Argentinosaurus* (one of the largest titanosaurs, at about 40 m long and weighing 83 tons)—they would have likely had more constrained ranges of motion in the joints of their limbs to be able to support and move such a large body. With those parameters implemented, however, Sellers et al. (2013) showed that *Argentinosaurus* would have been capable of locomotion at slower speeds.

Indeed, there is abundant variation in muscular reconstruction across taxa, and it is crucial to understand these variations in modeling, especially when dealing with animals of enormous size, including wide-gauge gait in titanosaurs (e.g., Ibiricu et al., 2013, 2018; Ullmann et al., 2017; Klinkhamer et al., 2018, 2019; Otero, 2018; Voegele et al., 2020, 2021). Ullmann et al. (2017) described how the gracile humeri of titanosauriforms allowed more excursion in locomotion (with some decrease in abduction/adduction of the shoulder), whereas the broadened femora created greater leverage for abduction and adduction of the hindlimbs. Klinkhamer et al. (2019) demonstrated with 3D musculoskeletal modeling of the forelimb that the titanosaur *Diamantinasaurus* still shows high leverage in shoulder adduction in wide-gauge stance with a laterally flared humerus that would have been less prone to ground-reaction forces. Shoulder extension leverage was also found to be high in both *Diamantinasaurus* as well as *Giraffatitan* for greater propulsion forward in locomotion (with the added benefit of an elongate olecranon process in *Diamantinasaurus* for better elbow extension and flexion as well). Klinkhamer et al. (2018) identified a reduction in hip extensors in *Diamantinasaurus*, indicating a more forelimb-dominant locomotion. Hip adductors had high leverages, however, as in the adductors of the forelimbs. Additionally, *Giraffatitan* showed high leverage in knee flexion and extension but *Diamantinasaurus* did not, even though these titanosaurs had broader femoral condyles. Voegele et al. (2020, 2021) also indicated that adductor muscles of the limbs in titanosaurs like *Dreadnoughtus* may have been important in wide-gauge titanosaurs, used to counteract the torque of the scapula and pelvis.

Ibiricu et al. (2013, 2018) reconstructed pelvic and hindlimb musculature in titanosaurs and noted differences in m. caudofemoralis lengths between different titanosaurs. They indicated that this difference in morphology would have affected surrounding muscles and would have had larger influences in locomotion as a whole, such as in the amount of hindlimb extension and lateral flexion of the tail (Ibiricu et al., 2013, 2018). Vidal and Diaz (2017) showed that the titanosaur tail (specifically *Lirainosaurus*) was regionalized in function, with more restrictions in the dorsoventral plane within the proximal and middle regions (although variation is possible among titanosaurs). Although most sauropods are thought to have had a more horizontal neutral pose of the tail, Vidal et al.'s (2020c) study of *Aeolosaurus* suggested a more ventrally arched, sigmoidal curvature of the tail in neutral pose based on the arching of the cranial-most caudal vertebrae. They hypothesized that this orientation would have increased torque in femoral rotation with the contraction of m. caudofemoralis.

As for the tail of brachiosaurids, Díaz et al. (2020) created 3D models of musculature in the tail of extremely tall *Giraffatitan*. They showed that hypaxial musculature as well as the musculature associated with the hindlimbs, especially m. caudofemoralis and m. ilioischiocaudalis, were extremely large and well developed (roughly weighing a total of 2,500 kg) and that this would have functionally made up for the consequently shorter muscles of a shorter tail when it came to forward propulsion of the hindlimbs and tail stabilization. Furthermore, Mallison (2011b), using kinetic-dynamic modeling, hypothesized that, compared to diplodocids, brachiosaurids would have had a harder time rearing up on their hindlimbs with support of the tail to reach even higher vegetation, although given their extreme height, this likely would not have been to its detriment.

CHAPTER 10

Heterodontosaurids and Early Thyreophorans

Introducing Ornithischians

The diversity of ornithischian dinosaurs is, for lack of a better term, staggering. Ornithischia consists of Heterodontosauridae and Genasauria, with Genasauria (meaning "cheeked reptile") including the clades Thyreophora (e.g., the armored and spiked stegosaurs and ankylosaurs) and Neornithischia, which itself includes the major clade Cerapoda (e.g., ornithopods, such as the broad-billed hadrosaurs and other relatives, and marginocephalians, like the dome-headed pachycephalosaurs and the horned and frilled ceratopsians) as well as thescelosaurines and other closely related forms.

Each of the major clades within Ornithischia has its own set of unique traits that determine their feeding style. One mandibular trait that is unique to ornithischians (and shared by all of them) is the predentary bone—a small, unpaired, midline symphyseal bone located rostral to the two rostral ends of the dentaries. Nabavizadeh and Weishampel (2016) described three separate functional predentary-dentary joint (PDJ) morphotypes present in ornithischians. (1) PDJ Morphotype I consists of two mediolaterally compressed rostral ends of the dentaries that contact a cup or saddle-shaped caudal surface of the predentary. (2) PDJ Morphotype II is described as a ventromedially curved rostral end of the dentaries (symphyseal processes) coming together at a U-shaped symphysis, allowing a greater range of mobility against the broader caudal surface of the predentary. (3) PDJ Morphotype III is defined as the rostral ends of the dentaries that are completely enveloped by caudal processes of the preden-

tary, creating a functionally secondarily "fused" symphysis. Other traits shared by most ornithischians are a laterally bowed jugal flange (with medially infolded caudal margin of the maxilla), a jaw joint lowered relative to the level of the tooth row, and an emarginated (i.e., inset) tooth row, mainly along the dentary but many times along the maxillary tooth row as well. A retroverted pubis with at least a small preacetabular process is also a key feature of ornithischians, resulting in a reorganization of hindlimb musculature (Maidment and Barrett, 2011) as well as an increase in volume of the torso and abdominal region for a strong ventilation system and an even larger gut fermentation system owing to the amount of space provided as result of this retroversion (Farlow, 1987; Macaluso and Tschopp, 2018; Radermacher et al., 2021).

Heterodontosauridae

For reference in understanding relationships and clade affiliation of early ornithischians discussed in this chapter, see the phylogeny (fig. 10.1) and taxon list (table 10.1) below.

According to phylogenetic analyses by Butler et al. (2008), Heterodontosauridae is among the earliest-diverging clades within Ornithischia. Heterodontosaurids ranged from the Early Jurassic to Early Cretaceous and had nearly worldwide distribution. The skulls of heterodontosaurids (e.g., *Heterodontosaurus* [fig. 10.2], *Tianyulong, Echinodon, Pegomastax, Fruitadens, Abrictosaurus,* and *Manidens*) are generally no larger than the

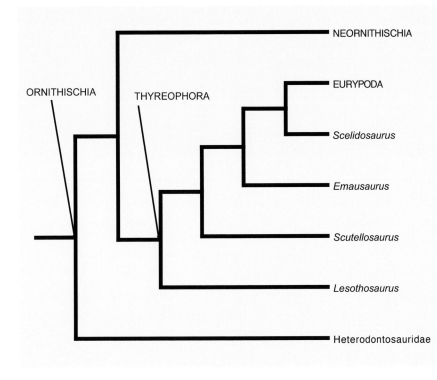

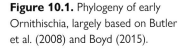

Figure 10.1. Phylogeny of early Ornithischia, largely based on Butler et al. (2008) and Boyd (2015).

Table 10.1.
Heterodontosaurid and early-diverging thyreophoran genera organized by clade

Major clade	Subclade	Genus
Heterodontosauridae		
		Heterodontosaurus
		Abrictosaurus
		Tianyulong
		Manidens
		Fruitadens
		Pegomastax
		Echinodon
Early-diverging Thyreophora		
		Lesothosaurus (?)
		Scelidosaurus
		Emausaurus
		Scutellosaurus

palm of your hand and possess a deep occiput with a midline sagittal crest that extends from just behind the orbits to the caudal edge of the skull roof. Heterodontosaurids have a triangular snout with large orbits and antorbital fossae. A dorsally arched notch is present at the premaxillary-maxillary junction, where a caniniform (i.e., "canine-like") fang from the corresponding region of the dentary rests inside at occlusion. Otherwise, the tooth-bearing regions of the maxilla and dentary sit parallel to one another at occlusion. Typically, the heterodontosaurid dentition consists of small premaxillary teeth, one caniniform tooth at the mesial end on the dentary resting just caudal to those premaxillary teeth at occlusion, and a row of precisely occluding "hypsodont" cheek teeth (adapted for constant wearing) with flat oblique occlusal surfaces (Norman et al., 2011; see below). Their temporal region (and supratemporal fenestra) was rostrocaudally broad and transversely narrow, and the postorbital adductor region and infratemporal fenestra are subrectangular and vertically oriented. The palate of heterodontosaurids is broad, with a moderately sized pterygoid flange.

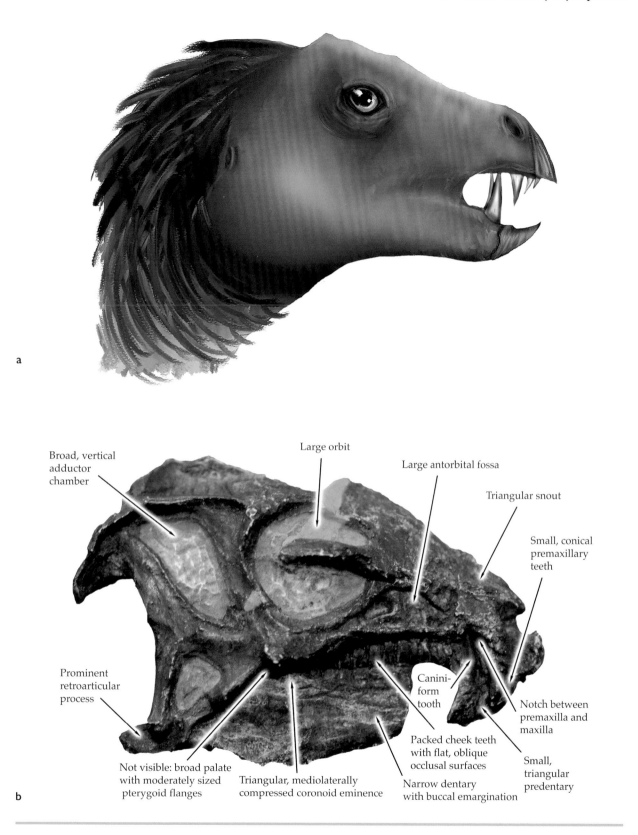

Figure 10.2. *Heterodontosaurus* (a) reconstructed head and (b) skull in lateral view (length approximately 10 to 11 cm).

In the mandible, the predentary is small and triangular (categorized as PDJ Morphotype I; Nabavizadeh and Weishampel, 2016), with some being longer than tall (*Heterodontosaurus* and *Abrictosaurus*) and others taller than long (*Pegomastax*). The dentaries are narrow and elongate, the triangular coronoid eminence is raised and mediolaterally compressed, the mandibular fenestra is reduced in size, and there is a prominent retroarticular process caudal to the jaw joint.

The morphology of the namesake heterodont dentition in heterodontosaurids (meaning teeth that are morphologically and functionally different along the tooth row) is highly critical for understanding of feeding mechanisms (fig. 10.3); consequently, they deserve more in-

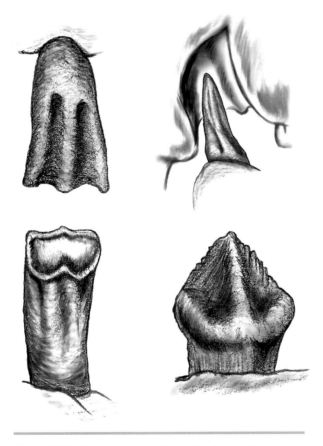

Figure 10.3. Heterodontosaurid dentition in lateral views: (*clockwise from top left*) *Heterodontosaurus* maxillary tooth, *Heterodontosaurus* caniniform dentary tooth, *Fruitadens* dentary tooth (note spade-like shape), and *Heterodontosaurus* dentary tooth.

depth description. The premaxilla bears three conical, shallowly caudally recurved, and mediolaterally compressed teeth on either side (two in *Tianyulong*; Zheng et al., 2009), each successively larger in size caudally with the third the largest and caniniform (Norman et al., 2011; Sereno, 2012). This premaxillary caniniform tooth is reduced in size in *Abrictosaurus* and is located in the maxilla in *Echinodon* (Sereno, 2012).

The dentary possesses one large, conical, caniniform tooth that, during occlusion, rests in the embayment formed at the premaxilla-maxilla suture diastema described above (Norman et al., 2011; Sereno, 2012). This tooth, slightly recurved distally at its apex, sits in an alveolus that is raised relative to the rest of the cheek dentition distal to it. In *Abrictosaurus*, this caniniform tooth is relatively much smaller (just as in the premaxilla), which along with the much reduced premaxillary caniniform suggests a reduction or elimination of caniniform use in general for this taxon (Sereno, 2012).

While there are gaps between maxillary teeth in some heterodontosaurids (i.e., *Abrictosaurus*, *Echinodon*, *Fruitadens*, and *Tianyulong*), the maxillary dentition in *Heterodontosaurus* is tightly packed and roughly level and flush ventrally in lateral view as well as slightly medially bowed. Each tooth crown is columnar and gradually mesiodistally expanded at the distal end relative to the narrower alveolar end. A single median ridge extends apicobasally in lateral view on the enamel crown, forming a pointed occlusal edge. The first four teeth also have an additional small ridge mesial and distal to the prominent median ridge. The mesial-most teeth are more mediolaterally compressed, forming a thinner cutting occlusal edge for shearing vegetation, whereas the more distal teeth are thicker and form a broader, flatter, butterfly-shaped occlusal surface, likely for crushing vegetation with more surface area (Norman et al., 2011; Sereno, 2012).

Within the dentary, a short diastema extends between the caniniform tooth and the rest of the dentition. The dentary tooth row is tightly packed and gradually raised dorsally as the height of the dentary rises in lateral view (Norman et al., 2011). The entire tooth row is also slightly bowed so that its dorsal occlusal surface is uneven, oriented 35 to 40 degrees to the horizontal plane (Weishampel, 1984a), although certain portions of the tooth row are more even with each other than others. The occlusal surfaces of dentary teeth face laterally, and the medial surface of each tooth resembles the lateral surface of each maxillary tooth, with a median ridge running dorsoventrally along the enamel crown. The mesial and distal dentary cheek teeth together form a planar occlusal surface; however, the middle teeth form more of a cupped, steplike wear facet. This wear pattern likely indicates more crushing or grinding action at this part of the lower jaw (see below).

Additionally, the enamel of maxillary teeth in *Heterodontosaurus* is thicker labially while the enamel of dentary is thicker lingually, which means the occlusal surfaces are much more amenable to wear, ultimately creating a sharper cutting edge. Some other heterodontosaurids do not show this asymmetry of enamel coating, however, as seen in *Echinodon*, which has teeth that occlude (Norman et al., 2011). Analysis of enamel microstructure in *Manidens* has shown that heterodontosaurid teeth are structurally the simplest among ornithischian groups, showing similarities with the sauropodomorph *Plateosaurus* (Becerra and Pol, 2020). Furthermore, heterodontosaurid tooth eruption and replacement patterns are largely comparatively slower than in other ornithischian dinosaurs. While some have concluded that tooth replacement was uniquely episodic (Hopson, 1980; Butler et al., 2008; Porro, 2009; Porro et al., 2010; Norman et al., 2011), others have argued that at least some taxa may have been replacing teeth continuously (Sereno, 2012; Becerra et al., 2021). Becerra et al. (2021) noted a differential wear pattern from mesial to distal, showing that tooth replacement was likely tied to the different types of novel feeding among heterodontosaurid taxa (see below).

Outside of the dental morphology in the best-studied heterodontosaurid, *Heterodontosaurus*, dental morphology shows a range of morphological variation across other taxa. *Fruitadens*, a Late Jurassic heterodontosaurid (Butler et al., 2012), has a more leaf-shaped

dentary tooth morphology with denticles and a pointed apex that differs from what is seen in most other heterodontosaurids. *Echinodon* shows symmetrical distribution of denticles of maxillary teeth, as seen in *Manidens*, with signs of a shearing mechanism. Interestingly, the dentary in *Manidens* is asymmetrical, like that in *Pegomastax* (Becerra et al., 2018). Unique swollen rims at the base of the teeth in *Manidens* called "cingular entolophs" form a platform on which opposing teeth occlude with its apical crown, forming a steplike "double occlusion" to increase shearing efficiency in feeding (Becerra et al., 2018).

As the most basal major clade known within Ornithischia, Heterodontosauridae is likely to include taxa with a primitive, strictly orthal (dorsoventral) feeding cycle and isognathous jaw spacing owing to its simplicity in such a small animal. Despite their basal position among ornithischians, however, some heterodontosaurids, such as *Heterodontosaurus*, exhibited a much more complex apparatus in which an orthal action is merely just one aspect (fig. 10.4). The heterodontosaurid skull itself is likely akinetic and similar in overall shape to most small, bipedal, and herbivorous basal Ornithopoda, a clade in which heterodontosaurids were originally placed phylogenetically until recently (e.g., Butler et al., 2008). Additionally, a characteristic of heterodontosaurid jaws is the kinetic nature of the mandibular elements relative to one another, with the predentary and dentaries acting as separate entities with presumably highly mobile joints between them (Weishampel, 1984a; Crompton and Attridge, 1986; Norman et al., 1991, 2011; Porro, 2007, 2009; Sereno, 2012; Nabavizadeh and Weishampel, 2016).

Thulborn (1971, 1974, 1978) suggested that, owing to the continuous occlusal surface in the distal dentition, a propalinal (mesiodistally oriented) jaw action was used by *Heterodontosaurus*. The dentition would have constantly worn away in a mesiodistally planar surface, a mechanism that was also supported by Barrett (1998). Hopson (1980), however, suggested a different jaw mechanism based on the mammal-like nature of the dentition, with an orthal power stroke coupled with

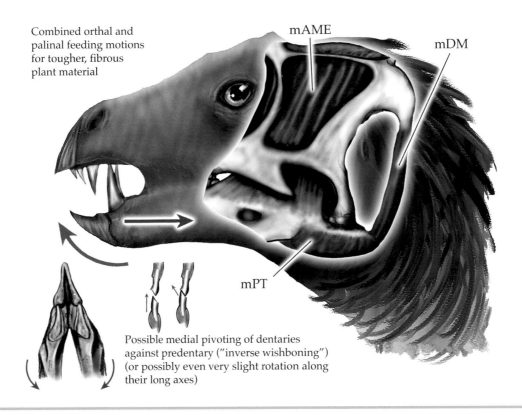

Combined orthal and palinal feeding motions for tougher, fibrous plant material

mAME

mDM

mPT

Possible medial pivoting of dentaries against predentary ("inverse wishboning") (or possibly even very slight rotation along their long axes)

Figure 10.4. *Heterodontosaurus* head showing feeding function with jaw muscles. Red arrows show various jaw actions, with a dorsal view of predentary-dentary junction on bottom left next to a view of tooth occlusion (redrawn from Sereno, 2012). Abbreviations are as follows: mAME, m. adductor mandibulae externus; mDM, m. depressor mandibulae; mPT, m. pterygoideus.

transverse movement of the entire mandible against the maxilla. Later, Weishampel (1984a) also investigated the skull of *Heterodontosaurus* and was the first to interpret the potentially mobile nature of its mandibular elements into a hypothesis regarding their mechanism. He described a spheroidal (ball-in-cup) joint between the predentary and each dentary, with the rostral end of the dentary being ball-shaped and the articular surface of the predentary rather concave and cup-shaped. Weishampel (1984a) proposed a jaw mechanism involving orthal adduction of the mandible coupled with long-axis rotation of each of the dentaries separately against the predentary. This was the first suggestion of intramandibular kinesis for heterodontosaurids, and it set the stage for a significant number of subsequent arguments regarding this concept of predentary-dentary joint mobility, which could explain the functional purpose of the origin of the

predentary bone itself in Ornithischia, given the basal positioning of Heterodontosauridae.

Crompton and Attridge (1986) and Norman et al. (2011) reexamined Weishampel's (1984a) arguments, and while they agree with the predentary-dentary joint in *Heterodontosaurus* being spheroidal, they rejected long-axis rotation of the dentaries or any major mobility at that site. Arguments against long-axis rotation include (1) wear facets that are planar and that widen labiolingually toward the distal end of the tooth row and (2) a mandibular glenoid of the jaw joint that is transversely expanded. Instead, they proposed an orthal mechanism coupled with an "inverse wishboning" of the mandible in which each mandibular corpus would independently shift medially and simultaneously on both sides of the jaw against the predentary (fig. 10.4). Crompton and Attridge's (1986) mechanism takes both of the above arguments against Weishampel's (1984a) mechanism into account. Crompton and Attridge (1986) also noted that *Heterodontosaurus* was said to have a higher ratio of adductor muscle mass to tooth and beak surfaces than *Lesothosaurus*.

Further analysis by Porro (2007, 2009), using FEA force and stress analysis, and Norman et al. (2011), using morphological analysis, also proposed a jaw action for *Heterodontosaurus* similar to (although slightly modified from) that suggested by Crompton and Attridge (1986). Norman et al. (2011) proposed a slight palinal motion along with orthal action and inverse wishboning. They pointed out that any lateral excursion or long-axis rotation of the mandibular corpora would have been restricted because of the tight fit between the caniniform teeth, the medial walls of the diastema, and the elongate nature of the pterygoid flange and ventral jugal process. These form an aperture that would have directed jaw closure in a specific orientation. Norman et al. (2011) also demonstrated that the predentary-dentary joint was not spheroidal but was in fact more morphologically complex in nature. This joint probably inhibited long-axis rotation but would not have precluded any type of wishboning jaw action. They suggested that heterodontosaurids were likely able to chew tough vegetation and possibly even the flesh of small prey.

Sereno (2012), in a separate analysis, rejected the ability of muscle tissue slicing in heterodontosaurids owing to the outward angled nature of their caniniform dentition, which was ineffective for puncturing soft tissue. He also made the observation that the predentary-dentary joint in heterodontosaurids is more saddle-shaped rather than spheroidal, and also stated that constraints shown by Norman et al. (2011) are not significant owing to the skull being so small in nature. With these criteria, Sereno (2012) suggested that long-axis rotation of the individual mandibular corpora might have been permitted after all, resurrecting Weishampel's (1984a) original hypothesis, although through slightly different morphological observations. He observed low angle wear facets on the premaxillary teeth, and this suggested to him that medial inverse wishboning might

have occurred as well. He used the quadrate-articular jaw joint to further support these ideas about jaw motions, stating that this joint is actually a well-fitted rotary joint with the lateral condyle of the quadrate, offset ventral to the medial condyle that provided a curved articular surface for long-axis rotation.

The slight variations in craniodental features and associated feeding mechanisms across heterodontosaurid taxa have also led to the suggestion that there might have been niche partitioning among heterodontosaurids. For example, Porro et al. (2010) suggested that the seemingly more sophisticated feeding apparatus of *Heterodontosaurus* was well suited for fibrous and generally tougher plant material, while *Abrictosaurus* might have fed on softer, nutritious plant material or might have possibly even been omnivorous. Butler et al. (2012) reexamined heterodontosaurid jaw mechanics and concluded that the earlier *Heterodontosaurus* was able to occlude all of its teeth simultaneously with stronger jaw adduction with smaller gape angles (again, a good adaptation for fibrous diets), whereas later taxa, such as *Tianyulong* and *Fruitadens*, used more of a scissorlike jaw-closing mechanism with more triangular dentition. Butler et al. (2012) also suggested, with the support of morphology and lever arm mechanics, that later surviving heterodontosaurids, including *Fruitadens* and *Tianyulong*, were adapted to a puncture-crushing jaw mechanism with more rapid biting, larger gape angles, and greater leverage of palatal musculature, possibly indicating omnivory with a diet of fruit, soft vegetation, and invertebrates.

Cranial musculature has been reconstructed in heterodontosaurids in a handful of studies (e.g., Crompton and Attridge, 1986; Porro, 2009; Norman et al., 2011; Butler et al., 2012; Sereno, 2012; Nabavizadeh, 2016, 2020a, 2020b). The temporal m. adductor mandibulae externus (mAME) complex would have been rostrocaudally expanded at its origin, with the muscle oriented vertically within a vertical adductor chamber and inserted onto a subtriangular coronoid eminence of the mandible. This vertical fan of temporal musculature was ideal for a forceful orthal feeding stroke and slight pali-

Figure 10.5. *Heterodontosaurus* skeleton.

nal motion. Furthermore, the expansion of m. adductor mandibulae externus superficialis (mAMES) on the lateral aspect of the surangular would have acted in transverse mandibular wishboning or even slight long-axis hemi-mandibular rotation (Weishampel, 1984a; Crompton and Attridge, 1986; Porro, 2009; Norman et al., 2011; Sereno, 2012; Nabavizadeh, 2020a, 2020b; Nabavizadeh and Weishampel, 2016).

In a comprehensive study of the evolution of ornithischian jaw mechanics using two-dimensional lever arm mechanics, Nabavizadeh (2016) found that, in *Heterodontosaurus*, the mandibular mechanical advantage value relative to other ornithischians (therein referred to as the "relative bite force") is focused more at the predentary and mesially along the tooth row, with a moderate gradation of mechanical advantage more distally. These results, in addition to showing a general focus of mesial bite efficiency, may also be indicative of an adaptation accompanying the formation of the caniniform tooth just distal to the smaller premaxillary teeth that over-

hangs the lateral aspect of the keratinous rhamphotheca of the predentary. Previous studies have suggested that the caniniform tooth was a possible sign of combat (Thulborn, 1974; Molnar, 1977) or even an omnivorous diet in *Heterodontosaurus*, potentially acting in prey capture and dismemberment (Porro, 2009; Norman et al., 2011; Farke, 2014). Although this has been disputed by Sereno (2012), who suggested the caniniform teeth were mainly for display, the greater mesial mechanical advantage suggests that these teeth could very well have acted in feeding (Nabavizadeh, 2016).

Heterodontosaurids were small, bipedal animals with a somewhat S-shaped neck with nine cervical vertebrae (fig. 10.5). They possessed long, robust, and strong forelimbs with large hands capable of grasping, indicated by a longer second digit and small but sharp claws (possibly for pulling tough plants, pulling their roots or tubers, or maybe even for opportunistic predatory behavior as well if heterodontosaurids were indeed omnivorous; see above) (Sereno, 1997, 2012; Norman et al., 2011).

Their hindlimbs were elongate and gracile, built for a cursorial lifestyle, making escape from predators relatively easy and energetically sound (Norman et al., 2011; Sereno, 2012). Reconstructed musculature of the forelimbs shows more similarities overall to crocodilians and the common ancestor of archosaurs than to birds, and that the pectoral muscles likely acted in protraction and retraction of the humerus, as expected in a bipedal animal using its forelimbs (Maidment and Barrett, 2011). Hindlimb musculature shows a lateral migration of crucial femoral protractor muscles owing to an elongate preacetabular process and reduction of the protractor m. puboischiofemoralis because of the retroverted pubis specific to ornithischians (Maidment and Barrett, 2011). The fourth trochanter of the femur (the attachment site for m. caudofemoralis) is more proximally placed relative to larger quadrupedal ornithischians, which creates a quicker, shorter-distance retraction of the femur in faster locomotion (Persons and Currie, 2020).

Early-Diverging Thyreophora

Thyreophora is the first of the two genasaurian clades described here and includes early-diverging forms as well as the major clade Eurypoda, comprising the much larger and derived Stegosauria and Ankylosauria (see chap. 11).

The skulls of early-diverging thyreophorans (e.g., *Lesothosaurus* [fig. 10.6], *Scelidosaurus* [fig. 10.7], *Emausaurus*, and *Scutellosaurus*, with ranges largely from the Early to Middle Jurassic of Europe and Africa) are somewhat more elongate in lateral view relative to heterodontosaurids. *Lesothosaurus* (which may alternatively be an early-diverging neornithischian rather than a thyreophoran according to recent studies; e.g., Baron et al., 2017a) has the most dorsoventrally shorter skull of the four taxa (as well as being rostrocaudally elongate and narrow), while *Scelidosaurus* and *Emausaurus* have a relatively deeper occiput caudally with a narrower, triangular snout (implicating selective feeding) with leaf-shaped teeth. All early-diverging thyreophorans possess a rostrocaudally widened temporal region and supratemporal fenestra; however, while the temporal region in *Lesothosaurus* is narrowed, it is wider in *Scelidosaurus* and *Emausaurus*. They possess subrectangular and vertical postorbital adductor chambers and infratemporal fenestrae, and their palates are of moderate breadth. Like most ornithischians, the jugal is slightly flared with an infolded maxilla in most early-diverging thyreophorans except *Lesothosaurus*, in which the maxilla-jugal junction is straight and laterally continuous. A buccal emargination with medially inset teeth is seen in all taxa, which, with neurovascular foramina, is suggested to have been indicative of "lip"-like extra-oral structures in *Lesothosaurus* (Knoll, 2008).

In the mandible, the predentary is small and triangular- to cup-shaped (categorized as PDJ Morphotype I; Nabavizadeh and Weishampel, 2016), with an elongate ventral process bracing the ventral aspect of the symphysis (as seen in *Lesothosaurus*). The rostral extension of the dentary is narrow, and the coronoid eminence is low to nearly nonexistent in *Lesothosaurus* but raised and more triangular in *Scelidosaurus* and *Emausaurus*. The mandibular fenestra is of variable size (relatively larger in *Lesothosaurus*), and the retroarticular process is blunt but reasonably prominent. For a detailed look at the three-dimensional (3D) intricate anatomy of the skull of *Lesothosaurus*, see Porro et al. (2015).

Among early-diverging thyreophorans, *Lesothosaurus*, *Scutellosaurus*, and *Emausaurus* are all known to possess premaxillary teeth, with six tooth positions in *Lesothosaurus* and *Scutellosaurus* and five tooth positions in *Emausaurus*. The premaxillary teeth are conical, lacrimiform, and slightly recurved distally (fig. 10.8). In life, they rested against the lateral edge of the keratinous rhamphotheca of the predentary. The maxillary dentition in early-diverging thyreophorans is closely packed with mesiodistally leaf-shaped teeth, with apicobasal ridges running along the buccal and lingual surfaces. These ridges form denticles at the apical ridges of the teeth. The wear facets on the teeth of *Scelidosaurus* are small at the apex of the teeth, unlike the dentary teeth

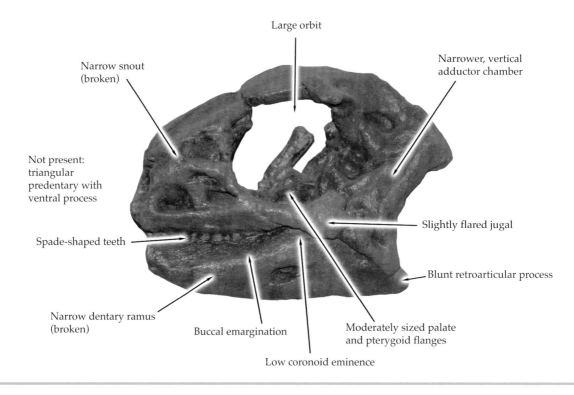

Figure 10.6. *Lesothosaurus* skull in lateral view (length of this incomplete skull approximately 40 to 50 mm).

described below (Barrett, 2001). *Lesothosaurus* had about 15 to 16 maxillary teeth, *Scelidosaurus* and *Scutellosaurus* (Colbert, 1981) had as many as 18 teeth, and *Emausaurus* (Haubold, 1990) had about 21 teeth.

The dentary teeth in early-diverging thyreophorans often retain the primitive appearance of ornithischian teeth, unlike the unusually derived morphology of the more basal heterodontosaurids. They are typically leaf-shaped in labial and lingual views, with a mesiodistally expanded base that extends apically to a point; ridges run apicobasally along the lingual and labial surfaces that create round-to-pointed denticles at the apical ridges (Colbert, 1981; Haubold, 1990; Sereno, 1991; Barrett, 2001), usually with a single denticle present at the apex of the tooth crown. The tooth root is narrow and columnar. In *Scelidosaurus*, the apex of the tooth is somewhat asymmetrical, and the teeth possess a shelf due to a buccal expansion at its base (Barrett, 2001). The dentary tooth count in early-diverging thyreophorans is variable. *Leso-*

thosaurus has between 11 and 14 teeth (Sereno, 1991), *Scelidosaurus* has at least 16 teeth (Barrett, 2001), *Scutellosaurus* (Colbert, 1981) has 18 teeth, and *Emausaurus* (Haubold, 1990) has 21 teeth.

The evolution of jaw mechanisms in ornithischians starts with (at least) an orthal component in the most early-diverging members, followed by the evolution of various other jaw actions in various subsequent clades. *Lesothosaurus* has been universally accepted to have implemented a primarily orthal component to their mechanism. Various hypotheses of orthal, isognathous slicing or shearing (Thulborn, 1971; Galton, 1986) and orthal pulping (Weishampel and Norman, 1989) have been suggested for *Lesothosaurus*. Its quadrate-glenoid jaw joint is offset ventral to the level of the maxillary tooth row, with a transversely broad ventral condyle of the quadrate (Thulborn, 1974; Galton, 1978; Cooper, 1985). According to Thulborn (1971), Weishampel (1984a), and Crompton and Attridge (1986), the dentition of

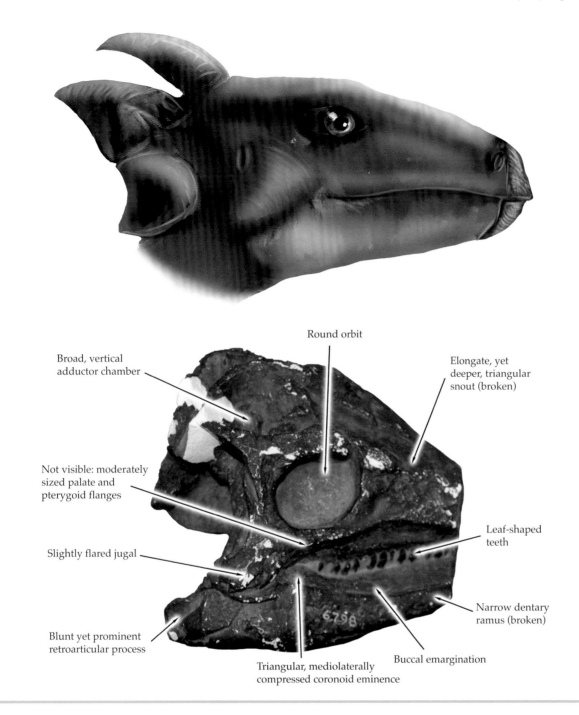

a

b

Round orbit

Broad, vertical
adductor chamber

Elongate, yet
deeper, triangular
snout (broken)

Not visible: moderately
sized palate and
pterygoid flanges

Leaf-shaped
teeth

Slightly flared jugal

Narrow dentary
ramus (broken)

Blunt yet prominent
retroarticular process

Buccal emargination

Triangular, mediolaterally
compressed coronoid eminence

Figure 10.7. *Scelidosaurus* (a) reconstructed head and (b) skull in lateral view (length of this incomplete skull approximately 20 cm).

Lesothosaurus exhibits two oblique wear facets, both on the mesial and distal sides of the occlusal surfaces, labially on the dentary teeth and lingually on the maxillary teeth, indicating an alternating, interlocking occlusion typical of many cases of orthal biting. Sereno (1991) rejected this observation on the basis of another specimen, stating that the wear is irregular on different teeth. He also indicated, similarly to Crompton and Attridge

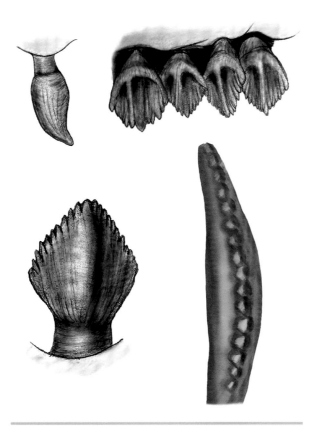

Figure 10.8. Early thyreophoran dentition: (*clockwise from top left*) *Lesothosaurus* premaxillary tooth, *Lesothosaurus* maxillary teeth, dorsal view of *Scelidosaurus* dentary tooth row showing medially bowed morphology, and *Scelidosaurus* dentary tooth.

(1986), that *Lesothosaurus* possesses a loose ball-and-socket articulation at the predentary-dentary junction, suggesting a slightly mobile joint, as predicted in *Heterodontosaurus* (Weishampel, 1984a; Crompton and Attridge, 1986). This would imply an orthal cropping mechanism with bilateral long-axis rotation of the dentaries against the predentary, although not to the extent seen in *Heterodontosaurus* (Crompton and Attridge, 1986; Sereno, 1991; Norman et al., 2011; Sereno, 2012). This finding, along with what is seen in other early-diverging ornithischians such as *Heterodontosaurus*, might shed some light on the original function of the predentary at its origin at the base of Ornithischia. Crompton and Attridge (1986) noted differences in muscle body sizes among ornithischians. For example,

they noted that, in *Lesothosaurus*, adductor muscle size relative to tooth surface area was only slightly greater than in prosauropods (although they were still weak). Additionally, ornithopods possessed much larger adductor muscle mass relative to stegosaurs, ankylosaurs, "fabrosaurids," and *Scelidosaurus*, with much smaller muscle volume to biting area ratio. The paucity of attritional wear on *Lesothosaurus* teeth, despite arid habitat, indicates a soft diet (Sciscio et al., 2017) and suggests that it and other early thyreophorans like *Scutellosaurus* used their narrow bills to select high-quality foodstuffs such as shoots, fruits, and invertebrates (Barrett, 2001; Sciscio et al., 2017).

Scelidosaurus has been suggested to have had a primarily orthal jaw action, although not quite like that of *Lesothosaurus* (Barrett, 2001). Through qualitative observations of dental micro- and mesowear, Barrett (2001) observed signs of direct tooth-tooth contact between the maxillary and dentary tooth rows, with the lingual surface of the maxillary dentition occluding with the labial surface of the dentary teeth, rather than an alternating occlusion as seen in *Lesothosaurus* (see above). The maxillary dentition exhibits small, apical wear facets, and the dentary tooth counterparts generally exhibit large, bowlike facets. Additionally, the tooth rows are also medially bowed. All of these attributes suggest an orthal puncture-crushing jaw mechanism, much like what is seen in a mortar-and-pestle action (Barrett, 2001) (fig. 10.9). Norman (2020a, 2021), in further examination, suggested the possibility of a difference in function across the tooth row, noting that mesial dentition may have been used in orthal slicing, while the distal dentition may have used a combined orthal pulping and crushing mechanism.

With no known predentary, it is difficult to discern the extent to which the individual dentaries rotated around their long axes against the predentary at occlusion; however, it is likely that rotation was slightly greater than that seen in *Lesothosaurus* simply looking at the generally larger size of the skull and curved nature of the tooth row (Barrett, 2001; Nabavizadeh, 2016; Nabavizadeh and Weishampel, 2016; Norman 2020a, 2021).

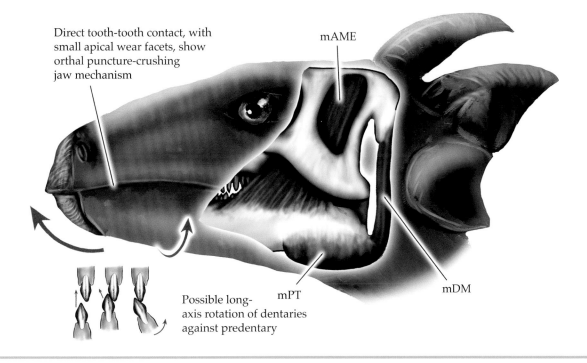

Direct tooth-tooth contact, with small apical wear facets, show orthal puncture-crushing jaw mechanism

mAME

Possible long-axis rotation of dentaries against predentary

mPT

mDM

Figure 10.9. *Scelidosaurus* head showing feeding function with jaw muscles. Red arrows show various jaw actions, with a view of possible tooth occlusions on bottom left. Abbreviations: mAME, m. adductor mandibulae externus; mDM, m. depressor mandibulae; mPT, m. pterygoideus.

Norman (2021) also suggested the potential for the quadrate head to move slightly at its articulation with the squamosal, potentially adding some wishboning movement of the mandible as well. Other early-diverging thyreophorans observed by Barrett (2001) include *Emausaurus*, which he suggests implemented a puncture mechanism similar to that inferred for *Scelidosaurus* (despite the slightly broader skull in *Emausaurus* and near absence of a medially bowed dentition) and *Scutellosaurus*, which he likened to the orthal slicing mechanism of *Lesothosaurus* based on the apparent presence of double wear facets.

Cranial muscle anatomy in early-diverging thyreophorans has been described by Crompton and Attridge (1986), Holliday (2009), Nabavizadeh (2016, 2020a, 2020b) and Norman (2020a, 2021). The mAME complex in early-diverging thyreophorans was rostrocaudally expanded in the temporal origin, with an overall vertical muscle body running down the adductor cham-

ber to insert onto low (e.g., *Lesothosaurus* and *Emausaurus*) or triangular (e.g., *Scelidosaurus*) coronoid eminences. This orientation creates a vertical muscle vector ideal for a primarily orthal feeding mechanisms with slight hemi-mandibular long-axis rotation in all early-diverging thyreophorans (Crompton and Attridge, 1986; Galton, 1986; Weishampel and Norman, 1989; Norman et al., 1991; Barrett, 2001; Nabavizadeh, 2016, 2020a, 2020b; Nabavizadeh and Weishampel, 2016; Norman, 2020a, 2021). Overall mAME mass in *Scelidosaurus* is also relatively larger, which would have given it even greater strength in its orthal feeding mechanism (Barrett, 2001; Nabavizadeh, 2016, 2020b). Palatal musculature (mainly m. pterygoideus, or mPT) was moderate in size and assisted in the orthal feeding mechanism as well.

Nabavizadeh (2016) found that *Lesothosaurus* and *Emausaurus* showed similar RBF values along the tooth row, slightly more mesial in *Emausaurus* and slightly more distal in *Lesothosaurus*. These results are likely a product of *Lesothosaurus* having a relatively slightly longer snout, creating a longer output lever arm to the predentary. An overall longer tooth row extends more distally, creating a shorter output lever arm to the distal-most tooth. Among all early-diverging thyreophorans studied, *Scelidosaurus* presented with the highest mechanical advantage across the tooth row because of the lower mAME muscle vector angle (owing to the heightened, triangular coronoid eminence) increasing the moment arm and mechanical advantage (Nabavizadeh, 2016).

Thyreophorans are the first of the three major ornithischian groups to transition from bipedal to quadrupedal locomotion (the others being Ornithopoda and Ceratopsia), with nearly all its other members having been quadrupedal. Although the true affinity of *Lesothosaurus* as an early-diverging thyreophoran or an early-diverging neornithischian is unclear, it is a good example of an early bipedal taxon in a clade whose members transition to quadrupedality thereafter. *Lesothosaurus* was a small, bipedal, cursorial ornithischian with long, gracile lower limbs, long tail, and relatively shorter forelimbs that may have had grasping function not seen in quadrupedal animals (Thulborn, 1972; Sereno, 1991; Carrano, 1999; Bates et al., 2012b; Maidment and Barrett, 2014; Baron et al., 2017b).

The locomotor function of *Lesothosaurus* was tested by Bates et al. (2012b) and later Maidment et al. (2014a) using 3D musculoskeletal modeling of the hindlimbs. Their findings supported muscular reconstructions hypothesized by Maidment and Barrett (2011) in that the m. iliofemoralis complex maintains the early dinosaurian condition of showing greater leverage in hip abduction in lateral limb support. They also find that *Lesothosaurus* likely stood in a more upright posture and that the proximal placement of the fourth trochanter of the femur, for m. caudofemoralis attachment, allowed for quicker retraction of the femur when contracted during fast running (Bates et al., 2012b; Maidment et al., 2014a), a conclusion further supported by morphological analysis by Persons and Currie (2020). *Scutellosaurus* has also been considered as a bipedal cursor (or possibly faculta-

tively quadrupedal), although pelvic material is lacking to confirm this hypothesis (Colbert, 1981; Carrano, 1999; Maidment et al., 2014b).

Just as *Lesothosaurus* is a good example of an early bipedal ornithischians, *Scelidosaurus* is a good example of the transition to quadrupedality (Norman, 2020b). Although *Scelidosaurus* was larger than *Lesothosaurus*, it was still a relatively small-bodied animal compared to most eurypodan thyreophorans (i.e., stegosaurs and ankylosaurs) and likely a low-browsing animal. With its narrow beak, *Scelidosaurus* (and likely other early-diverging thyreophorans) probably selectively fed on low-growing plants like cycads, ferns, and horsetails (Norman, 2021). *Scelidosaurus* is covered in thick osteoderms (scutes) along its back and tail (Norman, 2020c)—a prerequisite to the osteoderms and spikes of stegosaurs and ankylosaurs.

The locomotor style of *Scelidosaurus* seems to be a "small-bodied intermediate" form between being cursorial and graviportal (Carrano, 1999; Maidment et al., 2014a, 2014b; Norman, 2020b, 2021). Its forelimbs are slightly shorter than its hindlimbs, and with detailed reconstruction and modeling of its locomotor musculature and center of mass estimates, *Scelidosaurus* has been deemed either a quadruped or a facultative quadruped (owing to its transitional anatomy) (Colbert, 1981; Maidment and Barrett, 2011, 2012; Maidment et al., 2014a, 2014b; Norman, 2021). Good indications of quadrupedality in *Scelidosaurus* are its transversely expanded ilium and a femur that is longer than its tibia (Maidment and Barrett, 2014; Maidment et al., 2014a; Barrett and Maidment, 2017).

Three-dimensional musculoskeletal modeling of *Scelidosaurus* pelvic and hindlimb musculature by Maidment et al. (2014a) showed some of the higher muscular moment arms, with flexion helped by the preacetabular process pulling flexor muscle vectors cranially, and extension helped by the postacetabular process pulling extensor muscle vectors caudally—an important adaptation in quadrupeds. *Scelidosaurus* also shows generally high moment arms in adduction at the hip joint; however, it shows middle to lower moment arms for hip abduction. Although *Scelidosaurus* shows signs of a transversely broadening ilium (often affiliated with quadrupedality), it is still not enough to create higher muscular moment arms in range for effective abduction, as seen in stegosaurs and ankylosaurs (Maidment and Barrett, 2012; Maidment et al., 2014a). Also, lateral rotators of the femur and the femoral retraction function of m. caudofemoralis show increased moment arms with increased hip extension, and medial rotation shows a lower moment arm. Generally, these muscular adaptations show a range of similarities between both bipedal and quadrupedal animals, speaking to the transitional nature of *Scelidosaurus* locomotor strategy (Maidment et al., 2014a).

CHAPTER 11

Eurypodans

Stegosauria

For reference in understanding relationships of eurypodans discussed in this chapter, see the phylogeny in figure 11.1.

Stegosauria includes huayangosaurids (e.g., *Huayangosaurus*, *Paranthodon*, and *Tuojiangosaurus*) and stegosaurids (e.g., *Stegosaurus*, *Kentrosaurus*, and *Miragaia*) (fig. 11.2). They ranged from the Middle Jurassic to Early Cretaceous with largely worldwide distribution (although with relatively less known diversity). (For reference in understanding clade affiliation of stegosaurs discussed in this chapter, see the taxon list in table 11.1.)

Table 11.1.
Stegosaurian genera organized by clade

Major clade	Subclade	Genus
Stegosauria		
	Huayangosauridae	
		Huayangosaurus
		Tuojiangosaurus
		Paranthodon
	Stegosauridae	
		Stegosaurus
		Kentrosaurus
		Miragaia

Figure 11.1. Phylogeny of Eurypoda, based on Sereno (1986).

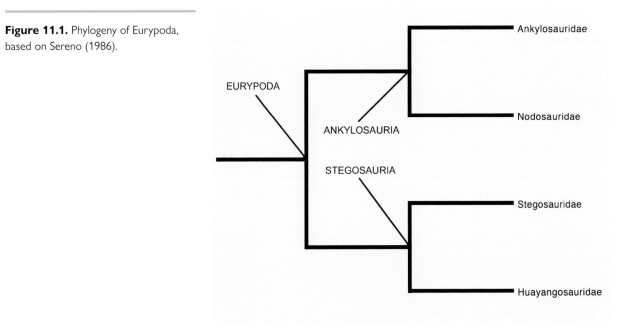

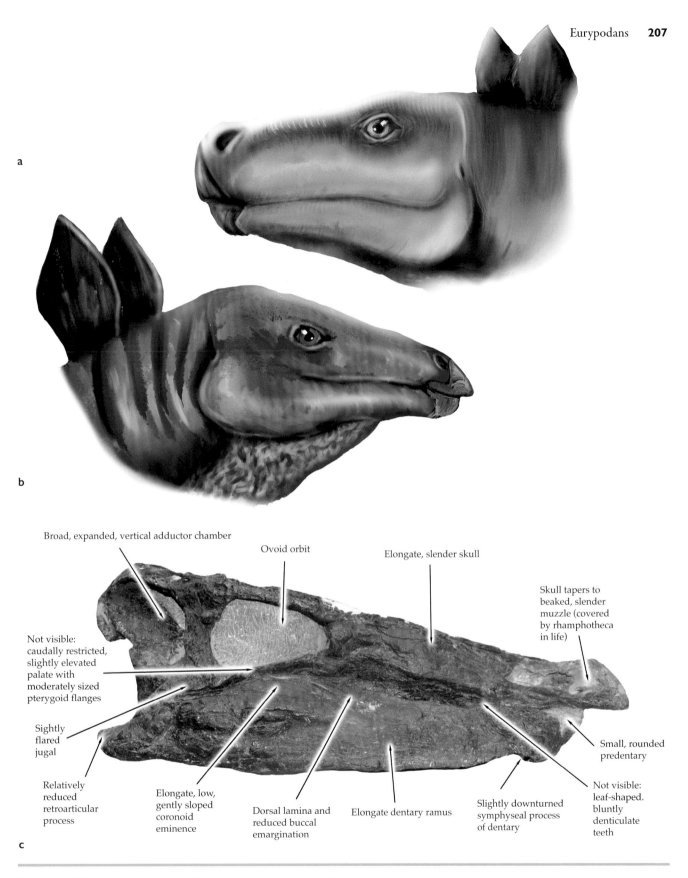

a

b

Broad, expanded, vertical adductor chamber

Ovoid orbit

Elongate, slender skull

Skull tapers to beaked, slender muzzle (covered by rhamphotheca in life)

Not visible: caudally restricted, slightly elevated palate with moderately sized pterygoid flanges

Sightly flared jugal

Relatively reduced retroarticular process

Elongate, low, gently sloped coronoid eminence

Dorsal lamina and reduced buccal emargination

Elongate dentary ramus

Slightly downturned symphyseal process of dentary

Small, rounded predentary

Not visible: leaf-shaped. bluntly denticulate teeth

c

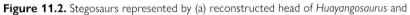

Figure 11.2. Stegosaurs represented by (a) reconstructed head of *Huayangosaurus* and (b) reconstructed head and (c) skull of *Stegosaurus* in lateral view (length approximately 46 cm).

Their skulls are miniscule compared to the size of their large postcrania. They are generally elongate and slender in lateral view as well, tapering to a beaked snout rostrally, where the ventral angle is abruptly heightened at the level of the external nares. The general shape of the muzzle is slender in the derived *Stegosaurus* compared to the more rectangular, broader muzzle of the more early-diverging *Huayangosaurus*. Dorsally, the rostrum of stegosaurs tapers to a point, signaling a more selective feeding habit (Nabavizadeh and Weishampel, 2016). Throughout stegosaur evolution, premaxillary dentition became reduced and eventually was lost in more derived stegosaurs such as *Stegosaurus*. The dentition of the lower jaw in stegosaurs became reduced toward the rostral end as well. Their teeth were leaf-shaped and bluntly denticulate. The orbits are of moderate size, and the antorbital fenestrae are present in the stegosaur *Huayangosaurus* and highly reduced to nonexistent in another stegosaur, *Stegosaurus*.

The temporal regions are somewhat variable in morphology, with some (e.g., *Stegosaurus*) possessing a transversely and rostrocaudally expanded temporal region and others (e.g., *Huayangosaurus*) whose temporal region is caudally restricted (still with transverse expansion). The supratemporal fenestrae in these taxa are small and circular. The postorbital adductor region and infratemporal fenestrae are subrectangular and vertically oriented, being relatively larger in *Stegosaurus* and somewhat smaller in *Huayangosaurus*. All stegosaurs present with the typical ornithischian slight flaring of the jugal, although it is more prominently flared in *Huayangosaurus*. The palate is caudally restricted and slightly elevated, with moderately sized pterygoid flanges.

The mandible possesses a small, rounded predentary resting against slightly downward curved symphyseal processes of the rostral dentaries, categorized as predentary-dentary joint (PDJ) Morphotype II (Nabavizadeh and Weishampel, 2016). The dentaries are elongate, with *Huayangosaurus* possessing teeth completely down the jawline, while *Stegosaurus* has a diastema (or gap) between the predentary and the mesial-most tooth. This reduction of the tooth series in derived stegosaurs

coincides with the development of a dorsally projecting eminence, or dorsal lamina, in the coronoid region that became more prominent, hiding the dentary dentition laterally at the distal end and projecting along the lateral edge of the tooth row. This could have acted as a mode of keeping vegetation in the oral cavity. A dorsally projecting ridge along the medial side of the tooth row rostral to the dentition served as a cropping mechanism that acted as a continuation of the dentition from its caudal end (Berman and McIntosh, 1986; Czerkas and Gillette, 1999; Barrett, 2001).

The tooth row in *Stegosaurus* is also slightly bowed medially. The coronoid eminence in *Stegosaurus* is elongate, low, and gently sloped, in contrast to the more triangular and caudally displaced coronoid eminence in *Huayangosaurus*. A well-developed buccal emargination can be seen extending rostrolabially (lateral to the distal dentition) in *Huayangosaurus* and *Paranthodon*, but it is much more reduced in the derived *Kentrosaurus* and *Stegosaurus*. This buccal emargination in *Huayangosaurus* may correspond to a rostrally extended adductor muscle attachment (Nabavizadeh, 2020a, 2020b; see below). The mandibular fenestra is larger in *Stegosaurus* than in *Huayangosaurus*. In all stegosaurs, the retroarticular process is reduced in size.

Among stegosaurs, *Huayangosaurus* (Sereno and Zhimin, 1992) bears seven roughly conical premaxillary teeth just caudal to the rostral edge of the premaxilla in lateral view. At occlusion, they rest against the lateral rim of the rhamphotheca of the predentary, overhanging it. This rostral edge was likely used for puncturing vegetation during food acquisition. In contrast, in all other known stegosaur taxa, the premaxilla is edentulous and rested against the dorsal rim of the predentary at occlusion. The maxillary and dentary teeth are generally leaf-shaped with ridges that extend apicobasally along its buccal and lingual surfaces (fig. 11.3). They are buccolingually expanded and bulbous at the cingulum (especially in *Paranthodon*; Galton and Coombs, 1981), with thin, cylindrical roots at the level of the alveoli. The apicobasal ridges create a variable number of blunted denticles (depending on the taxon) at the apical ridge of each tooth.

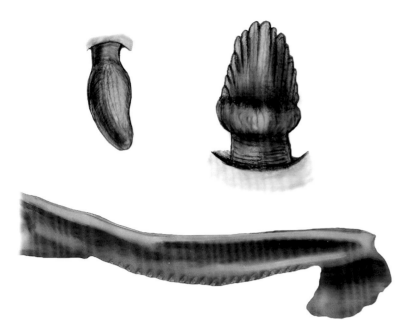

Figure 11.3. Stegosaur dentition: (*clockwise from top left*) premaxillary tooth of *Huayangosaurus*, dentary tooth of *Stegosaurus*, and tooth row (dorsal view) in *Stegosaurus*.

These denticles are much blunter and less pointed than those seen in ankylosaur teeth (see below). The tooth row consists of 20 to 23 tightly packed teeth in *Stegosaurus*, whereas in the earlier-diverging *Huayangosaurus* (Sereno and Zhimin, 1992), the tooth row has roughly 28 teeth (Sereno and Zhimin, 1992), meaning the tooth row shortened evolutionarily.

Unfortunately, stegosaur skulls are rare, like many early-diverging ornithischians, and the only partial to relatively complete skulls found of more derived stegosaurian taxa (not including *Huayangosaurus*) belong to *Kentrosaurus*, *Paranthodon* (a maxilla), *Tuojiangosaurus*, and *Stegosaurus*, with *Stegosaurus* being the only derived form known from complete lower jaw material. As such, the availability of stegosaur dentitions is therefore also limited, limiting the chances of analyzing true wear direction in stegosaurs. Dental wear patterns therefore had not been well known in stegosaurs until recently, as isolated stegosaur teeth are rare, and most teeth known are in situ and tightly packed to the point where wear is not visible.

With their elongate, narrow snouts, akinetic skulls, leaf-shaped denticulate teeth (similar to that of ankylosaurs, early-diverging ornithopods, and early-diverging marginocephalians), and pointed beaks with a predentary shaped much like that of the early-diverging thyreophoran *Lesothosaurus*, stegosaurs have long been thought to have simply used an orthal feeding mechanism (Weishampel and Norman, 1989; Barrett, 2001; Nabavizadeh, 2016; Button and Zanno, 2020). Barrett (2001) noted that the overall flattened shape of the teeth as well as slight macrowear on the apical rim of the teeth indeed suggest an orthal feeding mechanism. Barrett (2001) also noted that the jaw joint in stegosaurs is ventrally offset relative to the tooth row, indicating an herbivorous lifestyle with the maxillary and dentary tooth rows occluding simultaneously. He describes the rostrocaudal length of the slightly concave glenoid as about equal to the rostrocaudal breadth of the ventral condyles of the quadrate that articulates with it, which he indicates would have prohibited any fore-aft propalinal movement of the lower jaw and in turn would have constrained stegosaurs to a strictly orthal feeding mechanism. Nabavizadeh and Weishampel (2016), in further inspection of other stegosaur specimens, noted that the rostrocaudal breadth of the glenoid was actually slightly broader rostrocaudally than the ventral quadrate. They hypothesized that there might have been a slight palinal component to stegosaur feeding, in addition to orthal motions as well as slight hemi-mandibular rotation against the predentary (supported by PDJ morphology and slight medial curvature of the tooth rows). Woodruff et al. (2019) and Skutschas et al. (2021) confirmed this mechanism with stegosaur dental microwear analyses showing orthal and slight palinal feeding motions along with puncturing motions, like a mortar-and-pestle jaw action, providing evidence of generally more complicated feeding mechanisms as a whole.

Cranial muscle anatomy in stegosaurs has been described by Holliday (2009), Lautenschlager et al. (2016), and Nabavizadeh (2016, 2020a, 2020b). Palatal musculature (m. pterygoideus, or mPT) in stegosaurs was moderately sized overall, both in palatal origin and insertion along the retroarticular process, both articular and angular. This muscle body would have added strength in jaw closure, but this strength would have been secondary to that of temporal musculature in these animals. Comparisons of temporal muscle reconstructions between huayangosaurids (e.g., *Huayangosaurus*; fig. 11.4) and stegosaurids (e.g., *Stegosaurus*; fig. 11.5) present a good example of how slight morphological variations of the skull can affect the overall function of the feeding apparatus. For instance, the smaller-skulled yet broader-

muzzled *Huayangosaurus* would have had a rostrocaudally shortened yet transversely broad, temporal m. adductor mandibulae externus (mAME) origin. The m. adductor mandibulae externus superficialis (mAMES) would have fanned out rostroventrally and, with room given by a clear, laterally flaring jugal, would have inserted along the labial dentary ridge as well as the lateral aspect of the surangular and triangular coronoid eminence. This expansion of muscle would have increased mechanical advantage of the mAMES, possibly aiding in increased palinal motion and slight hemi-mandibular long-axis rotation against the predentary (Nabavizadeh, 2020a, 2020b), although tooth wear is not yet known in *Huayangosaurus* to confirm this.

In *Stegosaurus*, although the origins and insertions were mostly the same as in *Huayangosaurus*, the overall architecture was slightly different. The temporal mAME origin was both rostrocaudally expanded as well as transversely broadened (creating a relatively larger muscle body origin) with a more vertically oriented adductor chamber and less prominently flared jugal that would not have allowed as much rostrolabial expansion of mAMES as that in *Huayangosaurus*. The mAME would have in-

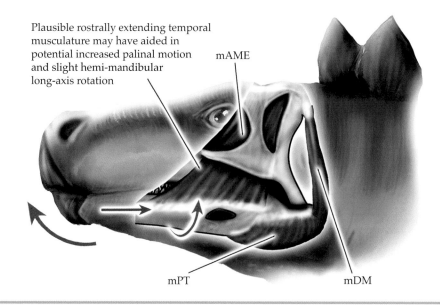

Plausible rostrally extending temporal musculature may have aided in potential increased palinal motion and slight hemi-mandibular long-axis rotation

mAME

mPT

mDM

Figure 11.4. *Huayangosaurus* head showing feeding function with jaw muscles. Red arrows show various jaw actions. Abbreviations: mAME, m. adductor mandibulae externus; mDM, m. depressor mandibulae; mPT, m. pterygoideus.

serted around the low coronoid eminence, with mAMES mostly being restricted on the lateral aspect of the surangular. Although slight palinal feeding motions in stegosaurids have been suggested by tooth wear (Woodruff et al., 2019) and osteological examination (Nabavizadeh and Weishampel, 2016), isognathous orthal feeding with slight hemi-mandibular long-axis rotation against the predentary was a major component of their overall feeding mechanisms, and the muscular anatomy described above would have been well suited for these jaw motions (Barrett, 2001; Galton and Upchurch, 2004; Nabavizadeh, 2016, 2020a, 2020b; Nabavizadeh and Weishampel, 2016; Woodruff et al., 2019).

Nabavizadeh (2016) found that mandibular mechanical advantage of stegosaurs varied with jaw shape as well, specifically when looking at the mechanical advantage of temporal musculature. The broad-muzzled *Huayangosaurus* possessed a relatively longer tooth row that stretched the entire length of the oral margin, extending sufficiently distally to create a greater distal mechanical advantage. The tooth row of *Stegosaurus*, however, showed higher mechanical advantage at the mesial end of the tooth row in their elongated snout, which is likely

a consequence of its rostrally displaced, shorter tooth row relative to that of *Huayangosaurus*. The muscular line of action in *Stegosaurus* is decreased in angle compared to that of *Huayangosaurus* because of the more rostrally displaced apex of its coronoid eminence. This would have assisted in improving leverage in its bite by creating a larger moment arm (Nabavizadeh, 2016).

The first time finite element analysis (FEA) was used in a stegosaur was in a study by Reichel (2010b). With digital three-dimensional (3D) reconstruction of a *Stegosaurus* tooth model and applied metrics of cranial anatomy overall, bite forces calculated in stegosaur teeth ranged from 140.1 N (mesially), 183.7 N (at the middle tooth position), and 275 N (distally). Reichel (2010b) explains that the homodont, denticulate nature of stegosaur teeth would most likely have helped dissipate stresses in biting smaller branches and leaves, although there was not much of a difference in distributions of force between serrated and unserrated tooth models. Reichel (2010b) also suggested that *Stegosaurus* likely was not using its full bite force all of the time during feeding, possibly because of greater use of the beak in foraging behavior.

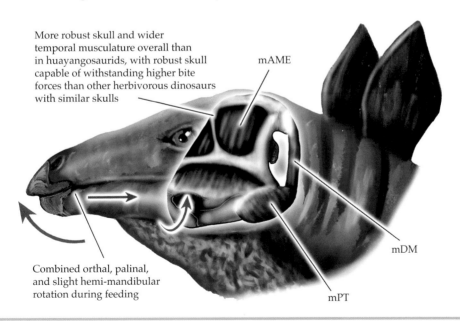

Figure 11.5. *Stegosaurus* head showing feeding function with jaw muscles. Red arrows show various jaw actions. Abbreviations: mAME, m. adductor mandibulae externus; mDM, m. depressor mandibulae; mPT, m. pterygoideus.

Lautenschlager et al. (2016), in a study using both FEA and multibody dynamics analysis (MDA) on skulls scanned by computed tomography (CT), found that the skull of *Stegosaurus* was better built for more powerful feeding compared to that of the therizinosaur *Erlikosaurus* and the "prosauropod" *Plateosaurus* (both of which possessed similarly gracile-looking skull morphotypes). They found bite forces similar to those of Reichel (2010b), ranging from 166 to 321 N across the tooth row in their scaled model. Cranial stress distributions also show an interesting pattern, with stresses primarily being focused on the rostral and antorbital region as well as throughout the jaw (especially the postdentary and jaw joint region). These findings corroborate with Nabavizadeh (2016) in finding that bite forces were likely focused more mesially relative to other herbivorous dinosaurs. It also stands to reason that the skull is suited for stresses in the antorbital region, as *Stegosaurus* lacks antorbital fenestrae, a characteristic that in turn adds more bone to that region for stresses to transmit. The addition of a modeled keratinous beak also shows mitigation of stresses to the bones themselves, although distribution of stresses is still similar.

Stegosaurs were large, quadrupedal, and graviportal animals, with short forelimbs compared to hindlimb length (fig. 11.6). Because most stegosaurs usually did not possess an extremely long, sauropod-like neck that would aid in reaching plant material that was higher off the ground (except for the anomalous *Miragaia*, which unusually possessed 17 cervical vertebrae), they were likely browsers on lower-growing vegetation. Some have suggested stegosaurs may have been capable of standing up on their hindlimbs with their tail as a support to reach up for higher vegetation in a tripodal posture (Bakker, 1978; Weishampel and Norman, 1989). Their necks, backs, and elongate, horizontal tails were adorned with two parallel, longitudinally arranged rows of plates and spikes of various shapes, sizes, and patterns, depending on the genus. Stegosaurs had unusually small heads for the size of their bodies, hardly reaching 5% of its total body length. Because of this and the unsophisticated feeding apparatus compared to other ornithischians,

much of the processing and fermentation of vegetation among stegosaurs likely occurred in a complex and extended gut system, implicated by an enlarged abdominal region (Weishampel and Norman, 1989; Button and Zanno, 2020). The body mass of the largest stegosaur, *Stegosaurus*, has been estimated at around 1,560 kg with the use of volumetric computer modeling (Brassey et al., 2015), and the center of mass was relatively more caudal than in other ornithischians, being at the level of the hindlimb close to the hip, possibly allowing for the previously hypothesized "tripodal" feeding posture (on hindlimbs and supported by the tail; see above) (Maidment et al., 2014b; Mallison, 2014).

Stegosaurs show multiple adaptations in the limbs that make them obligate quadrupeds. Their humerus shows an elongate deltopectoral crest, and the acromial process of the scapula is oriented caudally, creating an m. deltoideus vector less mechanically advantageous in retraction and abduction of the forelimb than in bipedal, early-diverging ornithischians. Instead, m. deltoideus may have acted in lateral rotation of the humerus to rotate the elbow toward the body for better control in quadrupedal motion (Maidment and Barrett, 2011, 2012; Barrett and Maidment, 2017). Stegosaurs also possessed transversely expanded ilia, which create a larger moment arm for hip abductor musculature (Maidment and Barrett, 2011, 2012, 2014; Maidment et al., 2014a).

Three-dimensional musculoskeletal modeling of *Kentrosaurus* pelvic and hindlimb musculature by Maidment et al. (2014a) showed improved adaptations for obligate quadrupedality. Among the ornithischians examined, *Kentrosaurus* showed some of the higher muscular moment arms across the range of flexion and extension angles of the hip joint, as in *Scelidosaurus*. Unlike *Scelidosaurus*, however, it showed among the largest hip abduction moment arms at the expense of adductor moments, owing to the lateral expansion of the trunk and pelvis. Although it showed higher relative moment arms for hip extension (retraction of the hindlimb), the fourth trochanter in stegosaurs is reduced in size, possibly indicating at least a partial decoupling of m. caudofemoralis from the femur to focus the muscle control more within

Figure 11.6. Skeleton of *Stegosaurus*.

the tail itself. This may have allowed the animal to swing its spiked tail at predators for self-defense (also supported by biomechanical analysis of the tail of *Kentrosaurus* by Mallison, 2011b) or even could have been used in intraspecific combat (Maidment et al., 2014a). Variation likely occurred in stegosaurian locomotion as well. For instance, the larger-bodied *Stegosaurus* shows a larger m. caudofemoralis moment arm than that seen in *Kentrosaurus*, for more power hindlimb extension (Brassey et al., 2015).

As stated, stegosaurs likely fed on lower-growing vegetation (about a meter off the ground; Weishampel and Norman, 1989)—most likely ferns, cycads, and horsetails, among other similarly low-growing plants. Unfortunately, for the most part, we do not have much evidence pertaining to the specific plant species stegosaurs might have eaten. One particular animal, *Isaberrysaura* (which has recently been placed within Stegosauria; Han et al., 2018), has been found with permineralized seeds in the gut region, the larger ones thought to be in the cycad family. The ecological connection between

stegosaurs and cycads specifically has been discussed previously, and although their diversification is positively correlated with each other (Butler et al., 2009a), there are no significant spaciotemporal patterns between the two (Butler et al., 2009b). It is safe to say that stegosaurs were not as specialized as previously thought and ate a broader range of plant life, although they still were likely selective eaters with their narrow snouts (Nabavizadeh and Weishampel, 2016).

Ankylosauria

Ankylosaurs are mostly divided into two main armored clades: the non-club-tailed and large-spiked Nodosauridae (e.g., *Panoplosaurus* [fig. 11.7], *Edmontonia*, *Silvisaurus*, *Gargoyleosaurus*, *Tatankacephalus*, *Pawpawsaurus*, *Struthiosaurus*, *Sauropelta*, and *Hungarosaurus*) and the largely club-tailed Ankylosauridae (e.g., *Ankylosaurus*, *Euoplocephalus* [fig. 11.8], *Pinacosaurus*, *Saichainia*, *Minmi* [possibly], *Crichtonsaurus*, and *Shamosaurus*).

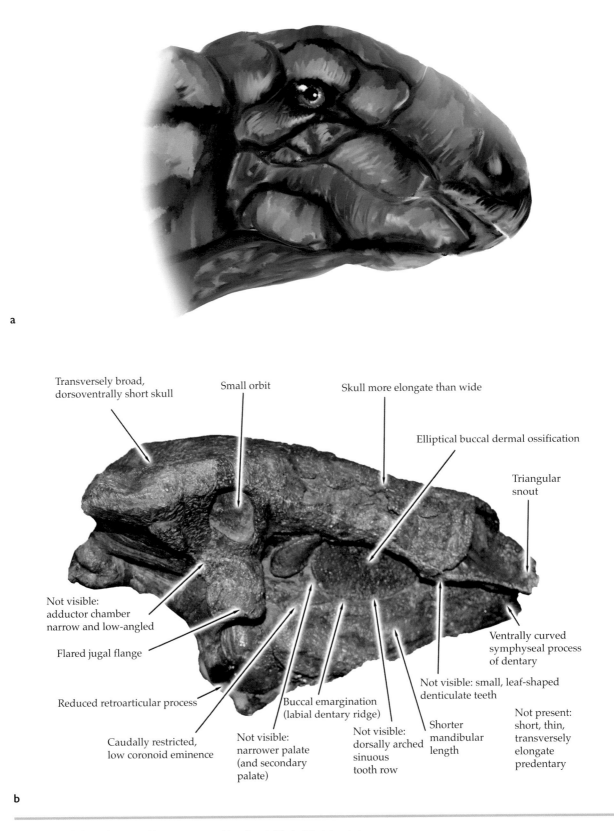

Figure 11.7. *Panoplosaurus* (a) reconstructed head and (b) skull in lateral view (length approximately 36 cm).

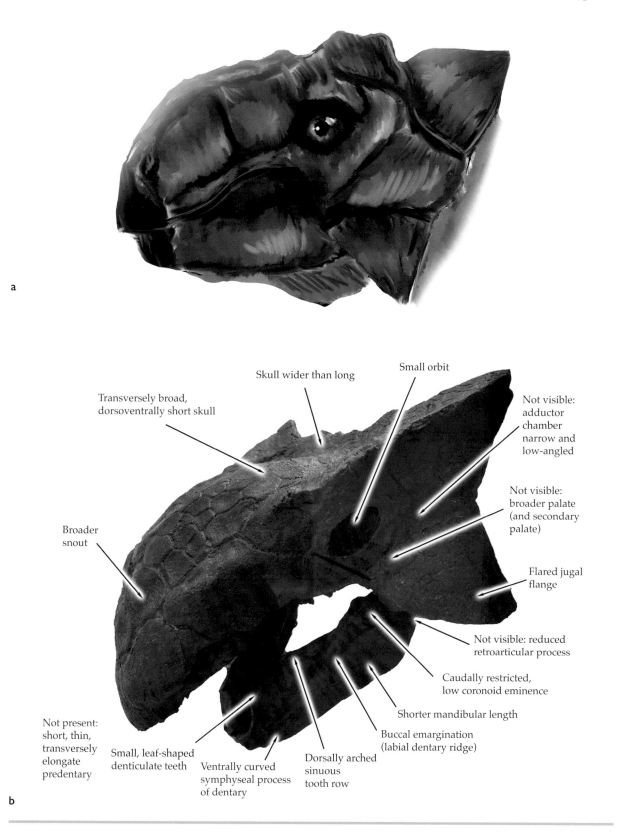

a

b

Figure 11.8. Ankylosaurids represented by (a) reconstructed head of *Euoplocephalus* and (b) skull of *Ankylosaurus* in lateral view (length reaching up to approximately 55 to 67 cm).

They ranged from the Middle Jurassic to Late Cretaceous with largely worldwide distribution. (For reference in understanding clade affiliation of ankylosaurs discussed in this chapter, see the taxon list in table. 11.2.)

Ankylosaur skulls are generally described as transversely broad and dorsoventrally short (Coombs, 1978). In dorsal view, the skull is approximately triangular, with a much broader rostral end and occiput in ankylosaurids than in nodosaurids, the latter typically (but not always) having much narrower skulls with more pointed snouts than the former. Ankylosaurid skulls are transversely wider than they are rostrocaudally long, whereas the opposite is seen in nodosaurids skulls (Coombs, 1971, 1978). The skull is rostrocaudally short and squared-off laterally in earlier-diverging ankylosaurids, such as *Minmi* (Molnar, 1996), *Crichtonsaurus* (Dong, 2002), and *Shamosaurus* (Tumanova, 1983, 1985), but becomes more elongate in more derived genera. Ankylosaur temporal

regions are both rostrocaudally and transversely broad. Uniquely, the supratemporal fenestrae in ankylosaurs is not visible owing to the robust surrounding dermal ossifications of various shapes and sizes tightly on the dorsal surface of the skull (Coombs, 1978). The postorbital adductor chamber is rostrocaudally narrow and rotated caudally at a lower angle in relation to the horizontal plane of the skull. In ankylosaurids, an ossified plate surrounding the quadratojugal region hides each infratemporal fenestra and quadrate in lateral view as well; however, in some nodosaurids, the infratemporal fenestrae are still exposed to some extent as caudally slanted, slender rectangles (Vickaryous et al., 2001; Vickaryous and Russell, 2003; Vickaryous, 2006). They possess a laterally flared jugal flange (which is accentuated by rugose dermal armor) and corresponding subjugal opening with the infolded caudal end of the maxilla.

Ventrally, ankylosaurids have a broad palate, while nodosaurid palates are much narrower owing to their slimmer skulls. In nodosaurids, a secondary palate (when present) has been described as a single sheet of bone running horizontally between the medially bowed maxillary tooth row, and in *Sauropelta* it is unossified (Coombs, 1978). In ankylosaurids, the secondary palate is more elaborate, made of two horizontal sheets of bone between the maxillary tooth rows. The resulting nasal passages in ankylosaurids, with several sinuses intact, are more complex and sinuous than in nodosaurids (Coombs, 1978). Vickaryous (2006) showed that the nodosaurids do show paranasal sinuses, however, and Witmer and Ridgely (2008) and Bourke et al. (2018), with the help of CT scanning, found more complex, looping nasal passages in nodosaurids than were previously known.

The mandible in ankylosaurs is uniquely shaped and quite short rostrocaudally. The predentary is small, thin, and transversely elongate with small dorsal and ventral processes that loosely articulate with the rostral ends of the dentaries, which possess a ventrally curved symphyseal process (categorized as PDJ Morphotype II; Nabavizadeh and Weishampel, 2016). The dentaries hold a dorsally arched and sinuous tooth row with small leaf-shaped, denticulate teeth (fig. 11.9). This tooth row

Table 11.2.
Ankylosaurian genera organized by clade

Major clade	Subclade	Genus
Ankylosauria		
	Ankylosauridae	
		Ankylosaurus
		Euoplocephalus
		Pinacosaurus
		Tarchia
		Saichania
		Minmi (?)
	Nodosauridae	
		Edmontonia
		Gastonia
		Sauropelta
		Panoplosaurus
		Silvisaurus
		Gargoyleosaurus
		Tatankacephalus
		Pawpawsaurus
		Struthiosaurus
		Hungarosaurus

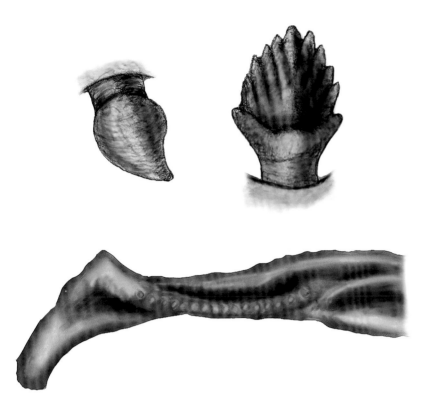

Figure 11.9. Ankylosaur dentition: (*clockwise from top left*) premaxillary tooth, dentary tooth, and curved tooth row in dorsal view.

in many ankylosaurs forms a large, sigmoidal curve in lateral view while at the same time bowing medially in dorsal view, with the maxillary tooth row likewise bowing medially in ventral view (Barrett, 2001).

The coronoid eminences in ankylosaurs are low and slightly rounded to triangular and caudally restricted with a rostral extension, creating a labial dentary ridge lateral to the dentition forming a buccal emargination. In such taxa as *Panoplosaurus* and *Edmontonia*, this buccal emargination is surmounted by an elliptical dermal ossification, likely indicating the presence of dermal soft tissue in this region (possibly functioning as a dermal "cheek" helping keep vegetation in the oral cavity) (Galton, 1973; Coombs, 1978; Barrett, 2001; Vickaryous, 2006; Nabavizadeh, 2020a). There is no mandibular fenestra, and the retroarticular process is reduced in size.

The quadrate-glenoid jaw joint is ventrally offset from the level of the maxillary tooth row, as in most ornithischians. The mandibular glenoid is variable in shape. According to Barrett (2001), in forms such as *Sauropelta* and *Panoplosaurus*, the glenoid fossa is short-ened rostrocaudally and bounded by heightened ridges, just large enough to fit the ventral cotylus of the quadrate as a mostly hinge-like joint (fig. 11.10). In *Euoplocephalus*, however, the glenoid is more rostrocaudally lengthened, with no clear ridge that indicates boundary constraints of the quadrate and instead implies a more freely mobile joint (Barrett, 2001; Rybczynski and Vickaryous, 2001).

Premaxillary dentition is found only in the nodosaurids *Silvisaurus* (Eaton, 1960), *Gargoyleosaurus* (Kilbourne and Carpenter, 2005), *Tatankacephalus* (Parsons and Parsons, 2009), *Pawpawsaurus* (Lee, 1996), *Struthiosaurus* (Nopcsa, 1928), and *Sauropelta* (Coombs, 1978) and is typically peg-like to conical with a flattened edge, although seemingly much stubbier than in stegosaurs. Present at the tip of the snout are eight or nine premaxillary teeth, which flare laterally along the lateral edge before following the caudomedial ridge that connects with the maxillary tooth row. Most ankylosaurids lack premaxillary teeth (Barrett, 2001; Carpenter, 2004).

Both major groups of ankylosaurs had maxillary and

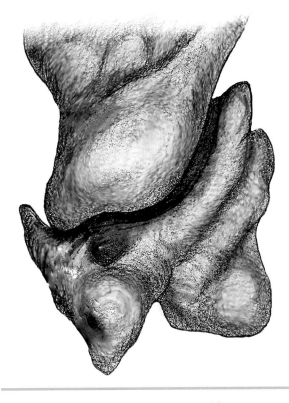

Figure 11.10. *Panoplosaurus* jaw joint in caudal view.

dentary teeth that were likely only used minimally, as indicated by the limited wear seen on isolated teeth (Barrett, 2001; Rybczynski and Vickaryous, 2001). For the most part, ankylosaur teeth are similar in shape to stegosaur teeth, although not entirely (fig. 11.9). Like stegosaurs, they are generally laterally compressed and leaf-shaped with ridges that extend apicobasally along its buccal and lingual surfaces. They are buccolingually expanded and bulbous at the base (cingulum) with thin, cylindrical roots at the level of the alveoli. Unlike stegosaurs, however, the apicobasal ridges create sharp, prominent denticles at the mesial and distal apical ridges of each tooth, rather than the more blunted denticles seen in stegosaurs. Typically, no more than two or three replacement teeth can be seen beneath each alveolus, and the teeth do not interlock. In ankylosaurids, the crown is small relative to the length of the roots, whereas in nodosaurids, the crown is much larger. Although all ankylosaur teeth are swollen at the base, nodosaurids typically have a more distinct cingulum. Variable wear facets

have been described from the small sample of ankylosaur teeth known, some with high-angle mesial and distal wear facets (similar to *Lesothosaurus* [see above], although this is rare in ankylosaurs) and others with large, planar wear facets covering the entire crown (similar to *Scelidosaurus*; see above) or just at the apex (Barrett, 2001; Rybczynski and Vickaryous, 2001; Mallon and Anderson, 2014b; Ősi et al., 2014, 2017).

Ankylosaurs, like stegosaurs, also possessed small heads for their body size; however, ankylosaurs had much more broadened skulls of various shapes and sizes. Ankylosaur skulls are generally dorsoventrally compressed. Fusion of many intracranial elements along with fusion of plates of dermal bone surrounding the skull would have prevented any form of cranial kinesis during feeding. The skulls of nodosaurids, particularly *Edmontonia*, *Panoplosaurus*, *Silvisaurus*, and *Gastonia*, were triangular in dorsal view and came to more of a point rostrally, a morphology likely adapted for more selective feeding. Alternatively, ankylosaurids such as *Euoplocephalus* and *Ankylosaurus* had skulls that are much squarer and had broad snouts, likely adapted for more generalized feeding behavior (Coombs, 1978; Mallon and Anderson, 2013, 2014a).

To test degrees of selectivity in feeding, Mallon and Anderson (2014a) performed a quantitative morphometric analysis of the shapes of muzzles and beaks of ornithischian dinosaurs from the Dinosaur Park Formation of Alberta, Canada, including those of the nodosaurid *Panoplosaurus* (fig. 11.11) and the ankylosaurid *Euoplocephalus* as well as various hadrosaurids and ceratopsids. Concurrent with similar beak shape studies in herbivorous mammals (e.g., Solounias et al., 1988; Solounias and Moelleken, 1992, 1993), Mallon and Anderson (2014a) used metrics that captured the shape of the beak from dorsal view to see how broad or narrow and of what shape the beak would have been. They found that the broad muzzle of the ankylosaurid *Euoplocephalus* would have implied it was more of a bulk feeder and that the less broad yet still rounded beak of the nodosaurid *Panoplosaurus* implied that it could have been more of a mixed feeder (still more selective than *Euoplo-*

cephalus). Another example of variable feeding style in ankylosaurs of a given ecosystem is that of two Mongolian ankylosaurids: the more generalist-feeding, broader-muzzled *Talarurus* and the comparatively more selective, rounder-muzzled *Tsagantegia* (Park et al., 2020). In general, ankylosaurids and nodosaurids show numerous characteristics that are typical of herbivorous dinosaurs and have shown similar trends to other clades, speaking to the importance of these characters for more complex oral processing (Button and Zanno, 2020).

Ankylosaur jaw mechanisms have been studied for a long time. Nopcsa (1928) suggested the possibility that ankylosaurs were insectivorous, based on the weak nature of their jaws. It is now generally accepted that they were in fact herbivorous (Haas, 1969; Coombs, 1971; Galton, 1986; Weishampel and Norman, 1989; Barrett, 2001; Rybczynski and Vickaryous, 2001). Russell (1940) and Haas (1969) described possible movements of jaws in ankylosaurs, suggesting that the lower jaw must have adducted lateral to the maxillary tooth row. Haas (1969) indicated that the jaw musculature was too weak for a powerful bite force and suggested that ankylosaurs likely fed on soft vegetation. Macrowear and qualitative studies of different bone and joint morphologies have previously suggested an orthal component in jaw mechanisms of many ankylosaurs (Galton, 1986; Weishampel and Norman, 1989), with a possibility of a puncture-crushing mechanism in some, similar to that in *Scelido-*

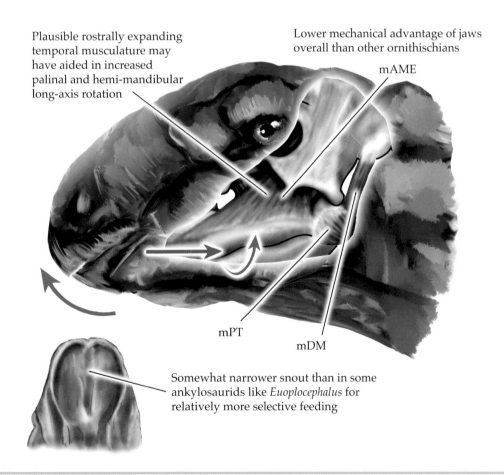

Figure 11.11. *Panoplosaurus* head showing feeding function with jaw muscles, with ventral view of the skull's snout on bottom left. Red arrows show various jaw actions. Abbreviations: mAME, m. adductor mandibulae externus; mDM, m. depressor

saurus (Barrett, 2001). But interlocking tooth occlusion (like that seen in *Lesothosaurus*) was not possible in ankylosaurs (Barrett, 2001). Coombs (1971) suggested that although dental microwear seemed to suggest otherwise, a strictly orthal jaw mechanism was unlikely to be a result of craniomandibular morphology as well as size and orientation of the inferred jaw adductor musculature. Instead, he suggested that there was a propalinal movement of the jaw during occlusion that coincided with an orthal movement during the power stroke. Moderate transverse motion of the jaw was also suggested owing to the unusual shape of the tooth row (Coombs, 1971). Coombs and Maryańska (1990) described wear facets in ankylosaur teeth that indicated a propalinal component, further validating this inference.

Later, an analysis of the craniomandibular complex of the ankylosaurid *Euoplocephalus* (fig. 11.12) by Rybczynski and Vickaryous (2001) suggested palinal motion coupled with a medial "pivoting" of each mandibular corpus against the quadrate at the jaw joint on either side during the power stroke (also suggested by Coombs, 1971). Evidence for this motion lies in the rostrocaudally and somewhat transversely widened, slightly concave expansions of the glenoid articular surfaces, providing a large range of movement. Rybczynski and Vickaryous (2001) described two regions of quadratic stability at the jaw joint that are at an angle, one slightly elevated rostrolateral region and one transversely expanded caudomedial region (which is at more of an oblique angle relative to the long axis of the tooth row). This jaw mechanism was also characterized by a highly mobile predentary-dentary and dentary-dentary symphyseal joint because of the peculiar morphology and absence of a firm, clasping junction between the predentary and dentary. But the likelihood of any sort of rotation of the mandibular corpora around their long axes was precluded in this study, although a clear explanation was not given as to why this might be so (Rybczynski and Vickaryous, 2001).

Nabavizadeh (2011, 2016), Ősi et al. (2014, 2017), and Nabavizadeh and Weishampel (2016) presented a slightly different mechanism, suggesting that, along with orthal and palinal motions of the jaw, ankylosaurs would have incorporated hemi-mandibular long-axis rotation against the predentary, with reference to the exaggerated curvature of the tooth row and the loose articular morphology between the predentary and rostral end of the dentaries. Nabavizadeh and Weishampel (2016) noted that *Euoplocephalus* would have likely had an even greater range of motion at the predentary joint than other ankylosaurids (like *Pinacosaurus*) and nodosaurids. This is because of the greater degree of sinusoidal tooth row curvature in *Euoplocephalus* and the fact that its predentary is more transversely straight and does not have the more restricting caudolateral inflections on either side that give the predentary more of a bracket shape in other taxa. Nabavizadeh and Weishampel (2016) also noted that the ventrally curved symphyseal process and the rostrocaudally expanded dentary symphysis would have allowed an even greater range of motion at the predentary joint in most ankylosaurs.

Dental microwear combined with morphological analyses of jaws have given even greater insights into the feeding mechanisms of ankylosaurs. Ősi et al. (2014), looking further in depth at the dentary and curved tooth row morphology as well as extensive dental microwear of the nodosaurid *Hungarosaurus*, found that it would have incorporated orthal and palinal feeding motions with hemi-mandibular rotation against the predentary; albeit a predentary was missing in the specimen they examined. Furthermore, they proposed that lateral excursion of the dentaries existed in tandem with long-axis rotation. Further analysis of dental wear morphology and microwear by Mallon and Anderson (2014b) found palinal along with orthal feeding strokes in both the ankylosaurid *Euoplocephalus* and the nodosaurid *Panoplosaurus* and that dietary overlap was possible between the two taxa, given the similarities in their microwear textures. Mallon and Anderson (2014b) noted, however, that given their overall morphology and size, ankylosaurid teeth were better suited for eating fruits and softer food items, contrary to nodosaurid teeth, which they suggest were better suited for a foliage-rich diet.

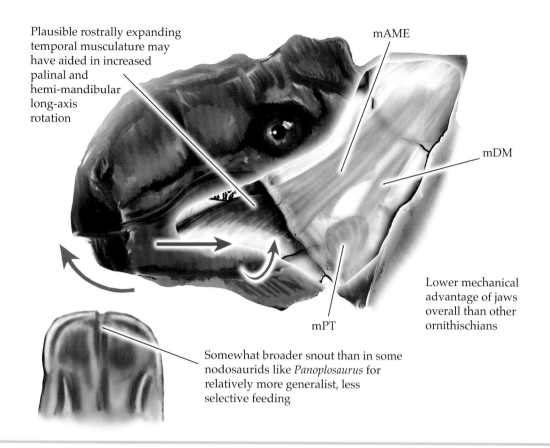

Plausible rostrally expanding temporal musculature may have aided in increased palinal and hemi-mandibular long-axis rotation

mAME

mDM

mPT

Lower mechanical advantage of jaws overall than other ornithischians

Somewhat broader snout than in some nodosaurids like *Panoplosaurus* for relatively more generalist, less selective feeding

Figure 11.12. *Euoplocephalus* head showing feeding function with jaw muscles, with ventral view of the skull's snout on bottom left. Red arrows show various jaw actions. Abbreviations: mAME, m. adductor mandibulae externus; mDM, m. depressor mandibulae; mPT, m. pterygoideus.

Ősi et al. (2017) investigated the evolution of feeding adaptations across Ankylosauria as a whole, examining osteology, tooth wear, and muscular anatomy. Narrower-snouted early-diverging ankylosaurs such as *Kunburrasaurus* and the early-diverging nodosaurid *Gargoyleosaurus* were limited to simple orthal occlusion, and the few cases of tooth-tooth contact (as in *Gargoyleosaurus*) were incidental. Early Cretaceous nodosaurids then developed more pronounced muscle attachment sites (both for temporal and palatal musculature) and more variable muzzle shapes. The early nodosaurid *Sauropelta* shows the first signs of dental occlusion and possible biphasic jaw motion (with implementation of palinal movement of the lower jaw in addition to the initial orthal motion). Independently, many nodosaurids and ankylosaurids had more precise occlusion across the tooth row with a more

powerful shearing bite. Furthermore, they indeed found that many nodosaurids (e.g., *Sauropelta*, *Edmontonia*, *Panoplosaurus*, and *Hungarosaurus*) and some ankylosaurids (e.g., *Euoplocephalus* and *Ankylosaurus*) independently evolved a biphasic feeding mechanism, with an orthal and palinal feeding stroke that also incorporated hemi-mandibular long-axis rotation against the predentary (as noted above) (Rybczynski and Vickaryous, 2001; Ősi et al., 2014, 2017; Nabavizadeh and Weishampel, 2016). Asian ankylosaurid taxa were thought to have generally been orthal pulpers without tooth occlusion (Ősi et al., 2017); however, a recent study showed that the early-diverging Asian ankylosaurid *Jinunpelta* also shows biphasic wear patterns in their teeth, providing further evidence of a global distribution of this feeding mechanism across Ankylosauria (Kubo et al., 2021).

Cranial musculature in ankylosaurs has been described in numerous studies, including Haas (1969), Holliday (2009), Carpenter et al. (2011), Ősi et al. (2014, 2017), and Nabavizadeh (2016, 2020a, 2020b). Their palatal musculature was relatively small- or medium-sized, given the moderately sized pterygoid flanges and reduced retroarticular process and surrounding mPT attachment sites, suggesting it was not as useful in jaw closure and was secondary to the temporal musculature in this regard (Ősi et al., 2014, 2017; Nabavizadeh, 2020a, 2020b). But the pterygoids and corresponding mPT complex origins are more expanded in later nodosaurids (e.g., *Silvisaurus*, *Pawpawsaurus*, and *Edmontonia*), plausibly indicating larger mPT muscle bodies in these taxa (Ősi et al., 2017).

Temporal musculature in ankylosaurs is more difficult to assess compared to other dinosaurs because of the fusion of dermal bone that surrounds it, covering up the entire temporal region. In general, most ankylosaurs possessed a caudally displaced and rostrocaudally short postorbital temporal region (relative to the length of their skull). Their tunnel-like adductor chamber was oriented rostroventrally at a low angle relative to the horizontal plane of the skull. The jugal flared laterally, creating a subjugal opening at the maxilla-jugal junction, which allowed a large fan of muscle (mAMES) to extend and attach rostrolabially along the labial dentary ridge lateral to its dentition (at varying lengths depending on the taxon). With this reconstruction, mAMES increased the mechanical advantage of adductor musculature as it braced the sides of the mandible, aiding in both palinal retraction of the jaw at occlusion as well as facilitating in hemi-mandibular long-axis rotation (Nabavizadeh, 2020a, 2020b).

Although ankylosaur cranial biomechanics studies have been scarce, recent lever arm studies—using methodology initially used by Ostrom (1961, 1964a, 1966)—have gained insights into the mechanical advantage of ankylosaur jaw systems, at least with respect to temporal musculature. Mallon and Anderson (2015), in investigating mandibular lever arm mechanics of all ornithischian megaherbivores of the Dinosaur Provincial Park,

found that the nodosaurid *Panoplosaurus* showed a generally higher mechanical advantage than the ankylosaurid *Euoplocephalus*, although this does not present a direct distinction between nodosaurids and ankylosaurids. With the addition of more taxa, Nabavizadeh (2016) found there to be more variation within each grouping. For instance, *Panoplosaurus*, although again showing higher mechanical advantage values than *Euoplocephalus*, was found to have lower mechanical performance across the tooth row than the ankylosaurids *Ankylosaurus* and *Pinacosaurus*. This discrepancy can be explained in a couple of ways. The mandible of other ankylosaurids like *Ankylosaurus* is relatively longer than that of *Euoplocephalus*. The coronoid eminence is also the lowest and displaced more caudally along the length of the jaw in *Euoplocephalus* relative to other ankylosaurids, especially *Ankylosaurus*. Among the nodosaurids studied, *Edmontonia* showed generally higher mechanical advantage than *Panoplosaurus*, especially at the distal-most tooth position. These two taxa have previously been seen as indistinct from each other in many ways, which makes it even more interesting that key characters in their jaws seem to be morphologically different. *Panoplosaurus* shows a more rounded coronoid eminence and slightly higher jaw joint, while *Edmontonia* shows a triangular coronoid eminence and more ventrally positioned jaw joint relative to the level of the tooth row. These characters have a direct impact on the overall mechanical advantage of the jaw system because of their influence on the moment arm length as well as muscle vector angle. Further study is needed to incorporate the mechanical advantage of the newly hypothesized rostrolabially expanded mAMES reconstruction to determine whether this feature affects the overall comparisons across all ankylosaurs.

Lastly, ankylosaurs are known to have large hyobranchial apparatus. Hill et al. (2015) found exceptionally large triangular hyobranchial bones that form a large platform with clear muscle scars, implicating enhanced tongue muscle attachment. The ankylosaur tongue was likely highly mobile and used for rotation and processing of vegetation during feeding, especially with an otherwise generally weaker mandibular feeding system than most other ornithischians.

Ankylosaurs were large, heavily built, graviportal animals with fusiform shapes, including shorter necks, transversely wide trunks, and robust limbs with shorter forelimbs relative to their hindlimbs (fig. 11.13). Their trunk was dorsoventrally flattened with fused dorsal vertebrae and a broad, reinforced, transversely board ribcage with wide ribs covered dorsally by an enormous sheet of variably sized osteoderms and different arrangements of spikes, creating what looks like an enormous carapace for ample protection. Their broad abdominal cavities (especially those of ankylosaurids) likely would have housed an extensive digestive tract for longer-term gut retention for elaborate fermentation of vegetation. Despite their large body sizes, however, ankylosaurs stood low to the ground, likely eating vegetation no higher than about 1 m high (Mallon et al., 2013).

Ankylosaurs generally showed adaptations to obligate quadrupedality somewhat similar to stegosaurs. The center of mass in ankylosaurs was generally at or just cranial to the hip joint (Maidment et al., 2014b). In the forelimb, they possess an elongate deltopectoral crest on the humerus, as in stegosaurs; however, the acromial process of the scapula is laterally everted, creating m. deltoideus muscle vectors that retain moment arms for humeral abduction (Maidment and Barrett, 2011, 2012; Barrett and Maidment, 2017). It is likely that their overall stance was different from that of stegosaurs, with feet held slightly lateral to the glenoid (as suggested by trackways; McCrea et al., 2001) and elbows likely adducted, meaning adductive ground reaction forces would have caused similar forces at the glenoid that would have required counteracting m. deltoideus abductive forces for extra control in stance (Maidment and Barrett, 2012; Barrett and Maidment, 2017). Their digits (unguals) likely had hooflike structures to help in weight-bearing (Barrett and Maidment, 2017). Like stegosaurs, ankylosaurs had transversely expanded hips, creating a larger moment arm for the abductor musculature (Maidment and Barrett, 2011, 2012, 2014; Maidment et al., 2014a). Recent speculation has suggested that ankylosaurids were competent at digging out roots from the ground, especially with specializations of the forelimb that include a large acromion of the scapula, robust humeri with well-developed deltopectoral crest, a prominent

Figure 11.13. Skeleton of *Gargoyleosaurus*.

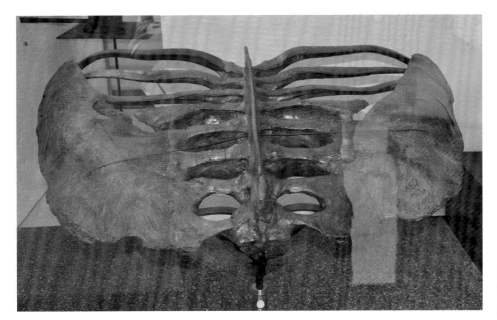

Figure 11.14. Widened pelvis of *Edmontonia*.

olecranon process of a shorter ulna, and short or absent pubic symphysis (Park et al., 2021).

As in stegosaurs, 3D musculoskeletal modeling of *Dyoplosaurus* pelvic and hindlimb musculature by Maidment et al. (2014a) showed adaptations indicative of obligate quadrupedality, although in somewhat different ways compared to stegosaurs. Among the ornithischians examined, *Dyoplosaurus* had the highest overall muscular moment arms in extension of the hip. Additionally, it also scored among the highest moment arms in flexion as the other tested thyreophorans (*Scelidosaurus* and *Stegosaurus*). *Dyoplosaurus* had high total abductor moment arms as well (similar to *Stegosaurus*), with lateral expansion of the trunk and pelvis, but it also had some of the highest adductor moment arm values (similar to *Scelidosaurus*), including with the addition of m. caudofemoralis longus acting as an adductor (probably owing to the lateral extension with much wider hips; fig. 11.14). *Dyoplosaurus* also had among the highest lateral femoral rotator moment arms among ornithischians owing to the large number of muscles dedicated to this function in ankylosaurs (Maidment et al., 2014a).

Many ankylosaurid ankylosaurs also possessed a club-like tail with large knobs capable of significant damage when swung at either a predator or another ankylosaur

(Arbour, 2009; Arbour and Snively, 2009). Despite having higher relative moment arms for hip extension (mostly owing to muscle origins on the postacetabular process), the fourth trochanter is less pronounced in ankylosaurs (as in stegosaurs) and may indicate a possible partial decoupling of m. caudofemoralis from the femur to be co-opted for more control in tail movement, thus allowing for subtleties in tail-swinging during fighting and self-defense. Arbour (2009), in modeling the tail in segments, showed that lateral motion (and not vertical motion) was the most likely mode of tail-swinging in at least some ankylosaurids. Arbour (2009) described pelvic muscle scars indicative of enlarged m. longissimus caudae that would have induced more lateral motion when contracted, with ossified tendons running along these muscles (possibly with m. spinalis) helping to keep the tail structurally sound. Furthermore, in examining impact stress distribution with FEA modeling, Arbour and Snively (2009) found that long prezygapophyses and neural spines in the modified distal caudal vertebrae (known as "handle" vertebrae) were helpful in creating greater distribution of forces from impact, with variability in knob size showing different overall values.

Isotopic evidence places ankylosaurs in the range of a clearly herbivorous lifestyle overlapping especially with ceratopsids in dietary niche signatures (Cullen et al., 2020, 2022). As stated above, ankylosaurs were low-level browsers (about 1 m off the ground) and as such may have fed on ferns, cycads, low-growing plants, fruits, and seeds. Although their jaws were not necessarily weak, ankylosaurs still had relatively less mechanical advantage compared to the more sophisticated feeding apparatuses of some of their ornithischian contemporaries (i.e., hadrosaurs and ceratopsids) (Mallon and Anderson, 2015; Nabavizadeh, 2016). It therefore stands to reason that they might have preferred to feed on softer plant tissues. With evidence from ankylosaur cololites (stomach contents) in *Kunburrasaurus* (Molnar and Clifford, 2000, 2001; therein referred to *Minmi*), at least some ankylosaurs are known to have fed on fruits, spherical seeds, fern sporangia, and vascular bundles of plant tissue. More recently, Brown et al. (2020) described cololites from the nodosaurid *Borealopelta* as showing 88% leaf tissue, with much more leptosporangiate ferns (85%) than cycad-cycadophyte (3%), and trace signs of coni-

fers, inferring primarily selective feeding on ferns in these nodosaurids. Larger olfactory bulbs in some ankylosaurs (Kuzmin et al., 2020) and the complex, convoluted nasal passages in some ankylosaurs (Witmer and Ridgely, 2008) are suggestive that some of them used their sense of smell for finding specific food sources. Ji et al. (2016) have also suggested the possibility of fish-eating in ankylosaurs, with fish skeletons associated with a specimen of the small, presumed semiaquatic ankylosaurid *Liaoningosaurus*, although it is difficult to confirm whether they were eating the fish or just preserved alongside each other. With reference to their having a larger vagus nerve pathway than might be expected, Alifanov and Saveliev (2019) suggested ankylosaurs may have had more complex digestive function owing to an omnivorous lifestyle. They also suggested that ankylosaurs may have had larger pharyngeal muscles used for filter-feeding capabilities by widening the oropharynx, which, if true, would corroborate with the hypothesized enlarged tongue indicated by Hill et al. (2015).

CHAPTER 12

Early Neornithischians and Ornithopods

Early-Diverging Neornithischia and Thescelosaurinae

For reference in understanding relationships and clade affiliation of early neornithischians and ornithopods discussed in this chapter, see the phylogeny (fig. 12.1) and taxon list (table 12.1) below.

Neornithischia is the other major clade of Genasauria. It consists of cerapodans, thescelosaurines, and other more basal taxa. Non-cerapodan neornithischians included early-diverging neornithischians like *Agilisaurus* as well as parksosaurids (e.g., *Parksosaurus* and *Orycto-dromeus*), jeholosaurids (e.g., *Jeholosaurus*), and thescelosaurines (e.g., *Thescelosaurus*; fig. 12.2). The latter three

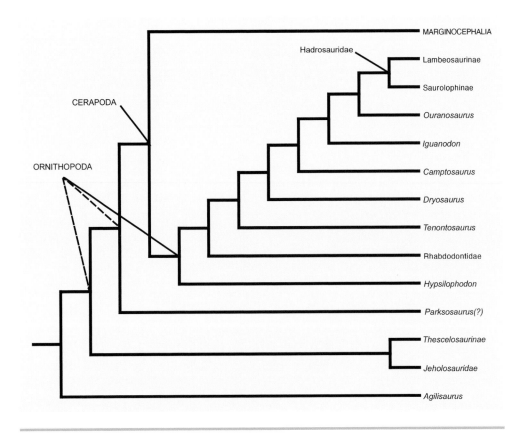

Figure 12.1. Phylogeny of Early Neornithischia and Ornithopoda, based largely on Prieto-Márquez (2010) and Boyd (2015).

Table 12.1.
Early neornithischian and ornithopod genera organized by clade

Major clade	Subclade	Genus
Early-diverging Neornithischia		
		Agilisaurus
		Orodromeus
	Jeholosauridae	
		Jeholosaurus
		Haya
		Changchunsaurus
	Thescelosaurinae	
		Thescelosaurus
	Parksosauridae	
		Parksosaurus
		Oryctodromeus
Ornithopoda		
		Hypsilophodon
	Non-hadrosaurid Iguanodontia	
		Bactrosaurus
		Camptosaurus
		Dryosaurus
		Gasparinisaura
		Tenontosaurus
		Rhabdodon
		Zalmoxes
		Mochlodon
		Iguanodon
		Ouranosaurus
	Lambeosaurine Hadrosaurids	
		Corythosaurus
		Hypacrosaurus
		Parasaurolophus
		Lambeosaurus
		Tsintaosaurus
	Saurolophine Hadrosaurids	
		Edmontosaurus
		Gryposaurus
		Brachylophosaurus
		Maiasaura
		Prosaurolophus
		Saurolophus
		Shantungosaurus

groupings have also been deemed early-diverging ornithopods (see Yang et al., 2020, for the latest) but will be treated separately herein because of the as yet controversial nature of their relationships (e.g., Boyd, 2015; Herne et al., 2019; Yang et al., 2020). They ranged mainly from the Middle Jurassic to Late Cretaceous with largely worldwide distribution. Early neornithischians show laterally wedge-shaped skulls that are more elongate in thescelosaurines and jeholosaurids than other forms, which come to a point at the snout. Their teeth are leaf-shaped and denticulate, and they have a large orbit and smaller antorbital fenestra. Their temporal region is rostrocaudally expanded and transversely narrow, and their postorbital adductor region is moderately sized and subrectangular. They possess the typical ornithischian slight jugal flare and corresponding small subjugal opening. The mandible has a small triangular predentary (which is rostrally extended in thescelosaurines and jeholosaurines) articulating with a vertical dentary symphysis, categorized as predentary-dentary joint (PDJ) Morphotype I (Nabavizadeh and Weishampel, 2016). The dentary is long and thin in lateral view with a slightly bowed tooth row. The coronoid eminence is broad and triangular with no mandibular fenestra seen ventral to it and a blunt and rugose retroarticular process.

Early neornithischians had peg-like, sometimes recurved (e.g., *Agilisaurus*) teeth in the premaxilla and denticulate, worn, leaf-shaped teeth in the maxillae and dentary, with a diastema between the premaxillary and maxillary teeth in thescelosaurines (Liyong et al., 2010; Boyd, 2014). Some, including *Haya*, had more flattened apices of the cheek teeth, likely a specialization for more efficient feeding (Barta and Norell, 2021). With slightly oblique occlusal facets, basal neornithischians, especially thescelosaurines, were well equipped for orthal isognathous feeding mechanisms (e.g., Weishampel, 1984a). In a histological study, Chen et al. (2018) found that the thescelosaurine *Changchunsaurus* could maintain a continuous chewing surface along the tooth row while its teeth were replaced because the developing teeth did not invade the pulp cavity until it was almost completely formed. Chen et al. (2018) also found that *Changchun-*

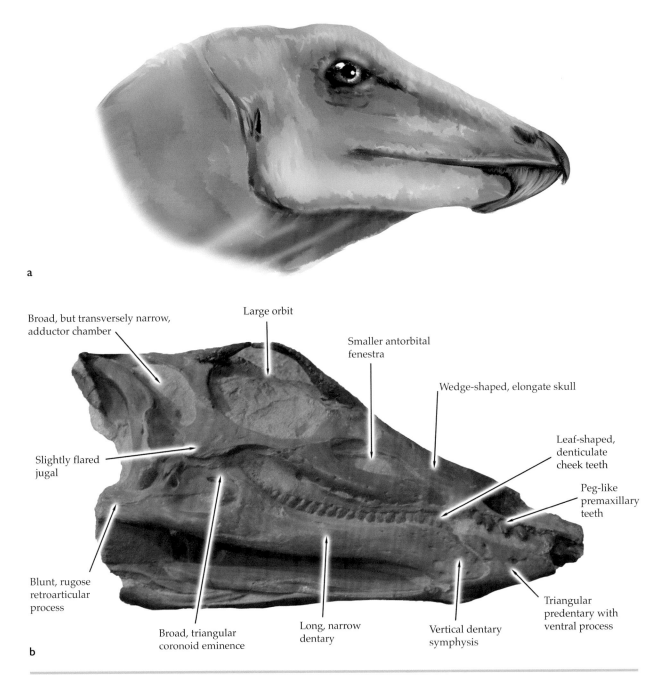

a

Broad, but transversely narrow,
adductor chamber

Large orbit

Smaller antorbital
fenestra

Wedge-shaped, elongate skull

Leaf-shaped,
denticulate
cheek teeth

Slightly flared
jugal

Peg-like
premaxillary
teeth

Blunt, rugose
retroarticular
process

Broad, triangular
coronoid eminence

Long, narrow
dentary

Vertical dentary
symphysis

Triangular
predentary with
ventral process

b

Figure 12.2. *Thescelosaurus* (a) reconstructed head and (b) skull in lateral view
(length approximately 34 to 35 cm).

saurus shows the earliest signs of wavy enamel in ornithischians (thought to be specific to the complex dentition of hadrosaurs), suggesting it aided in the development of a shearing dentition. Furthermore, slight hemi-mandibular long-axis rotation was likely a result of the predentary-dentary articulation (Nabavizadeh and Weishampel, 2016). The long and pointed predentary was probably useful in selective feeding strategies.

Non-cerapodan neornithischians generally show the typical early configuration of cranial musculature (Holli-

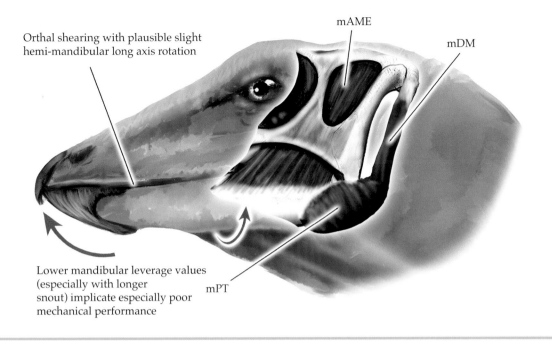

Orthal shearing with plausible slight hemi-mandibular long axis rotation

mAME

mDM

Lower mandibular leverage values (especially with longer snout) implicate especially poor mechanical performance

mPT

Figure 12.3. *Thescelosaurus* head showing feeding function with jaw muscles. Red arrows show various jaw actions. Abbreviations: mAME, m. adductor mandibulae externus; mDM, m. depressor mandibulae; mPT, m. pterygoideus.

day, 2009; Nabavizadeh, 2016, 2020a, 2020b) (fig. 12.3). The origin of the temporal m. adductor mandibulae externus (mAME) complex was moderately rostrocaudally expanded. The adductor musculature was likely well suited for orthal feeding, as the adductor chamber is oriented more vertically. The mAME inserted around a low, rounded, or subtriangular coronoid eminence of the mandible. The palatal m. pterygoideus (mPT) was likely moderately sized and acted secondary to the power of temporal musculature during the chewing stroke. Nabavizadeh (2016) found that non-cerapodan neornithischians show lower mechanical advantage values across the tooth row relative to other ornithischians, implicating generally poor mechanical performance of the jaw. *Thescelosaurus* and *Jeholosaurus* show especially poor mechanical performance, with *Agilisaurus* and *Parksosaurus* showing slightly greater values, likely owing to shorter and slightly more robust skulls (Nabavizadeh, 2016; Button and Zanno, 2020).

Non-cerapodan neornithischians were small- to medium-sized, bipedal, mostly cursorial animals with a flexible neck, slender torso, elongate lower limbs, and long tails (fig. 12.4). They were generally fast runners, with strong hip extensor musculature owing to a caudally elongated postacetabular process of the ilium (a synapomorphy of Neornithischia) pulling the extensors origins caudally to create high moment arms for femoral retraction (Maidment et al., 2014a). Some, such as *Oryctodromeus*, show enlarged scapular adaptations for larger forelimb muscle attachment indicative of strong digging and burrowing capacity, including a strong m. deltoideus and m. teres major (although it had a relatively weaker m. triceps) (Fearon and Varricchio, 2016). Owing to their relatively weaker feeding apparatus, smaller early neornithischians and thescelosaurines likely selectively ate fruits, seeds, and roots with their narrow beaks. Some specimens of *Haya* have been found with associated gastroliths, however, indicating the potential processing of tough vegetation in the gut as well (Makovicky et al., 2011; Barta and Norell, 2021).

Figure 12.4. *Parksosaurus* skeleton.

Ornithopoda

Ornithopoda includes early-diverging forms (e.g., *Hypsilophodon* [fig. 12.5] and others) as well as the more derived group Iguanodontia, including taxa such as rhabdodontids (e.g., *Rhabdodon, Zalmoxes,* and *Mochlodon*), *Bactrosaurus, Gasparinisaura, Dryosaurus, Camptosaurus, Tenontosaurus, Iguanodon* (fig. 12.6), *Ouranosaurus,* and of course the major derived "duck-billed" clade Hadrosauridae (including the crested Lambeosaurinae [e.g., *Parasaurolophus, Corythosaurus, Lambeosaurus, Hypacrosaurus, Tsintaosaurus*] and non-crested Saurolophinae [also called Hadrosaurinae; e.g., *Edmontosaurus, Shantungosaurus, Gryposaurus, Brachylophosaurus, Maiasaura, Prosaurolophus, Saurolophus*]). They ranged from the Middle Jurassic to Late Cretaceous with worldwide distribution.

Ornithopod skulls are generally wedge-shaped in lateral view (without the cranial ornamentation seen in many cases), rostrocaudally blunter relatively in more basal forms, and more elongate in the derived iguanodontians. Ornithopod skulls in general are triangular in dorsal view. The derived hadrosauroids evolved a dorsoventrally compressed, transversely broadened rostrum, creating what is commonly described as an edentulous "duck-bill," although from what is known of the overlying rhamphotheca preserved in a select few hadrosaurs, the bill curves ventrally in life (Morris, 1970; Farke et al., 2013), which is different from what is seen in many flat-billed ducks. The external naris is typically small and circular or oval. Of note are the elaborate and diverse hollow crests seen on the skull roof of lambeosaurine hadrosaurs, such as the rounded crests of *Corythosaurus* (fig. 12.7) and the long, tube-like crest of *Parasaurolophus* (fig. 12.8), to name a few. In ornithopods, the infratemporal fenestra is small and circular in earlier diverging taxa and much larger and rectangular in iguanodontians (although variable in shape across taxa), with a thin upper temporal bar dorsally. A small, rounded antorbital fenestra is seen in basal ornithopods, and it becomes either highly reduced or disappears in the derived iguanodontians. In lambeosaurine hadrosaurids, the nasal, along with other dorsal cranial elements, is integrated into an elaborate hollow crest, with each taxon possess-

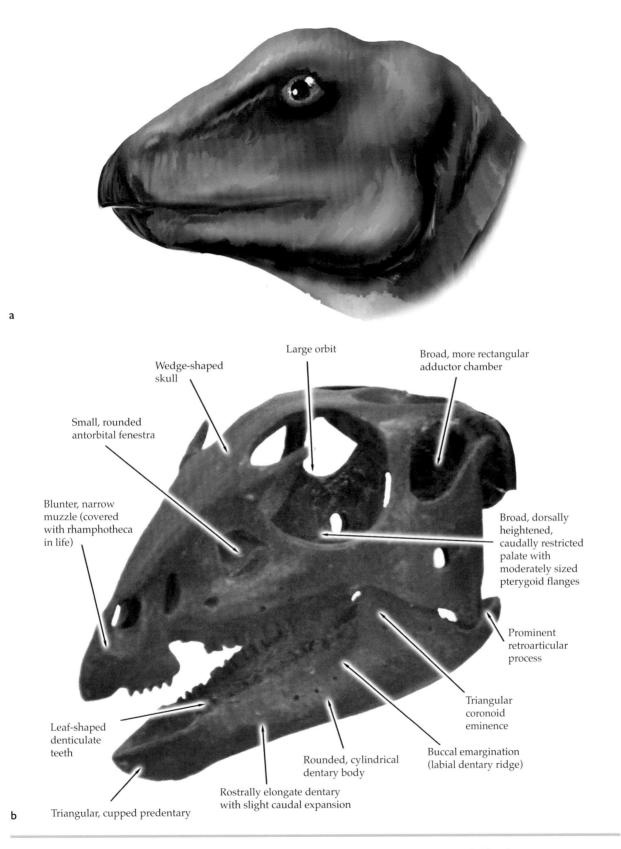

Figure 12.5. *Hypsilophodon* (a) reconstructed head and (b) skull in lateral view (length approximately 12 cm).

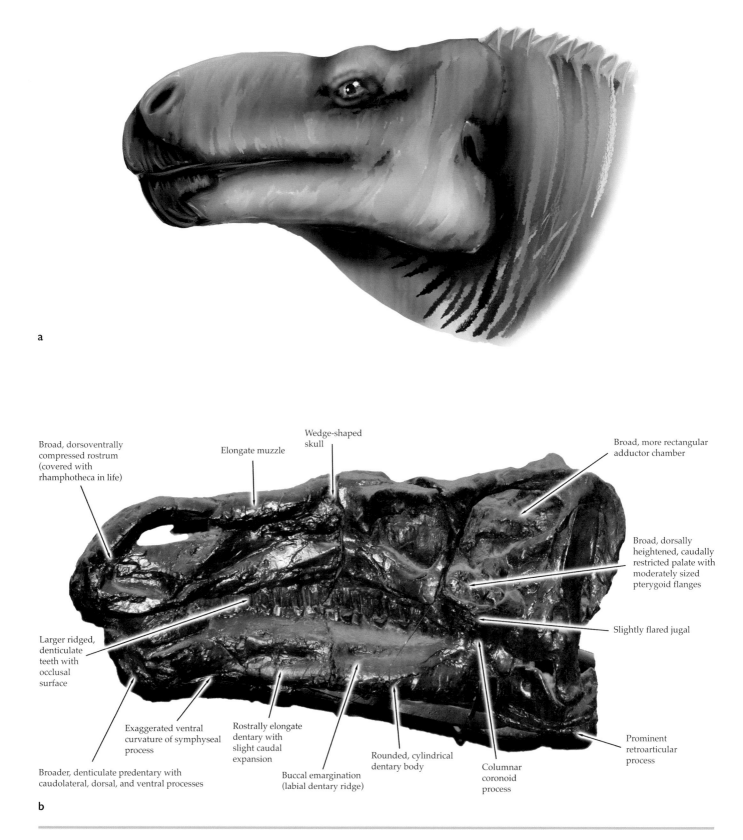

a

b

Broad, dorsoventrally compressed rostrum (covered with rhamphotheca in life)

Elongate muzzle

Wedge-shaped skull

Broad, more rectangular adductor chamber

Broad, dorsally heightened, caudally restricted palate with moderately sized pterygoid flanges

Larger ridged, denticulate teeth with occlusal surface

Slightly flared jugal

Exaggerated ventral curvature of symphyseal process

Rostrally elongate dentary with slight caudal expansion

Rounded, cylindrical dentary body

Columnar coronoid process

Prominent retroarticular process

Broader, denticulate predentary with caudolateral, dorsal, and ventral processes

Buccal emargination (labial dentary ridge)

Figure 12.6. *Iguanodon* (a) reconstructed head and (b) skull in lateral view (length approximately 54 cm).

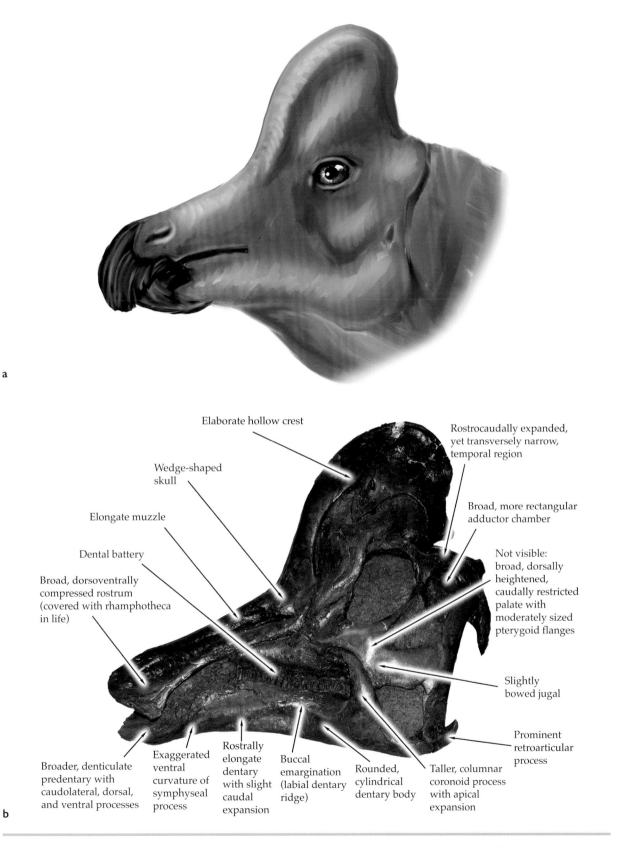

Figure 12.7. *Corythosaurus* (a) reconstructed head and (b) skull in lateral view (length approximately 55 to 75 cm).

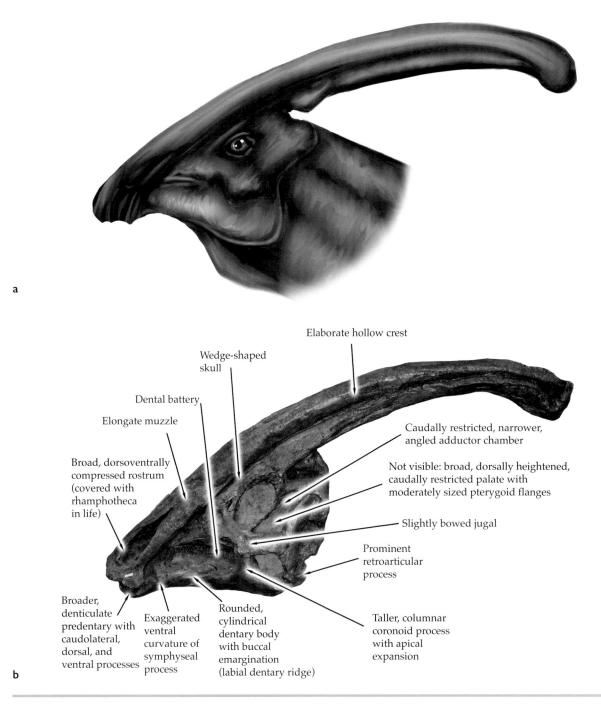

a

Elaborate hollow crest

Wedge-shaped
skull

Dental battery

Elongate muzzle

Caudally restricted, narrower,
angled adductor chamber

Broad, dorsoventrally
compressed rostrum
(covered with
rhamphotheca
in life)

Not visible: broad, dorsally heightened,
caudally restricted palate with
moderately sized pterygoid flanges

Slightly bowed jugal

Prominent
retroarticular
process

Broader,
denticulate
predentary with
caudolateral,
dorsal, and
ventral processes

Exaggerated
ventral
curvature of
symphyseal
process

Rounded,
cylindrical
dentary body
with buccal
emargination
(labial dentary ridge)

Taller, columnar
coronoid process
with apical
expansion

b

Figure 12.8. *Parasaurolophus* (a) reconstructed head and (b) skull in lateral view
(length approximately greater than 1.5 m; approximately 75 cm without crest).

ing its own unique morphology from a long tube-like structure, such as in *Parasaurolophus*, to a hump-like crest, such as in *Corythosaurus*, to other extravagant crests, such as in *Lambeosaurus*.

Ornithopod temporal morphology is variable. Early-diverging ornithopods, non-hadrosaurid iguanodontians, and some saurolophine hadrosaurids (e.g., *Gryposaurus* [fig. 12.9] and *Brachylophosaurus*) have both transversely

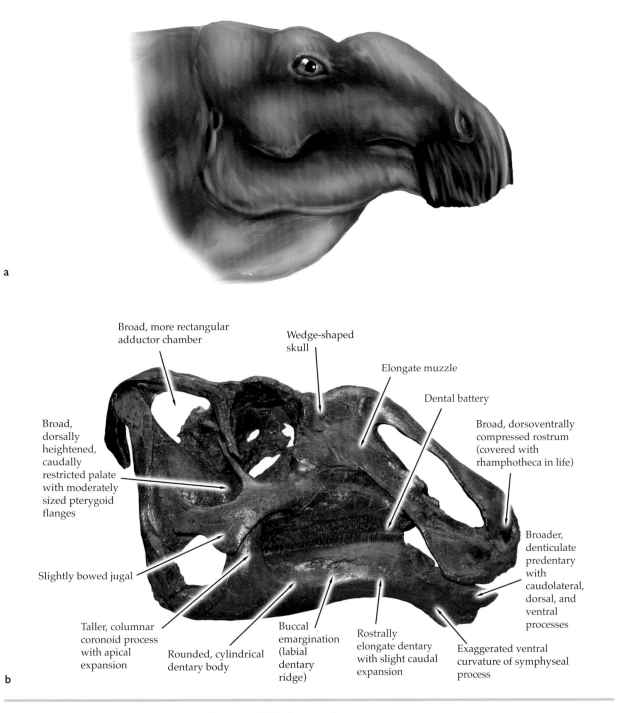

a

b

Broad, more rectangular adductor chamber

Wedge-shaped skull

Elongate muzzle

Dental battery

Broad, dorsally heightened, caudally restricted palate with moderately sized pterygoid flanges

Broad, dorsoventrally compressed rostrum (covered with rhamphotheca in life)

Broader, denticulate predentary with caudolateral, dorsal, and ventral processes

Slightly bowed jugal

Taller, columnar coronoid process with apical expansion

Rounded, cylindrical dentary body

Buccal emargination (labial dentary ridge)

Rostrally elongate dentary with slight caudal expansion

Exaggerated ventral curvature of symphyseal process

Figure 12.9. *Gryposaurus* (a) reconstructed head and (b) skull in lateral view (length approximately up to 98 cm).

and rostrocaudally expanded temporal region. Some other saurolophines (e.g., *Edmontosaurus* [fig. 12.10], *Shantungosaurus*, and *Saurolophus*) and the lambeosaurine *Parasaurolophus* possess caudally restricted and trans-

versely narrowed temporal regions. Most other lambeosaurines (e.g., *Corythosaurus*, *Hypacrosaurus*, *Lambeosaurus*, and *Tsintaosaurus*) and the saurolophines *Maiasaura* and *Prosaurolophus* have a rostrocaudally expanded and

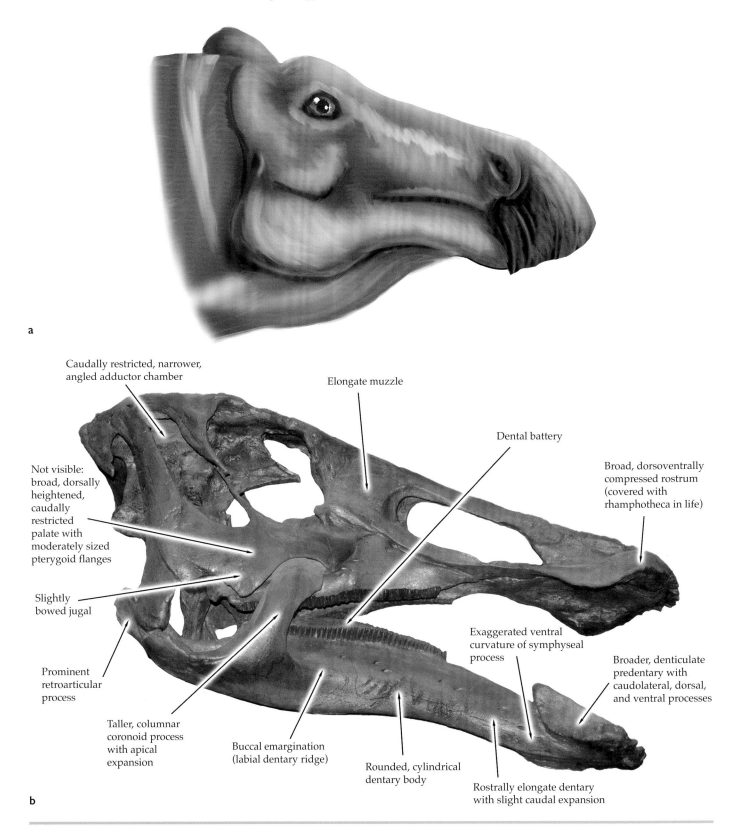

a

b

Caudally restricted, narrower, angled adductor chamber

Elongate muzzle

Dental battery

Broad, dorsoventrally compressed rostrum (covered with rhamphotheca in life)

Not visible: broad, dorsally heightened, caudally restricted palate with moderately sized pterygoid flanges

Slightly bowed jugal

Prominent retroarticular process

Taller, columnar coronoid process with apical expansion

Buccal emargination (labial dentary ridge)

Rounded, cylindrical dentary body

Exaggerated ventral curvature of symphyseal process

Broader, denticulate predentary with caudolateral, dorsal, and ventral processes

Rostrally elongate dentary with slight caudal expansion

Figure 12.10. *Edmontosaurus* (a) reconstructed head and (b) skull in lateral view (length approximately 80 to more than 100 cm).

transversely narrowed temporal regions. The postorbital adductor region and infratemporal fenestrae are larger and subrectangular in basal ornithopods, non-hadrosaurid iguanodontians, some lambeosaurines (e.g., *Corythosaurus*, *Hypacrosaurus*, *Lambeosaurus*, and *Tsintaosaurus*), and some saurolophine hadrosaurids (e.g., *Gryposaurus*, *Brachylophosaurus*, *Maiasaura*, and *Prosaurolophus*). The lambeosaurine hadrosaurid *Parasaurolophus* and some saurolophine hadrosaurids (e.g., *Edmontosaurus*, *Saurolophus*, and *Shantungosaurus*) conversely possess low-angled adductor regions. All ornithopods share the typical ornithischian slightly flared jugal (or, in the case of hadrosaurids, a bowed jugal) with corresponding subjugal opening. The palate is dorsally heightened, caudally restricted within the skull, and broad, with moderately sized pterygoid flanges.

Ornithopod mandibles are quite variable as well. The predentary varies from small rounded, cup-shaped, or triangular forms in basal ornithopods (categorized as PDJ Morphotype I; Nabavizadeh and Weishampel, 2016)

to much broader, elaborately denticulate predentaries with caudolateral, dorsal, and ventral processes articulating with the exaggerated ventral curvature and rostrocaudally expanded symphysis of the rostral ends of the dentaries in iguanodontians (categorized as PDJ Morphotype II; Nabavizadeh and Weishampel, 2016) (fig. 12.11). The dentary is elongate rostrally and expanded slightly caudally. In iguanodontians (especially hadrosaurids), the rostral symphyseal process of the dentary curves ventrally rostral to the dentition and then curves medially to meet its counterpart. In some hadrosaurids, this symphyseal process morphology forms a sort of shovel-like lower bill (e.g., Prieto-Márquez et al., 2020a). The body of the dentary in iguanodontians is rounded and cylindrical along its long axis.

The tooth row is slightly medially curved in early-diverging ornithopods and even non-hadrosaurid iguanodontians, with leaf-shaped, denticulate teeth. In hadrosaurids, a dental battery is formed with hundreds of diamond-shaped teeth packed together in columns that

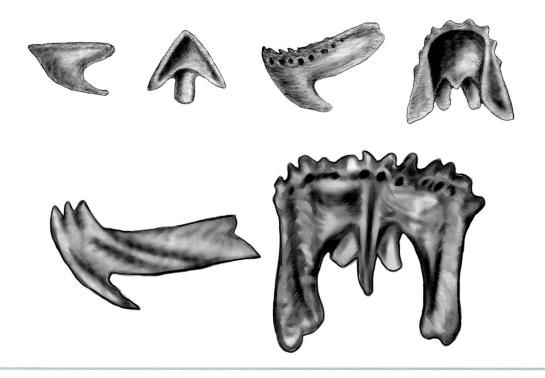

Figure 12.11. Comparison of (*left*) lateral and (*right*) dorsal views of predentaries in (*clockwise from top left*) *Hypsilophodon*, *Iguanodon*, and the hadrosaurid *Lambeosaurus*.

are worn down as one mesiodistally elongate occlusal surface that occludes with the maxillary dentition. The coronoid elevations are also variable, with basal ornithopods and the non-hadrosaurid iguanodontian *Dryosaurus* possessing a raised, triangular coronoid eminence, all other non-hadrosaurid iguanodontians possessing a more columnar coronoid process, and all hadrosaurids raising the coronoid process higher (increasing mechanical advantage of temporal muscles; see below), curving it more medially and expanding the apex (fig. 12.12). The rostroventral margin of the coronoid process in iguanodontians extends slightly rostrolabially lateral to the dentition, continuing as the labial dentary ridge. There is no mandibular fenestra, and the retroarticular process is prominent.

Basal ornithopods have roughly five conical premaxillary teeth that are sometimes lacrimiform and have a pointed apex, with a cupped inner enamel surface (e.g., *Hypsilophodon*). These teeth run along the lingual margin of the premaxilla on either side and were suitable for puncturing vegetation at the initial bite. Premaxillary dentition is subsequently lost in more derived ornithopods (i.e., iguanodontians). The maxillary teeth are rounded in basal forms (e.g., *Hypsilophodon*) to diamond-shaped (e.g., iguanodontians; fig. 12.13), apicobasally elongate, and have thin ridges running along the outer surface (usually one median ridge with one or two ridges parallel on either side). Hadrosaurids have a single thin midline ridge on the lingual side on an elongate, diamond-shaped tooth crown that runs the height of the tooth

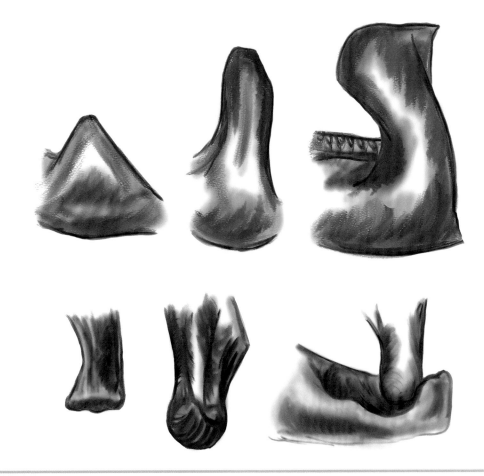

Figure 12.12. Illustrations of coronoid elevations in (*top, left to right*) *Dryosaurus*, *Iguanodon*, and *Parasaurolophus* as well as jaw ventral quadrate shapes in (*bottom, left to right*) a non-hadrosaurid ornithopod, hadrosaurid, along with a depiction of a hadrosaurid jaw joint.

(fig. 12.13). In iguanodontians, especially hadrosaurids, the tooth row in the maxilla does not contain nearly as many teeth, with only a tightly packed row of single teeth seen in lateral view. The continuous occlusal surface is oriented ventromedially, occluding with the dorsolateral occlusal surface of the dentary tooth battery described below.

The most distinctive feature of the ornithopod dentary is the tooth row, which is formed into what is known as a dental battery in hadrosauroids (i.e., many columns of multiple teeth packed together; fig. 12.13). It covers more than half the length of the dentary. The tooth row begins at the mesial-most oral margin of the dentary in more basal ornithopods, but just caudal to the diastema in most iguanodontians (Kubota and Kobayashi, 2009). The dentary contains roughly 10 to 15 teeth in basal forms and 40 or more tightly packed columns of teeth in hadrosaurids. The number of tooth positions varies with ontogeny and species, each extending dorsally from an elongate row of alveoli. The tooth row extends even further caudally than the coronoid process, which lies lateral to it, especially in hadrosauroids. Positioning of teeth in basal ornithopods is a typical reptilian tooth row with teeth in line with each other. Non-hadrosauroid iguanodontians have a tooth row with fewer teeth that are larger with respect to the size of the dentary itself and have visible replacement teeth, although only with two or three replacement teeth per tooth position (Weishampel and Jianu, 2011). The tooth row in non-hadrosauroid iguanodontians is unique in that, much like that seen in ankylosaurs, the tooth row has a sinusoidal configuration with alveoli that are oriented slightly dorsolaterally at the rostral end, and as the tooth row extends caudally, it curves medially and the alveoli are gradually oriented dorsally and, in the most extreme cases such as *Ouranosaurus* (Taquet, 1976), slightly dorsomedially. In hadrosauroids, in which dental batteries have many more teeth, all tooth columns jointly form a medially convex configuration extending dorsally to form a flat, smooth occlusal surface. This occlusal surface faces dorsolaterally and occludes with the ventrolateral orientation of the dentition of the maxilla

dorsally. In medial view, the alveoli are V-shaped, with both rostral and caudal ends closest to the dorsal surface with the fewest teeth in the column. A thin, rugose layer of bone conceals the ventral half of the medial surface of the dental battery. Individual dentary teeth, like maxillary teeth, are rounded with a straight and thin apex in basal forms (e.g., *Hypsilophodon*) to diamond-shaped and apicobasally elongate (e.g., iguanodontians). A thin midline ridge on the lingual side runs the height of the tooth in hadrosaurids. Basal ornithopods possess many more apicobasally oriented ridges along the outer surface of

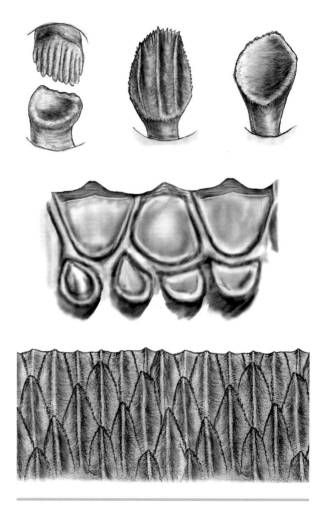

Figure 12.13. Dentition in ornithopods: (*top left*) *Hypsilophodon* maxillary and dentary teeth coming to occlusion, *Iguanodon* dentary tooth in (*top middle*) non-occlusal and (*top right*) occlusal views, (*middle*) hadrosaurid tooth occlusal surface, and (*bottom*) medial view of dentary tooth battery in a hadrosaurid.

the teeth that form rounded denticle-like protrusions on the apical ridge of the dentition in basal ornithopods, but not typically in iguanodontians. A worn edge along the lingual edge of the teeth is present at variably acute angles. Non-hadrosauroid iguanodontians, such as *Iguanodon*, possess a much more mesiodistally expanded apical ridge with a denticle formed apically by the large midline ridge and a large wear facet laterally. The midline ridge in the derived hadrosauroids, when reaching the continuous occlusal surface mentioned above, creates a serrated edge that runs the mesiodistal length of the lingual side of the tooth row when combined with the rest of the tooth columns. This serrated edge forms a saw-blade configuration of the tooth row that is bowed dorsally in the center that would help strip vegetation.

The packed dentition in hadrosaurids comes from a unique form of tooth development. A study by LeBlanc et al. (2016) performed a tissue-level ontogenetic study of hadrosaurid teeth showing that as hadrosaurid teeth grew in columns, development of older tooth generations was halted, retained, and packed together as it was pushed upward from the dentary. This meant that the packed teeth could be worn all the way painlessly, including the dead root, as new teeth developed in the alveolus near the jawline (a few teeth down each column). A highly modified type of dinosaurian gomphosis with tiny dental tissues (interconnected ligaments) also helped keep the packed teeth together into a solid structure with slight flexibility. The occlusal surface that formed at the apex of the packed dental battery created the longitudinal platform for effectively processing highly fibrous vegetation like conifers in the oral cavity before swallowing (LeBlanc et al., 2016). Furthermore, Bramble et al. (2017) made serial sections through an adult hadrosaurid dental battery and inferred gradual ontogenetic tooth migration, with mesiodistal displacement of later tooth generations along with signs of a great deal of remodeling. Again, the interconnected ligamentous tissue helped hold these teeth in place as they erupted in this unique way. They noted the possibility that these changes in direction of tooth eruption could be further indication of changes in diet through ontogeny (Bramble et al., 2017).

Hadrosaurid dental tissue makeup is enormously complex. Erickson et al. (2012) examined the microstructure of the teeth of *Edmontosaurus* and found that the structural composition of hadrosaurid teeth transformed the normal reptilian two-tissue dental composition to a six-tissue tooth composition, creating an occlusal surface capable of withstanding more complex feeding strategies (see below). The dental batteries of various hadrosaurids were used for different modes of feeding, and the dental microstructure and wear are indicative of this difference. For instance, Erickson et al. (2012) showed how some hadrosaurids had flatter grinding surfaces on their teeth, while others show more steep angled wear indicative of slicing action (also shown through mesowear studies by Weishampel, 1984a). Erickson and Zelenitsky (2014) also found this tooth microstructure in *Hypacrosaurus*, with tooth morphologies in different age groups indicating varied diet. Embryos and neonates showed cuplike teeth, subadults showed shearing plane dentition, and adults showed dual slicing and crushing surfaces. Although this is only one taxon, the fact that ontogenetic changes are visible gives a glimpse into the complexity of hadrosaur paleoecology.

Dental wear patterns (both microwear and microwear) are relatively well known in ornithopods. Early-diverging ornithopods and non-hadrosaurid iguanodontians show similar wear patterns. In early-diverging ornithopods like *Hypsilophodon*, wear is slightly cupped, while the larger, non-hadrosaurid iguanodontians show a flatter wear plane (Galton, 1974; Norman, 1980, 1984; Weishampel, 1984a). The maxillary tooth row possesses lingually facing occlusal surfaces of teeth that contact the labially facing occlusal surfaces of dentary teeth. The wear facets generally indicate a likely orthal, isognathous feeding motion of the mandible in these animals. Dental wear patterns in hadrosaurids have been studied numerous times over the past few decades (Ostrom, 1961; Weishampel, 1983, 1984a; Williams et al., 2009; Mallon and Anderson, 2014b; Nabavizadeh, 2014; Rivera-Sylva et al., 2019). Wear patterns have consistently shown scratches in a range between both the mesiodistal and buccolingual direction (to varying degrees in different taxa), as has been shown statistically by Williams et al. (2009), Mallon and Anderson (2014b), and Rivera-Sylva et al. (2019). This is a clear indication of a combination of both orthal and palinal feeding strokes (with intermediate wear angles) in hadrosaurid feeding mechanisms (see below). Possible signs of hemi-mandibular rotation against the predentary have also been described, with rounded apical wear in the maxillary dentition described by Nabavizadeh (2014) as resulting from rotational movements of the dentary teeth during the power stroke, supporting numerous other hypotheses of dentary rotation along their long axes (Nopcsa, 1900; Versluys, 1923; Kripp, 1933; Cuthbertson et al., 2012; Nabavizadeh, 2014; Nabavizadeh and Weishampel, 2016; see below).

The general shape of hadrosaurid skulls can be described as relatively large and narrow with a ventrally recurved snout, with saurolophine skulls slightly larger than lambeosaurine skulls (Mallon and Anderson, 2013). The intermediate size and shape of the broadened hadrosaurid beak is suggestive of a more mixed diet (Mallon and Anderson, 2014a). A recent ontogenetic study of *Edmontosaurus* showed that although some cranial elements developed isometrically, the dentary itself did undergo significant changes throughout growth, with the dental battery exceptionally variable across specimens (Wosik et al., 2019). Using cranial allometry and tooth microwear, Wyenberg-Henzler et al. (2022) showed that there were ontogenetic changes in hadrosaurid feeding selectivity, with juvenile hadrosaurids being relatively more selective feeders than adults, and likely used lateral movements of the neck to eat softer, low-growing vegetation. Morphometric analyses have also confirmed that ontogenetic changes in skull morphology likely resulted in intraspecific variability in most taxa, including, for example, the saurolophine hadrosaurid *Gryposaurus* (Lowi-Merri and Evans, 2020). Dentary variation is also common in hadrosauroids, with various fluctuating morphologies, again, affecting ontogenetic changes in food selection and feeding mechanics, as suggested by Söderblom (2017) by means of morphometric analysis.

Strickson et al. (2016) and Stubbs et al. (2019), through estimates of evolutionary rate and disparity analyses of morphological characteristics, found that cranial disparity in hadrosaurids was common by means of numerous morphological innovations of the skull, particularly of those pertaining to the feeding apparatus. Both studies found that feeding characters show low variance but high evolutionary rates throughout the ornithopod (especially hadrosaurid) subclades, meaning the complexity of the feeding apparatus was a key factor in the evolutionary diversity of hadrosaurids (Strickson et al., 2016; Stubbs et al., 2019). When compared to all herbivorous dinosaurs, ornithopods as a whole seem to trend evolutionarily in the same trajectory as ceratopsians, ankylosaurs, and heterodontosaurids, especially with the evolution of their complex feeding apparatus, dental packing, and coronoid process, increasing muscular mechanical advantage (see below) (Button and Zanno, 2020).

The functional anatomy and evolution of ornithopod feeding mechanisms have been studied extensively for well over century (Dollo, 1883, 1884; Marsh, 1893; Nopcsa, 1900; Versluys, 1910, 1912, 1923; Lambe, 1920; Kripp, 1933; Lull and Wright, 1942; Ostrom, 1961; Thulborn, 1971; Galton, 1974; Hopson, 1980; Sues, 1980; Norman, 1984; Weishampel, 1984a; Norman and Weishampel, 1985; Rybczynski et al., 2008; Bell et al., 2009; Williams et al., 2009; Cuthbertson et al., 2012; Erickson and Zelenitsky, 2014; Mallon and Anderson, 2014b, 2015; Nabavizadeh, 2014, 2016, 2020a, 2020b; Nabavizadeh and Weishampel, 2016; Rivera-Silva et al., 2019). Ornithopod cranial material is unique among both dinosaurs and large mammalian herbivores, making it difficult to pinpoint a modern analogue, although hadrosaurids have been somewhat likened to modern large ungulates owing to feeding adaptations such as flat, grinding occlusal surfaces of their teeth (Norman and Weishampel, 1985; Carrano and Janis, 1999).

Many of the earliest hypotheses of ornithopod feeding mechanisms, particularly those of hadrosaurids, have involved different forms of cranial or intramandibular kinesis. Both Marsh (1893) and Nopcsa (1900) suggested the quadrate might have rocked rostrocaudally against its articulation with the squamosal dorsally, causing the mandibular end of the quadrate to shift in a fore-aft motion against the articular condyle of the mandible. Although Versluys (1910) suggested an akinetic skull in hadrosaurids, he later agreed that the quadrate might have been slightly mobile against the squamosal (Versluys, 1912, 1923). Lambe (1920) and Lull and Wright (1942) suggested a vertical shearing mechanism in hadrosaurids, especially with the observation of vertical striations by the latter study. Furthermore, Nopcsa (1900), Versluys (1923), and Kripp (1933) suggested that the individual mandibular rami had the potential to rotate around their long axes on each side against the predentary, a mechanism suggested in numerous other ornithischian groups (see chapters 10, 11, and 13). Although these authors suggested different modalities of long-axis rotation, the general thinking stems from multiple lines of evidence, including the curvature of the coronoid process, mobility at the predentary-dentary joint, and a ball-and-socket jaw joint. Ostrom (1961) rejected the notion of long-axis rotation of the mandibular rami, calling it "erroneous" owing to a differing perspective of mobility of different elements. Instead, Ostrom (1961) proposed a propalinal feeding motion in which the mandible would itself move in a fore-aft motion during the power stroke along the flat occlusal surfaces. An orthal to possible buccolingual (transverse) feeding mechanism as well as possible slight cranial kinesis (both at the skull roof and the maxillae) have also been suggested for early-diverging ornithopods, such as *Hypsilophodon* (Galton, 1974) and *Zephyrosaurus* (Sues, 1980), based on craniodental morphology.

For the past three decades, one of the prominently known proposed mechanisms for most ornithopods has been pleurokinesis (named by Norman and Weishampel, 1985, but the mechanism was described in detail prior to that study; Weishampel, 1983, 1984a; Norman, 1984). Pleurokinesis is a complicated masticatory mechanism that uses a unique form of cranial kinesis involving many different cranial elements in motion with each other at various joints (fig. 12.14).

The cranial joint movements involved in pleurokinesis are summarized as follows. Initially, the ventral end of the quadrate moves caudolaterally, with its dorsal end

rocking in its cotylus against the squamosal (i.e., the otic joint). The mandible moves with the quadrate, although only slight movement is necessary. This movement is associated with mobility at the basipterygoid-pterygoid junction (i.e., the basal joint). The pterygoid-quadrate joint is immobile because it is broad and squamous with many ridges, which requires the pterygoid to move with the quadrate. This in turn also moves the pterygoid against the basipterygoid process of the braincase. Additionally, there would likely be mobility at the palatine-pterygoid joint, as it is a thin junction. The quadratojugal is mobile against the quadrate, bearing a relatively slim contact with it, especially in non-hadrosaurian ornithopods such as in *Iguanodon*, where it meets the dorsal and ventral half of the embayment on the rostral side of the quadrate but not inside. In hadrosauroids, the quadratojugal sits inside an expanded embayment. The quadratojugal-jugal junction is a broadly scarfed joint where the jugal overlaps the quadratojugal laterally, but it is likely immobile. The maxilla-jugal joint was immobile as well, and both elements would have acted as one unit. The maxilla-premaxilla joint, however, is considered highly mobile. In non-lambeosaurine iguanodontians, there are two premaxillary processes that contact the dorsal margin of the maxilla, forming a joint between the premaxilla and the maxilla. Lambeosaurines instead have a broad maxillary shelf in this region with which the premaxillae articulate. The jugal-lacrimal joint is continuous with the maxilla-premaxilla joint. Just rostral to the orbit, the lacrimal forms a buttress that contacts the jugal, wrapping around it at the rostral orbital margin, making the jugal-lacrimal contact mobile. This causes the ventral half of the orbit to rotate laterally while the maxilla is in occlusion. The postorbital-jugal joint is thin, delicate, and highly mobile as well, as the two bones are butted against each other and sometimes overlap, with the dorsal process of the jugal resting caudal to the ventral process of the postorbital. All of these movements of elements, starting with the quadrate moving against the squamosal, create a domino effect between all of these kinetic cranial elements that ultimately causes the maxillae to be pushed and rotate laterally as

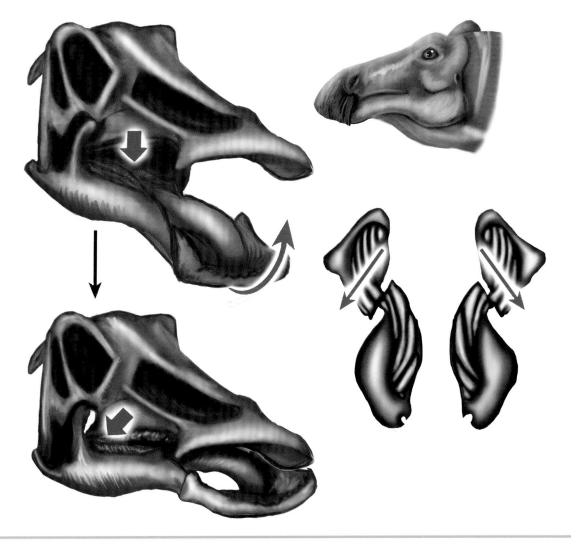

Figure 12.14. Pleurokinetic model in a hadrosaurid skull (*Edmontosaurus*), with rostral view of tooth occlusion shown on right.

the lower dentition comes into occlusion with the upper dentition (Weishampel, 1983, 1984a; Norman, 1984; Norman and Weishampel, 1985; Norman et al., 1991).

Although complex, the pleurokinetic model accounts for the unusual transverse (i.e., buccolingual or apicobasal) tooth wear patterns seen in hadrosaurs, contradicting Ostrom's (1961) hypothesis of a solely propalinal (fore-aft) jaw mechanism. Weishampel (1984a) also pointed out other indications from tooth wear that are supportive of pleurokinesis. The distribution of enamel on the labial surface of maxillary dentition and lingual surface of dentary dentition and shallowly concave wear

facets with flush enamel-dentine interfaces are two examples.

In terms of ornithopod feeding diversity, Weishampel (1984a) noted that early-diverging ornithopods and non-hadrosaurid iguanodontians show angled wear facets (with non-hadrosaurid iguanodontians like *Iguanodon* showing steeper wear angles) that indicate a primarily orthal, isognathous feeding motion (in congruence with pleurokinesis). This orthal feeding motion was also suggested by Galton (1974) and Sues (1980) for early-diverging ornithopods (see above) and Norman (1980, 1984) for non-hadrosaurid iguanodontians. More recently, Virág and Ősi (2017) found parallel orthal to slightly orthopalinal feeding motions with dental microwear in the rhabdodontid *Mochlodon* as well, further confirming the consistency in this wear pattern.

Among hadrosaurids, Weishampel (1984a) noticed variation in feeding strategies. In *Prosaurolophus*, for instance, the teeth wore at the steepest angles among hadrosaurids, creating a more vertical slicing action (although this condition could not be verified later by Mallon and Anderson, 2014b). In comparison, flatter or slightly concave wear facets were found in other hadrosaurids like *Lambeosaurus*, *Corythosaurus*, and *Edmontosaurus*. The variation in wear across taxa could broadly be linked to variable jaw morphology as well as dental battery structure. For instance, lambeosaurines generally possess fewer tooth families than saurolophines because of their overall smaller upper and lower jaws (Weishampel, 1984a).

Recent microwear analysis of the hadrosaurid *Edmontosaurus* quantified the transverse tooth wear patterns with statistical analysis of orientation of tooth wear on the occlusal surfaces and suggested that the tooth wear supports the pleurokinetic model (Williams et al., 2009). In addition to transverse wear, their analysis showed signs of Ostrom's (1961) proposed propalinal jaw action, suggesting a combination of the two in a more complex feeding mechanism.

Some recent studies have challenged the pleurokinetic model (Holliday and Witmer, 2008; Rybczynski et al., 2008; Bell et al., 2009; Cuthbertson et al., 2012). Holliday and Witmer (2008) argued that although there was potential mobility at both the basal and otic joints (in addition to the presence of protractor musculature), there were not enough permissible linkages within the cranium to make primary motions of pleurokinesis to occur and that there were endocranial spatial constraints as well. As an alternative, they posited a possible mobility at the predentary-dentary junction. Rybczynski et al. (2008) modeled joints within the skull of *Edmontosaurus* in three dimensions to test the movements of bones that would be involved in pleurokinesis. They found there may have been too much separation of some of the joints in secondary movements within the palate and face in the pleurokinetic mechanism. They also described potential intracranial spatial constraints (including those from soft tissues) that they proposed would have further hindered pleurokinetic movements. Additionally, Rybczynski et al. (2008) compared their analyses to movement of individual elements observed by Weishampel (1984a) of the lambeosaurine *Corythosaurus*. Whereas in *Corythosaurus* the maxilla and quadrate would rotate laterally 10 degrees and 7 degrees, respectfully (Weishampel, 1984a), in *Edmontosaurus*, both elements would only rotate 3 degrees (Rybczynski et al., 2008).

Bell et al. (2009) performed finite element analysis (FEA) on the mandible of a lambeosaurine hadrosaurid and compared it to that of a ceratopsid. Their results showed that joint reaction forces induced torsion in postdentary elements (with force vectors showing counterclockwise twisting), and the authors speculated the likelihood of long-axis rotation of the mandibular rami with a loose articulation with the predentary as being essential to the feeding mechanism. This assertion mirrors the original inferences of Nopcsa (1900), Versluys (1923), and Kripp (1933). They note that m. pterygoideus ventralis would have helped stabilize the jaw in counteracting the action of long-axis rotation (which they deduced would have been induced mainly by m. adductor mandibulae externus medialis). Additionally, they found increased tensile forces at the dentary-surangular junction at the level of the coronoid process, which likely dampened the stresses in that area by increasing the area of ligamentous attachment between the bones. When a transverse, isognathous feeding mechanism was tested, it resulted in high shear and buccolingually directed reaction forces from food along the occlusal surface (along with the torsional stresses of postdentary elements noted above).

Cuthbertson et al. (2012) performed a similar analysis to Rybczynski et al. (2008) with 3D animation of the laser-scanned skulls of both *Brachylophosaurus* and *Edmontosaurus*. They tested two feeding mechanisms. For the first mechanism, they tested the kinetic limitations of a pleurokinetic model. Overall, they came to a similar conclusion as Rybczynski et al. (2008), with greater depth regarding intracranial kinetic limitations affecting the capacity for a laterally swinging maxilla in pleurokinesis in the hadrosaurid skull. For the second mechanism, they re-created the hemi-mandibular long-axis rotation hypothesis posed by Bell et al. (2009) by first adducting the jaw in orthal closure and second incorporating a somewhat limited long-axis rotation of the mandibular rami along the transverse curvature of the longitudinal occlusal surface. Overall, Cuthbertson et al. (2012) suggested that hemi-mandibular long-axis rotation was the more plausible mechanism when animated. They further showed how there were both lingual and buccal wear zones along the tooth row, indicating differences in how the tooth row was used in orthal, palinal, and rotational chewing strokes.

Mallon and Anderson (2014b), in examining dental microwear of ornithischians of the Dinosaur Park Formation, found that hadrosaurids statistically show signs of a range across apicobasal (buccolingual), mesiodistal, and intermediately angled tooth wear, indicating combined orthal to palinal feeding strokes. These results largely agree with those of Williams et al. (2009); however, instead of claiming that these wear patterns are due to pleurokinesis, Mallon and Anderson (2014b) indicate that such tooth scratches are just as likely to occur in an orthopalinal feeding mechanism. They also found that the tooth batteries of hadrosaurids were suited for both shearing and crushing for a broader diet overall, with slight interspecific differences between and among lambeosaurines and saurolophines, indicating diverse feeding strategies even within communities where these animals coexisted, as was also noted by Weishampel (1984a). Mallon and Anderson (2014b) suggested that previously hypothesized hemi-mandibular long-axis rotation might be a possible mechanism as well, and that the tooth scratches are no less indicative of hemi-mandibular rotation than they are of pleurokinesis.

To examine the possibility of hemi-mandibular long-axis rotation further, Nabavizadeh (2014) described the morphology of the predentary and dentary as well as the junction between the two in more depth in an effort to assess functional implications. He noted that the cau-

dolateral processes of the broad denticulate predentary in hadrosaurids sit upon the dorsal rim of the rostral extent of the dentary (with the smaller dorsal and elongate ventral midline processes surrounding the symphysis); however, there is a loose articulation between the two elements, with the predentary seemingly floating around the rostral extent of the dentaries.

Nabavizadeh (2014) interpreted this morphology as indicative of mobility at this joint, with the rostrocaudally elongate symphysis between the dentaries rotating dorsoventrally as each side of the jaw rotates around its long axis. Although recent previous studies inferred possible mobility at the predentary-dentary joint as well (e.g., Holliday and Witmer, 2008; Bell et al., 2009; Cuthbertson et al., 2012), Nabavizadeh (2014) was the first to examine the morphology of the predentary itself and its relation to the dentaries. Nabavizadeh (2014) also examined the other mandibular elements and how they would play into a rotational movement. For instance, the cylindrical, rounded body of the dentaries would have kept the jaw from bowing outward. Also, possible mobility was described between some elements, including inferred flexibility between the splenial and other elements. Furthermore, Nabavizadeh (2014) described how the spherical nature of the ventral head of the quadrate created a clear ball-and-cup morphology of the jaw joint that would have also allowed substantial rotation of the hemi-mandibles.

In addition to mandibular morphology, Nabavizadeh (2014) qualitatively described the tooth rows of hadrosaurids and found similar dental wear patterns to Williams et al. (2009) and Mallon and Anderson (2014b), with both mesiodistal and buccolingual scratches (noting that the mesiodistal scratches are more apparent along the cutting edge of the dentary tooth row). Additionally, he described a slightly rounded wear along the maxillary cutting edge with a slightly concave pit where dentary teeth occluded and rotated. This along with a longitudinal 20-degree offset angle of occlusion as shown by the dentary occlusal surface further confirmed the likelihood of long-axis rotation of the dentaries against the predentary at occlusion. A four-step isognathous feeding mechanism was described by Nabaviza-deh (2014), with initial orthal closure of the jaws, followed by palinal retraction of the mandible, then medial rotation of hemi-mandibles while in occlusion, creating a bolt-cutter-like cropping mechanism between the dentary and maxillary dentition, and finally a reversion of the hemi-mandibles back to vertical orientation in jaw-opening. The benefit of such a unique mechanism is that it allows the animal to process a large amount of vegetation on both sides of the jaw simultaneously, maximizing the amount of food intake at every bite with their dental batteries (Nabavizadeh, 2014). Additionally, although this mechanism could act as an alternative to the abovementioned pleurokinetic model (Weishampel, 1983, 1984a; Norman, 1984a; Norman and Weishampel, 1985), it is possible for both models to have worked in tandem for a more complex feeding mechanism if, in fact, pleurokinesis did occur to some extent.

Nabavizadeh and Weishampel (2016), in examining predentary-dentary joint diversity across all of Ornithischia, described how hemi-mandibular long-axis rotation was present in feeding mechanisms throughout Ornithopoda, although to different degrees. Early-diverging ornithopods likely only rotated their mandibular corpora slightly owing to their simpler PDJ morphology (PDJ Morphotype I) in a more or less straight orthal, isognathous feeding mechanism (as noted by Weishampel, 1984a). Their predentaries ranged from triangular to rounded or cup-shaped rostrally (as noted above), likely with a more selective feeding style. Non-hadrosaurid iguanodontians likely also incorporated long-axis rotation of the mandibular corpora more so than early-diverging ornithopods, exhibiting PDJ Morphotype II (as in hadrosaurids); however, their mandibular rotation was still likely less extreme than hadrosaurids, as shown by their high-angled dental wear patterns and simpler tooth batteries. Non-hadrosaurid iguanodontians likely also incorporated orthopalinal feeding motions in addition to long-axis rotation, similar to the more extreme case in hadrosaurids. Iguanodontians as a whole exhibited much broader predentary morphologies compared to early-diverging ornithopods, indica-

tive of more generalist feeding style in comparison. The ventromedially recurved symphyseal process in iguanodontians (and other PDJ Morphotype II ornithischians such as stegosaurs and ankylosaurs) would have increased the capacity for the dentaries to medially rotate against each other along the rostrocaudally elongate symphysis without the dentaries separating (Nabavizadeh and Weishampel, 2016).

Cranial musculature has been studied extensively in ornithopods (e.g., Dollo, 1884; Lull and Wright, 1942; Ostrom, 1961; Galton, 1974; Norman, 1984; Weishampel, 1984a; Holliday, 2009; Mallon and Anderson, 2015; Nabavizadeh, 2016, 2020a, 2020b). Cranial morphology of early-diverging ornithopods (fig. 12.15a) suggests that the mAME muscle complex was moderately expanded rostrocaudally at its temporal origin and oriented more vertically within the adductor chambers to insert at a low, subtriangular coronoid eminence of the mandible, acting in pulling the jaw in orthal occlusion (Nabavizadeh, 2020b). The same morphology is generally seen in non-hadrosaurid iguanodontans as well (e.g., *Iguanodon*; fig. 12.15b), although the taller coronoid process in these animals created a greater overall mechanical advantage of temporal musculature (see below). A possible slight rostrolabial expansion of mAMES could additionally aid in a slight palinal component of non-iguanodontian ornithopod feeding by increasing leverage of the temporal muscle complex as well as helping to pull each side of the mandible to rotate medially for more extreme hemi-mandibular rotation (Nabavizadeh, 2020a, 2020b).

Hadrosaurids show more variation in cranial muscle morphology. Some hadrosaurids (e.g., the lambeosaurine *Corythosaurus* [fig. 12.16a] and the saurolophine *Gryposaurus* [fig. 12.17a]) had rostrocaudally expanded temporal muscle (mAME) origins (with a more subrectangular adductor chamber), whereas other hadrosaurids (e.g., the lambeosaurine *Parasaurolophus* [fig. 12.16b] and the saurolophine *Edmontosaurus* [fig. 12.17b]) show rostrocaudally shorter and caudally displaced origins for this muscle complex (with a more triangular adductor chamber). The latter taxa therefore suggest a lower-

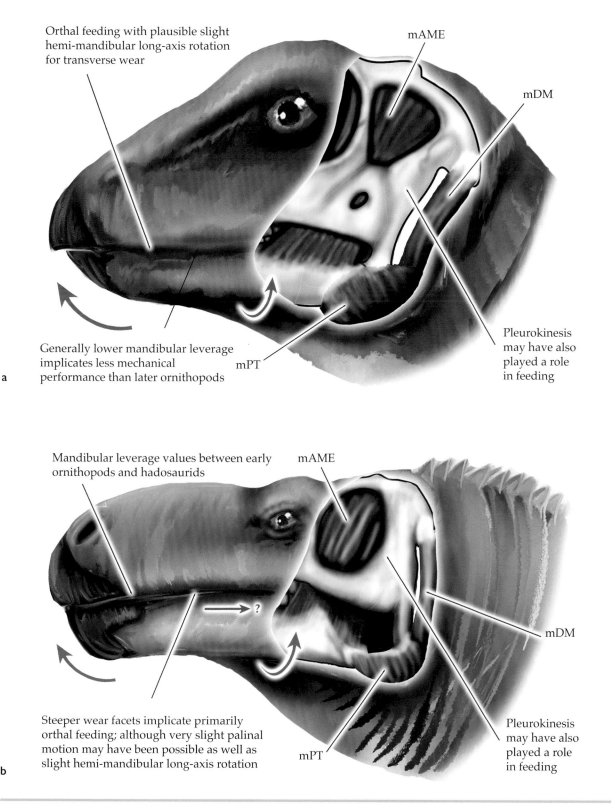

Figure 12.15. Heads of (a) *Hypsilophodon* and (b) *Iguanodon* showing feeding function with jaw muscles. Red arrows show various jaw actions. Abbreviations: mAME, m. adductor mandibulae externus; mDM, m. depressor mandibulae; mPT, m. pterygoideus.

angled morphology of adductor musculature with respect to the horizontal plane compared to the more vertical adductor musculature inferred for the former taxa (Nabavizadeh, 2020b). The slightly laterally flared jugal flange has also led to the suggestion that a somewhat larger fan of temporal musculature may have extended to attach rostrolabially along the labial dentary ridge, covering the dentition in lateral view. This fan of muscle may have been mAMES, as suggested for other ornithischians such as ceratopsians and ankylosaurs, or possibly fibers of m. pseudotemporalis in hadrosaurids specifically, owing to the ambiguously restricted nature of

the bowed coronoid process against the inner margin of the jugal (Nabavizadeh, 2020a, 2020b). If a larger, rostrolabially expanded fan of temporal muscle indeed existed, it would have increased the overall mechanical advantage of the feeding apparatus and would have helped brace the jaw bilaterally to aid in a palinal feeding motion at occlusion. Along with palinal feeding, this muscle reconstruction would have also facilitated hemi-mandibular long-axis rotation against the predentary because of how the insertion of the muscle lateral to the bowed coronoid process would make it pull each dentary medially when contracted (Nabavizadeh, 2020a, 2020b).

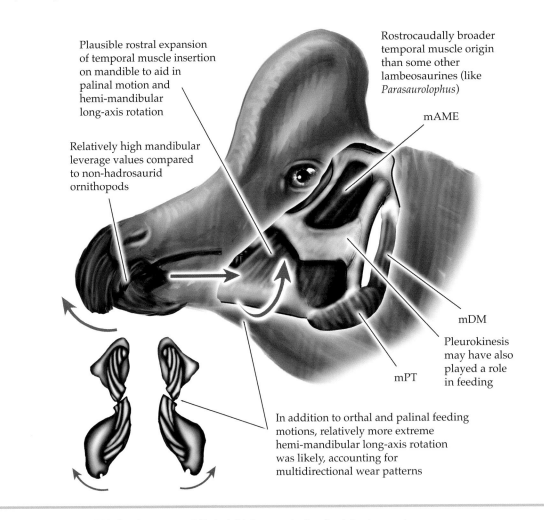

Figure 12.16. Heads of (a) *Corythosaurus* and (facing) (b) *Parasaurolophus* (both lambeosaurine hadrosaurids) showing feeding function with jaw muscles and rostral views of tooth occlusion showing hemi-mandibular long-axis rotation (elements of tooth occlusion redrawn from Ostrom, 1961). Red arrows show various jaw actions. Abbreviations: mAME, m. adductor mandibulae externus; mDM, m. depressor mandibulae; mPT, m. pterygoideus.

Palatal muscle attachment sites are smaller- to moderate-sized in early-diverging ornithopods and would have added more strength to jaw closure secondary to the strength of temporal musculature. In all iguanodontians, the dorsally heightened and moderately expanded palate would have directed larger mPT muscle bodies more vertically for more forceful orthal feeding strokes (Nabavizadeh, 2020b). Additionally, it is possible that mPT may have been larger in iguanodontian taxa with relatively reduced temporal musculature (Nabavizadeh, 2016). As Bell et al. (2009) noted, the mPT insertion around the retroarticular process and angular would have

helped allowed this muscle to counteract rotational forces during hemi-mandibular long-axis rotation.

Using a concept and methodology for lever arm mechanics analysis initially proposed for hadrosaurids and ceratopsids by Ostrom (1961, 1964a, 1966), Mallon and Anderson (2015) analyzed the relative bite force with respect to the muscle vector of m. adductor mandibulae externus profundus (mAMEP, inserting at the apex of the coronoid elevation) at three points along the tooth rows (predentary, mesial tooth, and distal tooth) of a large sample of four hadrosaurid genera from the Dinosaur Park Formation. They compared the individuals sta-

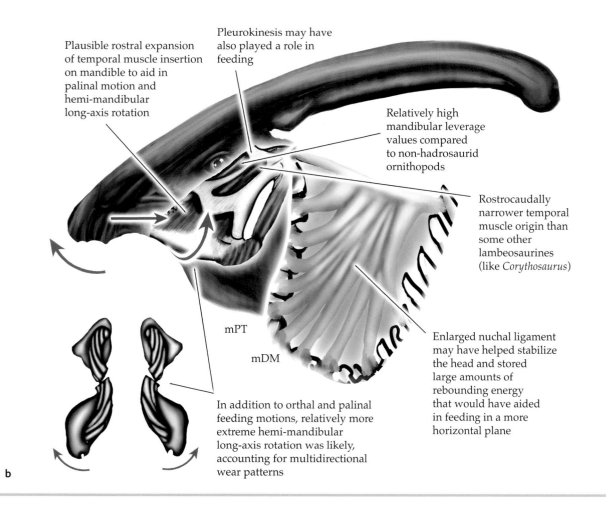

Plausible rostral expansion of temporal muscle insertion on mandible to aid in palinal motion and hemi-mandibular long-axis rotation

Pleurokinesis may have also played a role in feeding

Relatively high mandibular leverage values compared to non-hadrosaurid ornithopods

Rostrocaudally narrower temporal muscle origin than some other lambeosaurines (like *Corythosaurus*)

mPT

mDM

In addition to orthal and palinal feeding motions, relatively more extreme hemi-mandibular long-axis rotation was likely, accounting for multidirectional wear patterns

Enlarged nuchal ligament may have helped stabilize the head and stored large amounts of rebounding energy that would have aided in feeding in a more horizontal plane

b

tistically with each other as well as with other cohabitating ornithischians. They found no significant difference between individuals of the same genus and also no significant difference between hadrosaurid genera. Additionally, although they did find significant difference in relative bite forces (RBFs) between hadrosaurids and ankylosaurids, they found no significant differences between hadrosaurids and ceratopsids. Both hadrosaurids and ceratopsids showed much higher RBF values, especially at the most distal tooth positions (Mallon and Anderson, 2015).

Nabavizadeh (2016), using the same lever arm methodology at four points along the tooth row (predentary,

mesial tooth, middle tooth, and distal tooth), compared jaw adductor mechanics across all of Ornithischia. With respect to ornithopods, results showed variably significant differences evolutionarily. Early-diverging ornithopods showed generally lower mechanical advantage values that gradually increased along the tooth row from the predentary to the distal tooth positions. Non-hadrosaurid iguanodontians show a slightly significant increase in RBF values when compared to early-diverging ornithopods (with *Iguanodon* and *Altirhinus* scoring at the higher end), a trend that continues on throughout ornithopod evolution with the enlargement and lengthening of the snout, heightening of the coronoid process, and the ex-

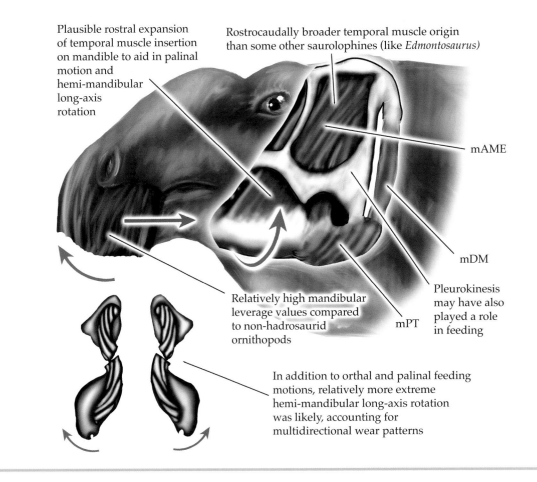

Figure 12.17. Heads of (a) *Gryposaurus* and (facing) (b) *Edmontosaurus* (both saurolophine hadrosaurids) showing feeding function with jaw muscles and rostral views of tooth occlusion showing hemi-mandibular long-axis rotation (elements of tooth occlusion redrawn from Ostrom, 1961). Red arrows show various jaw actions. Abbreviations: mAME, m. adductor mandibulae externus; mDM, m. depressor mandibulae; mPT, m. pterygoideus.

tension of the tooth row medial and caudal to the coronoid process (Nabavizadeh, 2016), which creates shorter output lever arms for the distal-most tooth position. This is taken to the extreme in hadrosaurids and creates a transition in the jaw apparatus from a third-class lever to a second-class lever, convergent with ceratopsids (Ostrom, 1961, 1964a, 1966; Tanoue et al., 2009b; Mallon and Anderson, 2015; Nabavizadeh, 2016). Indeed, because of this transition in anatomy and bite performance, the most significant change in mechanical advantage is seen in hadrosaurids, which score significantly higher in relative bite force values than in all other ornithopods. This indicates a much more mechanically advantageous

feeding apparatus in hadrosaurids, especially at the middle and distal-most tooth positions, which makes sense given the incredibly complex nature of their feeding mechanisms described above.

No significant difference is seen among hadrosaurid taxa, however, including between taxa belonging to both Lambeosaurinae and Saurolophinae, where Nabavizadeh (2016) found substantial overlap in RBF values in agreement with Mallon and Anderson (2015). When observed more closely, some saurolophines such as *Edmontosaurus* show lower leverage at the predentary, likely because of their highly elongate diastemata stretching the bill much further rostrally than in other hadrosaurs,

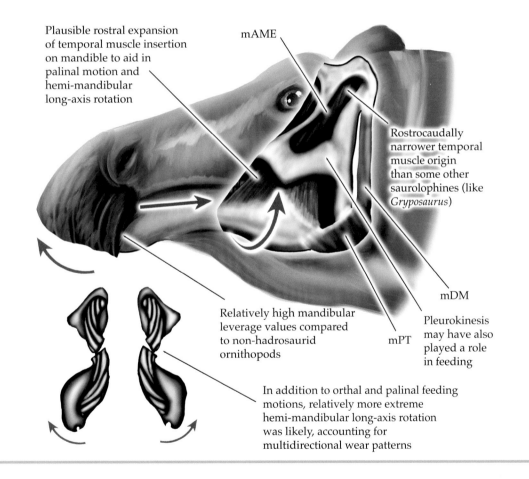

Plausible rostral expansion of temporal muscle insertion on mandible to aid in palinal motion and hemi-mandibular long-axis rotation

mAME

Rostrocaudally narrower temporal muscle origin than some other saurolophines (like *Gryposaurus*)

Relatively high mandibular leverage values compared to non-hadrosaurid ornithopods

mDM

mPT

Pleurokinesis may have also played a role in feeding

In addition to orthal and palinal feeding motions, relatively more extreme hemi-mandibular long-axis rotation was likely, accounting for multidirectional wear patterns

b

such as lambeosaurines with shorter diastemata. Interestingly, *Edmontosaurus* also showed the highest mechanical advantage at the distal-most tooth position, suggesting a more distally focused biting capacity overall. Minor fluctuations in RBF values may also indicate variations in food sources, such as ferns, palms, cycads, conifers, and angiosperms (Nabavizadeh, 2016).

The evolutionary trends seen between ornithopods and marginocephalians are especially interesting because, although there was significant difference found between hadrosaurids and earlier-diverging ornithopods, hadrosaurids showed no significant difference with ceratopsids in mechanical advantage values (as shown by Mallon and Anderson, 2015; see above). Likewise, early-diverging ornithopods showed no significant difference with early-diverging marginocephalians, indicating a convergence in heightened mechanical advantage brought on by the heightening of the coronoid process in both hadrosau-

rids and ceratopsids (Nabavizadeh, 2016). Despite this superficial convergence in mechanical advantage, however, hadrosaurids seemed to increase leverage with increased adductor muscle angle compared to earlier ornithopods, whereas ceratopsids did so with a decrease in adductor angle, highlighting how many factors in cranial anatomy can influence jaw mechanics in different ways and can still lead to similar results overall (Nabavizadeh, 2016).

Ornithopoda is the second of three ornithischian groups discussed here that portray an evolutionary transition from bipedality in early-diverging, smaller forms (fig. 12.18) to quadrupedality in the more derived iguanodontians (fig. 12.19). Ornithopods (especially iguanodontians) generally have curved cervical vertebrae, shorter forelimbs than hindlimbs, and a large, elongate tail. The necks of ornithopods are typically of moderate length, but muscle scars on the caudal aspect of the occiput are

Figure 12.18. Skeleton of *Dryosaurus* (being chased by *Ceratosaurus*).

Figure 12.19. Skeleton of *Iguanodon*.

indicative of strong cervical musculature (Lull and Wright, 1942; Ostrom, 1961; Bertozzo et al., 2020). The series of cervical vertebrae is recurved ventrally in quadrupedal stance of hadrosaurs, and recent analysis of pathological dorsal vertebrae in *Parasaurolophus* by Bertozzo et al. (2021) has described the likely presence of a tall and rostrocaudally expanded nuchal ligament stretching from the occiput to the neural spines of dorsal vertebrae in hadrosaurs (fig. 12.16b). The strong elastic nature and girth of such a large ligament would have effectively stabilized the head and stored large amounts of rebounding energy helpful in feeding in a horizontal plane vegetation within its range of height.

The vertebral column as a whole is especially strongly built in iguanodontians, which show signs of strong epaxial musculature and lattice-like ossified tendons spanning throughout the dorsal, sacral, and caudal vertebrae that help stabilize the back and tail. Organ's (2006a, 2006b) reconstruction of epaxial musculature based on extant archosaurs has helped elucidate the origins and functional implications of these tendons as well, showing them to make the tail stiffer and thereby increasing rigidity of the attachment of m. caudofemoralis and improving its ability to retract the hindlimb forcefully. Additionally, the ossified tendons have been suggested as helping support a horizontal, quadrupedal stance and as a source of stored elastic energy in gait (Organ, 2006a, 2006b). Notably, there was likely still some mobility in locomotion between some mid to distal caudal vertebrae (where ossified tendons are more lacking), as interpreted owing to signs of mechanical stress recently described in the tail of *Edmontosaurus* (Siviero et al., 2020). Slight differences in tailbone and inferred muscular morphology have been found in other hadrosauroids as well (such as in *Tethyshadros*), suggesting further variability in locomotor strategies (Dalla Vecchia, 2020).

Appendicular musculature and locomotor function in ornithopods has been reconstructed or discussed numerous times, giving valuable insight into the diversity and complexities of locomotor style throughout the clade (Romer, 1927; Lull and Wright, 1942; Ostrom, 1964b; Galton, 1969, 1970, 1974; Norman, 1980, 1986; Wright, 1999; Dilkes, 2000, 2001; Carpenter and Wilson, 2008; Maidment and Barrett, 2011, 2014; Maidment et al., 2012, 2014a, 2014b; Persons and Currie, 2014, 2020; Barrett and Maidment, 2017). Many of the more recent works listed (especially from the 1980s to present) have determined some degree of quadrupedality in hadrosaurids and other ornithopods. Additionally, there is a divergence seen in rates of postcranial changes (particularly the appendicular adaptations) between lambeosaurine and saurolophine hadrosaurids (Stubbs et al., 2019), possibly highlighting the complexity of diverse locomotor strategies.

The transition from bipedality to graviportal quadrupedality among ornithopods is revealed by osteological characteristics that are demonstrative of mechanical reconfiguration of limb musculature. Early-diverging ornithopods do not typically show characteristics associated with quadrupedality. Hadrosaurids (fig. 12.20), however, show the five major characteristics, including a rostrolateral process of the ulna, hooflike unguals on the hands, a transversely broadened ilium, femora longer than tibia, and a reduced fourth trochanter. Interestingly, and perhaps unsurprisingly, numerous non-hadrosaurid iguanodontians show a middle ground between the two morphotypes, portraying a mosaic of some of these characteristics that implicates various methods of facultative quadrupedality depending on the taxon (Maidment and Barrett, 2014). There has, however, still been speculation on the facultative or obligate nature of quadrupedality in hadrosaurids for various reasons (e.g.,

Sellers et al., 2009; Maidment et al., 2014b), including how the center of mass is positioned directly dorsal to the hind feet, allowing them to stand in both positions (Bates et al., 2009; Maidment et al., 2014b). Still, other studies have shown characteristics for more obligate quadrupedality (Dilkes, 2001; Maidment et al., 2012; Maidment and Barrett, 2012, 2014; Barrett and Maidment, 2017).

In the forelimb, hadrosaurids show enhanced characteristics for humeral abductor moment arms, with a laterally projected acromial process and large m. deltoideus attachment sites, with their manus placed near the midline and elbows tucked inward (Maidment and Barrett, 2012, 2014; Barrett and Maidment, 2017). A slender preacetabular process (with smaller muscles associated with femoral protraction) and enlarged ischium (with caudal movement of adductors co-opting them for assisting in more powerful femoral retraction) are also adaptations for quadrupedality (Maidment and Barrett, 2012). In addition, ontogenetic shifts from bipedality in juvenile hadrosaurids to quadrupedality in adult hadrosaurids have been proposed (Norman, 1980; Dilkes, 2001; Maidment and Barrett, 2014; Barrett and Maidment, 2017).

The bipedal early-diverging ornithopods such as *Hypsilophodon* possess a more proximally positioned fourth trochanter of the femur, allowing for more rapid retraction at the expense of mechanical advantage—a trait seen in other small early-diverging ornithischians as well (Maidment and Barrett, 2014; Barrett and Maidment, 2017; Persons and Currie, 2020). Maidment et al.'s (2014a) biomechanical study found that *Hypsilophodon*, which has a small ilium and short postacetabular process, shows relatively some of the lowest moment arms of most functional muscle groups in the hindlimbs, implicating less developed locomotor ability despite a cursorial *Bauplan*.

Contrarily, derived iguanodontians—especially hadrosaurids—show a more distally positioned, reduced fourth trochanter of the femur for increased mechanical advantage at the expense of speed (Maidment and Barrett, 2014; Persons and Currie, 2014, 2019). This adap-

Figure 12.20. Skeleton of *Edmontosaurus*.

tation is ideal for sustained locomotion and endurance, especially for running away from predators like tyrannosaurids (Persons and Currie, 2014). Biomechanics of the hindlimb of the saurolophine hadrosaurid *Edmontosaurus* show high moment arms in muscle groups involved in flexion and extension as well as medial and lateral rotation of the femur. Although hadrosaurids show a laterally everted ilium, it is not as exaggerated as in the wider-hipped derived thyreophorans and ceratopsids. Consequently, abductor moment arms show intermediate results between bipedal and quadrupedal groups (with high abduction moments in the wider-hipped quadrupeds; Maidment and Barrett, 2014; Maidment et al., 2014a; Barrett and Maidment, 2017).

The smaller, bipedal ornithopods likely used their small beaks for eating small, high-energy plant material like stems, shoots, fruits, and seeds within a 2-m-high range (Weishampel and Norman, 1989). Gastroliths have been found associated with *Gasparinisaura*, which may indicate the need to digest moderately tougher plant material (Cerda, 2008). Although early-diverging bipedal ornithopods were quite small, iguanodontians—particularly hadrosaurids—often grew to be truly gigantic, in some cases even competing with sauropods in terms of size (e.g., *Shantungosaurus*, which grew to be up to 16.5 m in length; Zhao et al., 2007; Słowiak et al., 2020). Recent histological evidence has suggested that the massive sizes of hadrosaurids may have been the result of an evolutionary transition from a cyclic to a continuous, uninterrupted growth cycle (Słowiak et al., 2020).

Isotopic evidence confirms that hadrosaurids were clearly in range of an herbivorous lifestyle (Cullen et al., 2020), which was also likely the case with the rest of Ornithopoda. Cullen et al. (2022) found that hadrosaurid dietary niche signatures differed from that of other ornithischians owing to height and range differences. With such large sizes in iguanodontians also comes taller feeding height ranges. Non-hadrosaurid iguanodontians likely had a more generalized diet feeding on vegetation at heights of between 2 and 4 m (Weishampel and Norman, 1989). Ostrom (1964b) early on proposed that hadrosaurids were active terrestrial foragers, browsing on conifers, deciduous trees, and shrubs in lowlands, coastal plains, and forests. Mallon et al. (2013) found that hadrosaurids of the Dinosaur Park Formation would have been capable of feeding up to 2 m high in quadrupedal stance and up to 5 m high in bipedal stance, allowing them to reach shrubs or low-growing trees. With the broad range in their feeding envelopes, hadrosaurid diets (and possibly those of other iguanodontian) were likely more generalized, consisting of tougher leaves and stems of conifers, ferns, horsetails, and angiosperms, with likely consumption of fruits and seeds in the process (Mallon and Anderson, 2014b). Coprolites associated with the saurolophine have shown traces of a leaf-dominated diet, with uniform and finely processed leaf remains affirming the complex feeding mechanisms of hadrosaurids described above (Tweet et al., 2008). Furthermore, properties of amber associated with a recently described *Prosaurolophus* jaw implicated that it fed in a coniferous environment (McKellar et al., 2019). Chin et al. (2017) also described hadrosaurid coprolites showing signs of rotted coniferous woody material along with crustaceous cuticles, possibly indicating seasonal dietary shifts related to oviparous breeding activities (gaining proteins and calcium). This finding shows the ability for large herbivorous dinosaurs like hadrosaurids to gain a more flexible diet as the need arose (Chin et al., 2017).

Mallon et al. (2013) also suggested that coexisting hadrosaurids may have used different feeding heights and postures to avoid competition. For instance, with some saurolophines like *Prosaurolophus* showing finer, grittier, more abundant microwear scratches than the lambeosaurine *Lambeosaurus*, there may have been differences in the heights at which they reached vegetation (with the former feeding on lower vegetation with more grit particles; Mallon and Anderson, 2014b). Indeed, an earlier study by Carrano and Janis (1999) also posited differences in feeding environments between saurolophines (which are suggested to have fed in more open environments) and lambeosaurines (which are suggested to have fed in closed environments). Furthermore, Carrano and Janis (1999) suggested that, as a whole, hadrosaurids shared numerous features with modern-day ungulates in terms of feeding adaptation and behavior, and that parallels can be easily drawn between them ecologically.

CHAPTER 13

Marginocephalians

Pachycephalosauria

For reference in understanding relationships of marginocephalians discussed in this chapter, see the phylogeny in figure 13.1.

Marginocephalia includes the dome-headed Pachycephalosauria and the horned and frilled Ceratopsia. Pachycephalosaurs (e.g., *Pachycephalosaurus*, *Stegoceras*, *Prenocephale*, *Tylocephale*, and *Homalocephale*) were mainly from the Late Cretaceous with distribution throughout North America and Asia. (For a list of pachycephalo-

saurs discussed in this chapter, see the taxon list in table 13.1.)

The pachycephalosaur skull (fig. 13.2) presents with a prominent thickening of the skull roof into a dome. Along with this thickening, most articulations of skull elements are co-ossified and akinetic. A lateral and caudal expansion of the squamosals that overhangs the occiput as well as a dorsoventral heightening of the occiput itself also characterize pachycephalosaurs as a clade. Pachycephalosaur skulls have many rounded or conical osteoderms, usually on the squamosals and on the nasal

Figure 13.1. Phylogeny of Early Marginocephalia, based largely on You et al. (2005),

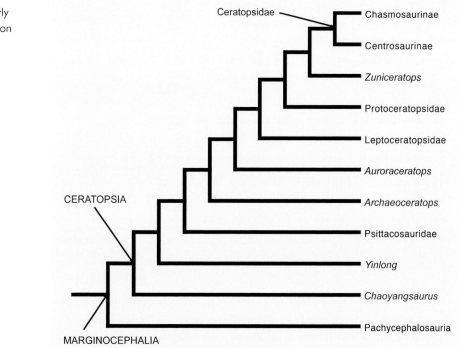

257

**Table 13.1.
Pachycephalosaur genera**

Major clade	Subclade	Genus
Pachycephalosauria		
		Homalocephale
		Pachycephalosaurus
		Stegoceras
		Prenocephale
		Tylocephale
		Goyocephale

region (fig. 13.3). Earlier-diverging pachycephalosaurs, like *Homalocephale* and *Stegoceras*, possess circular supratemporal fenestrae, albeit they are much smaller and nearly closed in *Stegoceras*. Many other derived pachycephalosaur taxa, however, do not have a supratemporal fenestra because it has been closed off because of skull thickening. The infratemporal fenestra is present but rostrocaudally compressed as a thin oval to a crescentic opening. Pachycephalosaurs usually have no antorbital fenestra, except for *Prenocephale*, in which it is small. The dome of the pachycephalosaur skull is made of the frontal and parietal as well as the supraorbital and postorbital along its lateral margins, and it has a variable thickness depending on the taxon (Gilmore, 1924; Maryańska and Osmólska, 1974). Skull thickening was previously thought to have protected the small brain and basicranium from damage during head-butting. But histological studies have shown internal structures inconsistent with head-butting behavior, instead suggesting that the dome was used for species recognition or sexual display (Goodwin and Horner, 2004). Still, it is possible that they could have used their heads in a defense mechanism of some form, including flank-butting (Carpenter, 1997; Farke, 2014).

The temporal region can vary in breadth, with some forms (e.g., *Stegoceras*, *Prenocephale*, and *Tylocephale*) showing caudally restricted and mediolaterally smaller temporal regions, and others (e.g., *Pachycephalosaurus* and *Homalocephale*) exhibiting both transversely and rostrocaudally expanded temporal regions. In *Homalo-*

cephale, by contrast, the postorbital adductor region and infratemporal fenestrae are relatively larger and subrectangular (although at a slight angle), and those of all most other pachycephalosaurs are narrower and lower angled relative to the horizontal plane. Pachycephalosaurs present with the typical ornithischian slight jugal flaring with corresponding subjugal opening. The palate in pachycephalosaurs is notable in that the pterygoid process is relatively quite large. Their palate as a whole is broad, angled more vertically, and restricted more caudally within the skull. The mandible of pachycephalosaurs is slender rostrally, and although no predentary has been reported in the literature, the vertical symphyseal morphology suggests that this element was likely small and cup-shaped or triangular, categorized as predentary-dentary joint (PDJ) Morphotype I (Nabavizadeh and Weishampel, 2016). The dentary is narrow in profile with leaf-shaped, denticulate teeth along the tooth row. The coronoid eminence is raised and triangular with only a slight rostral extension of the coronoid into the labial dentary ridge (LDR). Furthermore, there is no mandibular fenestra, and the retroarticular process is especially long relative to that of other ornithischians.

Pachycephalosaurs typically have three conical, caudoventrally recurved caniniform teeth in the premaxilla, each one larger than the one mesial to it. These teeth are seen in *Prenocephale* (Maryańska and Osmólska, 1974), *Dracorex* (Bakker et al., 2006), *Goyocephale* (Perle et al., 1982), and *Stegoceras* (Gilmore, 1924) but likely existed in other pachycephalosaurs, possibly acting in either plucking plant material or indicative of omnivorous feeding behavior. They are denticulate on their distal margin and possess vertical wear facets (Sues and Galton, 1987; Varriale, 2011), indicative of an orthal tooth-tooth movement. The maxilla contains between 15 and 20 teeth that project ventrally at its ventrolateral margin, although *Goyocephale* had a relatively smaller tooth row (Perle et al., 1982). These maxillary teeth differ from those of the premaxilla in being small, triangular to leaf-shaped, and mediolaterally compressed, with a slightly bulbous cingulum. The crown surfaces have apicobasally oriented ridges that form marginal denticles at the apex,

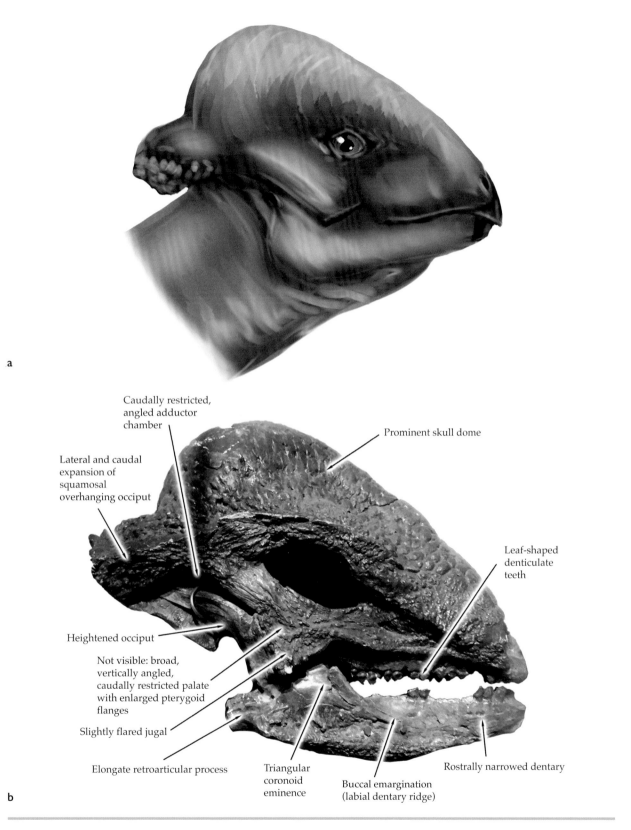

Caudally restricted, angled adductor chamber

Prominent skull dome

Lateral and caudal expansion of squamosal overhanging occiput

Leaf-shaped denticulate teeth

Heightened occiput

Not visible: broad, vertically angled, caudally restricted palate with enlarged pterygoid flanges

Slightly flared jugal

Elongate retroarticular process

Triangular coronoid eminence

Buccal emargination (labial dentary ridge)

Rostrally narrowed dentary

a

b

Figure 13.2. *Stegoceras* (a) reconstructed head and (b) skull in lateral view (length approximately 20 cm).

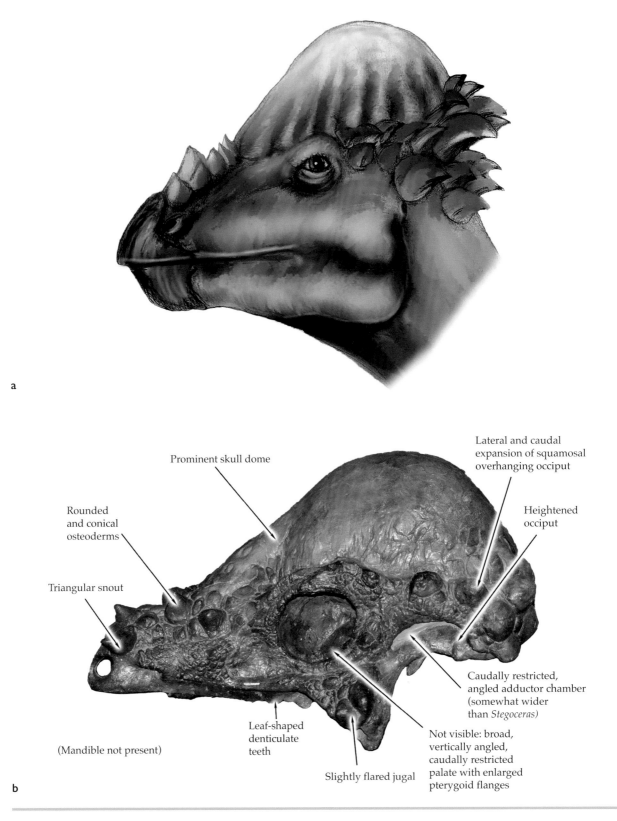

a

b

Prominent skull dome

Lateral and caudal
expansion of squamosal
overhanging occiput

Heightened
occiput

Rounded
and conical
osteoderms

Triangular snout

Caudally restricted,
angled adductor chamber
(somewhat wider
than *Stegoceras*)

(Mandible not present)

Leaf-shaped
denticulate
teeth

Slightly flared jugal

Not visible: broad,
vertically angled,
caudally restricted
palate with enlarged
pterygoid flanges

Figure 13.3. *Pachycephalosaurus* (a) reconstructed head and (b) skull in lateral view
(length approximately 65 cm).

with the largest denticles in the center of the tooth. The teeth are almost in line with each other mesiodistally, with the mesial end of each more caudal tooth overlapped labially by the tooth mesial to it (Maryańska et al., 2004). The tooth crown is slightly convex lingually, while the labial side is concave (Maryańska et al., 2004). The tooth roots are long and cylindrical.

The dentition of the dentary is similar to the maxillary dentition (fig. 13.4). The crown is triangular to leaf-shaped and projects dorsally from a bulbous base. The mesial dentition is typically larger in size than the more distal teeth. The crowns are ridged apicobasally, with the middle ridges the most pronounced. Wear is on the labial side of the dentition, meaning that the dentary teeth occluded with the maxillary teeth lingually (Maryańska et al., 2004).

Pachycephalosaur jaw mechanisms have not been studied in great detail, mainly owing to scarcity of craniodental material with which to infer feeding styles. Gilmore (1924) described the teeth in *Stegoceras*, which, as in most pachycephalosaurs, were leaf-shaped, laterally compressed, and had denticulations directed toward the apex of each tooth, which was rather sharp and curved distally. Most pachycephalosaurs possess a larger canini-

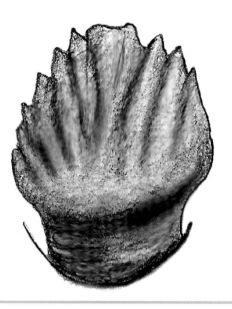

Figure 13.4. *Stegoceras* dentary tooth.

form tooth. Pachycephalosaurs had teeth in the premaxilla and the maxilla, as well as the dentary, and Gilmore (1924) suggested that they (specifically *Stegoceras*) fed on soft vegetation. Maryańska and Osmólska (1974) indicated differences in dental wear and tooth size in *Homalocephale*, *Prenocephale*, and *Tylocephale*, suggesting that each pachycephalosaur, although possessing similar looking teeth, had its own particular feeding style. In some cases, such as *Goyocephale* and *Tylocephale*, the wear on all teeth combined form one occlusal plane, indicative of uniform use of the tooth row as a whole in chewing (Maryańska et al., 2004). Maryańska and Osmólska (1974) also suggested the feeding apparatus of pachycephalosaurs was best equipped for feeding on softer vegetation, such as fruits, leaves, and seeds (with the possible addition of insects as well).

Sues and Galton (1987) interpreted pachycephalosaurs—*Stegoceras* specifically—as having a strictly orthal isognathous feeding mechanism, restricted by a simple hinge joint at the jaw joint that would have prevented palinal feeding motions (fig. 13.5). A mesial convergence of the upper and lower tooth rows and double wear facets on the maxillary and dentary teeth imply shearing capabilities. Sues and Galton (1987) also suggested a potential slight hemi-mandibular long-axis rotation against the predentary (although a predentary is yet unknown among pachycephalosaurs, as noted above). In a comprehensive statistical study investigating tooth wear in marginocephalians with scanning electron microscopy, Varriale (2011) verified Sues and Galton's (1987) hypothesis of orthal feeding motions in pachycephalosaurs. Primarily vertical orthal feeding strokes are clearly seen across the tooth row and, in conjunction with slight hemi-mandibular rotation against the predentary, would have been the most prevalent feeding mechanism in pachycephalosaurs (Sues and Galton, 1987; Varriale, 2011; Nabavizadeh, 2016; Nabavizadeh and Weishampel, 2016).

Pachycephalosaur cranial musculature has only been reconstructed in a handful of studies (Maryańska and Osmólska, 1974; Sues and Galton 1987; Holliday, 2009; Nabavizadeh 2020a, 2020b). The origin of temporal adductor musculature (m. adductor mandibulae externus,

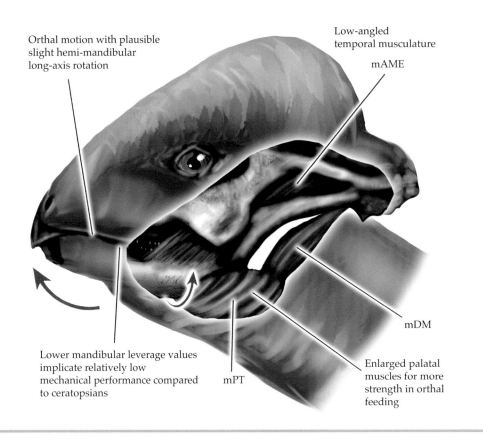

Orthal motion with plausible slight hemi-mandibular long-axis rotation

Low-angled temporal musculature

mAME

Lower mandibular leverage values implicate relatively low mechanical performance compared to ceratopsians

mPT

mDM

Enlarged palatal muscles for more strength in orthal feeding

Figure 13.5. *Stegoceras* head showing feeding function with jaw muscles. Red arrows show various jaw actions. Abbreviations: mAME, m. adductor mandibulae externus; mDM, m. depressor mandibulae; mPT, m. pterygoideus.

or mAME) at the supratemporal bar varies from rostrocaudally short and caudally displaced with respect to the length of the skull (e.g., *Stegoceras*) to rostrocaudally expanded for a larger muscle body (e.g., *Homalocephale* and *Pachycephalosaurus*). These muscle bodies travel within a tunnel-like adductor chamber positioned at a low angle with respect to the horizontal plane and insert around the coronoid eminence of the mandible. The caudal displacement of the temporal origin and the low angle of the adductor musculature positions the jaw joint more rostrally so that, when contracted, these muscles would have elevated the lower jaw while braced near the jaw joint to allow a primarily isognathous orthal feeding motion (as seen in diplodocoids and ornithomimosaurs). The m. adductor mandibulae posterior (mAMP) is also oriented parallel to mAME complex, helping to give supporting orthally directed force (Maryańska and Osmól-

ska, 1974). Additionally, m. adductor mandibulae externus superficialis (mAMES) would have acted in slight medial rotation of the mandibular ramus against the predentary. The rostral displacement of the jaw joint in pachycephalosaurs also positions the insertion of the palatal m. pterygoideus (mPT) musculature closer to its origin just rostrodorsal to it. The enlarged palate creates an equally large mPT muscle origin. The mPT then extends down, and each of its two main muscle bodies inserts around the rostrocaudally elongate retroarticular process. This position of mPT creates a more vertical muscle vector for the mPT muscle complex, and in turn the mechanical advantage of this muscle body was improved. It is therefore likely that mPT would have been the dominant muscle body acting in oral processing, with temporal musculature more so used for the initial closure of the oral cavity (again, like that seen in diplodocoids).

Quantitative biomechanical analyses of pachycephalosaur jaw systems are nearly nonexistent. Nabavizadeh (2016), with two-dimensional (2D) lever arm mechanics on m. adductor mandibulae externus profundus (mAMEP), found that the pachycephalosaur *Stegoceras*, with its low, triangular coronoid eminence, showed relatively low relative bite force (RBF) values (similar to many non-ceratopsoid ceratopsians) compared to other more derived ornithischians, with variable overlap throughout the tooth row in terms of mechanical advantage. As stated above, however, mPT was likely the more dominant muscle body in terms of the actual feeding mechanism in pachycephalosaurs, so these results are likely not representative of the actual feeding mechanics in pachycephalosaurs.

Pachycephalosaurs were bipedal ornithischians with short slender forelimbs, long slender hindlimbs, and a long tail (Maryańska and Osmólska, 1974; Sues and Galton, 1987) (fig. 13.6). They show seemingly intermediate locomotor adaptations that reduce likelihood of cursoriality, such as a femur longer than the tibia (Carrano, 1999; Maidment and Barrett, 2014). Still, they are considered bipedal because their forelimbs are too slender to bear any weight (Maidment and Barrett, 2014).

Pachycephalosaurs show an exceptionally large pelvis (possibly an indication of an enlarged gut system) (Maryańska and Osmólska, 1974; Sues and Galton, 1987), and they also show relatively high moment arm length in hip extension owing to muscular origination from an elongate postacetabular process (Maidment et al., 2014a). It has been suggested that they likely had a more horizontal stance with the head, neck, and tail all held horizontally (Maryańska and Osmólska, 1974). Part of the reasoning for this is the inferred presence of well-developed, strong neck extensor musculature, as shown by exceptionally large occipital muscle scars (especially of m. spinalis capitis and m. longissimus capitis; Maryańska and Osmólska, 1974; Sues and Galton, 1987; Tsuihiji, 2010). There is also evidence for the presence of a strong nuchal ligament, adding extra support to the hypothesis of an erect cervical series, because these features would have been ideal for strength in stabilization of the head (Maryańska and Osmólska, 1974; Sues and Galton, 1987; Tsuihiji, 2010).

A "double ridge-in-groove" articulation is seen between the zygopophyses of some dorsal vertebrae and has been interpreted as preventing major lateral movement of the vertebrae. This form of vertebral articular

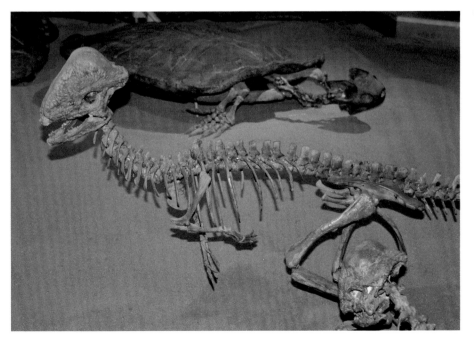

Figure 13.6. Skeleton of *Stegoceras*.

morphology has been suggested as having created a more rigid vertebral column, possibly for easy transmission of impact forces running from the cervical through dorsal vertebrae as a result of some form of head-ramming behavior to prevent buckling (Maryańska and Osmólska, 1974; Sues and Galton, 1987). This adaptation may have been used as a defense against predators in addition to an action in intraspecific combat, although again this behavioral interpretation is still debated (Goodwin and Horner, 2004; Farke, 2014).

Pachycephalosaur caudal vertebrae possess ossified tendons and long caudal ribs, creating extra stability in a horizontally held tail. Tripodal stance using the tail has been proposed for pachycephalosaurs (Maryańska and Osmólska, 1974; Sues and Galton, 1987), possibly to reach for higher vegetation. This may have been a rare occurrence, however, because, as stated previously, pachycephalosaurs likely only fed on softer vegetation such as leaves, fruits, and seeds (Maryańska and Osmólska, 1974), which are usually found low to the ground.

Ceratopsia

Ceratopsians arguably have some of the most elaborate cranial anatomy of all dinosaurian clades. Ceratopsians are a spectacularly diverse clade, ranging from the Late Jurassic to Late Cretaceous with distribution in North America, Europe, and Asia. (For reference in understanding clade affiliation of ceratopsians discussed in this chapter, see the taxon list in table 13.2.)

Ceratopsians included early-diverging taxa (e.g., *Chaoyangsaurus* and *Yinlong*), psittacosaurids (e.g., *Psittacosaurus* [fig. 13.7] and *Hongshanosaurus*), and neoceratopsians (including *Archaeoceratops* [fig. 13.8], *Auroraceratops*, leptoceratopsids [e.g., *Leptoceratops*, *Prenoceratops*, and *Udanoceratops*], protoceratopsids [e.g., *Protoceratops*, *Bagaceratops*, and *Magnirostris*], and ceratopsoids—including early-diverging taxa such as *Zuniceratops*, *Turanoceratops*, and *Ajkaceratops* as well as the derived ceratopsids). Ceratopsidae consisted of the major division between Chasmosaurinae (e.g., *Chasmosaurus*,

Table 13.2.
Ceratopsian genera organized by clade

Major clade	Subclade	Genus
Ceratopsia		
		Yinlong
	Chaoyangsauridae	
		Chaoyangsaurus
	Psittacosauridae	
		Hongshanosaurus
		Psittacosaurus
	Early-diverging Neoceratopsia	
		Archaeoceratops
		Auroraceratops
		Liaoceratops
	Leptoceratopsidae	
		Leptoceratops
		Prenoceratops
		Udanoceratops
	Protoceratopsidae	
		Protoceratops
		Magnirostris
		Bagaceratops
	Non-ceratopsid Ceratopsoidea	
		Zuniceratops
		Turanoceratops
		Ajkaceratops
	Centrosaurine Ceratopsids	
		Centrosaurus
		Einiosaurus
		Monoclonius
		Pachyrhinosaurus
		Nasutoceratops
		Diabloceratops
	Chasmosaurine Ceratopsids	
		Anchiceratops
		Arrhinoceratops
		Chasmosaurus
		Pentaceratops
		Spiclypeus
		Styracosaurus
		Torosaurus
		Triceratops

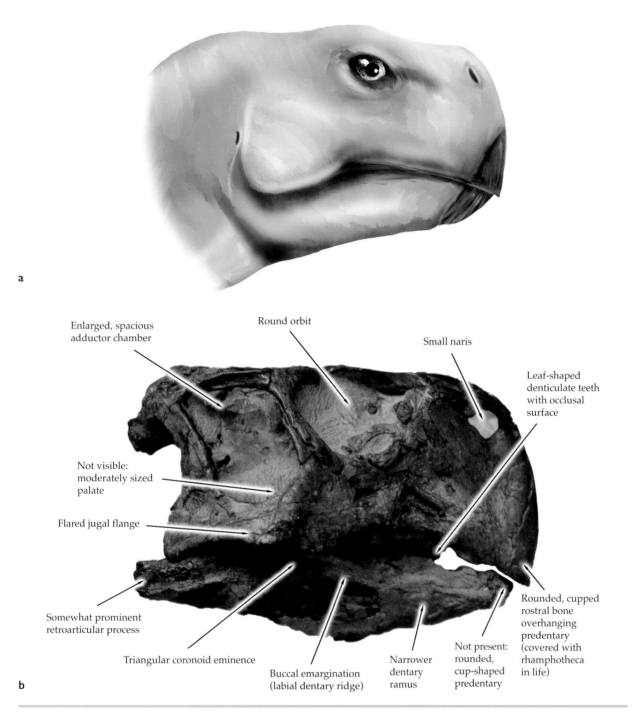

a

b

Enlarged, spacious
adductor chamber

Round orbit

Small naris

Leaf-shaped
denticulate teeth
with occlusal
surface

Not visible:
moderately sized
palate

Flared jugal flange

Somewhat prominent
retroarticular process

Triangular coronoid eminence

Buccal emargination
(labial dentary ridge)

Narrower
dentary
ramus

Not present:
rounded,
cup-shaped
predentary

Rounded, cupped
rostral bone
overhanging
predentary
(covered with
rhamphotheca
in life)

Figure 13.7. *Psittacosaurus* (a) reconstructed head and (b) skull in lateral view
(variably lengths between approximately 12 and 20 cm).

Pentaceratops, *Arrhinoceratops*, *Anchiceratops*, *Spiclypeus*,
and *Triceratops*) and Centrosaurinae (e.g., *Centrosaurus*,
Styracosaurus, *Pachyrhinosaurus*, *Einiosaurus*, *Nasutoceratops*, and *Diabloceratops*).

Neoceratopsians are known primarily for their distinctive, caudodorsally expanded parietosquamosal frill (figs.
13.9 and 13.10), which is otherwise diminutive in non-
neoceratopsians like the earliest-diverging ceratopsians

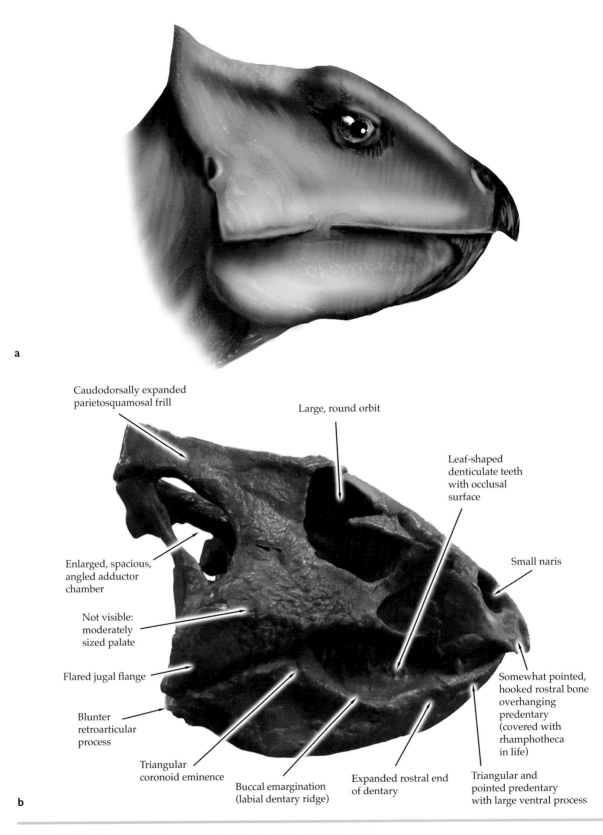

a

Caudodorsally expanded
parietosquamosal frill

Large, round orbit

Leaf-shaped
denticulate teeth
with occlusal
surface

Small naris

Enlarged, spacious,
angled adductor
chamber

Not visible:
moderately
sized palate

Flared jugal flange

Somewhat pointed,
hooked rostral bone
overhanging
predentary
(covered with
rhamphotheca
in life)

Blunter
retroarticular
process

Triangular
coronoid eminence

Buccal emargination
(labial dentary ridge)

Expanded rostral end
of dentary

Triangular and
pointed predentary
with large ventral process

b

Figure 13.8. *Archaeoceratops* (a) reconstructed head and (b) skull in lateral view
(length approximately 15 cm).

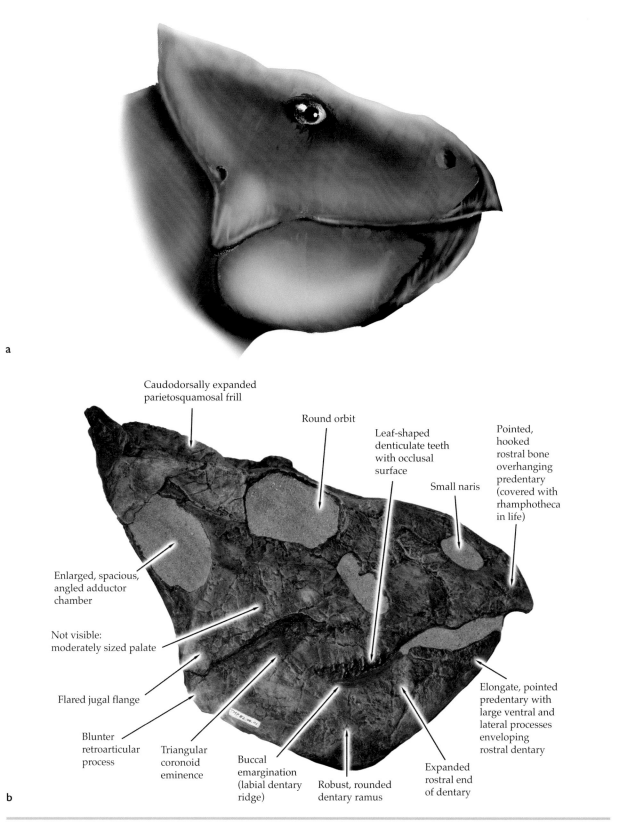

Caudodorsally expanded
parietosquamosal frill

Round orbit

Leaf-shaped
denticulate teeth
with occlusal
surface

Small naris

Pointed,
hooked
rostral bone
overhanging
predentary
(covered with
rhamphotheca
in life)

Enlarged, spacious,
angled adductor
chamber

Not visible:
moderately sized palate

Flared jugal flange

Blunter
retroarticular
process

Triangular
coronoid
eminence

Buccal
emargination
(labial dentary
ridge)

Robust, rounded
dentary ramus

Expanded
rostral end
of dentary

Elongate, pointed
predentary with
large ventral and
lateral processes
enveloping
rostral dentary

a

b

Figure 13.9. *Leptoceratops* (a) reconstructed head and (b) skull in lateral view
(length approximately 45 cm).

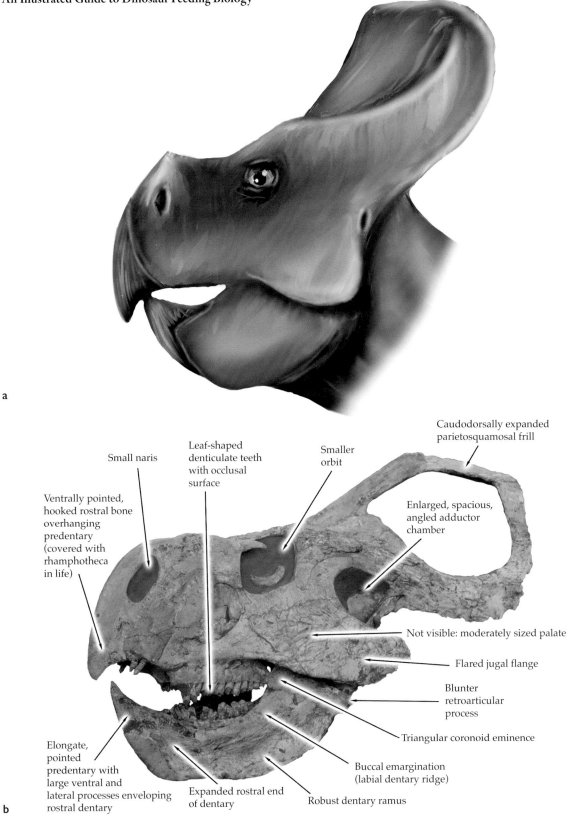

a

b

Caudodorsally expanded
parietosquamosal frill

Small naris

Leaf-shaped
denticulate teeth
with occlusal
surface

Smaller
orbit

Ventrally pointed,
hooked rostral bone
overhanging
predentary
(covered with
rhamphotheca
in life)

Enlarged, spacious,
angled adductor
chamber

Not visible: moderately sized palate

Flared jugal flange

Blunter
retroarticular
process

Triangular coronoid eminence

Elongate,
pointed
predentary with
large ventral and
lateral processes enveloping
rostral dentary

Expanded rostral end
of dentary

Buccal emargination
(labial dentary ridge)

Robust dentary ramus

Figure 13.10. *Protoceratops* (a) reconstructed head and (b) skull in lateral view
(length approximately 60 to 70 cm with frill).

and psittacosaurids. This frill is formed by a caudal extension of the squamosal caudolaterally and the parietal medially, creating a median parietal bar, and often possesses a series of small, bony protrusions called epoccipitals (more specifically, episquamosals and epiparietals) that surround the distal edge of the frill circumference. The frill usually has an enlarged fenestra on either side of the dorsal surface surrounding the parietal bar, except in *Triceratops*, which simply has a solid sheet of bone throughout the frill (Hatcher et al., 1907; Dodson, 1996; Forster, 1996). It is possible that the frill had the structural properties capable of withstanding predatory bites, as shown with three-dimensional (3D) modeling of a solid *Triceratops* frill (Farke et al., 2010). Among ceratopsids, chasmosaurines tend to have much more caudally elongate frills, whereas centrosaurines mostly have relatively shorter frills (figs. 13.11 and 13.12). Rostrally, the frill is sutured to the jugal on either side of the skull, which is represented by a caudoventrally oriented, triangular bony protrusion that slightly enlarges at its distal end.

Ceratopsians are also known for their sharp beak, with a unique rostral bone articulating in front of the premaxillae that is ventrally curved and variably rounded and cupped (in early-diverging ceratopsians) to larger and ventrally pointed and hooked (in derived neoceratopsians), overhanging the equally pointed and recurved predentary rostrodorsally (fig. 13.13). The external naris is relatively small in early ceratopsians and grows in relative rostrocaudal breadth to extremely large size in derived ceratopsids (typically more circular in centrosaurines and more ovoid in chasmosaurines). This enlargement of the nares correlates with the reduction in size of the subtriangular antorbital fenestra in early-diverging ceratopsians to its obliteration in derived taxa.

The signature long and pointed horns of ceratopsians are not particularly evident until the appearance of Ceratopsoidea. Vascular, keratinous sheaths, like those of bovid mammals, surrounded ceratopsian horn cores (Dodson, 1996). Horn cores are known to be of different shapes and sizes, depending on the taxon. The chasmosaurine *Bauplan* typically includes two long brow horn cores above the orbits with a smaller nasal horn core, although this is not always the case (e.g., *Vagaceratops* and some specimens of *Chasmosaurus*). Centrosaurines, however, typically possess a long nasal horn core and two small brow horn cores, again with exceptions (e.g., *Albertaceratops* and *Nasutoceratops*). In life, the horns would have been used either for defense or display (Farke, 2004, 2014; Farke et al., 2009) and are what set apart most ceratopsians from the rest of Ornithischia. Ornamentation like horns and epoccipitals was generally highly diverse across ceratopsians overall; however, there is no evidence that these display features were necessarily for species recognition between sympatric species (Knapp et al., 2018).

Ceratopsian skulls are quite variable, even when you do not count the ornamentation of the head (Dodson, 1993). The temporal region in ceratopsians is expanded both rostroventrally and transversely in relation to the frill attachment just caudal to it. The supratemporal fenestra in early ceratopsians (e.g., *Yinlong*, psittacosaurids, and early-diverging neoceratopsians) is relatively larger and of circular to rectangular shape, as in the most early-diverging ornithischians. In more derived ceratopsians (e.g., leptoceratopids, protoceratopsids, and ceratopsids), however, it is reduced in size to a small opening, sometimes as small as a V-shaped slit at the rostral border of either side of the frill, and is only visible in dorsal view. Numerous closed sinuses (possibly of pneumatic origin) are present within the cranium of ceratopsians rostrodorsal to the adductor chamber, likely for structural integrity with greater loads (Farke, 2010). Also of note is the dorsoventral expansion of the supratemporal bar in ceratopsids, which displaces the infratemporal fenestra ventrally, causing it to reduce in size as well. The adductor chamber is consequently large and subrectangular, and it is rotated caudally at a lower angle relative to the horizontal plane of the skull. As in most ornithischians, the laterally flaring jugal flange is present in ceratopsians, with ceratopsids taking it to an extreme with an even greater flaring as well as infolding of the caudal margin of the maxilla. This condition creates a large subjugal opening ideal for allowing a large

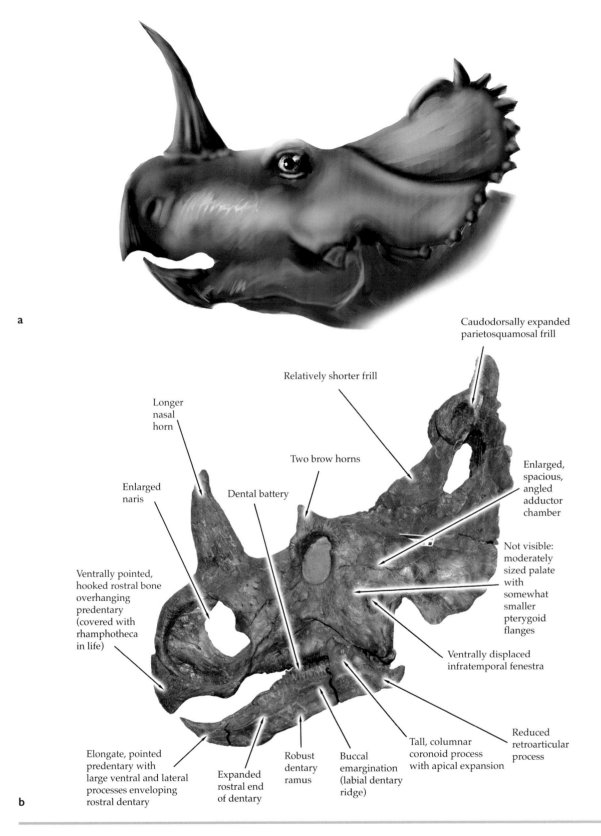

a

Caudodorsally expanded
parietosquamosal frill

Relatively shorter frill

Longer
nasal
horn

Two brow horns

Enlarged,
spacious,
angled
adductor
chamber

Enlarged
naris

Dental battery

Not visible:
moderately
sized palate
with
somewhat
smaller
pterygoid
flanges

Ventrally pointed,
hooked rostral bone
overhanging
predentary
(covered with
rhamphotheca
in life)

Ventrally displaced
infratemporal fenestra

Reduced
retroarticular
process

Tall, columnar
coronoid process
with apical expansion

Buccal
emargination
(labial dentary
ridge)

Robust
dentary
ramus

Elongate, pointed
predentary with
large ventral and lateral
processes enveloping
rostral dentary

Expanded
rostral end
of dentary

b

Figure 13.11. *Centrosaurus* (a) reconstructed head and (b) skull in lateral view
(length approximately 100 to 110 cm with frill).

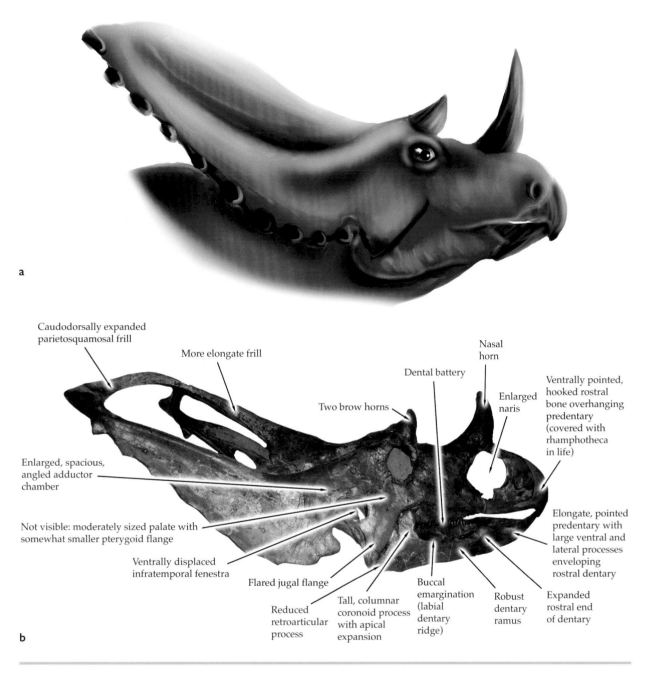

a

b

Caudodorsally expanded
parietosquamosal frill

More elongate frill

Nasal
horn

Dental battery

Enlarged
naris

Ventrally pointed,
hooked rostral
bone overhanging
predentary
(covered with
rhamphotheca
in life)

Two brow horns

Enlarged, spacious,
angled adductor
chamber

Not visible: moderately sized palate with
somewhat smaller pterygoid flange

Elongate, pointed
predentary with
large ventral and
lateral processes
enveloping
rostral dentary

Ventrally displaced
infratemporal fenestra

Flared jugal flange

Reduced
retroarticular
process

Tall, columnar
coronoid process
with apical
expansion

Buccal
emargination
(labial
dentary
ridge)

Robust
dentary
ramus

Expanded
rostral end
of dentary

Figure 13.12. *Chasmosaurus* (a) reconstructed head and (b) skull in lateral view
(length approximately 1.5 m with frill).

muscle body to exit the adductor chamber (see below). The palate in ceratopsians is variable in morphology but is mostly moderately sized in early ceratopsian taxa (with variability in the extent to which the palate is located rostrocaudally within the skull (Dodson et al., 2010). Of note is the fact that, in the derived ceratop-

sids, the pterygoid flange of the palate is substantially smaller relative to other ornithischians.

The mandible in ceratopsians is variable morphologically as well (Tanoue et al., 2010). The predentary ranges from rounded or cup-shaped in psittacosaurids and earlier-diverging ceratopsians, to triangular and

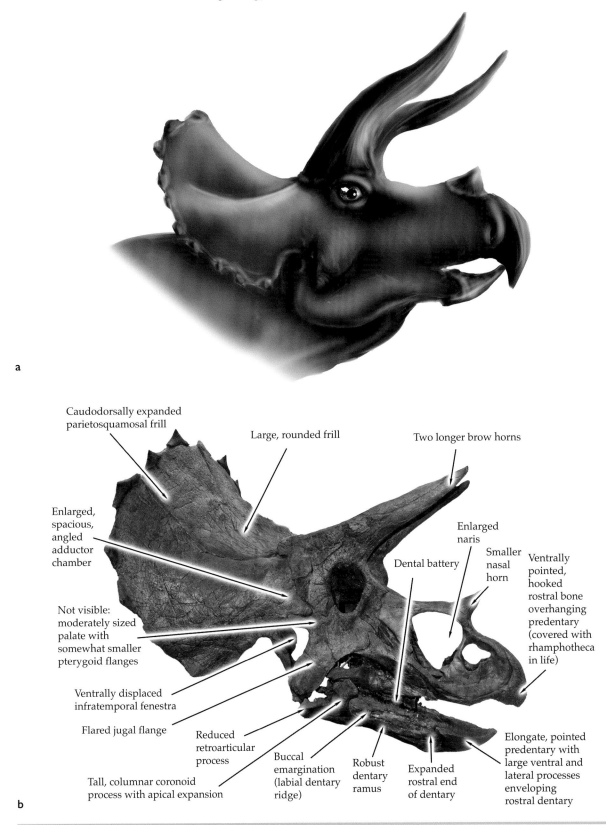

a

Caudodorsally expanded
parietosquamosal frill

Large, rounded frill

Two longer brow horns

Enlarged,
spacious,
angled
adductor
chamber

Enlarged
naris

Smaller
nasal
horn

Ventrally
pointed,
hooked
rostral bone
overhanging
predentary
(covered with
rhamphotheca
in life)

Dental battery

Not visible:
moderately sized
palate with
somewhat smaller
pterygoid flanges

Ventrally displaced
infratemporal fenestra

Flared jugal flange

Reduced
retroarticular
process

Buccal
emargination
(labial dentary
ridge)

Robust
dentary
ramus

Expanded
rostral end
of dentary

Elongate, pointed
predentary with
large ventral and
lateral processes
enveloping
rostral dentary

Tall, columnar coronoid
process with apical expansion

b

Figure 13.13. *Triceratops* (a) reconstructed head and (b) skull in lateral view
(length reaching approximately up to 2 m or more with frill).

pointed rostrally as in early-diverging neoceratopsians (Tanoue et al., 2010)—mostly categorized as PDJ Morphotype I (Nabavizadeh and Weishampel, 2016). In the more derived ceratopsians (e.g., leptoceratopsids, protoceratopsids, and ceratopsids), however, the predentary grows more pointed and sometimes recurved, and the caudal process envelops the rostral aspects of the dentary, which is expanded dorsoventrally at its rostral end (categorized as PDJ Morphotype III; Nabavizadeh and Weishampel, 2016) (fig. 13.14). The dentary is narrower in early-diverging taxa (which have leaf-shaped teeth with occlusal surfaces) (Tanoue et al., 2010) but is more robust especially in ceratopsids (which have dental batteries made of many teeth that form a common mesiodistally elongate occlusal surface, as in hadrosaurids). Leptoceratopsids also show a dorsoventrally highly robust, rounded mandible.

The coronoid elevation is also variable (fig. 13.14), with more early-diverging ceratopsians, leptoceratopsids, and protoceratopsids possessing a triangular coronoid eminence (tallest in protoceratopsids) and ceratopsids possessing dorsally elevated coronoid processes (for greater mechanical advantage of temporal musculature as seen in hadrosaurids; see below) that is expanded apically with a slight medial curvature. The rostroventral margin of the coronoid eminence and process extend rostrolabially along a prominent ridge making up the LDR. A mandibular fenestra is absent beneath the coronoid elevation, although a deep fossa is found ventrolateral to it. The retroarticular process is mildly prominent in all early ceratopsians, but the retroarticular process in the derived ceratopsids is highly reduced to nearly nonexistent.

Ceratopsian premaxillary dentition is only present in *Yinlong*, basal neoceratopsians, and protoceratopsids.

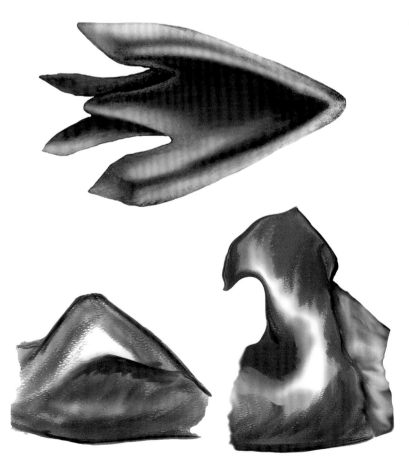

Figure 13.14. Various ceratopsian elements, including (*top*) a *Triceratops* predentary in dorsal view and (*bottom left*) a comparison of *Archaeoceratops* coronoid eminence with (*bottom right*) a *Triceratops* coronoid process.

They are large (larger than the maxillary teeth), lacrimiform, and bulbous at the base of the crown in *Yinlong*, with the labial side of the crown showing vertical wear against the predentary (Varriale, 2011). They are much more cylindrical and peg-like in basal neoceratopsians (including protoceratopsids) and fewer in number (between two and four depending on the taxon) (Tanoue et al., 2009a). Neoceratopsian premaxillary teeth were likely not as crucial in feeding as in *Yinlong* because of a caudal transition in tooth use and therefore disappear in more derived taxa (Tanoue et al., 2009a; Varriale, 2011). The maxillary dentition is similar to the dentary dentition in ceratopsians, except the occlusal surface is on the lingual side instead of the buccal side, as it occludes with the buccal occlusal surface of dentary teeth in this orientation. Depending on the genus, the angle of the maxil-

lary occlusal surface is oriented dorsomedial to ventrolateral from base to apex. The teeth have a denticulate edge along the oral apical margin of the tooth crowns.

Early-diverging ceratopsians, such as *Yinlong* and *Archaeoceratops* among others, have leaf-shaped teeth with a prominent buccal medial apicobasal ridge and a less prominent lingual medial ridge on the occlusal side. The dentition of the dentary is similar to that of the maxilla in early ceratopsians. The entire apical margin of the tooth crown is denticulate. In basal ceratopsians, such as psittacosaurids, the tooth crown is leaf-shaped and projects dorsally from the inset oral margin of the dentary (fig. 13.15). The lingual surface of the crown has a prominent ridge apicobasally, and in basal neoceratopsians, less prominent ridges on either side of the primary ridge are present. The buccal surface of the dentition also has a medial apicobasal ridge; however, it is less pronounced owing to the wear of the lingual surface of the maxillary tooth surface. The crowns in early-diverging ceratopsian teeth are slightly swollen at their base, with a cylindrical root (differing from the double-root in ceratopsids, described below). Furthermore, there is typically only one replacement tooth per tooth position in some early-diverging ceratopsians (Tanoue et al., 2009a), but signs of an increase in replacement teeth have been found in some taxa such as *Liaoceratops* and others that may hint at the beginnings of the formation of the dental battery in ceratopsids (see below) (He et al., 2018; Hu et al., 2022).

Enamel is present on both sides of the teeth in most non-ceratopsoid ceratopsians (Tanoue et al., 2009a). The crown and single root in early ceratopsian teeth are at a steep angle relative to one another, resulting in an orthal occlusal pattern, which is also seen in dental microwear (Varriale, 2011). Psittacosaurid wear patterns termed "clinolineal" consist of oblique, concave facets (Sereno et al., 2010; Varriale, 2011). Protoceratopsids (e.g., *Protoceratops*) show more oblique occlusion and wear facets (with teeth butting up against each other at occlusion), and this has been considered the preceding occlusal type to the more derived ceratopsids that show vertical slicing occlusion, with teeth vertically running

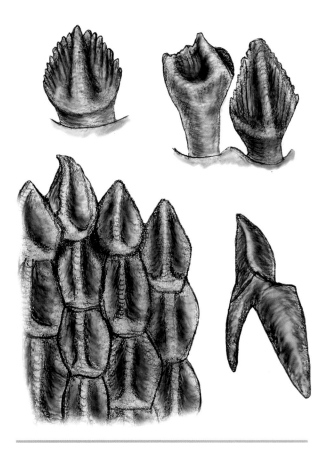

Figure 13.15. Ceratopsian dentition: (*clockwise from top left*) *Psittacosaurus* dentary tooth, *Leptoceratops* dentary teeth, and a ceratopsid dentary tooth battery and individual tooth.

past each other when the mouth is closed (Ostrom, 1966). Interestingly, the occlusal plane in leptoceratopsids (e.g., *Leptoceratops*) has been suggested to sometimes show an excavated shelflike morphology (labial shelf) on the buccal surface of dentary teeth where the maxillary dentition occluded, although this is likely more variable and depends on the individual (Ostrom, 1966; Varriale, 2016). Rounded tooth scratches indicate that the dentaries of *Leptoceratops* rotated palinally in a circular manner with initial orthal motions (a mechanism termed "circumpalinal"; Varriale, 2016).

Ceratopsoids possess a dental battery (fig. 13.15), superficially similar to that in hadrosaurid ornithopods, in which up to 30 to 40 columns of 3 to 5 teeth are tightly packed across the mesiodistal extent of the tooth row (Ostrom, 1966). The dental battery creates a large, mesiodistally elongate convexity along the medial surface of the dentary, and an arched series of alveolar foramina is visible ventral to the tooth columns. Studies in ontogenetic trends have shown that the number of tooth positions increases as ceratopsians grow, as in other ornithischian clades (Lehman, 1989; Dodson et al., 2004), but tooth size increases with it as well. Ceratopsid tooth roots are different from all other dinosaurs in that they are bifurcated (split into two conical roots) and arranged buccolingually instead of a cylindrical root in more basal forms.

The buccal occlusal surface is vertical, creating an apicobasal plane against which the maxillary teeth occlude. Although enamel is present on the non-occluding side of ceratopsid teeth, the occluding sides (i.e., the labial side of dentary teeth and the lingual side of maxillary teeth) do not show any enamel, which helped facilitate the heavy wearing patterns produced in a range of varying degrees from orthal to palinal feeding strokes in an orthopalinal slicing mechanism known in ceratopsids, as shown by tooth wear (Varriale, 2004, 2011; Mallon and Anderson, 2014b). Furthermore, 3D biomechanical wear analysis of the teeth of *Triceratops* by Erickson et al. (2015) showed that it had five different tissue types in its teeth, arranged so that the teeth would be built to reduce friction during sophisticated feeding strokes, similar to that seen in hadrosaurid teeth with six tissue types.

As mentioned, the ceratopsian skull shows considerable variability, including various reorientations of bones throughout the skull that affect the shape of the skull such as frill angle, quadrate angle, jugal morphology, and ventral squamosal expansion, among others (Dodson, 1993). The overall shape differences of skulls and frills across Ceratopsia have led to numerous studies that address ecological resource partitioning. The shapes of ceratopsian skulls overall are generally considered large and narrow with a rostrally extended snout (Mallon and Anderson, 2013). Henderson (2010) argued, using biomechanical analyses incorporating beam theory, that the dorsoventrally deeper skulls of centrosaurines were more resistant to forces in stronger bites than the dorsoventrally shorter skulls of chasmosaurines (which would have been relatively weaker). This has led to the suggestion that there was variation in foraging behavior between the two (Henderson, 2010). Furthermore, variation among members within each group shows further plasticity in cranial shape (Henderson, 2010). Although Mallon and Anderson (2013) tentatively agreed with this assessment, they highlighted the importance of intraspecific variability and how there are many factors that could hinder the ability to reduce this concept as a dichotomy between the two groups. Otherwise, Mallon and Anderson (2013) found no significant differences in cranial shape across the ceratopsians of the Dinosaur Park Formation. Mallon and Anderson (2014a), in measuring beak shapes of large ornithischians of Dinosaur Park Formation, suggested that the narrow snouts in ceratopsids indicate more selective diets (compared to the broader snouts of hadrosaurids and especially ankylosaurs) and that their larger size and sophisticated feeding apparatuses may have allowed for a diet of low-growing flowering plants and shrubs, similar to the selective diet of the modern narrow-lipped black rhinoceros (Mallon and Anderson, 2014a).

Recent geometric morphometric analysis by Prieto-Márquez et al. (2020b) indicates peramorphic evolutionary trends in ceratopsian frills. The ceratopsian frill varies widely in shape, size, and ornamentation, especially by clade, and the evolution of this progression of anatomical diversity was accelerated in early neoceratopsians, with later-diverging ceratopsians occupying a wider range of morphospace. They suggest that the decoupling of the cranial musculature from the frill may have been facilitated by peramorphic evolution of the frill and that decoupling explains how ceratopsian frills would have then been free to develop a wide range of morphologies across Ceratopsidae (Prieto-Márquez et al., 2020b). Additionally, although the frill was likely not indicative of sexual dimorphism in some ceratopsians like *Protoceratops*, recent 3D morphometric analyses have shown that they possibly functioned to some degree as sociosexual signaling (Knapp et al., 2021).

Feeding mechanisms in ceratopsians have been widely studied for more than a century. Hatcher et al. (1907), in observing tooth occlusal surfaces, suggested that ceratopsians (mainly ceratopsids) fed using a vertical (orthal) feeding motion (with no ability to produce transverse feeding strokes) owing to the morphology of the combined serrated edge of the tooth apices throughout the dental battery, creating a scissorlike vertical slicing mechanism along the vertical plane of the occlusal surfaces (with the labial side of the dentary teeth contacting the lingual side of the maxillary teeth). In addition to a vertical feeding stroke, Lull (1908) proposed a slight caudally oriented (palinal) feeding stroke (referring to oblique striations on the occlusal surface); however, this statement is not consistent with his later assertion of a strictly vertical feeding stroke with no fore-aft or transverse motions (Lull, 1933).

As mentioned above, Ostrom (1964a) (examining *Triceratops*) and Ostrom (1966) (examining a larger range of ceratopsians) agreed for the most part with previous studies in proposing a mainly orthal slicing mechanism in ceratopsids, based on a number of factors including dentition and musculature, and he additionally noted that any other motion (including any fore-aft motion of the jaw) would have likely been impeded by restrictions at the transversely expanded, mostly hinge-like quadrate-articular jaw joint. Ostrom (1966) also dis-

cussed feeding mechanisms with reference to tooth occlusion in two earlier-diverging ceratopsians, *Leptoceratops* and *Protoceratops*. As noted above, *Leptoceratops* sometimes shows a labial shelf on the dentary teeth where maxillary teeth likely occluded. Ostrom (1966) suggested this might be for a sort of crushing mechanism (rather than slicing as in ceratopsids). He also noted that the more oblique occlusion in *Protoceratops* was likely a primordial crushing mechanism, but one that paved the way to additional vertical shearing action in ceratopsids. Weishampel and Norman (1989) and Tanoue et al. (2009b) further suggested that these were the feeding mechanisms in ceratopsians.

A few hypotheses have been put forth for feeding mechanisms in the early-diverging psittacosaurid, including propalinal movement (Sereno, 1987; Norman et al., 1991) and the potential for anisognathous (unilateral) bites, atypical of most reptiles (Norman et al., 1991). In examining dental microwear of a larger *Psittacosaurus* species, Sereno et al. (2010) coined the term "clinolineal" for its feeding mechanism, meaning an isognathous jaw closure with combined orthal and palinal motion that created tooth wear in the form of inclined arcs (fig. 13.16). The distally diverging tooth rows would have accommodated such jaw movements. This style of feeding indicates mechanisms where higher bite forces, especially more mesially, were induced to crop and crush shelled seeds and nuts (Sereno et al., 2010).

Varriale (2004, 2008, 2011) analyzed dental microwear of an extensive sample of marginocephalian (mainly ceratopsian) taxa spanning the entire clade with statistical quantitative methods, revealing the evolution of tooth wear orientations in ceratopsian feeding mechanisms. Varriale (2011) documented a progression of chewing orientations that largely follows a progression toward the ceratopsid method of feeding. First, strictly orthal mechanisms were seen in the earliest-diverging ceratopsian taxa, including *Yinlong* and members of Chaoyangsauridae. Psittacosauridae also showed orthal wear patterns, although, as described by Sereno et al. (2010), there are also signs of palinal wear, creating the arcs of "clinolineal" feeding and making psittacosaurids

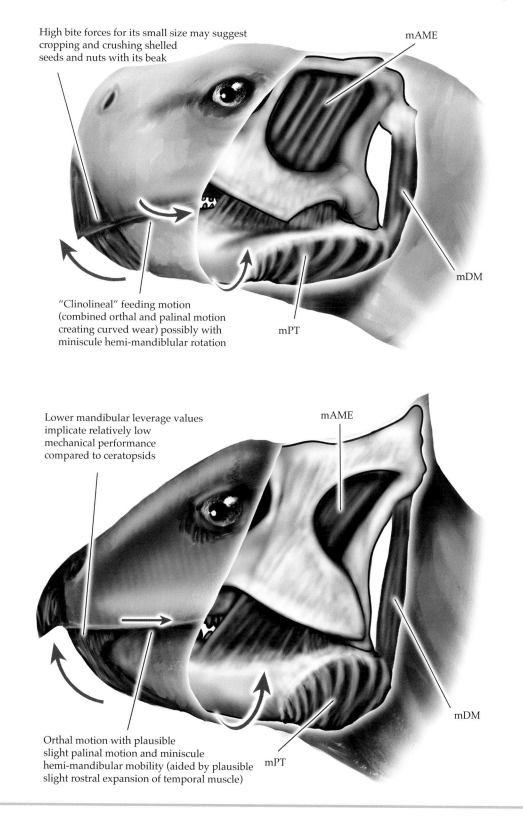

High bite forces for its small size may suggest cropping and crushing shelled seeds and nuts with its beak

mAME

mDM

"Clinolineal" feeding motion (combined orthal and palinal motion creating curved wear) possibly with miniscule hemi-mandiblular rotation

mPT

a

Lower mandibular leverage values implicate relatively low mechanical performance compared to ceratopsids

mAME

mDM

Orthal motion with plausible slight palinal motion and miniscule hemi-mandibular mobility (aided by plausible slight rostral expansion of temporal muscle)

mPT

b

Figure 13.16. Heads of (a) *Psittacosaurus* and (b) *Archaeoceratops* showing feeding function with jaw muscles. Red arrows show various jaw actions. Abbreviations: mAME, m. adductor mandibulae externus; mDM, m. depressor mandibulae; mPT, m. pterygoideus.

an intermediary group between feeding orientations. More curvilinear (recurved) palinal feeding motions were more seen in non-ceratopsoid neoceratopsians, such as *Liaoceratops* (Varriale, 2008, 2011). As noted above, a more recent detailed dental microwear study by Varriale (2016) further described the jaw motions of the leptoceratopsid *Leptoceratops* as circular motions in palinal feeding strokes (therein termed "circumpalinal" motion) (fig. 13.17). Furthermore, these circumpalinal microwear patterns were also described by Virág and Ősi (2017) in the early-diverging ceratopsoid *Ajkaceratops*.

Finally, Varriale (2004, 2011) suggested that the more derived ceratopsids show multidirectional wear patterns in which there is frequent orthal wear (with apicobasally oriented wear facets) along with some palinal microwear to varying degrees across ceratopsid taxa that indicate retractive feeding components—a mechanism therein termed orthopalinal feeding (figs. 13.18 and 13.19). Owing to the divergent nature of the tooth rows in ceratopsids, Varriale (2004, 2011) suggested that any palinal motion of the jaw was conducted anisognathously, since the lower jaw would have had to follow the maxillary tooth row on one side at a time.

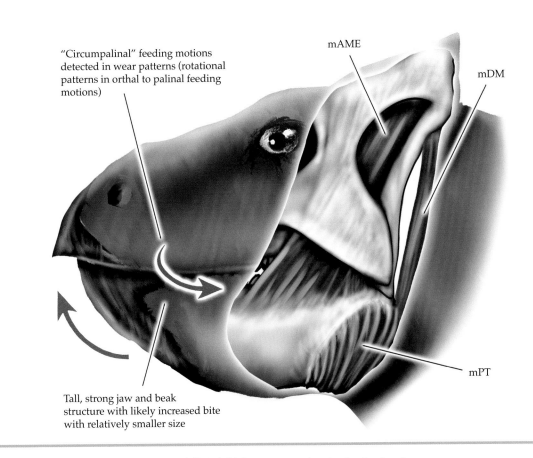

"Circumpalinal" feeding motions detected in wear patterns (rotational patterns in orthal to palinal feeding motions)

mAME

mDM

mPT

Tall, strong jaw and beak structure with likely increased bite with relatively smaller size

a

Figure 13.17. Heads of (a) *Leptoceratops* and (*facing*) (b) *Protoceratops* showing feeding function with jaw muscles. Red arrows show various jaw actions. Abbreviations: mAME, m. adductor mandibulae externus; mDM, m. depressor mandibulae; mPT, m. pterygoideus.

Although some degree of palinal motion in ceratopsid jaws had been reported previously (e.g., Lull, 1908; Sampson, 1993; Barrett, 1998), Varriale (2004, 2011) was the first to document and quantify orientation angles of both orthal and palinal microwear in ceratopsids, showing that ceratopsid mechanisms were in fact considerably more complicated than previously thought, and that all previous studies suggesting that ceratopsids had strictly orthal feeding mechanisms underestimated the full scope of ceratopsid jaw function.

Dental microwear analysis by Mallon and Anderson (2014b) also showed a primarily orthopalinal mechanism in ceratopsid taxa from the Dinosaur Park Formation to varying degrees (e.g., the chasmosaurines *Chasmosaurus* and *Vagaceratops* and the centrosaurines *Centrosaurus*, *Styracosaurus*, and a pachyrhinosaur). Occasional protrusion of the jaw is noted as well, possibly indicating reversion of the jaw to begin another mandibular feeding cycle (or even rare incorporation into the feeding cycle). Some differences between centrosaurines and chasmosaurines were documented (with centrosaurines possibly eating somewhat more abrasive plant material). This was not a statistically significant finding, however. Furthermore, a later study also showed similar ortho-

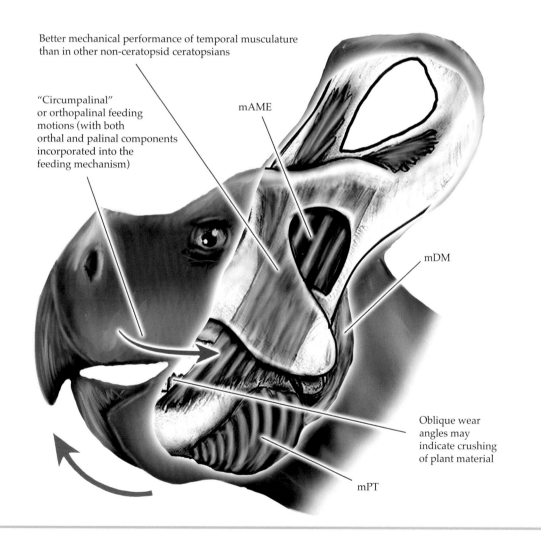

Better mechanical performance of temporal musculature than in other non-ceratopsid ceratopsians

"Circumpalinal" or orthopalinal feeding motions (with both orthal and palinal components incorporated into the feeding mechanism)

mAME

mDM

Oblique wear angles may indicate crushing of plant material

mPT

b

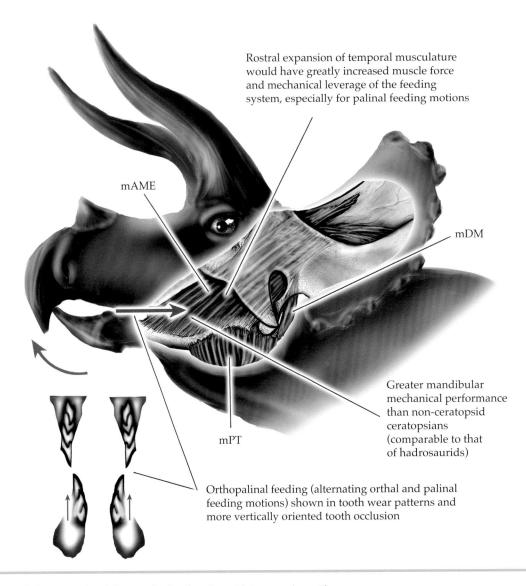

Rostral expansion of temporal musculature would have greatly increased muscle force and mechanical leverage of the feeding system, especially for palinal feeding motions

mAME

mDM

Greater mandibular mechanical performance than non-ceratopsid ceratopsians (comparable to that of hadrosaurids)

mPT

Orthopalinal feeding (alternating orthal and palinal feeding motions) shown in tooth wear patterns and more vertically oriented tooth occlusion

Figure 13.18. *Triceratops* head showing feeding function with jaw muscles, with rostral view of tooth occlusion on bottom left (redrawn from Ostrom 1966). Red arrows show various jaw actions. Abbreviations are as follows: mAME, m. adductor mandibulae externus; mDM, m. depressor mandibulae; mPT, m. pterygoideus.

palinal feeding in *Spiclypeus* (Mallon et al., 2016). In general, these findings indicated tougher woody browse in ceratopsids diets compared to those of other contemporaneous megaherbivores based on the heavy wear patterns in their dentition (Mallon and Anderson, 2014b).

Taking into account the multidirectional wear patterns in ceratopsians described above, Nabavizadeh and Weishampel (2016) examined the morphology of the

bones of ceratopsian mandible, with particular attention to the predentary and rostral extent of the dentary (the PDJ). Although the earliest-diverging ceratopsians are therein characterized as PDJ Morphotype I (with only miniscule mobility, if any, at the predentary-dentary junction), leptoceratopsids, protoceratopsids, and ceratopsids all clearly show a clasping junction, with the caudal processes of the predentary enveloping the rostral

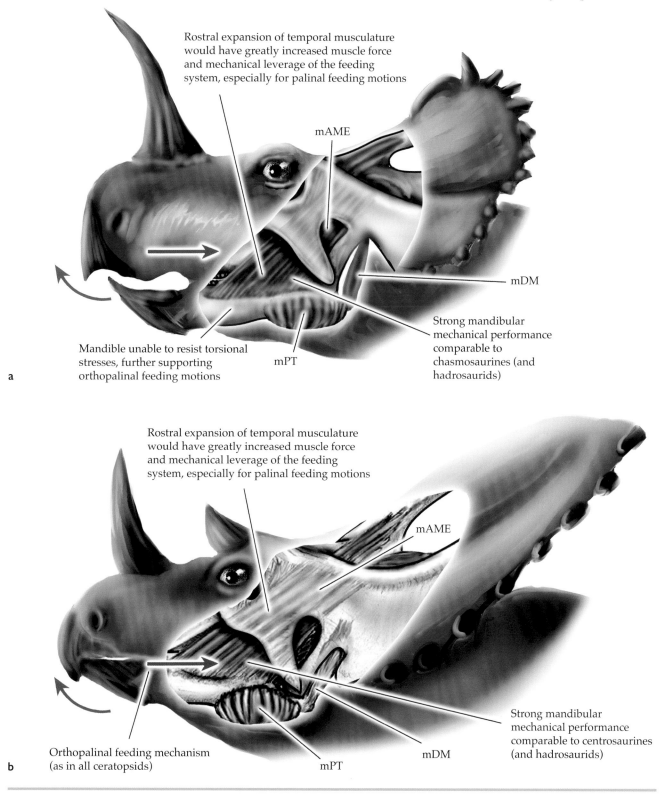

Rostral expansion of temporal musculature would have greatly increased muscle force and mechanical leverage of the feeding system, especially for palinal feeding motions

mAME

mDM

Strong mandibular mechanical performance comparable to chasmosaurines (and hadrosaurids)

Mandible unable to resist torsional stresses, further supporting orthopalinal feeding motions

mPT

a

Rostral expansion of temporal musculature would have greatly increased muscle force and mechanical leverage of the feeding system, especially for palinal feeding motions

mAME

Strong mandibular mechanical performance comparable to centrosaurines (and hadrosaurids)

Orthopalinal feeding mechanism (as in all ceratopsids)

mPT

mDM

b

Figure 13.19. Heads of (a) *Centrosaurus* (a centrosaurine ceratopsid) and (b) *Chasmosaurus* (a chasmosaurine ceratopsid) showing feeding function with jaw muscles. Red arrows show various jaw actions. Abbreviations: mAME, m. adductor mandibulae externus; mDM, m. depressor mandibulae; mPT, m. pterygoideus.

ends of the dentaries. This condition indicates a "secondary fusion" of the predentary with the dentary, suggesting lack of mobility at the PDJ, as described in most other ornithischians (especially ornithopods and ankylosaurs; Nabavizadeh and Weishampel, 2016). Additionally, examination of all bone morphologies in the mandible support a powerful feeding mechanism, and although previous restrictions to multidirectional wear have been suggested for ceratopsians (Ostrom, 1964a, 1966), slight fluctuations in jaw movement seen in microwear data show that orthal and palinal orientations (Varriale, 2004, 2008, 2011, 2016; Mallon and Anderson, 2014b) would have still been possible owing to the larger size of ceratopsian (especially ceratopsid) skulls and soft tissues surrounding the joints (Nabavizadeh and Weishampel, 2016).

Cranial musculature has been described in ceratopsians for more than a century (Lull, 1908; Russell, 1935; Haas, 1955; Ostrom, 1964a, 1966; Holliday, 2009; Tanoue et al., 2009b; Sereno et al., 2010; Mallon and Anderson, 2015; Nabavizadeh, 2016, 2020a, 2020b; Taylor et al., 2017; Landi et al., 2021). Although the dorsal surface of the frill of ceratopsians was originally thought to have been a large platform for origination of a hyperenlarged mAME muscle body extending to the caudal extent of the frill (Lull 1908; Russell, 1935; Haas, 1955; Ostrom, 1964a, 1966), more in-depth morphological analysis has discounted this reconstruction, citing morphological inconsistencies across taxa and ontogenetic stages, properties of surface texture incompatible with muscular attachment, and general mechanical ineffectiveness (and needlessness) of such a large muscle in an herbivore (Dodson, 1993, 1996; Holliday, 2009; Mallon and Anderson, 2015; Nabavizadeh, 2016, 2020a, 2020b; Prieto-Márquez et al., 2020b). Still, the temporal origin of mAME itself in all ceratopsians was expanded both rostrocaudally and transversely, creating ample room for large mAME muscle bodies to originate, likely restricted to the base of the frill in ceratopsids. It is possible that the non-ceratopsid ceratopsians had tendinous attachment of mAME along their much smaller frill, as inferred by alleged muscle scarring along the rims of the frill of some non-ceratopsid neoceratopsians (Xu et al., 2002; Makovicky and Norell, 2006; Morcshhauser et al., 2018; Prieto-Márquez et al., 2020b); however, this muscle was likely decoupled from the frill in all ceratopsids (see above; Prieto-Márquez et al., 2020b).

Non-ceratopsoid ceratopsians typically show a more vertically oriented, subrectangular adductor chamber (with a vertical quadrate) where the temporal musculature would have stretched straight ventrally to attach around a low, rounded, or subtriangular coronoid eminence of the mandible (most pronounced in protoceratopsids for a relatively greater mechanical advantage). It is likely that a broad fan of muscle (mAMES) expanded along the lateral surface of the surangular and more rostrally along the labial dentary ridge (Nabavizadeh, 2020a, 2020b). The mAME as a whole would have facilitated in orthal as well as circumpalinal motions in non-ceratopsoid ceratopsian feeding mechanisms shown in microwear in studies by Varriale (2008, 2011) and Sereno et al. (2010). Additionally, this muscle would have helped facilitate possible slight hemi-mandibular long-axis rotation in the earliest-diverging ceratopsians with patent PDJs (Nabavizadeh and Weishampel, 2016). The mAMP would have also helped facilitate in palinal motion of the jaw (Varriale, 2016). Palatal origins and mandibular insertions of the mPT bellies were generally small to moderate in size in non-ceratopsoid ceratopsians. The mPT likely added to the strength in jaw closure but acted secondarily to the primary function of enhanced temporal musculature (Nabavizadeh, 2020b).

As noted above, the temporal origins of mAME in ceratopsoids were broad both rostrocaudally and transversely, indicating enlarged muscle bodies. The supratemporal bar is dorsoventrally expanded in ceratopsids, displacing the infratemporal fenestra ventrally and creating a much larger origination for mAMES relative to all other ornithischians. The low-angled adductor chamber combined with a strongly flared jugal and infolded maxilla at the maxilla-jugal suture has recently been interpreted as having allowed a large fan of temporal musculature (mainly mAMES) to expand rostrolabially to insert onto the prominent LDR. This muscle would have covered most of the dentition in lateral view and would have facilitated a powerful orthopalinal jaw action. It would have been relatively larger than that in both ankylosaurs and hadrosaurids, as shown by the greatly flared jugal flange (Nabavizadeh, 2020a, 2020b). Additionally, such a muscle would have helped retain food in the motion while feeding owing to its labial positioning relative to the teeth. The tall coronoid process in ceratopsids creates an even larger attachment area and greater mechanical advantage to mAME as a whole (see below; also Ostrom, 1964a, 1966; Mallon and Anderson, 2015; Nabavizadeh, 2016).

With the adductor chamber and quadrate oriented at a low angle relative to the horizontal plane, the jaw joint is consequently displaced rostrally, positioning it to be much closer to the palatal origin of mPT. This creates a more vertical vector for mPT (although it is still rostroventrally oriented), facilitating a more powerful orthal component in the ceratopsid orthopalinal feeding mechanism. The relatively small size of the pterygoids in ceratopsids and the reduced to nearly nonexistent nature of the retroarticular process lead to two interpretations: either mPTV was highly reduced and did not insert lateral to the jaw joint (or, if it did, it would have been minimal), or mPTV was larger, with a more tendon-like origin that wrapped ventrally around the ventral margin of the angular and inserted more rostrally on the lateral surface of the angular, as posited owing to the enlarged depression seen on the lateral surface of the angular (Nabavizadeh, 2020a, 2020b). Additionally, mAMP is also oriented more caudally and would have facilitated palinal chewing, as described by Varriale (2016).

Probably the most influential biomechanical studies of ceratopsian feeding were those of Ostrom (1964a, 1966), in which he introduced the use of lever arm mechanics methodology to evaluate the mechanical advantage of bite points throughout the jaw (see chap. 2). All specimens used in these studies were given a uniform value of 100 units of muscle exerted force because there was no way of estimating absolute mass of jaw adductor musculature. Ostrom (1964a, 1966) demonstrated that ceratopsids (as well as hadrosaurids; Ostrom, 1961)

show a unique adaptation in which the tall coronoid process oriented the moment arm in such a way that created a second-class lever. This is a result of the adductor force vector crossing the jaw line rostral to the distal bite points (rather than a third-class lever where the adductor force vector crosses caudal to the bite point, like most vertebrates). In turn, mechanical advantage is increased exponentially (e.g., with a roughly 300% increase in distal dentition in *Triceratops*) (Ostrom, 1964a). Consequently, however, this adaptation causes a substantial decrease in force at the mesial-most bite point (about 28% to 36% of what would be expected). This partially helps explain why the tooth row is restricted more distally with a large diastema mesial to it (Ostrom, 1964a).

Ostrom (1966), in describing evolutionary trends in mechanical advantage of ceratopsian jaw systems, described four evolutionarily successive taxa (based on previous evolutionary hypotheses) that show an incremental trend toward increased mechanical advantage and relative bite force, implicating a growing ability for ceratopsians to feed on more fibrous vegetation (e.g., cycads and palms). He also used the now-erroneous assertion that a large muscle mass extended all the way to the caudal rim of the frill to explain that bite forces would have been even greater than expected (Ostrom, 1966). With this muscle reconstruction in mind, he posited that ceratopsians with lower-angled frills, such as the chasmosaurine *Chasmosaurus*, would have had a smaller moment arm and therefore not as powerful of a bite overall than other ceratopsids, such as *Centrosaurus* or *Triceratops*. But this is now considered a moot point because of the decoupling of the frill from musculature in ceratopsids (see above as well as chap. 3) (Dodson, 1993, 1996; Holliday, 2009; Mallon and Anderson, 2015; Nabavizadeh, 2020a, 2020b; Prieto-Márquez et al., 2020b).

Biomechanical analysis of ceratopsian jaws using lever arm mechanics was not performed until more than four decades later. Tanoue et al. (2009b) invoked both 2D lever arm mechanics using Ostrom's (1964a, 1966) method and a special 3D dorsal lever arm mechanics methodology, first used by Greaves (1978) using dorsal views of a mammalian skull, to analyze jaw mechanics of early-diverging ceratopsians. Greaves's (1978) method presents a proxy for understanding the transfer of muscle forces through the mandibular symphysis to the opposite side of the jaw, which is feasible in ceratopsians since nearly all ceratopsians show a tight fusion of the predentary with the dentaries. Tanoue et al.'s (2009b) results presented interesting implications for bite mechanics in various ceratopsians. In psittacosaurids, more leverage was found at the mesial-most dentition and predentary compared to other early-diverging ceratopsians, possibly suggesting a more mesially focused bite. Additionally, Tanoue et al. (2009b) found that early-diverging neoceratopsian jaw systems could be considered second-class levers with respect to the distal-most teeth, showing a gradual evolutionary trajectory toward the much higher mechanical advantages in ceratopsid jaws with a tall coronoid process (as shown by Ostrom, 1964a, 1966). Additionally, the 3D analysis using a dorsal view found that the medial curvature of the tooth row in early-diverging ceratopsians would have influenced a greater mechanical advantage compared to a hypothetical straight tooth row, showing further that ceratopsians evolved numerous ways to increase jaw leverage for a better bite performance to be able to process tough vegetation (Tanoue et al., 2009b).

In their study statistically comparing jaw mechanics of megaherbivorous ornithischians from Dinosaur Park Formation, which also incorporated Ostrom's (1961, 1964a, 1966) lever arm methodology, Mallon and Anderson (2015) analyzed the mechanical advantage of jaw systems in a large sample of the centrosaurines *Centrosaurus* and *Styracosaurus* and the chasmosaurine *Chasmosaurus*. Mechanical advantages were calculated with respect to temporal musculature (stretching from the caudal edge of the supratemporal fenestra to the apex of the coronoid elevation) at three points along the tooth rows (predentary, mesial tooth, and distal tooth). No significant differences were found in mechanical advantage between individuals of the same genus or between ceratopsid genera. In comparison with other ornithischian groups, ceratopsids showed statistically significant differences with ankylosaurids (which showed lower me-

chanical advantage values) while exhibiting no significant difference with hadrosaurids. The convergently shared tall coronoid process of both ceratopsids and hadrosaurids again contributed to their exceptionally higher distal tooth mechanical advantage values (Mallon and Anderson, 2015).

Nabavizadeh (2016), again using the same lever arm methodology at four points along the tooth row (predentary, mesial tooth, middle tooth, and distal tooth), compared jaw adductor mechanics across Ornithischia, including all groupings of Ceratopsia, which showed varying significant differences evolutionarily. Early-diverging ceratopsians showed generally lower RBF values that gradually increased along the tooth row from the predentary to the distal tooth positions. In the derived ceratopsids, however, the caudal displacement of the tooth row allows for some increased leverage, and the distal teeth show a drastically improved mechanical advantage (again, owing to the tall coronoid process). These results show that there were significant differences found between the early-diverging non-ceratopsid ceratopsians and ceratopsids. Among ceratopsids themselves, no significant difference was found between centrosaurines and chasmosaurines, with variable overlap (similar to the results of Mallon and Anderson, 2015) (Nabavizadeh, 2016).

As stated in chapter 12, marginocephalians and ornithopods show parallel evolutionary trends toward the highest mechanical advantage among ornithischians. Whereas there was significant difference in mandibular mechanical advantage between ceratopsids and nonceratopsid ceratopsians, there was no significant difference between ceratopsids and hadrosaurids, both with the highest RBF values, especially at the distal tooth position. Likewise, early-diverging ceratopsians showed no significant difference with early-diverging ornithopods, again suggesting a convergence in heightened RBF brought on as a result of the heightening of the coronoid process in both ceratopsids and hadrosaurids. Still, as ceratopsids show a decreased adductor angle and hadrosaurids show an increased adductor angle, this convergence in RBF values was achieved through different means, highlighting the variable complexity of cranial characteristics that influence jaw mechanics to reach similar results (Nabavizadeh, 2016).

Ontogeny of feeding mechanics is rarely studied in ceratopsians. Recently, however, Landi et al. (2021) used 3D models of two specimens of *Psittacosaurus* (one hatchling and one adult) to explore whether there are differences in overall lever mechanics of the jaws that might relate to a difference in feeding ecologies, especially since these animals underwent a unique ontogenetic postural shift from quadrupedality to bipedality. Using the same 2D lever arm methods as described above, Landi et al. (2021) reconstructed and analyzed the mechanical advantage of all cranial musculature. The adult clearly showed higher mechanical advantages overall, deeming it better constructed for feeding on tougher plant material, although in general they found only slight differences between the two jaw systems, relatively speaking in terms of both shape and function. Further investigation of ontogenetic trends in ceratopsians would be greatly beneficial to our understanding of their fullscale feeding evolutionary ecology.

Computational biomechanical modeling methods have also been used in a few studies to investigate feeding in ceratopsians. For instance, Taylor et al. (2017) reconstructed all cranial musculature and performed FEA on the skull of *Psittacosaurus* and tested how the previously hypothesized presence of two additional, novel muscles seen in parrots (m. pseudomasseter, or mPSM, and m. adductor mandibular externus ventralis, or mAMEV (Sereno et al., 2010) would affect its feeding mechanics. Results showed that the rostrally positioned mAMEV and rostrodorsally inclined mPSM were found to have increased cranial stresses and deformation as well as increased bite force. Additionally, they suggested that in any case there was no clear evidence of the existence of these novel muscles, and morphological analysis showed that bite position had a greater influence on cranial deformation than muscle loads (Taylor et al., 2017).

Although FEA has rarely been used to assess ceratopsian feeding as a whole, few other instances have been used to investigate ceratopsian jaw mechanics. In an ef-

fort to better understand the mechanics of movement that would have been possible in large ceratopsid jaws, Bell et al. (2009) performed FEA on the 3D jaw of the ceratopsid *Centrosaurus* to model the force distributions throughout the jaw during feeding. Results showed that the mandibular ramus of *Centrosaurus* was not able to resist torsional stresses, indicating that the main type of jaw closure likely used in ceratopsids would have been an orthal, isognathous shearing mechanism, as previously proposed (Ostrom, 1964a, 1966). But a palinal component to the jaw mechanism was not precluded as a possibility, as described in dental microwear studies by Varriale (2004, 2008, 2011) and Mallon and Anderson (2014b). They also found that the labial dentary ridge (the leading edge of the coronoid process) showed higher stresses from jaw closure (Bell et al., 2009). This may also have implications for the hypothesized presence of mAMES attaching to and acting on this ridge (Nabavizadeh, 2020a, 2020b).

Although 3D FEA has yet to be performed on any additional ceratopsid jaw material, 2D FEA and geometric morphometrics (GMM) has recently been used to compare structure and function across a broad range of ceratopsian species. Maiorino et al. (2015) used these methods to compare across Ceratopsoidea specifically. GMM results found that ceratopsids that are considered Triceratopsini (e.g., *Triceratops*; *Torosaurus*, *Ojoceratops*) were found to have morphologically distinct mandibular morphology (i.e., larger absolute size, more robust dentary, larger coronoid process, and smaller angular) compared to other chasmosaurines and centrosaurines (which are more similar to each other morphologically) and early-diverging ceratopsoids. FEA results, in turn, showed that triceratopsins were relatively more stressed in the mandibles than other ceratopsids, possibly indicating dietary variation (Maiorino et al., 2015).

Later, Maiorino et al. (2018), using the same methods, investigated the structure and function of the feeding systems across all of Ceratopsia and found similar stress patterns throughout the jaws of taxa spanning the entire clade, with fluctuations in stress patterns specifically associated with morphological variation in the coronoid elevation (eminence or process). Most early-diverging ceratopsians and protoceratopsids show higher stress levels in their mandibles, suggestive of a diet of softer vegetation and fruits (although protoceratopsids still show adaptations between the dentary and post-dentary elements for withstanding higher stresses in feeding). The rostrocaudally shorter yet robust mandible (with a ventral flange) of psittacosaurids proves capable of withstanding relatively higher bite loads, even though their small coronoid eminence experiences more stress, and the highly robustly built jaws of leptoceratopsids also showed strong mandibular adaptations for stress dissipation. Of course, Late Cretaceous ceratopsoids showed the strongest jaws of all by comparison, with stresses focused in the caudal aspect of the jaw and coronoid process, especially with larger adductor muscles attaching there. The larger, longer dentary in the ceratopsid tribe Triceratopsini were most equipped for withstanding higher stress levels in feeding, especially with the high integration between the dentary and post-dentary elements to withstand high forces, as seen in protoceratopsids (Maiorino et al., 2018).

Ceratopsians possessed strongly built necks, with the first three cervical vertebrae (including the atlas and axis) fused together (termed "syncervicals") throughout Neoceratopsia (VanBuren et al., 2015). Tait and Brown (1928) interpreted these fused vertebrae as a reinforcement mechanism to help carry and support the massive heads of ceratopsids. The spherical ball-and-socket articulation of the head and neck allows for sufficient rotation of the head, especially along a rostrocaudal axis, for feeding purposes. Tait and Brown (1928) interpreted this feature as an adaptation to strip succulent plant material efficiently and effectively. Hypertrophied cervical musculature was also clearly present (particularly m. longissimus capitis and m. obliquus capitis magnus), with enlarged muscle attachment sites seen on the occiput, aiding in the support given to carry such a large head especially in ceratopsids (Lull, 1908; Russell, 1935; Tsuihiji, 2010).

VanBuren et al. (2015) tested whether the syncervicals initially evolved for supporting such a large head

(e.g., Marsh, 1891; Hatcher et al., 1907; Dodson, 1996; Campione and Holmes, 2006) as well as acting as a buttress for intraspecific or even antipredatory combat (e.g., Farlow and Dodson, 1975; Ostrom and Wellnhofer, 1986; Farke, 2004, 2014), but they found no correlation relating either explanation to the origin of the syncervicals, although this could mean the syncervicals were co-opted for these reasons with further modifications. This assertion is indeed confirmed with the presence of a syncervicals in small, bipedal early neoceratopsians like *Auroraceratops*, which do not bear large cranial frills or ornamentations (Li et al., 2018). In contrast, the early-diverging ceratopsian *Psittacosaurus* has an atlas-axis complex with slight flexibility, resembling that of ornithopods (Podlesnov, 2018).

Ceratopsia is the third ornithischian group that shows an evolutionary transition from cursorial bipedality to graviportal quadrupedality. The small, bipedal early-diverging ceratopsians, including *Yinlong*, chaoyangsaurids, and psittacosaurids, as well as early-diverging neoceratopsians, including *Archaeoceratops* and *Auroraceratops*, show longer, slender hindlimbs compared to the lengths of their forelimbs as well as numerous other adaptations for bipedality (Maidment and Barrett, 2014). For example, *Auroraceratops* has a laterally oriented process of the proximal ulna, dorsoventrally flattened manual unguals, shorter femur than tibia, transversely narrow pelvis, and a strongly pendant fourth trochanter of the femur with a more proximal position, all suggestive of obligate bipedality (Morschhauser et al., 2018). Senter (2007) performed a range-of-motion study with a manual manipulation method and found that forelimb mobility was highly limited in *Psittacosaurus* (making it an obligate biped) compared to that of *Protoceratops* and *Leptoceratops*. He concluded that *Psittacosaurus* could not reach its own mouth and could only tuck its arms in to clutch objects to its belly.

Quadrupedalism (whether facultative or obligate) has been generally accepted as having been the stance and locomotor strategies of both leptoceratopsids and protoceratopsids for anatomical and weight-bearing purposes (e.g., Russell, 1970; Tereschenko, 1996; Senter,

2007, Maidment and Barrett, 2014) (fig. 13.20). Russell (1970) argued that *Leptoceratops* was likely quadrupedal owing to the large size of the head, producing a forwardly positioned center of gravity. Tereschenko (1996) suggested that protoceratopsids were obligate bipeds with straight hindlimbs for faster locomotion and slightly splayed forelimbs for more weight-bearing locomotion. Tereschenko (2004, 2008) further examined vertebral adaptations, including those of the tail, in leptoceratopsids and protoceratopsids and broadly grouped them as potentially ranging from semiaquatic (*Protoceratops*) to facultatively aquatic (*Udanoceratops*) to more terrestrial (*Leptoceratops*), largely referring to the taller height of the middle caudal neural spines in the former taxa as a sign of a more aquatic lifestyle.

Regarding more specifics of stance, Senter (2007) deduced that *Leptoceratops* and *Protoceratops* would have tucked their elbows in for quadrupedal locomotion (although he speculated that they may have also splayed their humeri somewhat laterally for transverse pivoting display motions). The palms of *Protoceratops* would have

faced caudally owing to the radius crossing over the ulna, whereas those of *Leptoceratops* would have faced medially, with the digits creating caudally propulsive motions (Senter, 2007). Maidment and Barrett (2014) showed that both *Leptoceratops* and *Protoceratops* showed some bipedal characteristics, but that other characters such as a hooflike manual unguals (and those of front-heavy nature) point to quadrupedal or at least facultatively quadrupedal stance. Słowiak et al. (2019) used the presence of a bowed femur in lateral view as a sign of primary bipedality (as also posited by Chinnery, 2004, and Maidment and Barrett, 2014) and suggested that protoceratopsids and leptoceratopsids were likely quadrupedal owing to their straight femora.

Ceratopsids were clearly quadrupedal with graviportal stance, in part likely a result of the increased size and weight of the massive skulls pulling the center of mass more cranially (Maidment et al., 2014b; Barrett and Maidment, 2017) (fig. 13.21). Although some have posited that ceratopsids held their forelimbs straight and parasagittally (e.g., Bakker, 1975), more recent studies

Figure 13.20. Skeleton of *Protoceratops*.

have argued for a more splayed forelimb, with a caudo-ventrally oriented humerus with a slightly laterally flexed elbow (e.g., Johnson and Ostrom, 1995; Thompson and Holmes, 2007; Rega et al., 2010; Maidment and Barrett, 2012). The large deltopectoral crest suggests that a hypertrophied m. pectoralis assisted in bending the elbow in this interpretation (Johnson and Ostrom, 1995). The extent to which the forelimb is splayed has been debated with studies using manual manipulation, muscle reconstructions, and trackway evidence, among other methods, but the extent to which the humerus was laterally splayed was only slight, with further evidence for this shown by the presence of weight-bearing hoof-like unguals on the medial three digits instead of all five

(Thompson and Holmes, 2007; Rega et al., 2010; Maidment and Barrett, 2012).

It is widely accepted that, unlike the forelimbs, the hindlimbs of ceratopsids were held straight and their excursion was parasagittal (Maidment and Barrett, 2012; Maidment et al., 2014a). Although earlier studies have suggested ceratopsids were relatively fast runners (Bakker, 1968; Paul and Christiansen, 2000), no adaptations for cursoriality have been identified (Maidment and Barrett, 2012). Certain features of the hip suggest increased surface area and moment arms of flexion of the hip, such as an enlargement of the preacetabular process (Maidment and Barrett, 2012). But biomechanical analysis shows that overall moment arm in flexion was not as high as

Figure 13.21. Skeleton of *Chasmosaurus*.

would be expected (Maidment et al., 2014a). In fact, ceratopsids plot among the lowest in leverage of most muscles and functionality (e.g., flexion, extension, rotation, adduction) tested at the hip joint, implying relatively poor locomotor performance overall (Maidment et al., 2014a). Abduction at the hip is more midrange among ornithischians, possibly because of the transversely broadened ilium widening the hip (Maidment et al., 2014a).

As discussed above, diets were wide-ranging, especially when comparing that of the small early-diverging ceratopsians and the large, quadrupedal ceratopsids. As mentioned, psittacosaurids were likely able to eat low-growing vegetation—anything from softer plant material to tough, resistant vegetation as well as seeds and nuts—with their powerful beak and feeding mechanism (Osborn, 1924; Sereno et al., 2010). Many gastroliths have also been found associated with *Psittacosaurus* specimens, further implying a tougher diet (including nuts) in need of extra processing power (Xu, 1997; Sereno et al., 2010). Early-diverging neoceratopsians as well as leptoceratopsids and protoceratopsids could only reach low-growing vegetation, less than 1 m high (Weishampel and Norman,

1989), which, with feeding adaptations considered powerful for their size, was also likely resistant at least in some cases (consisting of conifers, ferns, cycads, and angiosperms, among other things).

Isotope analysis has, unsurprisingly, confirmed the herbivorous diet of the larger ceratopsids (Cullen et al., 2020) overlapping especially with ankylosaurs in dietary niche signatures (Cullen et al., 2022). Ceratopsids may have been capable of feeding at vegetative levels at or slightly above 1 m (Dodson, 1983; Mallon et al., 2013). Their highly powerful feeding apparatuses would have been capable of subdividing some of the toughest woody browse of conifers, including many leaves as well as likely a range of other plants like cycads, ferns, horsetails, and angiosperms (e.g., Mallon and Anderson, 2014b). Additionally, recent endocranial studies of ceratopsid skulls like that of *Triceratops* have shown relatively small olfactory bulbs (e.g., Sakagami and Kawabe, 2020), which implies a relatively low dependence on smell to differentiate among vegetation. Furthermore, the lateral semicircular canal of *Triceratops* implies that it held its head at about 45 degrees for feeding purposes (Sakagami and Kawabe, 2020).

CHAPTER 14

Shaping Dinosaur Ecosystems

Predator–Prey Relationships through Time

Even with all the knowledge that vertebrate paleontologists have gained over the past century and half in studying the biology of feeding in dinosaurs, much remains to be learned of their interactions with each other and with their environments. Throughout the Mesozoic, broadscale evolutionary patterns in skull morphology as well as postcranial adaptations and body size likely coincided with trophic standing (predator, omnivore, or herbivore) and individual taxonomic and geographic circumstances, including the types of prey available to predators in a certain setting or the kind of plant life available to herbivores in a certain environment (with omnivorous forms taking advantage of bits of both). All factors of an animal's anatomy, physiology, and ecological setting play a role in reconstructing how and what they ate to survive and how they lived their lives overall (Barrett and Rayfield, 2006).

Carnivorous dinosaurs were significantly outnumbered by their herbivorous counterparts worldwide (Bakker, 1975; Farlow, 1993; White et al., 1998; Barrett, 2014), which is to be expected given that a similar type of discrepancy is seen in the wild today between carnivorous mammals and herbivorous mammals. Unfortunately, given the relative paucity of specimen data from the fossil record, it is difficult to discern the effect of continuous population dynamics in predators that may have fed upon one or two particular species in their given ecological setting. In addition, few finds involve

two different animals that would give paleontologists a clue as to which predator might have interacted with which prey (e.g., the famous discovery of a *Velociraptor* and *Protoceratops* seemingly interlocked in combat; see chap. 7). The best ways to discern a relationship between a predatory dinosaur and its prey are by bite mark evidence, coprolite and cololite evidence, or simply circumstantial evidence (i.e., understanding what types of prey items, dinosaur or otherwise, may have lived contemporaneously with a particular predator).

Through time, the interactions between predatory dinosaurs (consisting of a vast majority of theropod taxa) and their prey (typically herbivorous dinosaurs of all kinds, though this is not always the case) put future offspring of these animals under pressures of natural selection. These pressures may turn into an evolutionary arms race between contemporaneous predators and prey of a given environmental setting, with both "one-upping" each other in terms of how well adapted they are to either capturing a specific kind of prey (in the case of the predator) or defending themselves against a specific kind of predator (in the case of the prey). With general knowledge from the fossil record regarding contemporaneous carnivorous and herbivorous dinosaurs in different settings, clues about predator–prey relationships can theoretically be pieced together accordingly.

One of the ways in which researchers have investigated possible predator–prey ecology is by examining the encephalization quotient (EQ) of the predatory and herbivorous dinosaurs (Hopson, 1977; Buchholtz, 2012). The EQ examines the relationship between brain size and body size to infer relative cognition. Hopson (1977) studied the EQs of dinosaurs of all different kinds and compared them with modern archosaurs and mammals. He inferred that dinosaurs with larger brain to body size ratios (>1.0) likely had a higher, more complex cognitive ability. Although this method has numerous setbacks to its interpretation (such as not incorporating individual functions of the many different parts of the brain), it still served as a basis to conclude that the higher EQs of predatory theropod dinosaurs generally indicated that they had higher cognitive ability than herbivorous dinosaurs, as would have been necessary for more strategic predation (Hopson, 1977).

Among herbivorous dinosaurs, ornithopods show relatively higher EQs than ceratopsians and especially thyreophorans and sauropods, likely an indicator that they required more cognitive ability to escape predators because they lacked the anatomical defense mechanisms these other animals had (in addition to the sometimes much larger body size than all except sauropods). Many herbivorous dinosaurs did show defensive structures—including the horns of ceratopsid heads, the domed head of pachycephalosaurs, the tail spikes of stegosaurs, the armor (and sometimes clubbed tail) of ankylosaurs, and the possible whiplike tail of sauropods, all of which would have been effective weapons and defense mechanisms used for protecting against predators—although evidence of use of these defensive structures is variable (Farke, 2004, 2014; Arbour, 2009; Arbour and Snively, 2009; Farke et al., 2009; Arbour and Zanno, 2020).

Predatory theropod dinosaurs possessed many skull and body plans as well as numerous methods of predation, all of which are discussed in chapters 4 through 7. Although not all theropods were carnivores, they did make up a vast majority—if not all—of the non-avialan predatory dinosaurs throughout the Mesozoic, and it is helpful to broadly summarize them in their chronologic and geographic contexts for a more understanding dinosaur paleoecology as a whole.

Evidence from the Late Triassic shows that the earliest predatory dinosaurs coexisted with, and likely competed with, crurotarsan archosaurs for about 30 million years, and ultimately some key adaptations (instead of mere superiority) helped these dinosaurs survive the end of the Triassic extinction and continue to evolve and diversify (Brusatte et al., 2008). The earliest predatory dinosaurs were smaller cursorial predators, including the tiny *Eoraptor* and the somewhat larger, more robustly built *Herrerasaurus*, both of Patagonia in Argentina. Their skulls were relatively weaker than larger theropods, and the grasping abilities of *Herrerasaurus* likely meant they were able to capture and feed upon small animals such as lepidosaurs and small mammals, along with their therapsid relatives. The small, 3-meter-long, gracile coelophysoids such as *Coelophysis*, as well as other small predators such as *Tawa*, were also prevalent throughout the Late Triassic. Found in abundance at Ghost Ranch in New Mexico, these animals were likely fast runners, and with their specialized forelimbs adapted for powerful grasping, they likely also preyed upon small animals like lepidosaurs, small mammals, insects, and even small crocodylomorphs (Nesbitt et al., 2006). The low strain and mechanical advantage of the long, gracile snouts of coelophysoids (Rayfield, 2005a; Sakamoto, 2010) would have allowed for faster biting capabilities—another adaptation for quick capture of small prey. Other slightly larger predators such as *Liliensternus* of Germany may have preyed upon larger animals such as early sauropodomorphs (Paul, 1988). It is difficult to imagine, however, that early sauropodomorphs like *Plateosaurus* would have had many predators (at least dinosaurian predators) owing to their relatively immense sizes compared to the small, gracile-bodied, weak-skulled predatory dinosaurs during the Late Triassic.

Throughout their diversification across the Triassic-Jurassic boundary, the body sizes of theropods started to increase gradually, giving them more opportunities to dispatch larger prey (Griffin and Nesbitt, 2020). Predatory theropods also developed more diverse adaptations in prey capture fast and powerful prey, which became useful with the evolution of not only sauropodomorphs but also the newly emerging ornithischians of that time. Predatory theropods of the Early Jurassic included dilophosaurids (*Dilophosaurus* of Arizona) and early-diverging tetanurans (*Cryolophosaurus* of Antarctica and *Sinosaurus* of China). *Dilophosaurus* was an agile predator that likely preyed on smaller dinosaurs of the Kayenta Formation, including small early ornithischians such as the bipedal early-diverging armored thyreophoran *Scutellosaurus* and possibly heterodontosaurids, and other smaller animals (possibly even fish) with fast slashing bites and strong two-handed prehension and clutching ability (Welles, 1984; Therrien et al., 2005; Milner and Kirkland, 2007; Senter and Sullivan, 2019), although bite mark evidence also suggests it preyed on larger "prosauropods" like *Sarahsaurus* (Marsh and Rowe, 2018, 2020), which may have been possible through pack-hunting behavior (Rowe and Gauthier, 1990). The larger-bodied early tetanurans likely did prey upon "prosauropods" with more ease, as evidenced by alleged *Sinosaurus* bite puncture marks in a specimen of *Lufengosaurus* (Xing et al., 2018).

Although dinosaur material is comparatively less known from the Middle Jurassic, this period saw the beginnings of increased diversification of medium-sized, agile, predatory theropods (mostly ranging between 5 and 9 m in length) that coincided with the continuing expansion of herbivorous sauropodomorphs and ornithischians worldwide. These theropods include early-diverging tetanurans (*Monolophosaurus* of China), some megalosauroids (*Piatnitzkysaurus* of Argentina, *Megalosaurus* of England, *Afrovenator* of Africa, and *Debreuillosaurus* of France), and the first signs of abelisaurid ceratosaurians (*Eoabelisaurus*) and allosauroids (*Asfaultovenator*), both from Argentina. These theropods all likely fed on small- to medium-sized ornithischians, such as stegosaurs (*Huayangosaurus* of China) and early neornithischians (*Agilisaurus* of China), as well as smaller (*Shunosaurus* of China) and much larger (*Jobaria* of northern Africa and *Omeisaurus* and *Datousaurus* of China) sauropods. *Monolophosaurus* and *Debreuillosaurus*, for instance, show intermediate stress patterns in the skull (Rayfield, 2011), potentially meaning they may have been capable of withstanding forces in preying upon prey of similar size.

The Late Jurassic saw even greater diversification of similar-sized to larger-bodied theropods, ranging between 5 and 13 m in length, as indicated by comparatively copious amounts of dinosaur material from this time. The relatively smaller end of these body sizes included ceratosaurids (the 5- to 7-m *Ceratosaurus* of the Morrison Formation of the United States), noasaurids (*Elaphrosaurus* of Tanzania), and megalosauroids (*Torvosaurus* of North America and Europe, *Eustreptospondylus* of England, and *Marshosaurus*, a piatnyzkysaurid from Colorado). Larger-bodied allosauroids diversified and expanded across the globe during this time as well, including the 9- to 12-m-long *Sinraptor* and *Yangchuanosaurus*, which were likely some of the most dominant predators of China, and the 10- to 13-m-long *Allosaurus* and *Saurophaganax*, likely the apex predators of the Morrison Formation of the United States.

Global distributions of both sauropods and ornithischians would have allowed for a broad range of prey options for theropods of varying sizes. In the Morrison Formation of the United States, for instance, *Ceratosaurus* and *Torvosaurus* likely preyed upon smaller- to medium-sized sauropods (possibly younger sauropods in particular) as well as ornithischians such as stegosaurs (*Stegosaurus* and *Hesperosaurus*), small early ankylosaurs (*Mymoorapelta* and the smaller *Gargoyleoaurus*), and small early ornithopods (such as *Dryosaurus*, *Uteodon*, *Camptosaurus*). The larger-bodied allosauroids of that same region (*Allosaurus* and *Saurophaganax*) likely also preyed upon some of these animals—including *Stegosaurus* (fig. 14.1), *Mymoorapelta*, and *Camptosaurus*—but also likely took the opportunity to hunt sauropod dinosaurs of different sizes (as evidenced by various feeding traces). Sauropods were highly diverse in the Late Jurassic of North America and included many enormous forms of diplodocoids (*Apatosaurus*, *Brontosaurus*, *Diplodocus*, *Barosaurus*, *Haplocanthosaurus*, *Supersaurus*) and macronarians (*Camarasaurus* and *Brachiosaurus*). The larger sauropods may have required more help to take down, but it is also likely that allosauroids would have preferred to hunt younger sauropods of more comparable size ranges, using their slashing head strike abilities to dispatch them quickly.

Figure 14.1. Depiction of a Late Jurassic predator–prey interaction in North America (Morrison Formation), showing (*left*) *Allosaurus* facing off against (*right*) *Stegosaurus*.

Similar situations would have also likely been the case in the Late Jurassic of other parts of the world. The Tendaguru Formation of Africa was home to similarly sized allosauroids (including early carcharodontosaurids), megalosauroids, and noasaurids (*Elaphrosaurus*), which would have also preyed upon stegosaurs (*Kentrosaurus*), early ornithopods (*Dysalotosaurus*), and sauropods, also including both diplodocoids (*Dicraeosaurus*) and large macronarians (*Giraffatitan*). The Late Jurassic of China saw many large predatory dinosaurs (*Sinraptor* and *Yangchuanosaurus*) and smaller ones, including the earliest tyrannosauroids (the 3-m-long *Guanlong*), all living alongside similar herbivorous dinosaur groups as in North America and Africa, including numerous stegosaurs (*Chungkingosaurus*, *Chialingosaurus*, *Tuojiangosaurus*, and *Gigantospinosaurus*) and sauropods (*Mamenchisaurus*). Tiny predatory dinosaurs (about 1 m long or less) were also prevalent globally in this time, including early coelurosaurs, compsognathids, and scansoriopterygids, which would have fed on small animals like lizards and small early mammals as well as insects.

With the continuing separation of continents, even among Laurasian and Gondwanan landmasses, the Early Cretaceous saw a broad diversity of larger predatory dinosaurs becoming more restricted in terms of migration between landmasses, which sparks more competition in smaller areas worldwide. Early Cretaceous predatory dinosaurs include a broad range of theropod groups of similar size ranges as the Late Jurassic, albeit relatively less material is known from the Early Cretaceous than from the Late Jurassic or Late Cretaceous. These groups include abelisaurids (*Spectrovenator* of Brazil), allosauroids (the over 11-m-tall *Acrocanthosaurus* of North America and *Neovenator* of the United Kingdom), and numerous spinosaurids in England (*Baryonyx*), Brazil (*Irritator*), and Niger (*Suchomimus*, the largest from the Early Cretaceous reaching 9–11 m). All of these animals had their choice of a wide range of sauropods and ornithischians, although the options were comparatively less given the further separation of the continents. *Spectrovenator*, for instance, may have encountered the early Brazilian titanosaurian sauropod *Tapuiasaurus*. *Acrocanthosaurus* likely encountered gargantuan sauropods such as the macronarians *Astrodon*

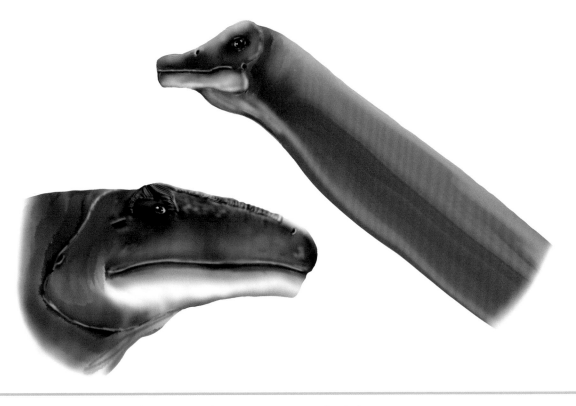

Figure 14.2. Depiction of a Late Cretaceous predator–prey interaction in South America (Candeleros Formation), showing (*left*) *Giganotosaurus* facing off against (*right*) a titanosaur.

and *Sauroposeidon* (depending on the region) as well as ornithopods like *Tenontosaurus*. Early Cretaceous spinosaurids may have been at least partially piscivorous, but they may have eaten other smaller animals near the water with elongate, narrow snouts. Dromaeosaurids also appeared in the Early Cretaceous and expanded both across North America and China (including the 3-m terrestrial *Deinonychus* of North America and the much smaller, tree-gliding *Microraptor* of China) with their fast biting capabilities, ideal for capturing smaller, faster prey. Evidence from isolated *Deinonychus* teeth associated with *Tenontosaurus* bones suggests a possibility of interaction between the two, for instance (Maxwell and Ostrom, 1995), although isotope evidence reveals a possible ontogenetic dietary shift from smaller to larger animals (Frederickson et al., 2020). Similar in size to dromaeosaurids were Early Cretaceous tyrannosauroids (*Dilong* of China), which were still much smaller than their Late Cretaceous evolutionary descendants.

Other smaller dinosaurs of the Early Cretaceous again included compsognathids from China and Italy as well as troodontids from China.

The Late Cretaceous was a time of the most extreme Mesozoic separation of continental landmasses, with each one home to its own dominant apex predator. In Gondwanan landmasses, there were many 7- to 9-m abelisaurids (including *Abelisaurus* and *Carnotaurus* of South America and *Rugops* and *Majungasaurus* of Africa and Madagascar) as well as much larger 10- to 13-m-long carcharodontosaurid allosauroids such as *Carcharodontosaurus* of northern Africa *Giganotosaurus* and *Mapusaurus* of South America. Arguably the largest of all Gondwanan theropods was the 15- to 16-m-long spinosaurid *Spinosaurus* of northern Africa, although multiple lines of evidence point to *Spinosaurus* (like other spinosaurids) feeding more on fish and other aquatic or shoreline prey items (Amiot et al., 2010; Hassler et al., 2018; Hone and Holtz, 2021), while African abelisaurids and carcharodontosaurids preyed on terrestrial animals (likely herbivorous dinosaurs; Hassler et al., 2018). The dominant herbivorous dinosaurs of the Late Cretaceous of Gondwana were rebbachisaurid and titanosaurian sauropods (including some of the largest of all like *Argentinosaurus* and *Patagotitan*), with relatively fewer ornithischians compared to Laurasian landmasses. Carcharodontosaurids and abelisaurids would have been well suited to prey upon sauropods of varying sizes and ages, with carcharodontosaurids especially capable of handling much larger prey (Henderson and Nicholls, 2015) (fig. 14.2). Since allosauroids like carcharodontosaurids are calculated to have lower agility (but higher rotational inertia) and are less efficient runners relative to tyrannosaurids, these animals would have been able to hold on to larger, slow-moving prey like sauropods (Snively et al., 2019; Dececchi et al., 2020a).

Tyrannosauroids were the apex predators throughout the Late Cretaceous of the northern Laurasian landmasses spanning from North America to Asia. This group grew to different sizes, ranging from 3 to 12 m and included *Suskityrannus*, *Alioramus*, *Qianzhousaurus*, *Tarbosaurus* of Asia as well as *Dryptosaurus*, *Albertosaurus*, *Gorgosaurus*, *Lythronax*, *Teratophoneus*, *Daspletosaurus*, and of course *Tyrannosaurus* of North America. Tyrannosauroids fed primarily on diverse ornithischian taxa of all shapes and sizes depending on how large of a prey their own body size could handle. Ornithischians of the Late Cretaceous included ankylosaurs, pachycephalosaurs, ceratopsians (especially ceratopsids), thescelosaurines, and hadrosaurids, all of which were distributed from North America to Asia (Farlow and Holtz, 2002). For instance, ornithischians of the Hell Creek Formation of North America included *Triceratops*, *Ankylosaurus*, *Thescelosaurus*, *Edmontosaurus*, and *Pachycephalosaurus*, all of which could have been prey for *Tyrannosaurus* (fig. 14.3). There were also a few titanosaurian sauropods in North America (*Alamosaurus*), but they were vastly outnumbered by ornithischians in this area.

The large heads with powerful bite forces, stronger overall build, and relatively higher agility in tyrannosaurids plausibly gave them a slight advantage over allosau-

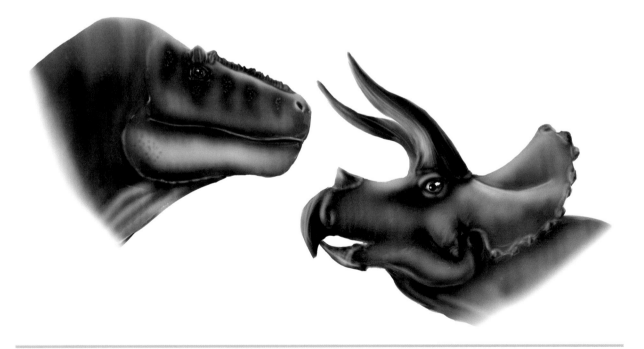

Figure 14.3. Depiction of a Late Cretaceous predator–prey relationship in North America (Hell Creek Formation), showing (*left*) *Tyrannosaurus* facing off against (*right*) *Triceratops*.

roids for attempting to take down large ornithischians, especially those with defense mechanisms like the horns of ceratopsids or the armor of ankylosaurs (Snively et al., 2019; Farlow and Holtz, 2002). But some prey could have more quickly outgrown their predators, such as those with higher growth rates than tyrannosaurids, like in the often enormous hadrosaurids (Cooper et al., 2008). Still, many predatory theropods may have preyed more so on the young, small individuals, as they were much easier to kill (Hone and Rauhut, 2010). It has also been shown that juvenile tyrannosaurids likely preyed on smaller herbivorous dinosaurs, which therefore explains the limited diversity of small- to medium-sized predatory dinosaurs in given areas (Schroeder et al., 2021; Holtz, 2021). Other relatively smaller predators (and omnivores) of the Late Cretaceous included 2- to 5-m-long dromaeosaurids (*Velociraptor*, *Bambiraptor*, *Dromaeosaurus*, *Buitreraptor*, and *Austroraptor*), noasaurids (*Masiakasaurus*), and troodontids, all of which likely preyed on smaller herbivorous dinosaurs or other smaller animals.

The Success of Dinosaur Herbivory

Herbivores are and always have been of utmost importance to ecosystems worldwide. Modern herbivores have acquired complex and often extreme evolutionary adaptations to be able to process and digest plant material of all kinds, and as discussed in previous chapters, the same was undeniably true for herbivorous dinosaurs (Weishampel and Norman, 1989; Smith, 1993; Fritz et al., 2011; Barrett, 2014). Representing more than 50% of all dinosaurs, herbivores likely played a large role in shaping the environments in which they lived by consuming large amounts of plant material as well as by acting as seed dispersers (Weishampel, 1984b; Tiffney, 1997; Farlow et al., 2010; Sander et al., 2010; Perry, 2021). Herbivores can really be thought of as the predators of plants, which is why plants show numerous structural and chemical defense mechanisms of their own (Weishampel, 1984b; Coley and Barone, 1996).

As such, the many adaptations of herbivorous dinosaurs were crucial for getting as much benefit as they could out of vegetation of variable nutritional value and fiber content (Weishampel, 1984b; Weishampel and Norman, 1989; Tiffney, 1997; Barrett, 2014). The more complex adaptations for herbivory that an animal possesses, the more likely that megaherbivores like hadrosaurs and ceratopsids were eating much tougher, high-fiber foods like bark, stems, and tough leaves rather than softer shoots, leaves, and fruits, which were typically eaten by smaller herbivores or larger megaherbivores with less complex feeding adaptations or even omnivores with a mix of adaptations for herbivory and carnivory (Barrett, 2014).

Evolutionary trends in the anatomy and ecology of herbivorous dinosaurs were often instigated by widespread competition based on the ability and efficiency of the dinosaurs in processing the available plant resources. These evolutionary trends also coincided with extinction events. Many times, these trends led to ecosystems in which herbivores as well as carnivores of diverse specializations and sizes showed patterns of dietary resource partitioning (or niche partitioning, as it is often referred to) so that there would be less chance of competition within a given environment (Upchurch and Barrett, 2000; Whitlock, 2011; Mallon and Anderson, 2013, 2014a, 2014b, 2015; Lautenschlager, 2017; Ma et al., 2019; Singh et al., 2021). Dietary partitioning occurred early in the evolution of dinosaurs, among not only the dinosaurs (of which there were very few) but also the other vertebrates around them, such as related non-dinosaurian dinosauromorphs and pseudosuchians, among others (Benton, 1983). Although the earliest dinosaurs were likely carnivorous or omnivorous (Barrett et al., 2011; Kubo, 2011; Singh et al., 2021), there were herbivorous animals living alongside them in the Late Triassic, including silesaurids, which were small, mostly herbivorous non-dinosaurian dinosauromorphs with denticulate leaf-shaped teeth similar to early ornithischian dinosaurs (Nesbitt, 2011) (although a recent phylogenetic analysis has suggested that silesaurids indeed may have been early-diverging ornithischians themselves; Müller and Garcia, 2020b).

Herbivory evolved multiple times within Dinosauria—in sauropodomorphs in the Late Triassic, in ornithischians in the Early Jurassic, and a few theropod groups throughout the Cretaceous, all likely by means of millions of years of evolutionary transition through omnivorous taxa (Weishampel and Norman, 1989; Barrett and Rayfield, 2006; Barrett, 2014). In many of these cases, herbivory can be inferred by the presence of complex craniodental feeding adaptations (including cranial shape, musculature, joint structures, and dental occlusion and replacement strategy) as well possible gastric mills in some cases. Other adaptations would include increased gut length with increased size, changes in browsing height, bulk feeding adaptations, and locomotor adaptations (Barrett and Rayfield, 2006; Zanno and Makovicky, 2011; Barrett, 2014). For detailed discussions of the diversity of all these adaptations throughout Dinosauria, see chapters 7 through 13.

Early, non-sauropod sauropodomorphs ("prosauropods") became the largest, most prominent herbivores of their ecosystems by the end of the Late Triassic Period (especially plateosaurids) and continued to diversify on through the Early Jurassic, especially. Ornithischians did not appear until the Early Jurassic, with such forms as heterodontosaurids and early-diverging thyreophorans, all of which were relatively rare compared to non-sauropod sauropodomorphs. With the exception of some heterodontosaurids, feeding mechanisms in all of these herbivores (and omnivores) were not particularly complex (Galton, 1985b; Barrett et al., 2011; Barrett, 2014). It was not until the Middle to Late Jurassic that herbivorous dinosaurs—particularly ornithischians like stegosaurs, ankylosaurs, and ornithopods—showed more complex adaptations suited for feeding on particular types of low- to middle-height vegetation (conifers, ferns, ginkgos, cycads, and horsetails). The Middle to Late Jurassic also saw the radiation of the long-necked sauropods, some of which reached enormous sizes that dwarfed the ornithischians of their time in both size and abundance (Barrett, 2014). By the end of the Late Jurassic, sauropod dinosaurs dominated the landscapes, with many types of macronarians (brachiosaurids and camarasaurids), diplodocoids (diplodocids and dicraeosaurids), and other sauropod groups (mamenchisaurids). Anatomical and paleoecological evidence has shown that there was clear partitioning of resources among sauropods feeding at different heights in different ecosystems (depending on how high they held their heads with their long necks). For instance, brachiosaurids and camarasaurids fed on higher, tougher vegetation, while diplodocoids were feeding on lower, softer vegetation (Calvo, 1994a; Upchurch and Barrett, 2000; Farlow et al., 2010; Whitlock, 2011; Button et al., 2014; MacLaren et al., 2017; see chaps. 8 and 9 for more details) (fig. 14.4). Craniodental evidence exists for partitioning between various ontogenetic stages of individual sauropod taxa (Woodruff et al., 2018). Stegosaurs (e.g., *Stegosaurus* and *Kentrosaurus*) also reached the peak of their global diversity in the Late Jurassic before going extinct sometime in the Early Cretaceous.

The Early Cretaceous saw a substantial decrease in sauropod diversity, with titanosaurs becoming the primary sauropods living through the Late Cretaceous, mainly in Gondwanan landmasses, but in relatively fewer numbers (Upchurch and Barrett, 2000; Barrett and Upchurch, 2005; Mannion et al. 2011; Barrett, 2014). Early ceratopsians, ornithopods, ankylosaurs, and even herbivorous theropods (therizinosaurs and ornithomimosaurs) started to develop, diversify, and disperse more broadly in the Early Cretaceous as well. The Late Cretaceous was when ceratopsids, hadrosaurids, ankylosaurs, pachycephalosaurs, and other ornithischians dominated North America, Europe, and Asia (as noted above). There were also some sauropods present in the northern continents and some ornithischians in the southern continents, but in each case, they were much rarer (Weishampel and Norman, 1989; Fastovsky and Smith, 2004; Zanno and Makovicky, 2011; Benson et al., 2013; Barrett, 2014).

Dietary partitioning was especially prevalent among Cretaceous ornithischians, and this feature has been especially well studied in the ornithischians of the Late Cretaceous of Dinosaur Park Formation in Alberta, Canada (fig. 14.5). These studies included quantitative data from skull morphology, snout shape, tooth shape and microwear, jaw biomechanics, feeding height stratification, and other general paleoecological factors. Broadly, hadrosaurids were mixed feeders that ate comparatively higher-level plant material that was generally composed of tough leaves and twigs; ceratopsians were more selective and fed somewhat lower on tough, woody browse; and ankylosaurs were mixed feeders (to varying degrees) and fed low to the ground on softer vegetation—although overlap did occur across diets as well (Mallon et al., 2012, 2013; Mallon and Anderson, 2013, 2014a, 2014b, 2015; Mallon, 2019; Wyenberg-Henzler et al., 2021, 2022; Wyenberg-Henzler, 2022; see chaps. 11–13 for more anatomical and paleobiological details on this topic). Similar paleoecological differentiation of plant diets has been discerned using isotopic data from ornithischians of the Hell Creek and

Figure 14.4. Three sauropods of the Late Jurassic of North America (Morrison Formation) that are of different heights, representing resource partitioning between taxa: (*top to bottom*) *Brachiosaurus*, *Camarasaurus*, and *Diplodocus*.

Oldman Formations as well, again with some dietary overlap (Fricke and Pearson, 2008; Cullen et al., 2020, 2022). Mallon (2019) concluded that competition among ornithischians of North America is what structured the communities, and this competition acted as a sort of structural principle that lasted through continued species turnover over time. With more megaherbivorous dinosaurs occupying broader ecomorphospaces and developing ecomorphological similarities (whether it was in terms of feeding anatomy and function, body size, or physiological adaptations), competition between these dinosaurs determined the extent of species richness across taxa in a given community (Mallon, 2019; Wyenberg-Henzler et al., 2021; Wyenberg-Henzler, 2022).

Global plant life changed drastically all throughout the Mesozoic (Sander et al., 2010; also see chap. 1), but the Cretaceous was by far the most unique compared to the Triassic and Jurassic. The Cretaceous was the time when angiosperms (flowering plants) grew, diversified, and dispersed rapidly across the globe, making them dominant in worldwide ecosystems, particularly those containing herbivorous dinosaurs (Wing and Sues, 1992;

Figure 14.5. Three ornithischians of the Late Cretaceous of North America (Dinosaur Park Formation) that are of different heights, representing resource partitioning between taxa: (*top to bottom*) *Gryposaurus*, *Chasmosaurus*, and *Euoplocephalus*.

Sander et al., 2010; Barrett, 2014). Owing to the evolution of complex feeding adaptations in Cretaceous herbivorous dinosaurs—particularly those of hadrosaurids, ceratopsids, and ankylosaurs, among others—some have suggested a coevolutionary relationship between dinosaurs and flowering plants (where reciprocal evolutionary pressures accelerated changes in both over time, fine-tuning their place in the ecosystem; Bakker, 1978; Coe et al., 1987; Wing and Tiffney, 1987; Weishampel and Norman, 1989; Tiffney, 1992, 2004; Taggart et al., 1997). The hypothesis (primarily proposed by Bakker, 1978) was based on general patterns of faunal and floral makeup of the Mesozoic. Whereas large-bodied, high-browsing sauropods dominated the gymnosperm-rich Jurassic ecosystems, the change to comparatively smaller ornithischian herbivores in the Early Cretaceous (following the extinction of many sauropod groups) created evolutionary pressures on lower-growing vegetation—specifically, conifers that grew slowly would have been eaten by these smaller herbivores before reaching taller heights. This process then made space for angiosperms, which are much better suited to thrive in low-growing environments disturbed by herbivores because they were capable of growing faster than gymnosperms. Additionally, the evolution of complex feeding mechanisms and the general paleoecology of many ornithischians were thought to have been greatly influenced by this coevolutionary relationship (Bakker, 1978; Wing and Tiffney, 1987; Weishampel and Norman, 1989; Tiffney, 1992).

Recent challenges to the dinosaur-angiosperm coevolution hypothesis by Barrett and Willis (2001) and Butler et al. (2009a, 2009b, 2010) have shown that the oversimplification inherent in the coevolutionary hypothesis does not take into account the full record of diversity of dinosaurs and flora in different environments, richness or abundance of different herbivorous taxa, or temporal or biostratigraphic patterns seen across Dinosauria and angiosperms in the Jurassic and Cretaceous. No qualitative relationship or statistically quantitative significance has been found to support the notion of an evolutionary relationship between dinosaurs and angiosperms (Barrett and Willis, 2001; Butler et al., 2009a, 2009b, 2010). This is not to say, however, that herbivorous dinosaurs did not ecologically interact with angiosperms. Quite the contrary: Dinosaurs assuredly fed on both angiosperms as well as gymnosperms. But a coevolutionary link is much more difficult to discern. Environmental conditions and interactions with pollinating insects were likely much stronger factors in the evolution of angiosperms, especially in the Late Cretaceous, when they were most abundant in the Mesozoic (Barrett and Willis, 2001; Barrett, 2014). Similar coevolutionary relationships have been proposed between cycads and sauropod and stegosaurian dinosaurs in the Jurassic owing to their potential as effective seed dispersers (Bakker, 1978; Mustoe, 2007); however, no statistical significance has been found in this relationship (Butler et al., 2010).

Abundance and spatial distributions of herbivorous dinosaurs were heavily influenced by the makeup of plant communities, depending on forest density, height stratification of plants, and their vegetative nutritional value (Weishampel and Norman, 1989; Butler and Barrett, 2008; Lyson and Longrich, 2011; Whitlock, 2011; Mallon et al., 2013; Mallon and Anderson, 2013, 2014a, 2014b, 2015; Barrett, 2014). Of course, what an herbivore eats depends on its physiological requirements to maintain their metabolism and body mass. Although body mass is reasonably predictable through various methods (see Campione and Evans, 2020), physiological and metabolic requirements are more difficult to discern, although these are being explored (Farlow, 1987; Colbert, 1993; Farlow et al., 1995; Hummel et al., 2008; McNab, 2009; Myhrvold, 2016). Dinosaur physiology, energy budgets, body size, and abundance are linked to plant productivity (Farlow et al., 1995; Barrett, 2014). Studies have suggested a possible association between high levels of CO_2 in the atmosphere and the abundance of large-bodied dinosaurs like sauropods owing to the likelihood of greater amounts of plant life in a given environment (Farlow et al., 1995; Burness et al., 2001). More recent studies have not found sufficient evidence for this assertion, however (Sander et al., 2011). It is possible that metabolic requirements for larger body size and gut length become less important and that therefore low, less nutritious plants with poor nitrogen content (like conifers, cycads, and other gymnosperms) would have been acceptable food sources, but more work is needed in this regard (Midgley et al., 2002; Engelmann et al., 2004; Farlow et al., 2010; Sander et al., 2011; Wilkinson and Ruxton, 2013; Barrett, 2014; Gill et al., 2018).

Nevertheless, the interplay between plants and herbivorous dinosaurs shaped entire ecosystems, and the importance of this relationship can be seen throughout the evolution of dinosaurs across the Mesozoic. The diversity of herbivorous dinosaurs (and dinosaur diversity in general) remained strong before the mass extinction at the end of the Cretaceous (Nordén et al., 2018; Chiarenza et al., 2019). An asteroid impact hitting earth about 66 million years ago was the most likely cause of the extinction of non-avian dinosaurs because it would have caused a global winter. The global winter would have blocked out the sun, thereby stopping photosynthesis and killing off food sources for herbivorous dino-

saurs, and by domino effect killing the herbivores and their predators thereafter (Alvarez et al., 1980; Chiarenza et al., 2019).

One group of theropod dinosaurs—the birds—did survive the event, however, and went on to evolve into an incredibly diverse array of forms of their own, with more than 10,000 species living today. Although many birds died in this mass extinction event, many neornithine ornithuromorphs survived because of key evolutionary innovations that allowed more room to evolve and diversify feeding strategies and ecologies in more difficult environments. These include cranial, gut, and flight adaptations (see chap. 7; O'Connor, 2019; Miller and Pittman, 2021). For instance, some seed-eating ornithuromorphs at the end of the Cretaceous would have likely been able to survive off seeds buried in the earth in harsh climates following asteroid impact (Larson et al., 2016). Following the Cretaceous, birds evolved and diversified in truly astounding ways, reaching overwhelmingly abundant populations all around the globe and making birds—living dinosaurs—one of the most successful major tetrapod groups alive on earth today.

References

Abler, William L. 1992. "The serrated teeth of tyrannosaurid dinosaurs, and biting structures in other animals." *Paleobiology* 18, no. 2: 161–83.

Abler, William L. 1999. "The teeth of the tyrannosaurs." *Scientific American* 281, no. 3: 50–51.

Abler, William L. 2001. "A kerf-and-drill model of tyrannosaur tooth serrations." In: D. H. Tanke and K. Carpenter, eds. *Mesozoic vertebrate life*. Bloomington: Indiana University Press, 84–89.

Abler, William L. 2013. "Internal structure of tooth serrations." In: R. E. Molnar and P. J. Currie, eds. *Tyrannosaurid paleobiology*. Bloomington: Indiana University Press, 81.

Adams, L. A. 1919. "A memoir on the phylogeny of the jaw muscles in recent and fossil vertebrates." *Annals of the New York Academy of Sciences* 28: 51–166.

Alexander, R. M. 1985. "Mechanics of posture and gait of some large dinosaurs." *Zoological Journal of the Linnean Society* 83, no. 1: 1–25.

Alifanov, V. R., and S. V. Saveliev. 2019. "The brain morphology and neurobiology in armored dinosaur *Bissekipelta archibaldi* (Ankylosauridae) from the Late Cretaceous of Uzbekistan." *Paleontological Journal* 53, no. 3: 315–21.

Alvarez, Luis W., Walter Alvarez, Frank Asaro, and Helen V. Michel. 1980. "Extraterrestrial cause for the Cretaceous-Tertiary extinction." *Science* 208, no. 4448: 1095–108.

Amiot, Romain, Eric Buffetaut, Christophe Lécuyer, Xu Wang, Larbi Boudad, et al. 2010. "Oxygen isotope evidence for semi-aquatic habits among spinosaurid theropods." *Geology* 38, no. 2: 139–42.

Apesteguía, Sebastián. 2004. "*Bonitasaura salgadoi* gen. et sp. nov.: a beaked sauropod from the Late Cretaceous of Patagonia." *Naturwissenschaften* 91, no. 10: 493–97.

Aranciaga Rolando, Alexis M., Mauricio A. Cerroni, and Fernando E. Novas. 2020. "Skull anatomy and pneumaticity of the enigmatic coelurosaurian theropod *Bicentenaria argentina*." *Anatomical Record* 303, no. 7: 1884–900.

Arbour, Victoria M. 2009. "Estimating impact forces of tail club strikes by ankylosaurid dinosaurs." *PLoS One* 4, no. 8: e6738.

Arbour, Victoria M., and Eric Snively. 2009. "Finite element analyses of ankylosaurid dinosaur tail club impacts." *Anatomical Record* 292, no. 9: 1412–26.

Arbour, Victoria M., and Lindsay E. Zanno. 2020. "Tail weaponry in ankylosaurs and glyptodonts: an example of a rare but strongly convergent phenotype." *Anatomical Record* 303, no. 4: 988–98.

Arden, Thomas M. S., Catherine G. Klein, Samir Zouhri, and Nicholas R. Longrich. 2019. "Aquatic adaptation in the skull of carnivorous dinosaurs (Theropoda: Spinosauridae) and the evolution of aquatic habits in spinosaurids." *Cretaceous Research* 93: 275–84.

Augustin, Felix J., Andreas T. Matzke, Michael W. Maisch, and Hans-Ulrich Pfretzschner. 2020. "A theropod dinosaur feeding site from the Upper Jurassic of the Junggar Basin, NW China." *Palaeogeography, Palaeoclimatology, Palaeoecology* 560: 109999.

Bailleul, Alida M., and Casey M. Holliday. 2017. "Joint histology in *Alligator mississippiensis* challenges the identification of synovial joints in fossil archosaurs and inferences of cranial kinesis." *Proceedings of the Royal Society B: Biological Sciences* 284, no. 1851: 20170038.

Bailleul, Alida M., Lawrence M. Witmer, and Casey M. Holliday. 2017. "Cranial joint histology in the mallard duck (*Anas platyrhynchos*): new insights on avian cranial kinesis." *Journal of Anatomy* 230, no. 3: 444–60.

Bailleul, Alida M., Zhiheng Li, Jingmai O'Connor, and Zhonghe Zhou. 2019. "Origin of the avian predentary and evidence of a unique form of cranial kinesis in Cretaceous ornithuromorphs." *Proceedings of the National Academy of Sciences* 116, no. 49: 24,696–706.

Ballell, Antonio, Emily J. Rayfield, and Michael J. Benton. 2022. "Walking with early dinosaurs: appendicular myology of the Late Triassic sauropodomorph *Thecodontosaurus antiquus.*" *Royal Society Open Science* 9, no. 1: 211356.

Bakker, Robert T. 1968. "*The superiority of dinosaurs.*" Discovery 3, no. 2: 11–22.

Bakker, Robert T. 1971. "Dinosaur physiology and the origin of mammals." *Evolution* 25: 636–58.

Bakker, Robert T. 1975. "Dinosaur renaissance." *Scientific American* 232, no. 4: 58–79.

Bakker, Robert T. 1978. "Dinosaur feeding behaviour and the origin of flowering plants." *Nature* 274, no. 5672: 661–63.

Bakker, Robert T. 1986. *The dinosaur heresies: new theories unlocking the mystery of the dinosaurs and their extinction.* New York: William Morrow.

Bakker, Robert T. 1998. "Brontosaur killers: Late Jurassic allosaurids as sabre-tooth cat analogues." *Gaia* 15: 145–58.

Bakker, Robert T., and Gary Bir. 2004. "Dinosaur crime scene investigations: theropod behavior at Como Bluff, Wyoming, and the evolution of birdness." In: Philip J. Currie, ed. *Feathered dragons: studies on the transition from dinosaurs to birds.* Bloomington: Indiana University Press, 301.

Bakker, Robert T., Robert M. Sullivan, Victor Porter, Peter Larson, and Steven J. Saulsbury. 2006. "*Dracorex hogwartsia,* n. gen., n. sp., a spiked, flat-headed pachycephalosaurid dinosaur from the Upper Cretaceous Hell Creek Formation of South Dakota." *New Mexico Museum of Natural History and Science Bulletin* 35: 331–45.

Ballell, Antonio, J. Logan King, James M. Neenan, Emily J. Rayfield, and Michael J. Benton. 2020. "The braincase, brain and palaeobiology of the basal sauropodomorph dinosaur *Thecodontosaurus antiquus.*" *Zoological Journal of the Linnean Society* 193, no. 2: 541–62.

Barker, Chris Tijani, Darren Naish, Elis Newham, Orestis L. Katsamenis, and Gareth Dyke. 2017. "Complex neuroanatomy in the rostrum of the Isle of Wight theropod *Neovenator salerii.*" *Scientific Reports* 7, no. 1: 1–8.

Baron, Matthew G., and Paul M. Barrett. 2017. "A dinosaur missing-link? *Chilesaurus* and the early evolution of ornithischian dinosaurs." *Biology Letters* 13, no. 8: 20170220.

Baron, Matthew G., David B. Norman, and Paul M. Barrett. 2017a. "A new hypothesis of dinosaur relationships and early dinosaur evolution." *Nature* 543, no. 7646: 501–6.

Baron, Matthew G., David B. Norman, and Paul M. Barrett. 2017b. "Postcranial anatomy of *Lesothosaurus diagnosticus* (Dinosauria: Ornithischia) from the Lower Jurassic of southern Africa: implications for basal ornithischian taxonomy and systematics." *Zoological Journal of the Linnean Society* 179, no. 1: 125–68.

Barrett, Paul M. 1998. "Herbivory in the non-avian Dinosauria." PhD diss., University of Cambridge.

Barrett, Paul M. 2000. "Prosauropods and iguanas: speculation on the diets of extinct reptiles." In H.-D. Sues, ed. *The evolution of herbivory in terrestrial vertebrates: perspectives from the fossil record.* Cambridge: Cambridge University Press, 42–78.

Barrett, Paul M. 2001. "Tooth wear and possible jaw action of *Scelidosaurus harrisonii* Owen and a review of feeding mechanisms in other thyreophoran dinosaurs." In: K. Carpenter, ed. *The armored dinosaurs.* Bloomington: Indiana University Press, 25–52.

Barrett, Paul M. 2005. "The diet of ostrich dinosaurs (Theropoda: Ornithomimosauria)." *Palaeontology* 48, no. 2: 347–58.

Barrett, Paul M. 2014. "Paleobiology of herbivorous dinosaurs." *Annual Review of Earth and Planetary Sciences* 42: 207–30.

Barrett, Paul M., and Susannah C. R. Maidment. 2017. "The evolution of ornithischian quadrupedality." *Journal of Iberian Geology* 43, no. 3: 363–77.

Barrett, Paul M., and E. J. Rayfield. 2006. "Ecological and evolutionary implications of dinosaur feeding behaviour." *Trends in Ecology and Evolution* 21, no. 4: 217–24.

Barrett, Paul M., and Paul Upchurch. 1994. "Feeding mechanisms of *Diplodocus.*" *Gaia* 10: 195–203.

Barrett, Paul M., and Paul Upchurch. 2005. "Phylogenetic and taxic perspectives on sauropod diversity." In: Kristina Curry Rogers and Jeffrey Wilson, eds. *The sauropods: evolution and paleobiology.* Berkeley: University of California Press, 125–56.

Barrett, Paul M., and Paul Upchurch. 2007. "The evolution of feeding mechanisms in early sauropodomorph dinosaurs." *Special Papers in Palaeontology* 77: 91–112.

Barrett, Paul M., and Katherine J. Willis. 2001. "Did dinosaurs invent flowers? Dinosaur–angiosperm coevolution revisited." *Biological Reviews* 76, no. 3: 411–47.

Barrett, Paul M., R. J. Butler, and S. J. Nesbitt. 2011. The roles of herbivory and omnivory in early dinosaur evolution. *Earth and Environmental Science Transactions of the Royal Society of Edinburgh* 101: 383–96.

Barsbold R. 1977. "Kinetism and specialty of the jaw apparatus of oviraptors (Theropoda, Saurischia)." *Transactions of the Joint Soviet-Mongolian Palaeontological Expedition* 4: 34–47.

Barsbold, R. 1983. "Carnivorous dinosaurs from the Cretaceous of Mongolia." *Transactions of the Joint Soviet-Mongolian Palaeontological Expedition* 19: 1–117.

Barta, Daniel, and Mark A. Norell. 2021. "The osteology of *Haya griva* (Dinosauria: Ornithischia) from the Late Cretaceous of Mongolia." *Bulletin of the American Museum of Natural History* 445. https://doi.org/10.1206/0003-0090 .445.1.1.

Bates, Karl T., and Peter L. Falkingham. 2012. "Estimating maximum bite performance in *Tyrannosaurus rex* using multi-body dynamics." *Biology Letters* 8, no. 4: 660–64.

Bates, Karl T., and Peter L. Falkingham. 2018a. "Correction to 'Estimating maximum bite performance in *Tyrannosaurus rex* using multi-body dynamics.'" *Biology Letters* 14, no. 4: 20180160.

Bates, Karl T., and Peter L. Falkingham. 2018b. "The importance of muscle architecture in biomechanical reconstructions of extinct animals: a case study using *Tyrannosaurus rex*." *Journal of Anatomy* 233, no. 5: 625–35.

Bates, Karl T., Phillip L. Manning, David Hodgetts, and William I. Sellers. 2009. "Estimating mass properties of dinosaurs using laser imaging and 3D computer modelling." *PLoS One* 4, no. 2: e4532.

Bates, Karl T., Roger B. J. Benson, and Peter L. Falkingham. 2012a. "A computational analysis of locomotor anatomy and body mass evolution in Allosauroidea (Dinosauria: Theropoda)." *Paleobiology* 38, no. 3: 486–507.

Bates, Karl T., Susannah C. R. Maidment, Vivian Allen, and Paul M. Barrett. 2012b. "Computational modelling of locomotor muscle moment arms in the basal dinosaur *Lesothosaurus diagnosticus*: assessing convergence between birds and basal ornithischians." *Journal of Anatomy* 220, no. 3: 212–32.

Bates, Karl T., Philip D. Mannion, Peter L. Falkingham, Stephen L. Brusatte, John R. Hutchinson, Alejandro Otero, William I. Sellers, Corwin Sullivan, Kent A. Stevens, and Vivian Allen. 2016. "Temporal and phylogenetic evolution of the sauropod dinosaur body plan." *Royal Society Open Science* 3, no. 3: 150636.

Becerra, Marcos G., and Diego Pol. 2020. "The enamel microstructure of *Manidens condorensis*: new hypotheses on the ancestral state and evolution of enamel in Ornithischia." *Acta Palaeontologica Polonica* 65, no. 1: 59–70.

Becerra, Marcos G., Kevin L. Gomez, and Diego Pol. 2017. "A sauropodomorph tooth increases the diversity of dental morphotypes in the Cañadón Asfalto Formation (Early–Middle Jurassic) of Patagonia." *Comptes Rendus Palevol* 16, no. 8: 832–40.

Becerra, Marcos G., Diego Pol, Gertrud E. Rössner, and Oliver W. M. Rauhut. 2018. "Heterodonty and double occlusion in *Manidens condorensis*: a unique adaptation in an Early Jurassic ornithischian improving masticatory efficiency." *Science of Nature* 105, no. 7: 1–5.

Becerra, Marcos G., Diego Pol, John A. Whitlock, and Laura B. Porro. 2021. "Tooth replacement in *Manidens condorensis*: baseline study to address the replacement pattern in dentitions of early ornithischians." *Papers in Palaeontology* 7, no. 2: 1167–93.

Behrensmeyer, Anna K., and Robert W. Hook. 1992. "Paleoenvironmental contexts and taphonomic modes." In: Anna K. Behrensmeyer, John D. Damuth, William A. DiMichele, et al., eds. *Terrestrial ecosystems through time: evolutionary paleoecology of terrestrial plants and animals*. Chicago: University of Chicago Press.

Bell, Phil R., and Philip J. Currie. 2009. "A tyrannosaur jaw bitten by a confamilial: scavenging or fatal agonism?" *Lethaia* 43, no. 2: 278–81.

Bell, Phil R., and Christophe Hendrickx. 2020. "Crocodile-like sensory scales in a Late Jurassic theropod dinosaur." *Current Biology* 30, no. 19: R1068–R1070.

Bell, Phil R., Eric Snively, and Lara Shychoski. 2009. "A comparison of the jaw mechanics in hadrosaurid and ceratopsid dinosaurs using finite element analysis." *Anatomical Record* 292, no. 9: 1338–51.

Benson, Roger B. J., and Jonah N. Choiniere. 2013. "Rates of dinosaur limb evolution provide evidence for exceptional radiation in Mesozoic birds." *Proceedings of the Royal Society B: Biological Sciences* 280, no. 1768: 20131780.

Benton, Michael J. 1983. "Dinosaur success in the Triassic: a noncompetitive ecological model." *Quarterly Review of Biology* 58, no. 1: 29–55.

Benton, Michael J. 1985. "Classification and phylogeny of the diapsid reptiles." *Zoological Journal of the Linnean Society* 84, no. 2: 97–164.

Berman, David S., and John Stanton McIntosh. 1986. "Description of the lower jaw of *Stegosaurus* (Reptilia, Ornithischia)." *Annals of the Carnegie Museum* 55: 29–40.

Berman, David S., and Bruce M. Rothschild. 2005. "Neck posture of sauropods determined using radiological imaging to reveal three-dimensional structure of cervical vertebrae." In: Virginia Tidwell and Kenneth Carpenter, eds. *Thunder-lizards: the sauropodomorph dinosaurs*. Bloomington: Indiana University Press, 233–47.

Bertozzo, Filippo, Fabio Manucci, Matthew Dempsey, Darren H. Tanke, David C. Evans, Alastair Ruffell, and Eileen Murphy. 2021. "Description and etiology of paleopathological lesions in the type specimen of *Parasaurolophus walkeri* (Dinosauria: Hadrosauridae), with proposed reconstructions of the nuchal ligament." *Journal of Anatomy* 238, no. 5: 1055–69.

Bhullar, Bhart-Anjan S., Michael Hanson, Matteo Fabbri, Adam Pritchard, Gabe S. Bever, and Eva Hoffman. 2016. "How to make a bird skull: major transitions in the evolution of the avian cranium, paedomorphosis, and the beak as a surrogate hand." *Integrative and Comparative Biology* 56, no. 3: 389–403.

Birn-Jeffery, Aleksandra, and Emily Rayfield. 2009. "Finite element analysis of pedal claws to determine mode of life in birds, lizards and maniraptoran theropods." *Journal of Vertebrate Paleontology* 29: 64A.

Bishop, Peter J. 2019. "Testing the function of dromaeosaurid (Dinosauria, Theropoda) 'sickle claws' through musculoskeletal modelling and optimization." *PeerJ* 7: e7577.

Bishop, Peter J., Christofer J. Clemente, R. E. Weems, D. F. Graham, L. P. Lamas, et al. 2017. "Using step width to compare locomotor biomechanics between extinct, non-avian theropod dinosaurs and modern obligate bipeds." *Journal of the Royal Society Interface* 14, no. 132: 20170276.

Bishop, Peter J., Andrew R. Cuff, and John R. Hutchinson. 2021. "How to build a dinosaur: musculoskeletal modeling and simulation of locomotor biomechanics in extinct animals." *Paleobiology* 47, no. 1: 1–38.

Blanco, R. Ernesto, and Gerardo V. Mazzetta. 2001. "A new approach to evaluate the cursorial ability of the giant theropod *Giganotosaurus carolinii.*" *Acta Palaeontologica Polonica* 46, no. 2.

Bock, Walter J. 1964. "Kinetics of the avian skull." *Journal of Morphology* 114, no. 1: 1–41.

Bonaparte, J. F., and O. Mateus. 1999. "A new diplodocid, *Dinheirosaurus lourinhanensis* gen. et sp. nov., from the Late Jurassic beds of Portugal." *Revista del Museo Argentino de Ciencias Naturales* 5, no. 2: 13–29.

Bonaparte, J. F., and José A. Pumares. 1995. "Notes on the first skull and jaws of *Riojasaurus incertus* (Dinosauria, Prosauropoda, Melanorosauridae), Late Triassic of La Rioja, Argentina." *Ameghiniana* 32, no. 4: 341–34.

Bonaparte, J. F., F. E. Novas, and R. A. Coria. 1990. "*Carnotaurus sastrei* Bonaparte, the horned, lightly built carnosaur from the Middle Cretaceous of Patagonia." *Contributions in Science* 416, no. 4: 1–41.

Bonnan, M. F., F. Matthew, and P. Senter. 2007. "Evolution and palaeobiology of early sauropodomorph dinosaurs." *Special Papers in Palaeontology* 77: 139–55.

Borsuk-Bialynicka, Magdalena. 1977. "A new camarasaurid sauropod *Opisthocoelicaudia skarzynskii* gen. n., sp. n. from the Upper Cretaceous of Mongolia." *Palaeontologia Polonica* 37, no. 5: 5–64.

Bouabdellah, Florian, Emily Lessner, and Julien Benoit. 2022. "The rostral neurovascular system of *Tyrannosaurus rex.*" *Palaeontologia Electronica* 25, no. 1: 1–21.

Bourke, Jason M., W. M. Ruger Porter, Ryan C. Ridgely, Tyler R. Lyson, Emma R. Schachner, Phil R. Bell, and Lawrence M. Witmer. 2014. "Breathing life into dinosaurs: tackling challenges of soft-tissue restoration and nasal airflow in extinct species." *Anatomical Record* 297, no. 11: 2148–86.

Bourke, Jason M., W. Ruger Porter, and Lawrence M. Witmer. 2018. "Convoluted nasal passages function as efficient heat exchangers in ankylosaurs (Dinosauria: Ornithischia: Thyreophora)." *PLoS One* 13, no. 12: e0207381.

Bout, Ron G., and Gart A. Zweers. 2001. "The role of cranial kinesis in birds." *Comparative Biochemistry and Physiology Part A: Molecular and Integrative Physiology* 131, no. 1: 197–205.

Boyd, C. A. 2014. "The cranial anatomy of the neornithischian dinosaur *Thescelosaurus neglectus.*" *PeerJ* 2: e669.

Boyd, C. A. 2015. "The systematic relationships and biogeographic history of ornithischian dinosaurs." *PeerJ* 3: e1523.

Bramble, Katherine, Aaron R. H. LeBlanc, Denis O. Lamoureux, Mateusz Wosik, and Philip J. Currie. 2017. "Histological evidence for a dynamic dental battery in hadrosaurid dinosaurs." *Scientific Reports* 7, no. 1: 1–13.

Brassey, Charlotte A., Susannah C. R. Maidment, and Paul M. Barrett. 2015. "Body mass estimates of an exceptionally complete *Stegosaurus* (Ornithischia: Thyreophora): comparing volumetric and linear bivariate mass estimation methods." *Biology Letters* 11, no. 3: 20140984.

Brett-Surman, Michael K., Thomas R. Holtz, and James O. Farlow, eds. 2012. *The complete dinosaur.* Bloomington: Indiana University Press.

Brink, K. S., R. R. Reisz, A. R. H. LeBlanc, R. S. Chang, Y. C. Lee, C. C. Chiang, T. Huang, and D. C. Evans. 2015. "Developmental and evolutionary novelty in the serrated teeth of theropod dinosaurs." *Scientific Reports* 5, no. 1: 1–12.

Brink, K. S., Yu-Cheng Chen, Ya-Na Wu, Wei-Min Liu, Dar-Bin Shieh, Timothy D. Huang, Chi-Kuang Sun, and Robert R. Reisz. 2016. "Dietary adaptions in the ultrastructure of dinosaur dentine." *Journal of the Royal Society Interface* 13, no. 125: 20160626.

Bronzati, Mario, Rodrigo T. Müller, and Max C. Langer. 2019. "Skull remains of the dinosaur *Saturnalia tupiniquim* (Late Triassic, Brazil): with comments on the early evolution of sauropodomorph feeding behaviour." *PLoS One* 14, no. 9: e0221387.

Brown, Caleb M., David R. Greenwood, Jessica E. Kalyniuk, Dennis R. Braman, Donald M. Henderson, Cathy L. Greenwood, and James F. Basinger. 2020. "Dietary palaeoecology of an Early Cretaceous armoured dinosaur (Ornithischia; Nodosauridae) based on floral analysis of stomach contents." *Royal Society Open Science* 7, no. 6: 200305.

Brownstein, Chase D. 2019. "*Halszkaraptor escuilliei* and the evolution of the paravian bauplan." *Scientific Reports* 9, no. 1: 1–16.

Brusatte, Stephen L., Michael J. Benton, Marcello Ruta, and Graeme T. Lloyd. 2008. "Superiority, competition, and opportunism in the evolutionary radiation of dinosaurs." *Science* 321, no. 5895: 1485–88.

Brusatte, Stephen L., Richard J. Butler, Albert Prieto-Márquez, and Mark A. Norell. 2012. "Dinosaur morphological diversity and the end-Cretaceous extinction." *Nature Communications* 3, no. 1: 1–8.

Brusatte, Steve. 2018. *The rise and fall of the dinosaurs: the untold story of a lost world*. New York: Mariner Books.

Bryant, Harold N., and Anthony P. Russell. 1992. "The role of phylogenetic analysis in the inference of unpreserved attributes of extinct taxa." *Philosophical Transactions of the Royal Society of London, Series B: Biological Sciences* 337, no. 1282: 405–18.

Bryant, Harold N., and Kevin L. Seymour. 1990. "Observations and comments on the reliability of muscle reconstruction in fossil vertebrates." *Journal of Morphology* 206, no. 1: 109–17.

Buchholtz, Emily. 2012. "Dinosaur paleoneurology." In: M. K. Brett-Surman, T. R. Holtz, and J. O. Farlow, eds. *The complete dinosaur*. Bloomington: Indiana University Press, 191–208.

Buffetaut, Eric, David Martill, and François Escuillié. 2004. "Pterosaurs as part of a spinosaur diet." *Nature* 430, no. 6995: 33–33.

Burch, Sara H. 2014. "Complete forelimb myology of the basal theropod dinosaur *Tawa hallae* based on a novel robust muscle reconstruction method." *Journal of Anatomy* 225, no. 3: 271–97.

Burch, Sara H. 2017. "Myology of the forelimb of *Majungasaurus crenatissimus* (Theropoda, Abelisauridae) and the morphological consequences of extreme limb reduction." *Journal of Anatomy* 231, no. 4: 515–31.

Burness, Gary P., Jared Diamond, and Timothy Flannery. 2001. "Dinosaurs, dragons, and dwarfs: the evolution of maximal body size." *Proceedings of the National Academy of Sciences* 98, no. 25: 14,518–523.

Busbey, Arthur B., III. 1989. "Form and function of the feeding apparatus of *Alligator mississippiensis*." *Journal of Morphology* 202, no. 1: 99–127.

Busbey, Arthur B. 1995. "The structural consequences of skull flattening in crocodilians." In: J. J. Thomason, ed. *Functional morphology in vertebrate paleontology*. Cambridge: Cambridge University Press, 173–92.

Butler, Richard J., and Paul M. Barrett. 2008. "Palaeoenvironmental controls on the distribution of Cretaceous herbivorous dinosaurs." *Naturwissenschaften* 95, no. 11: 1027–32.

Butler, Richard J., Paul Upchurch, and David B. Norman. 2008. "The phylogeny of the ornithischian dinosaurs." *Journal of Systematic Palaeontology* 6, no. 1: 1–40.

Butler, Richard J., Paul M. Barrett, Paul Kenrick, and Malcolm G. Penn. 2009a. "Diversity patterns amongst herbivorous dinosaurs and plants during the Cretaceous: implications for hypotheses of dinosaur/angiosperm co-evolution." *Journal of Evolutionary Biology* 22, no. 3: 446–59.

Butler, Richard J., Paul M. Barrett, Paul Kenrick, and Malcolm G. Penn. 2009b. "Testing co-evolutionary hypotheses over geological timescales: interactions between Mesozoic non-avian dinosaurs and cycads." *Biological Reviews* 84, no. 1: 73–89.

Butler, Richard J., Paul M. Barrett, Malcolm G. Penn, and Paul Kenrick. 2010. "Testing coevolutionary hypotheses over geological timescales: interactions between Cretaceous dinosaurs and plants." *Biological Journal of the Linnean Society* 100, no. 1: 1–15.

Butler, Richard J., Laura B. Porro, Peter M. Galton, and Luis M. Chiappe. 2012. "Anatomy and cranial functional morphology of the small-bodied dinosaur *Fruitadens haagarorum* from the Upper Jurassic of the USA." *PLoS One* 7, no. 4: e31556.

Button, David J., and Lindsay E. Zanno. 2020. "Repeated evolution of divergent modes of herbivory in non-avian dinosaurs." *Current Biology* 30, no. 1: 158–68.

Button, David J., Emily J. Rayfield, and Paul M. Barrett. 2014. "Cranial biomechanics underpins high sauropod diversity in resource-poor environments." *Proceedings of the Royal Society B: Biological Sciences* 281, no. 1795: 20142114.

Button, David J., Paul M. Barrett, and Emily J. Rayfield. 2016. "Comparative cranial myology and biomechanics of *Plateosaurus* and *Camarasaurus* and evolution of the sauropod feeding apparatus." *Palaeontology* 59, no. 6: 887–913.

Button, David J., Paul M. Barrett, and Emily J. Rayfield. 2017. "Craniodental functional evolution in sauropodomorph dinosaurs." *Paleobiology* 43, no. 3: 435–62.

Button, Khai, Hailu You, James I. Kirkland, and Lindsay Zanno. 2017. "Incremental growth of therizinosaurian dental tissues: implications for dietary transitions in Theropoda." *PeerJ* 5: e4129.

Cabreira, Sergio Furtado, Alexander Wilhelm Armin Kellner, Sérgio Dias-da-Silva, Lúcio Roberto da Silva, Mario Bronzati, et al. 2016. "A unique Late Triassic dinosauromorph assemblage reveals dinosaur ancestral anatomy and diet." *Current Biology* 26, no. 22: 3090–95.

Calvo, Jorge O. 1994a. "Jaw mechanics in sauropod dinosaurs." *Gaia* 10: 183–93.

Calvo, Jorge O. 1994b. "Gastroliths in sauropod dinosaurs." *Gaia* 10: 205–8.

Campione, Nicolás E., and David C. Evans. 2020. "The accuracy and precision of body mass estimation in non-avian dinosaurs." *Biological Reviews* 95, no. 6: 1759–97.

Campione, Nicolas E., and Robert Holmes. 2006. "The anatomy and homologies of the ceratopsid syncervical." *Journal of Vertebrate Paleontology* 26, no. 4: 1014–17.

Candeiro, Carlos Roberto A., Philip J. Currie, Caio L. Candeiro, and Lílian P. Bergqvist. 2017. "Tooth wear and microwear of theropods from the Late Maastrichtian Marília Formation (Bauru Group), Minas Gerais State, Brazil." *Earth and Environmental Science Transactions of the Royal Society of Edinburgh* 106, no. 4: 229–33.

Carballido, José L., Diego Pol, Alejandro Otero, Ignacio A. Cerda, Leonardo Salgado, Alberto C. Garrido, Jahandar Ramezani, Néstor R. Cúneo, and Javier M. Krause. 2017. "A new giant titanosaur sheds light on body mass evolution among sauropod dinosaurs." *Proceedings of the Royal Society B: Biological Sciences* 284, no. 1860: 20171219.

Carbone, Chris, Samuel T. Turvey, and Jon Bielby. 2011. "Intraguild competition and its implications for one of the biggest terrestrial predators, *Tyrannosaurus rex*." *Proceedings of the Royal Society B: Biological Sciences* 278, no. 1718: 2682–90.

Carpenter, Kenneth. 1997. "Agonistic behavior in pachycephalosaurs (Ornithischia, Dinosauria): a new look at head-butting behavior." *Rocky Mountain Geology* 32, no. 1: 19–25.

Carpenter, Kenneth. 1998. "Evidence of predatory behavior by carnivorous dinosaurs." *Gaia* 15: 135–44.

Carpenter, Kenneth. 2000. "Marsh's dinosaurs: the collections from Como Bluff." *Journal of Vertebrate Paleontology* 20, no. 4: 784–84.

Carpenter, Kenneth. 2002. "Forelimb biomechanics of nonavian theropod dinosaurs in predation." *Senckenbergiana Lethaea* 82, no. 1: 59–75.

Carpenter, Kenneth. 2004. "Redescription of *Ankylosaurus magniventris* Brown 1908 (Ankylosauridae) from the Upper Cretaceous of the western interior of North America." *Canadian Journal of Earth Sciences* 41, no. 8: 961–86.

Carpenter, Kenneth. 2013. "A closer look at the hypothesis of scavenging versus predation by *Tyrannosaurus rex*." In: R. E. Molnar and P. J. Currie, eds. *Tyrannosaurid paleobiology*. Bloomington: Indiana University Press, 265–78.

Carpenter, Kenneth, and Matt Smith. 2001. "Forelimb osteology and biomechanics of *Tyrannosaurus rex*." In: D. H. Tanke and K. Carpenter, eds. *Mesozoic vertebrate life*. Bloomington: Indiana University Press, 90–116.

Carpenter, Kenneth, and Yvonne Wilson. 2008. "A new species of *Camptosaurus* (Ornithopoda: Dinosauria) from the Morrison Formation (Upper Jurassic) of Dinosaur National Monument, Utah, and a biomechanical analysis of its forelimb." *Annals of Carnegie Museum* 76, no. 4: 227–63.

Carpenter, Kenneth, S. Hayashi, Y. Kobayashi, T. Maryańska, R. Barsbold, K. Sato, and I. Obata. 2011. "*Saichania chul-sanensis* (Ornithischia, Ankylosauridae) from the Upper Cretaceous of Mongolia." *Palaeontographica Abteilung A* 294, nos. 1–3: 1–61.

Carr, Thomas D. 2020. "A high-resolution growth series of *Tyrannosaurus rex* obtained from multiple lines of evidence." *PeerJ* 8: e9192.

Carr, Thomas D., David J. Varricchio, Jayc C. Sedlmayr, Eric M. Roberts, and Jason R. Moore. 2017. "A new tyrannosaur with evidence for anagenesis and crocodile-like facial sensory system." *Scientific Reports* 7, no. 1: 1–11.

Carrano, Matthew T. 1999. "What, if anything, is a cursor? Categories versus continua for determining locomotor habit in mammals and dinosaurs." *Journal of Zoology* 247, no. 1: 29–42.

Carrano, Matthew T. 2005. "The evolution of sauropod locomotion: morphological diversity of a secondarily quadrupedal radiation." In: Kristina Curry Rogers and Jeffrey Wilson, eds. *The sauropods: evolution and paleobiology*. Berkeley: University of California Press, 229–51.

Carrano, Matthew T., and Jonah Choiniere. 2016. "New information on the forearm and manus of *Ceratosaurus nasicornis* Marsh, 1884 (Dinosauria, Theropoda), with implications for theropod forelimb evolution." *Journal of Vertebrate Paleontology* 36, no. 2: e1054497.

Carrano, Matthew T., and John R. Hutchinson. 2002. "Pelvic and hindlimb musculature of *Tyrannosaurus rex* (Dinosauria: Theropoda)." *Journal of Morphology* 253, no. 3: 207–28.

Carrano, Matthew T., and Christine M. Janis. 1999. "Hadrosaurs as ungulate parallels: lost lifestyles and deficient data." *Acta Palaeontologica Polonica* 44, no. 3: 237–61.

Cau, Andrea. 2018. "The assembly of the avian body plan: a 160-million-year-long process." *Bollettino della Società Paleontologica Italiana* 57, no. 1: 1–25.

Cau, Andrea. 2020. "The body plan of *Halszkaraptor escuilliei* (Dinosauria, Theropoda) is not a transitional form along the evolution of dromaeosaurid hypercarnivory." *PeerJ* 8: e8672.

Cau, Andrea, Vincent Beyrand, Dennis F. A. E. Voeten, Vincent Fernandez, Paul Tafforeau, Koen Stein, Rinchen Barsbold, Khishigjav Tsogtbaatar, Philip J. Currie, and Pascal Godefroit. 2017. "Synchrotron scanning reveals amphibious ecomorphology in a new clade of bird-like dinosaurs." *Nature* 552, no. 7685: 395–99.

Cerda, Ignacio A. 2008. "Gastroliths in an ornithopod dinosaur." *Acta Palaeontologica Polonica* 53, no. 2: 351–55.

Cerda, Ignacio A., Fernando E. Novas, José Luis Carballido, and Leonardo Salgado. 2022. "Osteohistology of the hyperelongate hemispinous processes of *Amargasaurus cazaui* (Dinosauria: Sauropoda): implications for soft tissue reconstruction and functional significance." *Journal of Anatomy* 240: 1005–19.

Cerroni, Mauricio A., Juan I. Canale, Fernando E. Novas, and Ariana Paulina-Carabajal. 2020. "An exceptional neurovascular system in abelisaurid theropod skull: new evidence from Skorpiovenator bustingorryi." *Journal of Anatomy* 240, no. 4: 1–15.

Chapelle, Kimberley E. J., Paul M. Barrett, Jennifer Botha, and Jonah N. Choiniere. 2019a. "*Ngwevu intloko*: a new early sauropodomorph dinosaur from the Lower Jurassic Elliot Formation of South Africa and comments on cranial ontogeny in Massospondylus carinatus." *PeerJ* 7: e7240.

Chapelle, Kimberley E. J., Roger B. J. Benson, Josef Stiegler, Alejandro Otero, Qi Zhao, and Jonah N. Choiniere. 2019b. "A quantitative method for inferring locomotory shifts in amniotes during ontogeny, its application to dinosaurs and its bearing on the evolution of posture." *Palaeontology* 63, no. 2: 229–42.

Charig, Alan Jack, and Angela C. Milner. 1997. "*Baryonyx walkeri*, a fish-eating dinosaur from the Wealden of Surrey." *Bulletin of the Natural History Museum of London* 53: 11–70.

Chen, Jun, Aaron R. H. LeBlanc, Liyong Jin, Timothy Huang, and Robert R. Reisz. 2018. "Tooth development, histology, and enamel microstructure in *Changchunsaurus parvus*: implications for dental evolution in ornithopod dinosaurs." *PLoS One* 13, no. 11: e0205206.

Chiappe, Luis M., Bo Zhao, Jingmai K. O'Connor, Gao Chunling, Xuri Wang, Michael Habib, Jesus Marugan-Lobon, Qingjin Meng, and Xiaodong Cheng. 2014. "A new specimen of the Early Cretaceous bird *Hongshanornis longicresta*: insights into the aerodynamics and diet of a basal ornithuromorph." *PeerJ* 2: e234.

Chiarenza, Alfio Alessandro, Philip D. Mannion, Daniel J. Lunt, Alex Farnsworth, Lewis A. Jones, Sarah-Jane Kelland, and Peter A. Allison. 2019. "Ecological niche modelling does not support climatically-driven dinosaur diversity decline before the Cretaceous/Paleogene mass extinction." *Nature Communications* 10, no. 1: 1–14.

Chin, Karen, Timothy T. Tokaryk, Gregory M. Erickson, and Lewis C. Calk. 1998. "A king-sized theropod coprolite." *Nature* 393, no. 6686: 680–82.

Chin, Karen, David A. Eberth, Mary H. Schweitzer, Thomas A. Rando, Wendy J. Sloboda, and John R. Horner. 2003. "Remarkable preservation of undigested muscle tissue within a Late Cretaceous tyrannosaurid coprolite from Alberta, Canada." *Palaios* 18, no. 3: 286–94.

Chin, Karen, Rodney M. Feldmann, and Jessica N. Tashman. 2017. "Consumption of crustaceans by megaherbivorous dinosaurs: dietary flexibility and dinosaur life history strategies." *Scientific Reports* 7, no. 1: 1–11.

Chinnery, Brenda. 2004. "Morphometric analysis of evolutionary trends in the ceratopsian postcranial skeleton." *Journal of Vertebrate Paleontology* 24, no. 3: 591–609.

Christian, Andreas. 2002. "Neck posture and overall body design in sauropods." *Fossil Record* 5, no. 1: 271–81.

Christian, Andreas. 2010. "Some sauropods raised their necks—evidence for high browsing in *Euhelopus zdanskyi*." *Biology Letters* 6, no. 6: 823–25.

Christian, Andreas, and Gordon Dzemski. 2007. "Reconstruction of the cervical skeleton posture of *Brachiosaurus brancai* Janensch, 1914 by an analysis of the intervertebral stress along the neck and a comparison with the results of different approaches." *Fossil Record* 10, no. 1: 38–49.

Christian, Andreas, and Gordon Dzemski. 2011. "Neck posture in sauropods." In: N. Klein, K. Remes, C. T. Gee, and P. M. Sander, eds. *Biology of the sauropod dinosaurs: understanding the life of giants*. Bloomington: Indiana University Press, 251–60.

Christian, Andreas, and W.-D. Heinrich. 1998. "The neck posture of *Brachiosaurus brancai*." *Fossil Record* 1, no. 1: 73–80.

Christian, Andreas, and Holger Preuschoft. 1996. "Deducing the body posture of extinct large vertebrates from the shape of the vertebral column." *Palaeontology* 39, no. 4: 801–812.

Christian, Andreas, Dorothee Koberg, and Holger Preuschoft. 1996. "Shape of the pelvis and posture of the hindlimbs in *Plateosaurus*." *Paläontologische Zeitschrift* 70, nos. 3–4: 591–601.

Christian, Andreas, Guangzhao Peng, Toru Sekiya, Yong Ye, Marco G. Wulf, and Thorsten Steuer. 2013. "Biomechanical reconstructions and selective advantages of neck poses and feeding strategies of Sauropods with the example of *Mamenchisaurus youngi*." *PLoS One* 8, no. 10: e71172.

Christiansen, P. 1996. "The evidence for and implications of gastroliths in sauropods (Dinosauria, Sauropoda)." *Gaia* 12: 1–7.

Christiansen, P. 1999. "On the head size of sauropodomorph dinosaurs: implications for ecology and physiology." *Historical Biology* 13, no. 4: 269–97.

Christiansen, P. 2000. "Feeding mechanisms of the sauropod dinosaurs *Brachiosaurus*, *Camarasaurus*, *Diplodocus* and *Dicraeosaurus*." *Historical Biology* 14, no. 3: 137–52.

Chure, Daniel J. 1998. "On the orbit of theropod dinosaurs." *Gaia* 15: 233–40.

Cobb, Savannah Elizabeth, and William I. Sellers. 2020. "Inferring lifestyle for Aves and Theropoda: a model based on curvatures of extant avian ungual bones." *PLoS One* 15, no. 2: e0211173.

Cobley, Matthew J., Emily J. Rayfield, and Paul M. Barrett. 2013. "Inter-vertebral flexibility of the ostrich neck: implications for estimating sauropod neck flexibility." *PLoS One* 8, no. 8: e72187.

Coe, M. J., D. L. Dilcher, J. O. Farlow, D. M. Jarzen, and D. A. Russell. 1987. "Dinosaurs and land plants." In: *The origins of angiosperms and their biological consequences.* Cambridge: Cambridge University Press, 225–58.

Colbert, Edwin Harris. 1961. *Dinosaurs: their discovery and their world.* New York: Dutton.

Colbert, Edwin Harris. 1981. *A primitive ornithischian dinosaur from the Kayenta Formation of Arizona.* Bulletin Series 53. Flagstaff: Museum of Northern Arizona Press, 61 pp.

Colbert, Edwin Harris. 1989. *The Triassic dinosaur Coelophysis.* Bulletin Series 57. Flagstaff: Museum of Northern Arizona Press.

Colbert, Edwin H. 1993. "Feeding strategies and metabolism in elephants and sauropod dinosaurs." *American Journal of Science* 293, no. A: 1.

Colbert, Edwin Harris, and John H. Ostrom. 1958. "*Dinosaur stapes.*" *American Museum Novitates* 1900: 1–20.

Coley, Phyllis D., and J. A. Barone. 1996. "Herbivory and plant defenses in tropical forests." *Annual Review of Ecology and Systematics* 27, no. 1: 305–35.

Coombs, Walter Preston, Jr. 1971. *The Ankylosauria.* Columbia University.

Coombs, Walter Preston, Jr. 1975. "Sauropod habits and habitats." *Palaeogeography, Palaeoclimatology, Palaeoecology* 17, no. 1: 1–33.

Coombs, Walter Preston, Jr. 1978. "Theoretical aspects of cursorial adaptations in dinosaurs." *Quarterly Review of Biology* 53, no. 4: 393–418.

Coombs, Walter Preston, Jr., and T. Maryańska. 1990. "Ankylosauri." In D. B. Weishampel, P. Dodson, and H. Osmólska, eds. *The dinosauria*, 2nd ed. Berkeley: University of California Press, 456–83.

Cooper, Michael R. 1981. "The prosauropod dinosaur *Massospondylus carinatus* Owen from Zimbabwe: its biology, mode of life and phylogenetic significance." *Occasional Papers of the National Museums and Monuments Rhodesia* 6, no. 10: 690–840.

Cooper, Michael R. 1985. "A revision of the ornithischian dinosaur *Kangnasaurus coetzeei* Haughton, with a classification of the Ornithischia." *Annals of the South African Museum* 95, no. 8: 281–317.

Cooper, Lisa Noelle, Andrew H. Lee, Mark L. Taper, and John R. Horner. 2008. "Relative growth rates of predator and prey dinosaurs reflect effects of predation." *Proceedings of the Royal Society B: Biological Sciences* 275, no. 1651: 2609–15.

Cope, Edward Drinker. 1883. "On the characters of the skull in the Hadrosauridae." *Proceedings of the Academy of Natural Sciences of Philadelphia*: 97–107.

Cost, Ian N., Kevin M. Middleton, Kaleb C. Sellers, Michael Scott Echols, Lawrence M. Witmer, Julian L. Davis, and Casey M. Holliday. 2020. "Palatal biomechanics and its significance for cranial kinesis in *Tyrannosaurus rex.*" *Anatomical Record* 303, no. 4: 999–1017.

Cost, Ian N., Kaleb C. Sellers, Rachel E. Rozin, Anthony T. Spates, Kevin M. Middleton, and Casey M. Holliday. 2022. "2D and 3D visualizations of archosaur jaw muscle mechanics, ontogeny and phylogeny using ternary diagrams and 3D modeling." *Journal of Experimental Biology* 225: jeb243216.

Cracraft, Joel. 1971. "Caenagnathiformes: Cretaceous birds convergent in jaw mechanism to dicynodont reptiles." *Journal of Paleontology* 45, no. 5: 805–9.

Crompton, A. W., and J. Attridge. 1986. "Masticatory apparatus of the larger herbivores during Late Triassic and Early Jurassic times." In: Kevin Padian, ed. *The beginning of the age of the dinosaurs.* Cambridge: Cambridge University Press, 223–36.

Cuff, Andrew R., and Emily J. Rayfield. 2013. "Feeding mechanics in spinosaurid theropods and extant crocodilians." *PLoS One* 8, no. 5: e65295.

Cuff, Andrew R., and Emily J. Rayfield. 2015. "Retrodeformation and muscular reconstruction of ornithomimosaurian dinosaur crania." *PeerJ* 3: e1093.

Cullen, T. M., F. J. Longstaffe, U. G. Wortmann, L. Huang, F. Fanti, M. B. Goodwin, M. J. Ryan, and D. C. Evans. 2020. "Large-scale stable isotope characterization of a Late Cretaceous dinosaur-dominated ecosystem." *Geology* 48, no. 6: 546–51.

Cullen, Thomas M., Shuangquan Zhang, Joseph Spencer, and Brian Cousens. 2022. "Sr-O-C isotope signatures reveal herbivore niche-partitioning in a Cretaceous ecosystem." *Palaeontology* 65, no. 2: e12591.

Currie, Philip J., and David C. Evans. 2020. "Cranial anatomy of new specimens of *Saurornitholestes langstoni* (Dinosauria, Theropoda, Dromaeosauridae) from the Dinosaur Park Formation (Campanian) of Alberta." *Anatomical Record* 303, no. 4: 691–715.

Currie, Philip J., and Kevin Padian, eds. 1997. *Encyclopedia of dinosaurs.* New York: Elsevier.

Currie, Philip J., J. Keith Rigby Jr., and Robert E. Sloan. 1990. "Theropod teeth from the Judith River Formation of southern Alberta, Canada." In: K. Carpenter and P. J. Currie, eds. *Dinosaur systematics: approaches and perspectives.* Cambridge: Cambridge University Press, 107–25.

Currie, Philip J., Stephen J. Godfrey, and Lev Nessov. 1993. "New caenagnathid (Dinosauria: Theropoda) specimens from the Upper Cretaceous of North America and Asia." *Canadian Journal of Earth Sciences* 30, no. 10: 2255–72.

Cuthbertson, Robin S., Alex Tirabasso, Natalia Rybczynski, and Robert B. Holmes. 2012. "Kinetic limitations of intracranial joints in *Brachylophosaurus canadensis* and *Edmontosaurus regalis* (Dinosauria: Hadrosauridae), and their implications for the chewing mechanics of hadrosaurids." *Anatomical Record* 295, no. 6: 968–79.

Czerkas, S., and D. Gillette. 1999. "The beaked jaw of stegosaurs and their implications for other ornithischians." In: David Gillette, ed. *Vertebrate paleontology in Utah*. Salt Lake City: Utah Geological Survey, 143–50.

D'Amore, Domenic C. 2009. "A functional explanation for denticulation in theropod dinosaur teeth." *Anatomical Record* 292, no. 9: 1297–314.

D'Emic, Michael D., John A. Whitlock, Kathlyn M. Smith, Daniel C. Fisher, and Jeffrey A. Wilson. 2013. "Evolution of high tooth replacement rates in sauropod dinosaurs." *PLoS One* 8, no. 7: e69235.

D'Emic, Michael, D., Patrick M. O'Connor, Thomas R. Pascucci, Joanna N. Gavras, Elizabeth Mardakhayava, and Eric K. Lund. 2019. "Evolution of high tooth replacement rates in theropod dinosaurs." *PLoS One* 14, no. 11: e0224734.

Dalla Vecchia, Fabio Marco. 2020. "The unusual tail of *Tethyshadros insularis* (Dinosauria, Hadrosauroidea) from the Adriatic Island of the European Archipelago." *Rivista Italiana di Paleontologia e Stratigrafia* 126, no. 3.

Dalman, Sebastian G., and Spencer G. Lucas. 2018. "New evidence for predatory behaviour in Tyrannosaurid dinosaurs from the Kirtland Formation (Late Cretaceous, Campanian), northwestern New Mexico." *New Mexico Museum of Natural History Science Bulletin* 79: 113–24.

Dalman, Sebastian G., and Spencer G. Lucas. 2021. "New evidence for cannibalism in tyrannosaurid dinosaurs from the Upper Cretaceous (Campanian/Maastrichtian) San Juan Basin of New Mexico." *Fossil Record* 7, 82: 39.

Dal Sasso, Cristiano, and Marco Signore. 1998. "Exceptional soft-tissue preservation in a theropod dinosaur from Italy." *Nature* 392, no. 6674: 383–87.

Dal Sasso, Cristiano, and Simone Maganuco. 2011. "*Scipionyx samniticus* (Theropoda: Compsognathidae) from the Lower Cretaceous of Italy—Osteology, ontogenetic assessment, phylogeny, soft tissue anatomy, taphonomy and palaeobiology." *Memorie della Società Italiana de Scienze Naturali e del Museo Civico di Storia Naturale di Milano* 37, no. 1: 1–281.

Dececchi, T. Alexander, Aleksandra M. Mloszewska, Thomas R. Holtz Jr., Michael B. Habib, and Hans C. E. Larsson. 2020a. "The fast and the frugal: divergent locomotory strategies drive limb lengthening in theropod dinosaurs." *PLoS One* 15, no. 5: e0223698.

Dececchi, T. Alexander, Arindam Roy, Michael Pittman, Thomas G. Kaye, Xing Xu, Michael B. Habib, Hans C. E. Larsson, Xiaoli Wang, and Xiaoting Zheng. 2020b. "Aerodynamics show membrane-winged theropods were a poor gliding dead-end." *Iscience* 23, no. 12: 101574.

de Fabrègues, Claire Peyre, Shundong Bi, Hongqing Li, Gang Li, Lei Yang, and Xing Xu. 2020. "A new species of early-diverging Sauropodiformes from the Lower Jurassic Fengjiahe Formation of Yunnan Province, China." *Scientific Reports* 10, no. 1: 1–17.

de Souza, Geovane Alves, Marina Bento Soares, Luiz Carlos Weinschütz, Everton Wilner, Ricardo Tadeu Lopes, Olga Maria Oliveira de Araújo, and Alexander Wilhelm Armin Kellner. 2021. "The first edentulous ceratosaur from South America." *Scientific Reports* 11, no. 1: 1–15.

DePalma, Robert A., David A. Burnham, Larry D. Martin, Bruce M. Rothschild, and Peter L. Larson. 2013. "Physical evidence of predatory behavior in *Tyrannosaurus rex*." *Proceedings of the National Academy of Sciences* 110, no. 31: 12,560–64.

DePalma, Robert A., David A. Burnham, Larry D. Martin, Peter L. Larson, and Robert T. Bakker. 2015. "The first giant raptor (Theropoda: Dromaeosauridae) from the hell creek formation." *Paleontological Contributions* 2015, no. 14: 1–16.

Díaz, Verónica Díez, Oliver E. Demuth, Daniela Schwarz, and Heinrich Mallison. 2020. "The tail of the Late Jurassic sauropod *Giraffatitan brancai*: digital reconstruction of its epaxial and hypaxial musculature, and implications for tail biomechanics." *Frontiers in Earth Science* 8: 160.

Dilkes, David W. 2000. "Appendicular myology of the hadrosaurian dinosaur Maiasaura peeblesorum from the Late Cretaceous (Campanian) of Montana." *Earth and Environmental Science Transactions of the Royal Society of Edinburgh* 90, no. 2: 87–125.

Dilkes, David W. 2001. "An ontogenetic perspective on locomotion in the Late Cretaceous dinosaur *Maiasaura peeblesorum* (Ornithischia: Hadrosauridae)." *Canadian Journal of Earth Sciences* 38, no. 8: 1205–27.

Dodson, Peter. 1983. "A faunal review of the Judith River (Oldman) Formation, Dinosaur Provincial Park, Alberta." *Mosasaur* 1: 89–118.

Dodson, Peter. 1986. "A faunal review of the Judith River (Oldman) Formation, Dinosaur Provincial Park, Alberta." *The Mosasaur* 1: 89–118.

Dodson, Peter. 1993. "Comparative craniology of the Ceratopsia." *American Journal of Science* 293, no. A: 200–234.

Dodson, Peter. 1996. *The horned dinosaurs: a natural history*. Princeton, NJ: Princeton University Press.

Dodson, Peter, C. A. Forster, and S. D. Sampson. 2004. "Ceratopsidae." In: D. Weishampel, P. Dodson, and H. Osmólska, eds. *The Dinosauria*, 2nd ed. Berkeley: University of California Press, 494–513.

Dodson, Peter, H. You, and K. Tanoue. 2010. "Comments on the basicranium and palate of basal ceratopsians." In: Michael J. Ryan, Brenda J. Chinnery-Allgeier, and David E. Eberth, eds. *New perspectives on horned dinosaurs: the Royal Tyrrell Museum Ceratopsian Symposium*. Bloomington: Indiana University Press, 221–33.

Dollo, M. L. 1883. "Quatriéme note sur les dinosauriens de Bernissart." *Bulletin de Museé Royal d'Histoire Naturelle de Belgique* 2, no. 13: 223–48.

Dollo, L. 1884. "Cinquiéme note sur les dinosauriens de Bernissart." *Bulletin de Museé Royal d'Histoire Naturelle de Belgique* 3, no. 8: 129–46.

Dong, Zhiming. 2002. "A new armored dinosaur (Ankylosauria) from Beipiao Basin, Liaoning Province, northeastern China." *Vertebrata PalAsiatica* 40: 276–85.

Drongelen, V., and P. Dullemeijer. 1982. "The feeding apparatus of *Caiman crocodilus*, a functional-morphological study." *Anatomischer Anzeiger* 151, no. 4: 337–66.

Drumheller, Stephanie K., Julia B. McHugh, Miriam Kane, Anja Riedel, and Domenic C. D'Amore. 2020. "High frequencies of theropod bite marks provide evidence for feeding, scavenging, and possible cannibalism in a stressed Late Jurassic ecosystem." *PLoS One* 15, no. 5: e0233115.

Dullemeijer, P. 1974. *Concepts and approaches in animal morphology*. Assen, Netherlands: Van Gorcum.

Dzemski, Gordon, and Andreas Christian. 2007. "Flexibility along the neck of the ostrich (*Struthio camelus*) and consequences for the reconstruction of dinosaurs with extreme neck length." *Journal of Morphology* 268, no. 8: 701–14.

Eaton, Theodore H., Jr. 1960. "A new armored dinosaur from the Cretaceous of Kansas." *University of Kansas Paleontological Contributions* 8: 5–21.

Edinger, Tilly. 1929. "*Die fossilen Gehirne*." *Ergebnisse der Anatomie und Entwicklungsgeschichte* 28: 1–249.

Elzanowski, A., and L. M. Chiappe. 2002. "Archaeopterygidae (Upper Jurassic of Germany)." In: Luis M. Chiappe and Lawrence W. Witmer, eds. *Mesozoic birds: above the heads of dinosaurs*. Berkeley: University of California Press, 129–59.

Engelmann, George F., Daniel J. Chure, and Anthony R. Fiorillo. 2004. "The implications of a dry climate for the paleoecology of the fauna of the Upper Jurassic Morrison Formation." *Sedimentary Geology* 167, nos. 3–4: 297–308.

Erickson, Gregory M. 1996. "Incremental lines of von Ebner in dinosaurs and the assessment of tooth replacement rates using growth line counts." *Proceedings of the National Academy of Sciences* 93, no. 25: 14,623–27.

Erickson, Gregory M., and Kenneth H. Olson. 1996. "Bite marks attributable to *Tyrannosaurus rex*: preliminary description and implications." *Journal of Vertebrate Paleontology* 16, no. 1: 175–78.

Erickson, Gregory M., and Darla K. Zelenitsky. 2014. "Osteohistology and occlusal morphology of *Hypacrosaurus stebingeri* teeth throughout ontogeny." In: David A. Eberth and David C. Evans, eds. *Hadrosaurs*. Bloomington: Indiana University Press, 422–32.

Erickson, Gregory M., Samuel D. Van Kirk, Jinntung Su, Marc E. Levenston, William E. Caler, and Dennis R. Carter. 1996. "Bite-force estimation for *Tyrannosaurus rex* from tooth-marked bones." *Nature* 382, no. 6593: 706–8.

Erickson, Gregory M., Brandon A. Krick, Matthew Hamilton, Gerald R. Bourne, Mark A. Norell, Erica Lilleodden, and W. Gregory Sawyer. 2012. "Complex dental structure and wear biomechanics in hadrosaurid dinosaurs." *Science* 338, no. 6103: 98–101.

Erickson, Gregory M., Mark A. Sidebottom, David I. Kay, Kevin T. Turner, Nathan Ip, Mark A. Norell, W. Gregory Sawyer, and Brandon A. Krick. 2015. "Wear biomechanics in the slicing dentition of the giant horned dinosaur *Triceratops*." *Science Advances* 1, no. 5: e1500055.

Ezcurra, Martin D., and Stephen L. Brusatte. 2011. "Taxonomic and phylogenetic reassessment of the early neotheropod dinosaur *Camposaurus arizonensis* from the Late Triassic of North America." *Palaeontology* 54, no. 4: 763–72.

Fabbri, Matteo, Guillermo Navalón, Nicolas Mongiardino Koch, Michael Hanson, Holger Petermann, and Bhart-Anjan Bhullar. 2021. "A shift in ontogenetic timing produced the unique sauropod skull." *Evolution* 75, no. 4: 819–31.

Fabbri, Matteo, Guillermo Navalón, Roger B. J. Benson, Diego Pol, Jingmai O'Connor, et al. 2022. "Subaqueous foraging among carnivorous dinosaurs." *Nature* 603, no. 7903: 852–57.

Fairman, J. E. 1999. "Prosauropod and iguanid jaw musculature: a study on the evolution of form and function." Master's thesis, Johns Hopkins University.

Farke, Andrew A. 2004. "Horn use in Triceratops (Dinosauria: Ceratopsidae): testing behavioral hypotheses using scale models." *Palaeontologia Electronica* 7, no. 1: 10 pp.

Farke, Andrew A. 2010. "Evolution and functional morphology of the frontal sinuses in Bovidae (Mammalia: Artiodactyla), and implications for the evolution of cranial pneumaticity." *Zoological Journal of the Linnean Society* 159, no. 4: 988–1014.

Farke, Andrew A. 2014. "Evaluating combat in ornithischian dinosaurs." *Journal of Zoology* 292, no. 4: 242–49.

Farke, Andrew A., Ewan D. S. Wolff, and Darren H. Tanke. 2009. "Evidence of combat in *Triceratops*." *PLoS One* 4, no. 1: e4252.

Farke, Andrew, Ralph E. Chapman, and Art Anderson. 2010. "Modeling structural properties of the frill of *Triceratops*." In: Michael J. Ryan, Brenda J. Chinnery-Allgeier, and David E. Eberth, eds. *New perspectives on horned dinosaurs: the Royal Tyrrell Museum Ceratopsian Symposium*. Bloomington: Indiana University Press, 264–70.

Farke, Andrew A., Derek J. Chok, Annisa Herrero, Brandon Scolieri, and Sarah Werning. 2013. "Ontogeny in the tube-crested dinosaur *Parasaurolophus* (Hadrosauridae) and heterochrony in hadrosaurids." *PeerJ* 1: e182.

Farlow, James O. 1987. "Speculations about the diet and digestive physiology of herbivorous dinosaurs." *Paleobiology* 13, no. 1: 60–72.

Farlow, James O. 1993. "On the rareness of big, fierce animals; speculations about the body sizes, population densities, and geographic ranges of predatory mammals and large carnivorous dinosaurs." *American Journal of Science* 293, no. A: 167.

Farlow, James O., and Michael K. Brett-Surman, eds. 1997. *The complete dinosaur*. Bloomington: Indiana University Press.

Farlow, James O., and Peter Dodson. 1975. "The behavioral significance of frill and horn morphology in ceratopsian dinosaurs." *Evolution* 29, no. 2: 353–61.

Farlow, James O., and Thomas R. Holtz. 2002. "The fossil record of predation in dinosaurs." *Paleontological Society Papers* 8: 251–66.

Farlow, James O., Daniel L. Brinkman, William L. Abler, and Philip J. Currie. 1991. "Size, shape and serration density of theropod dinosaur lateral teeth." *Modern Geology* 16: 161.

Farlow, James O., Peter Dodson, and Anusuya Chinsamy. 1995. "Dinosaur biology." *Annual Review of Ecology and Systematics* 26, no. 1: 445–71.

Farlow, James O., Stephen M. Gatesy, Thomas R. Holtz Jr., John R. Hutchinson, and John M. Robinson. 2000. "Theropod locomotion." *American Zoologist* 40, no. 4: 640–63.

Farlow, James O., I. Dan Coroian, and John R. Foster. 2010. "Giants on the landscape: modelling the abundance of megaherbivorous dinosaurs of the Morrison Formation (Late Jurassic, western USA)." *Historical Biology* 22, no. 4: 403–29.

Fastovsky, David E., and Joshua B. Smith. 2004. "26. Dinosaur paleoecology." In: D. Weishampel, P. Dodson, and H. Osmólska, eds. *The Dinosauria*, 2nd ed. Berkeley: University of California Press, 614–26.

Fastovsky, David E., and David B. Weishampel. 2009. *Dinosaurs: a concise natural history*. Cambridge: Cambridge University Press.

Fearon, Jamie L., and David J. Varricchio. 2016. "Reconstruction of the forelimb musculature of the Cretaceous ornithopod dinosaur *Oryctodromeus cubicularis*: implications for digging." *Journal of Vertebrate Paleontology* 36, no. 2: e1078341.

Field, Daniel J., Michael Hanson, David Burnham, Laura E. Wilson, Kristopher Super, Dana Ehret, Jun A. Ebersole, and Bhart-Anjan S. Bhullar. 2018. "Complete *Ichthyornis* skull illuminates mosaic assembly of the avian head." *Nature* 557, no. 7703: 96–100.

Field, Daniel J., Juan Benito, Albert Chen, John W. M. Jagt, and Daniel T. Ksepka. 2020. "Late Cretaceous neornithine from Europe illuminates the origins of crown birds." *Nature* 579, no. 7799: 397–401.

Fiorillo, Anthony R. 1998. "Dental micro wear patterns of the sauropod dinosaurs *Camarasaurus* and *Diplodocus*: evidence for resource partitioning in the Late Jurassic of North America." *Historical Biology* 13, no. 1: 1–16.

Fiorillo, Anthony R. 1991. "Prey bone utilization by predatory dinosaurs." *Palaeogeography, Palaeoclimatology, Palaeoecology* 88, no. 3–4: 157–66.

Fiorillo, Anthony R. 2008. "On the occurrence of exceptionally large teeth of Troodon (Dinosauria: Saurischia) from the Late Cretaceous of northern Alaska." *Palaios* 23, no. 5: 322–28.

Fong, Raymond K. M., Aaron R. H. LeBlanc, David S. Berman, and Robert R. Reisz. 2016. "Dental histology of *Coelophysis bauri* and the evolution of tooth attachment tissues in early dinosaurs." *Journal of Morphology* 277, no. 7: 916–24.

Forster, Catherine A. 1996. "New information on the skull of *Triceratops*." *Journal of Vertebrate Paleontology* 16, no. 2: 246–58.

Fowler, Denver W., and Robert M. Sullivan. 2006. "A ceratopsid pelvis with toothmarks from the Upper Cretaceous Kirtland Formation, New Mexico: evidence of Late Campanian tyrannosaurid feeding behavior." *New Mexico Museum of Natural History and Science Bulletin* 35: 127–30.

Fowler, Denver W., Elizabeth A. Freedman, John B. Scannella, and Robert E. Kambic. 2011. "The predatory ecology of *Deinonychus* and the origin of flapping in birds." *PLoS One* 6, no. 12: e28964.

Fowler, Denver W., John B. Scannella, Mark B. Goodwin, and John R. Horner. 2012. "How to eat a *Triceratops*: large sample of toothmarks provides new insight into the feeding behavior of *Tyrannosaurus*." *Journal of Vertebrate Paleontology* 32, 96.

Franz, Ragna, Jürgen Hummel, Ellen Kienzle, Petra Kölle, Hanns-Christian Gunga, and Marcus Clauss. 2009. "Allometry of visceral organs in living amniotes and its implications for sauropod dinosaurs." *Proceedings of the Royal Society B: Biological Sciences* 276, no. 1662: 1731–36.

Franz, Ragna, Jürgen Hummel, Dennis W. H. Müller, Martin Bauert, Jean-Michel Hatt, and Marcus Clauss. 2011. "Herbivorous reptiles and body mass: effects on food intake, digesta retention, digestibility and gut capacity, and a comparison with mammals." *Comparative Biochemistry and Physiology Part A: Molecular and Integrative Physiology* 158, no. 1: 94–101.

Frazzetta, T. H., and Kenneth V. Kardong. 2002. "Prey attack by a large theropod dinosaur." *Nature* 416, no. 6879: 387–88.

Frederickson, J. A., M. H. Engel, and R. L. Cifelli. 2018. "Niche partitioning in theropod dinosaurs: diet and habitat preference in predators from the Uppermost Cedar Mountain Formation (Utah, USA)." *Scientific Reports* 8, no. 1: 1–13.

Frederickson, J. A., M. H. Engel, and R. L. Cifelli. 2020. "Ontogenetic dietary shifts in *Deinonychus antirrhopus* (Theropoda; Dromaeosauridae): insights into the ecology and social behavior of raptorial dinosaurs through stable isotope analysis." *Palaeogeography, Palaeoclimatology, Palaeoecology* 552: 109780.

Freimuth, William J., David J. Varricchio, Alexandria L. Brannick, Lucas N. Weaver, and Gregory P. Wilson Mantilla. 2021. "Mammal-bearing gastric pellets potentially attributable to Troodon formosus at the Cretaceous Egg Mountain locality, Two Medicine Formation, Montana, USA." *Palaeontology* 64, no. 5: 699–725.

Fricke, Henry C., and Dean A. Pearson. 2008. "Stable isotope evidence for changes in dietary niche partitioning among hadrosaurian and ceratopsian dinosaurs of the Hell Creek Formation, North Dakota." *Paleobiology* 34, no. 4: 534–52.

Fritz, Julia, Jürgen Hummel, Ellen Kienzle, Oliver Wings, W. Jürgen Streich, and Marcus Clauss. 2011. "Gizzard vs. teeth, it's a tie: food-processing efficiency in herbivorous birds and mammals and implications for dinosaur feeding strategies." *Paleobiology* 37, no. 4: 577–86.

Funston, Gregory F. 2020. "Caenagnathids of the Dinosaur Park Formation (Campanian) of Alberta, Canada: anatomy, osteohistology, taxonomy, and evolution." *Vertebrate Anatomy Morphology Palaeontology* 8: 105–53.

Funston, Gregory F., and Philip J. Currie. 2014. "A previously undescribed caenagnathid mandible from the late Campanian of Alberta, and insights into the diet of *Chirostenotes pergracilis* (Dinosauria: Oviraptorosauria)." *Canadian Journal of Earth Sciences* 51, no. 2: 156–65.

Funston, Gregory F., S. E. Mendonca, P. J. Currie, and R. Barsbold. 2018. "Oviraptorosaur anatomy, diversity and ecology in the Nemegt Basin." *Palaeogeography, Palaeoclimatology, Palaeoecology* 494: 101–20.

Funston, Gregory F., Ryan D. Wilkinson, D. Jade Simon, Aaron H. Leblanc, Mateusz Wosik, and Philip J. Currie. 2020. "Histology of caenagnathid (Theropoda, Oviraptorosauria) dentaries and implications for development, ontogenetic edentulism, and taxonomy." *Anatomical Record* 303, no. 4: 918–34.

Galton, Peter M. 1969. *The pelvic musculature of the dinosaur Hypsilophodon (Reptilia: Ornithischia)*. New Haven, CT: Peabody Museum of Natural History.

Galton, Peter M. 1970. "The posture of hadrosaurian dinosaurs." *Journal of Paleontology*: 464–73.

Galton, Peter M. 1971. "The prosauropod dinosaur Ammosaurus, the crocodile Protosuchus, and their bearing on the age of the Navajo Sandstone of northeastern Arizona." *Journal of Paleontology* 45, no. 5: 781–95.

Galton, Peter M. 1973. "The cheeks of ornithischian dinosaurs." *Lethaia* 6, no. 1: 67–89.

Galton, Peter M. 1974. "The ornithischian dinosaur *Hypsilophodon* from the Wealden of the Isle of Wight." *Bulletin of the British Museum (Natural History)* 25, no. 1: 3–152.

Galton, Peter M. 1978. "Fabrosauridae, the basal family of ornithischian dinosaurs (Reptilia: Ornithopoda)." *Paläontologische Zeitschrift* 52, no. 1: 138–59.

Galton, Peter M. 1984. "Cranial anatomy of the prosauropod dinosaur *Plateosaurus* from the Knollenmergel (Middle Keuper, Upper Triassic) of Germany, 1. Two complete skulls from Trossingen/Württ, with comments on the diet. *Geologica et Palaeontologica* 18: 139–71.

Galton, Peter M. 1985a. "Cranial anatomy of the prosauropod dinosaur *Plateosaurus* from the Knollenmergel (Middle Keuper, Upper Triassic) of Germany, 2. All the cranial material and details of soft-part anatomy." *Geologica et Palaeontologica* 19: 119–59.

Galton, Peter M. 1985b. "Diet of prosauropod dinosaurs from the Late Triassic and Early Jurassic." *Lethaia* 18, no. 2: 105–23.

Galton, Peter M. 1986. "Herbivorous adaptations of late Triassic and early Jurassic dinosaurs." In: Kevin Padian, ed. *The beginning of the age of dinosaurs: faunal change across the Triassic-Jurassic boundary*. Cambridge: Cambridge University Press, 203–21.

Galton, Peter M., and Walter P. Coombs Jr. 1981. "*Paranthodon africanus* (Broom) a stegosaurian dinosaur from the Lower Cretaceous of South Africa." *Geobios* 14, no. 3: 299–309.

Galton, Peter M., and Paul Upchurch. 2004. "Stegosauria." In: D. Weishampel, P. Dodson, and H. Osmólska, eds. *The*

Dinosauria, 2nd ed. Berkeley: University of California Press, 343–62.

Gauthier, Jacques. 1986. "Saurischian monophyly and the origin of birds." *Memoirs of the California Academy of Sciences* 8: 1–55.

Gee, Carole T. 2011. "Dietary options for the sauropod dinosaurs from an integrated botanical and paleobotanical perspective." In: N. Klein, K. Remes, C. T. Gee, and P. M. Sander, eds. *Biology of the sauropod dinosaurs: understanding the life of giants*. Bloomington: Indiana University Press, 34–56.

George, Ian D., and Casey M. Holliday. 2013. "Trigeminal nerve morphology in *Alligator mississippiensis* and its significance for crocodyliform facial sensation and evolution." *Anatomical Record* 296, no. 4: 670–80.

Gianechini, Federico A., Federico L. Agnolín, and Martín D. Ezcurra. 2011. "A reassessment of the purported venom delivery system of the bird-like raptor *Sinornithosaurus*." *Paläontologische Zeitschrift* 85, no. 1: 103–7.

Gianechini, Federico A., Marcos D. Ercoli, and Ignacio Díaz-Martínez. 2020. "Differential locomotor and predatory strategies of Gondwanan and derived Laurasian dromaeosaurids (Dinosauria, Theropoda, Paraves): inferences from morphometric and comparative anatomical studies." *Journal of Anatomy* 236, no. 5: 772–97.

Gignac, Paul M., and Gregory M. Erickson. 2017. "The biomechanics behind extreme osteophagy in *Tyrannosaurus rex*." *Scientific Reports* 7, no. 1: 1–10.

Gignac, Paul M., Peter J. Makovicky, Gregory M. Erickson, and Robert P. Walsh. 2010. "A description of *Deinonychus antirrhopus* bite marks and estimates of bite force using tooth indentation simulations." *Journal of Vertebrate Paleontology* 30, no. 4: 1169–77.

Gignac, Paul M., Nathan J. Kley, Julia A. Clarke, Matthew W. Colbert, Ashley C. Morhardt, et al. 2016. "Diffusible iodine-based contrast-enhanced computed tomography (diceCT): an emerging tool for rapid, high-resolution, 3-D imaging of metazoan soft tissues." *Journal of Anatomy* 228, no. 6: 889–909.

Gill, Fiona L., Jürgen Hummel, A. Reza Sharifi, Alexandra P. Lee, and Barry H. Lomax. 2018. "Diets of giants: the nutritional value of sauropod diet during the Mesozoic." *Palaeontology* 61, no. 5: 647–58.

Gilmore, Charles W. 1915. "On the fore limb of *Allosaurus fragilis*." *Proceedings of the United States National Museum* 49: 501–13.

Gilmore, Charles W. 1924. *On Troodon validus: an orthopodous dinosaur from the Belly River Cretaceous of Alberta, Canada*. Edmonton: University of Alberta Press.

Gishlick, Alan D. 2001. "The function of the manus and forelimb of *Deinonychus antirrhopus* and its importance for the origin of avian flight." In: J. Gauthier and L. F. Gall, eds. *New perspectives on the origin and early evolution of birds: proceedings of the international symposium in honor of John H. Ostrom; 1999 Feb 13–14 New Haven, Connecticut*. New Haven, CT: Peabody Museum of Natural History, Yale University, 301–18.

Gishlick, Alan D., and Jacques A. Gauthier. 2007. "On the manual morphology of *Compsognathus longipes* and its bearing on the diagnosis of Compsognathidae." *Zoological Journal of the Linnean Society* 149, no. 4: 569–81.

Goedert, Jean, Romain Amiot, Larbi Boudad, Eric Buffetaut, François Fourel, et al. 2016. "Preliminary investigation of seasonal patterns recorded in the oxygen isotope compositions of theropod dinosaur tooth enamel." *Palaios* 31, no. 1: 10–19.

Gold, Maria Eugenia Leone, Stephen L. Brusatte, and Mark A. Norell. 2013. "The cranial pneumatic sinuses of the tyrannosaurid *Alioramus* (Dinosauria: Theropoda) and the evolution of cranial pneumaticity in theropod dinosaurs." *American Museum Novitates* 3790: 1–46.

Gold, Maria Eugenia Leone, and Akinobu Watanabe. 2018. "Flightless birds are not neuroanatomical analogs of non-avian dinosaurs." *BMC Evolutionary Biology* 18, no. 1: 1–11.

Gong, Enpu, Larry D. Martin, David A. Burnham, and Amanda R. Falk. 2010. "The birdlike raptor *Sinornithosaurus* was venomous." *Proceedings of the National Academy of Sciences* 107, no. 2: 766–68.

Goodwin, Mark B., and John R. Horner. 2004. "Cranial histology of pachycephalosaurs (Ornithischia: Marginocephalia) reveals transitory structures inconsistent with head-butting behavior." *Paleobiology* 30, no. 2: 253–67.

Gow, C. E., J. W. Kitching, and M. A. Raath. 1990. "Skulls of the prosauropod dinosaur *Massospondylus carinatus* Owen in the collections of the Bernard Price Institute for Palaeontological Research." *Palaeontologia Africana* 27: 45–58.

Greaves, W. S. 1978. "The jaw lever system in ungulates: a new model." *Journal of Zoology* 184, no. 2: 271–85.

Gregory, William King. 1920. "A review of the evolution of the lacrymal bone of vertebrates with special reference to that of mammals." *Bulletin of the American Museum of Natural History* 42, 95–263.

Gregory, William King, and L. A. Adams. 1915. "The temporal fossae of vertebrates in relation to the jaw muscles." *Science*, n.s., 41: 763–65.

Griffin, Christopher T., and Sterling J. Nesbitt. 2020. "Does the maximum body size of theropods increase across the Triassic–Jurassic boundary? Integrating ontogeny, phylogeny, and body size." *Anatomical Record* 303, no. 4: 1158–69.

Guinard, Geoffrey. 2020. "Forelimb shortening of Carcharo-dontosauria (Dinosauria: Theropoda): an update on evolutionary anterior micromelias in non-avian theropods." *Zoology* 139: 125756.

Gunga, Hanns-Christian, Tim Suthau, Anke Bellmann, Andreas Friedrich, Thomas Schwanebeck, Stefan Stoinski, Tobias Trippel, Karl Kirsch, and Olaf Hellwich. 2007. "Body mass estimations for *Plateosaurus engelhardti* using laser scanning and 3D reconstruction methods." *Naturwissenschaften* 94, no. 8: 623–30.

Haas, Georg. 1955. "The jaw musculature in *Protoceratops* and in other ceratopsians. *American Museum Novitates* 1729: 1–24.

Haas, Georg. 1963. "A proposed reconstruction of the jaw musculature of *Diplodocus*." *Annals of the Carnegie Museum* 36: 139–57.

Haas, Georg. 1969. "On the jaw muscles of ankylosaurs." *American Museum Novitates* 2399: 1–11.

Hammer, William R., and William J. Hickerson. 1994. "A crested theropod dinosaur from Antarctica." *Science* 264, no. 5160: 828–30.

Han, Fenglu, Catherine A. Forster, Xing Xu, and James M. Clark. 2018. "Postcranial anatomy of *Yinlong downsi* (Dinosauria: Ceratopsia) from the Upper Jurassic Shishugou Formation of China and the phylogeny of basal ornithischians." *Journal of Systematic Palaeontology* 16, no. 14: 1159–87.

Hanai, Tomoya, and Takanobu Tsuihiji. 2019. "Description of tooth ontogeny and replacement patterns in a juvenile *Tarbosaurus bataar* (Dinosauria: Theropoda) using CT-scan data." *Anatomical Record* 302, no. 7: 1210–25.

Happ, John. 2008. "An analysis of predator-prey behavior in a head-to-head encounter between *Tyrannosaurus rex* and *Triceratops*." In: Peter L. Larson and Kenneth Carpenter, eds. *Tyrannosaurus Rex: the tyrant king*. Bloomington: Indiana University Press, 355–70.

Hassler, A., J. E. Martin, R. Amiot, T. Tacail, F. Arnaud Godet, R. Allain, and V. Balter. 2018. "Calcium isotopes offer clues on resource partitioning among Cretaceous predatory dinosaurs." *Proceedings of the Royal Society B: Biological Sciences* 285, no. 1876: 20180197.

Hatcher, John Bell, Othniel Charles Marsh, and Richard Swann Lull. 1907. *The Ceratopsia*. No. 310. Washington, DC: US Government Printing Office.

Haubold, H. 1990. "Ein neuer Dinosaurier (Ornithischia, Thyreophora) aus dem unteren Jura des nördlichen Mitteleuropa." *Revue de Paléobiologie* 9: 149–77.

He, Yiming, Peter J. Makovicky, Xing Xu, and Hailu You. 2018. "High-resolution computed tomographic analysis of tooth replacement pattern of the basal neoceratopsian *Liaoceratops yanzigouensis* informs ceratopsian dental evolution." *Scientific Reports* 8, no. 1: 1–15.

Heckeberg, Nicola S., and Oliver W. M. Rauhut. 2020. "Histology of spinosaurid teeth." *Palaeontologia Electronica* 23, no. 3: a48.

Heidweiller, J., and G. A. Zweers. 1990. "Drinking mechanisms in the zebra finch and the Bengalese finch." *Condor* 92, no. 1: 1–28.

Heidweiller, J., and G. A. Zweers. 1992. "Flexibility of the drinking mechanism in adult chickens (*Gallus gallus*) (Aves)." *Zoomorphology* (Berl) 111, no. 3: 141–59.

Henderson, Donald M. 2002. "The eyes have it: the sizes, shapes, and orientations of theropod orbits as indicators of skull strength and bite force." *Journal of Vertebrate Paleontology* 22, no. 4: 766–78.

Henderson, Donald M. 2006. "Burly gaits: centers of mass, stability, and the trackways of sauropod dinosaurs." *Journal of Vertebrate Paleontology* 26, no. 4: 907–21.

Henderson, Donald M. 2010. "Skull shapes as indicators of niche partitioning by sympatric chasmosaurine and centrosaurine dinosaurs." In: Michael J. Ryan, Brenda J. Chinnery-Allgeier, and David E. Eberth, eds. *New perspectives on horned dinosaurs: the Royal Tyrrell Museum Ceratopsian Symposium*. Bloomington: Indiana University Press, 293–307.

Henderson, Donald M. 2013. "Sauropod necks: are they really for heat loss?" *PLoS One* 8, no. 10: e77108.

Henderson, Donald M. 2018. "A buoyancy, balance and stability challenge to the hypothesis of a semi-aquatic Spinosaurus Stromer, 1915 (Dinosauria: Theropoda)." *PeerJ* 6: e5409.

Henderson, Donald M., and Robert Nicholls. 2015. "Balance and strength—estimating the maximum prey-lifting potential of the large predatory dinosaur *Carcharodontosaurus saharicus*." *Anatomical Record* 298, no. 8: 1367–75.

Henderson, Donald M., and David B. Weishampel. 2002. "Convergent evolution of the maxilla-dental-complex among carnivorous archosaurs." *Senckenbergiana Lethaea* 82, no. 1: 77–91.

Hendrickx, Christophe, S. A. Hartman, and O. Mateus. 2015a. "An overview of non-avian theropod discoveries and classification" *PalArch's Journal of Vertebrate Paleontology* 12: 1–73.

Hendrickx, Christophe, Octávio Mateus, and Ricardo Araújo. 2015b. "A proposed terminology of theropod teeth (Dinosauria, Saurischia)." *Journal of Vertebrate Paleontology* 35, no. 5: e982797.

Hendrickx, Christophe, Octávio Mateus, and Eric Buffetaut. 2016. "Morphofunctional analysis of the quadrate of Spinosauridae (Dinosauria: Theropoda) and the presence of *Spinosaurus* and a second spinosaurine taxon in the Cenomanian of North Africa." *PLoS One* 11, no. 1: e0144695.

Hendrickx, Christophe, O. Mateus, R. Araújo, and J. Choiniere. 2019. "The distribution of dental features in non-avian theropod dinosaurs: taxonomic potential, degree of homoplasy, and major evolutionary trends." *Palaeontologia Electronica* 22, no. 3: 1–110.

Hendrickx, Christophe, Josef Stiegler, Philip J. Currie, Fenglu Han, Xing Xu, Jonah N. Choiniere, and Xiao-Chun Wu. 2020. "Dental anatomy of the apex predator *Sinraptor dongi* (Theropoda: Allosauroidea) from the Late Jurassic of China." *Canadian Journal of Earth Sciences* 57, no. 9: 1127–47.

Hennig, W. 1966. *Phylogenetic systematics.* Urbana: University of Illinois Press.

Heredia, Arturo Miguel, Pablo José Pazos, Diana Elizabeth Fernández, Ignacio Díaz-Martínez, and Marcos Comerio. 2019. "A new narrow-gauge sauropod trackway from the Cenomanian Candeleros Formation, northern Patagonia, Argentina." *Cretaceous Research* 96: 70–82.

Herne, Matthew C., Jay P. Nair, Alistair R. Evans, and Alan M. Tait. 2019. "New small-bodied ornithopods (Dinosauria, Neornithischia) from the Early Cretaceous Wonthaggi Formation (Strzelecki Group) of the Australian-Antarctic rift system, with revision of *Qantassaurus intrepidus* Rich and Vickers-Rich, 1999." *Journal of Paleontology* 93, no. 3: 543–84.

Hieronymus, Tobin L., Lawrence M. Witmer, Darren H. Tanke, and Philip J. Currie. 2009. "The facial integument of centrosaurine ceratopsids: morphological and histological correlates of novel skin structures." *Anatomical Record* 292, no. 9: 1370–96.

Hill, Robert V., Michael D. D'Emic, G. S. Bever, and Mark A. Norell. 2015. "A complex hyobranchial apparatus in a Cretaceous dinosaur and the antiquity of avian paraglossalia." *Zoological Journal of the Linnean Society* 175, no. 4: 892–909.

Hinić-Frlog, S., and R. Motani. 2010. "Relationship between osteology and aquatic locomotion in birds: determining modes of locomotion in extinct Ornithurae." *Journal of Evolutionary Biology* 23, no. 2: 372–85.

Hofer, H. 1950. "Zur Morphologie der Kiefermuskulature der Vsgel." *Zoologische Jahrbücher* 70: 427–556.

Hohn, Bianca. 2011. "Walking with the shoulder of giants: biomechanical conditions in the tetrapod shoulder girdle as a basis for sauropod shoulder reconstruction." In: N. Klein, K. Remes, C. T. Gee, and P. M. Sander, eds. *Biology of the sauropod dinosaurs: understanding the life of giants.* Bloomington: Indiana University Press, 182–96.

Holland, William J. 1910. "A review of some recent criticisms of the restorations of sauropod dinosaurs existing in the museums of the United States, with special reference to that of Diplodocus carnegiei in the Carnegie Museum." *American Naturalist* 44, no. 521: 258–83.

Holland, William J., and Olof August Peterson. 1924. *The skull of Diplodocus.* Washington, DC: Board of Trustees of the Carnegie Institute.

Holliday, Casey M. 2009. "New insights into dinosaur jaw muscle anatomy." *Anatomical Record* 292, no. 9: 1246–65.

Holliday, Casey M., and Lawrence M. Witmer. 2007. "Archosaur adductor chamber evolution: integration of musculoskeletal and topological criteria in jaw muscle homology." *Journal of Morphology* 268, no. 6: 457–84.

Holliday, Casey M., and Lawrence M. Witmer. 2008. "Cranial kinesis in dinosaurs: intracranial joints, protractor muscles, and their significance for cranial evolution and function in diapsids." *Journal of Vertebrate Paleontology* 28, no. 4: 1073–88.

Holliday, Casey M., William Ruger Porter, Kent A. Vliet, and Lawrence M. Witmer. 2020. "The frontoparietal fossa and dorsotemporal fenestra of archosaurs and their significance for interpretations of vascular and muscular anatomy in dinosaurs." *Anatomical Record* 303, no. 4: 1060–74.

Holtz, Thomas R., Jr. 1994. "The phylogenetic position of the Tyrannosauridae: implications for theropod systematics." *Journal of Paleontology* 68, no. 5: 1100–1117.

Holtz, Thomas R., Jr. 1995. "The arctometatarsalian pes, an unusual structure of the metatarsus of Cretaceous Theropoda (Dinosauria: Saurischia)." *Journal of Vertebrate Paleontology* 14, no. 4: 480–519.

Holtz, Thomas R., Jr. 1998. "Spinosaurs as crocodile mimics." *Science* 282, no. 5392: 1276–77.

Holtz, Thomas R., Jr. 2003. "Dinosaur predation." In: Patricia H. Kelley, Michal Kowalewski, and Thor A. Hansen, eds. *Predator-prey interactions in the fossil record.* Boston, MA: Springer, 325–40.

Holtz, Thomas R., Jr. 2008. "A critical reappraisal of the obligate scavenging hypothesis for *Tyrannosaurus rex* and other tyrant dinosaurs." In: Peter L. Larson and Kenneth Carpenter, eds. *Tyrannosaurus Rex: the tyrant king.* Bloomington: Indiana University Press, 371–96.

Holtz, Thomas R., Jr. 2021. "Theropod guild structure and the tyrannosaurid niche assimilation hypothesis: implications for predatory dinosaur macroecology and ontogeny in later Late Cretaceous Asiamerica." *Canadian Journal of Earth Sciences* 99, no. 999: 1–18.

Holtz, Thomas R., Jr., Daniel L. Brinkman, and Christine L. Chandler. 1998. "Denticle morphometrics and a possibly omnivorous feeding habit for the theropod dinosaur *Troodon*." *Gaia* 15, no. 159: e166.

Holwerda, Femke M., Diego Pol, and Oliver W. M. Rauhut. 2015. "Using dental enamel wrinkling to define sauropod tooth morphotypes from the Cañadón Asfalto Formation, Patagonia, Argentina." *PLoS One* 10, no. 2: e0118100.

Holwerda, Femke M., Verónica Díez Díaz, Alejandro Blanco, Roel Montie, and Jelle W. F. Reumer. 2018. "Late Cretaceous sauropod tooth morphotypes may provide supporting evidence for faunal connections between North Africa and Southern Europe." *PeerJ* 6: e5925.

Hone, David W. E., and Thomas R. Holtz Jr. 2019. "Comment on: Aquatic adaptation in the skull of carnivorous dinosaurs (Theropoda: Spinosauridae) and the evolution of aquatic habits in spinosaurids. 93: 275–284." *Cretaceous Research* 93: 275–84.

Hone, David W. E., and Thomas R. Holtz Jr. 2021. "Evaluating the ecology of *Spinosaurus*: shoreline generalist or aquatic pursuit specialist?" *Palaeontologia Electronica* 24, no. 1: a03.

Hone, David W. E., and Oliver W. M. Rauhut. 2010. "Feeding behaviour and bone utilization by theropod dinosaurs." *Lethaia* 43, no. 2: 232–44.

Hone, David W. E., and Mahito Watabe. 2010. "New information on scavenging and selective feeding behaviour of tyrannosaurids." *Acta Palaeontologica Polonica* 55, no. 4: 627–34.

Hone, David W. E., Jonah Choiniere, Corwin Sullivan, Xing Xu, Michael Pittman, and Qingwei Tan. 2010. "New evidence for a trophic relationship between the dinosaurs *Velociraptor* and *Protoceratops*." *Palaeogeography, Palaeoclimatology, Palaeoecology* 291, no. 3–4: 488–92.

Hone, David W. E., W. Scott Persons, and Steven C. Le Comber. 2021. "New data on tail lengths and variation along the caudal series in the non-avialan dinosaurs." *PeerJ* 9: e10721.

Hopson, James A. 1977. "Relative brain size and behavior in archosaurian reptiles." *Annual Review of Ecology and Systematics* 8, no. 1: 429–48.

Hopson, James A. 1980. "Tooth function and replacement in early Mesozoic ornithischian dinosaurs: implications for aestivation." *Lethaia* 13, no. 1: 93–105.

Horner, John R., Mark B. Goodwin, and Nathan Myhrvold. 2011. "Dinosaur census reveals abundant *Tyrannosaurus* and rare ontogenetic stages in the Upper Cretaceous Hell Creek Formation (Maastrichtian), Montana, USA." *PloS one* 6, no. 2: e16574.

Hou, Lianhai, Luis M. Chiappe, Fucheng Zhang, and Cheng-Ming Chuong. 2004. "New Early Cretaceous fossil from China documents a novel trophic specialization for Mesozoic birds." *Naturwissenschaften* 91, no. 1: 22–25.

Hu, Han, Gabriele Sansalone, Stephen Wroe, Paul G. McDonald, Jingmai K. O'Connor, Zhiheng Li, Xing Xu, and Zhonghe Zhou. 2019. "Evolution of the vomer and its implications for cranial kinesis in Paraves." *Proceedings of the National Academy of Sciences* 116, no. 39: 19,571–78.

Hu, Jinfeng, Catherine A. Forster, Xing Xu, Qi Zhao, Yiming He, and Fenglu Han. 2022. "Computed tomographic analysis of the dental system of three Jurassic ceratopsians and implications for the evolution of tooth replacement pattern and diet in early-diverging ceratopsians." *Elife* 11: e76676.

Hughes, Stephen, John Barry, Jeremy Russell, Robert Bell, and Som Gurung. 2016. "Neck length and mean arterial pressure in the sauropod dinosaurs." *Journal of Experimental Biology* 219, no. 8: 1154–61.

Hummel, Jürgen, and Marcus Clauss. 2011. "Sauropod feeding and digestive physiology." In: N. Klein, K. Remes, C. T. Gee, and P. M. Sander, eds. *Biology of the sauropod dinosaurs: understanding the life of giants*. Bloomington: Indiana University Press, 11–33.

Hummel, Jürgen, Carole T. Gee, Karl-Heinz Südekum, P. Martin Sander, Gunther Nogge, and Marcus Clauss. 2008. "In vitro digestibility of fern and gymnosperm foliage: implications for sauropod feeding ecology and diet selection." *Proceedings of the Royal Society B: Biological Sciences* 275, no. 1638: 1015–21.

Hurum, J. H., and Phillip J. Currie. 2000. "The crushing bite of tyrannosaurids." *Journal of Vertebrate Paleontology* 20, no. 3: 619–21.

Hutchinson, John R. 2001a. "The evolution of pelvic osteology and soft tissues on the line to extant birds (Neornithes)." *Zoological Journal of the Linnean Society* 131, no. 2: 123–68.

Hutchinson, John R. 2001b. "The evolution of femoral osteology and soft tissues on the line to extant birds (Neornithes)." *Zoological Journal of the Linnean Society* 131: 169–97.

Hutchinson, John R. 2002. "The evolution of hindlimb tendons and muscles on the line to crown-group birds." *Comparative Biochemistry and Physiology Part A: Molecular and Integrative Physiology* 133, no. 4: 1051–86.

Hutchinson, John R. 2004. "Biomechanical modeling and sensitivity analysis of bipedal running ability, II. Extinct taxa." *Journal of Morphology* 262, no. 1: 441–61.

Hutchinson, John R., and Mariano Garcia. 2002. "*Tyrannosaurus* was not a fast runner." *Nature* 415, no. 6875: 1018–21.

Hutchinson, John R., Frank C. Anderson, Silvia S. Blemker, and Scott L. Delp. 2005. "Analysis of hindlimb muscle moment arms in *Tyrannosaurus rex* using a three-dimensional musculoskeletal computer model: implications for stance, gait, and speed." *Paleobiology* 31, no. 4: 676–701.

Hutchinson, John R., Karl T. Bates, Julia Molnar, Vivian Allen, and Peter J. Makovicky. 2011. "A computational analysis of limb and body dimensions in *Tyrannosaurus rex* with implications for locomotion, ontogeny, and growth." *PLoS One* 6, no. 10: e26037.

Huxley, Thomas Henry. 1868. "On the animals which are most nearly intermediate between birds and reptiles." *Annals and Magazine of Natural History* 2: 66–75.

Hwang, Sunny H. 2005. "Phylogenetic patterns of enamel microstructure in dinosaur teeth." *Journal of Morphology* 266, no. 2: 208–40.

Hwang, Sunny H. 2010. "The utility of tooth enamel microstructure in identifying isolated dinosaur teeth." *Lethaia* 43, no. 3: 307–22.

Hwang, Sunny H. 2011. "The evolution of dinosaur tooth enamel microstructure." *Biological Reviews* 86, no. 1: 183–216.

Ibiricu, Lucio M., Gabriel A. Casal, Rubén D. Martínez, Matthew C. Lamanna, Marcelo Luna, and Leonardo Salgado. 2013. "*Katepensaurus goicoecheai*, gen. et sp. Nov., a Late Cretaceous rebbachisaurid (Sauropoda, Diplodocoidea) from central Patagonia, Argentina." *Journal of Vertebrate Paleontology* 33, no. 6: 1351–66.

Ibiricu, Lucio M., Rubén D. Martínez, and Gabriel A. Casal. 2018. "The pelvic and hindlimb myology of the basal titanosaur Epachthosaurus sciuttoi (Sauropoda: Titanosauria)." *Historical Biology* 32, no. 6: 773–88.

Ibrahim, Nizar, Paul C. Sereno, Cristiano Dal Sasso, Simone Maganuco, Matteo Fabbri, David M. Martill, Samir Zouhri, Nathan Myhrvold, and Dawid A. Iurino. 2014. "Semiaquatic adaptations in a giant predatory dinosaur." *Science* 345, no. 6204: 1613–16.

Ibrahim, Nizar, Simone Maganuco, Cristiano Dal Sasso, Matteo Fabbri, Marco Auditore, et al. 2020. "Tail-propelled aquatic locomotion in a theropod dinosaur." *Nature* 581, no. 7806: 67–70.

Iordansky, N. N. 1964. "The jaw muscles of the crocodiles and some relating structures of the crocodilian skull." *Anatomischer Anzeiger* 115: 256–80.

Jacobsen, Aase Roland. 1998. "Feeding behaviour of carnivorous dinosaurs as determined by tooth marks on dinosaur bones." *Historical Biology* 13, no. 1: 17–26.

Janensch, W. 1929. "Die Wirbelsäule der Gattung *Dicraeosaurus*." *Palaeontographica* 2 (Suppl. 7): 39–133.

Janensch, W. 1935. "Die Schädel der Sauropoden *Brachiosaurus*, *Barosaurus* und *Dicraeosaurus* aus den Tendaguru-Schichten Deutsch-Ostafrikas." *Palaeontographica* 1 (Suppl. 7), no. 2: 145–298.

Janensch, W. J. 1950. "The vertebral column of *Brachiosaurus brancai*." *Palaeontographica* 3, no. 7: 27–93.

Jansen, Stig Olav Krekvik. 2008. "Beak morphology in oviraptorids, based on extant birds and turtles." Master's thesis, University of Oslo.

Ji, Q., P. J. Currie, M. A. Norell, and S.-A. Ji. 1998. "Two feathered dinosaurs from northeastern China." *Nature* 393: 753–61.

Ji, Q., S. Ji, J. Lu, H. You, W. Chen, Y. Liu, and Y. Liu. 2005. "First avialan bird from China (*Jinfengopteryx elegans* gen. et sp. Nov.)." *Geological Bulletin of China* 24, no. 3: 197–210.

Ji, Q., X. Wu, Y. Cheng, F. Ten, X. Wang, and Y. Ji. 2016. "Fish hunting ankylosaurs (Dinosauria, Ornithischia) from the Cretaceous of China." *Journal of Geology* 40, no. 2: 183–90.

Johnson, R. E., and J. H. Ostrom. 1995. "The forelimb of *Torosaurus* and an analysis of the posture and gait of ceratopsians." In: J. J. Thomason, ed. *Functional morphology in vertebrate paleontology*. Cambridge: Cambridge University Press, 205–18.

Johnson-Ransom, Evan D., Eric Snively, and Daniel Barta. 2021. "Biomechanical performance of the crania of tyrannosauroids and comparative implications for theropod feeding." Paper presented at the Society of Vertebrate Paleontology Virtual Meeting, November 1–5.

Kielan-Jaworowska, Zofia, and Rinchen Barsbold. 1972. "Narrative of the Polish–Mongolian palaeontological expeditions 1967–1971." *Palaeontologia Polonica* 27: 5–13.

Kilbourne, Brandon, and Kenneth Carpenter. 2005. "Redescription of *Gargoyleosaurus parkpinorum*, a polacanthid ankylosaur from the Upper Jurassic of Albany County, Wyoming." *Neues Jahrbuch für Geologie und Paläontologie, Abhandlungen* 237: 111–60.

King, J. Logan, Justin S. Sipla, Justin A. Georgi, Amy M. Balanoff, and James M. Neenan. 2020. "The endocranium and trophic ecology of Velociraptor mongoliensis." *Journal of Anatomy* 237, no. 5: 861–69.

Klein, Nicole, Andreas Christian, and P. Martin Sander. 2012. "Histology shows that elongated neck ribs in sauropod dinosaurs are ossified tendons." *Biology Letters* 8, no. 6: 1032–35.

Klinkhamer, Ada J., Heinrich Mallison, Stephen F. Poropat, George H. K. Sinapius, and Stephen Wroe. 2018. "Three-dimensional musculoskeletal modeling of the sauropodomorph hind limb: the effect of postural change on muscle leverage." *Anatomical Record* 301, no. 12: 2145–63.

Klinkhamer, Ada J., Heinrich Mallison, Stephen F. Poropat, Trish Sloan, and Stephen Wroe. 2019. "Comparative three-dimensional moment arm analysis of the sauropod forelimb: implications for the transition to a wide-gauge stance in titanosaurs." *Anatomical Record* 302, no. 5: 794–817.

Knapp, Andrew, Robert J. Knell, Andrew A. Farke, Mark A. Loewen, and David W. E. Hone. 2018. "Patterns of divergence in the morphology of ceratopsian dinosaurs: sympatry is not a driver of ornament evolution." *Proceedings of the Royal Society B: Biological Sciences* 285, no. 1875: 20180312.

Knapp, Andrew, Robert J. Knell, and David W. E. Hone. 2021. "Three-dimensional geometric morphometric analysis of the skull of *Protoceratops andrewsi* supports a socio-sexual signaling role for the ceratopsian frill." *Proceedings of the Royal Society B: Biological Sciences* 288, no. 1944: 20202938.

Knoll, Fabien. 2008. "Buccal soft anatomy in *Lesothosaurus* (Dinosauria: Ornithischia)." *Neues Jahrbuch fur Geologie und Palaontologie-Abhandlungen* 248, no. 3: 355–64.

Knoll, Fabien, Peter M. Galton, and Raquel López-Antoñanzas. 2006. "Paleoneurological evidence against a proboscis in the sauropod dinosaur *Diplodocus*." *Geobios* 39, no. 2: 215–21.

Kobayashi, Yoshitsugu, and J. C. Lu. 2003. "A new ornithomimid dinosaur with gregarious habits from the Late Cretaceous of China." *Acta Palaeontologica Polonica* 48, no. 2: 235–259.

Kobayashi, Yoshitsugu, Jun-Chang Lu, Zhi-Ming Dong, Rinchen Barsbold, Yoichi Azuma, and Yukimitsu Tomida. 1999. "Herbivorous diet in an ornithomimid dinosaur." *Nature* 402, no. 6761: 480–81.

Krauss, D., and J. Robinson. 2013. "The biomechanics of a plausible hunting strategy for *Tyrannosaurus rex*." In: R. E. Molnar and P. J. Currie, eds. *Tyrannosaurid paleobiology*. Bloomington: Indiana University Press, 251–64.

Kripp, D. 1933. "Die Kaubewegung und Lebensweise von Edmontosaurus spec. auf Grund der mechanisch-konstruktiven Analyse." *Palaeobiologica* 5: 409–22.

Kubo, Tai. 2011. "Evolution of bipedality and herbivory among Triassic dinosauromorphs." *Memoir of the Fukui Prefectural Dinosaur Museum* 10: 55–62.

Kubo, Tai, Wenjie Zheng, Mugino O. Kubo, and Xingsheng Jin. 2021. "Dental microwear of a basal ankylosaurine dinosaur, *Jinyunpelta*, and its implication on evolution of chewing mechanism in ankylosaurs." *PLoS One* 16, no. 3: e0247969.

Kubota, Katsuhiro, and Yoshitsugu Kobayashi. 2009. "Evolution of dentary diastema in iguanodontian dinosaurs." *Acta Geologica Sinica*, Eng. ed., 83, no. 1: 39–45.

Kundrát, Martin, Rodolfo A. Coria, Terry W. Manning, Daniel Snitting, Luis M. Chiappe, John Nudds, and Per E. Ahlberg. 2020. "Specialized craniofacial anatomy of a titanosaurian embryo from Argentina." *Current Biology* 30, no. 21: 4263–69.

Kuzmin, I., I. Petrov, A. Averianov, E. Boitsova, P. Skutschas, and H. D. Sues. 2020. "The braincase of *Bissektipelta archibaldi*—new insights into endocranial osteology, vasculature, and paleoneurobiology of ankylosaurian dinosaurs." *Biological Communications* 65, no. 2: 85–156.

Lakjer, T. 1926. *Studien über die Trigeminus-versorgte Kaumuskulatur der Sauropsiden*. Copenhagen: C. A. Reitsel Buchhandlung.

Lamanna, Matthew C., Hans-Dieter Sues, Emma R. Schachner, and Tyler R. Lyson. 2014. "A new large-bodied oviraptorosaurian theropod dinosaur from the latest Cretaceous of western North America." *PLoS One* 9, no. 3: e92022.

Lambe, Lawrence Morris. 1917. *The Cretaceous theropodus dinosaur Gorgosaurus. Memoir (Geological Survey of Canada)* 100: 1–84.

Lambe, Lawrence Morris. 1920. "The hadrosaur *Edmontosaurus* from the Upper Cretaceous of Alberta." *Memoir (Geological Survey of Canada)* 120: 1–79.

Landi, Damiano, Logan King, Qi Zhao, Emily J. Rayfield, and Michael J. Benton. 2021. "Testing for a dietary shift in the Early Cretaceous ceratopsian dinosaur *Psittacosaurus lujiatunensis*." *Palaeontology* 64, no. 3: 371–84.

Langer, Max C. 2003. "The pelvic and hind limb anatomy of the stem-sauropodomorph *Saturnalia tupiniquim* (Late Triassic, Brazil)." *PaleoBios* 23, no. 2: 1–40.

Langer, Max C., Marco A. G. Franca, and Stefan Gabriel. 2007. "The pectoral girdle and forelimb anatomy of the stem-sauropodomorph *Saturnalia tupiniquim* (Upper Triassic, Brazil)." *Special Papers in Palaeontology* 77: 113.

Larson, Derek W., Caleb M. Brown, and David C. Evans. 2016. "Dental disparity and ecological stability in bird-like dinosaurs prior to the end-Cretaceous mass extinction." *Current Biology* 26, no. 10: 1325–33.

Larsson, Hans C. E. 2008. "Palatal kinesis of *Tyrannosaurus rex*." In: Peter L. Larson and Kenneth Carpenter, eds. *Tyrannosaurus rex: the tyrant king*. Bloomington: Indiana University Press, 245–52.

Lautenschlager, Stephan. 2013. "Cranial myology and bite force performance of *Erlikosaurus andrewsi*: a novel approach for digital muscle reconstructions." *Journal of Anatomy* 222, no. 2: 260–72.

Lautenschlager, Stephan. 2014. "Morphological and functional diversity in therizinosaur claws and the implications for theropod claw evolution." *Proceedings of the Royal Society B: Biological Sciences* 281, no. 1785: 20140497.

Lautenschlager, Stephan. 2015. "Estimating cranial musculo-skeletal constraints in theropod dinosaurs." *Royal Society Open Science* 2, no. 11: 150495.

Lautenschlager, Stephan. 2017. "Functional niche partitioning in Therizinosauria provides new insights into the evolution of theropod herbivory." *Palaeontology* 60, no. 3: 375–87.

Lautenschlager, Stephan. 2020. "Multibody dynamics analysis (MDA) as a numerical modelling tool to reconstruct the function and palaeobiology of extinct organisms." *Palaeontology* 63, no. 5: 703–15.

Lautenschlager, Stephan, Lawrence M. Witmer, Perle Altangerel, and Emily J. Rayfield. 2013. "Edentulism, beaks, and biomechanical innovations in the evolution of theropod dinosaurs." *Proceedings of the National Academy of Sciences* 110, no. 51: 20,657–62.

Lautenschlager, Stephan, Charlotte A. Brassey, David J. Button, and Paul M. Barrett. 2016. "Decoupled form and function in disparate herbivorous dinosaur clades." *Scientific Reports* 6, no. 1: 1–10.

LeBlanc, Aaron R. H., Robert R. Reisz, David C. Evans, and Alida M. Bailleul. 2016. "Ontogeny reveals function and evolution of the hadrosaurid dinosaur dental battery." *BMC Evolutionary Biology* 16, no. 1: 1–13.

LeBlanc, Aaron R. H., Kirstin S. Brink, Thomas M. Cullen, and Robert R. Reisz. 2017. "Evolutionary implications of tooth attachment versus tooth implantation: a case study using dinosaur, crocodilian, and mammal teeth." *Journal of Vertebrate Paleontology* 37, no. 5: e1354006.

Lehman, Thomas M. 1989. "*Chasmosaurus mariscalensis*, sp. Nov., a new ceratopsian dinosaur from Texas." *Journal of Vertebrate Paleontology* 9, no. 2: 137–62.

Lee, Yuong-Nam. 1996. "A new nodosaurid ankylosaur (Dinosauria: Ornithischia) from the Paw Formation (late Albian) of Texas." *Journal of Vertebrate Paleontology* 16: 232–45.

Lee, Yuong-Nam, Rinchen Barsbold, Philip J. Currie, Yoshitsugu Kobayashi, Hang-Jae Lee, Pascal Godefroit, François Escuillié, and Tsogtbaatar Chinzorig. 2014. "Resolving the long-standing enigmas of a giant ornithomimosaur Deinocheirus mirificus." *Nature* 515, no. 7526: 257–60.

Li, D.-Q., C. Peng, H.-L. You, M. C. Lamanna, J. D. Harris, K. J. Lacovara, and P. Zhang. 2007. "A large therizinosauroid (Dinosauria: Theropoda) from the Early Cretaceous of northwestern China." *Acta Geologica Sinica*, Eng. ed., 81: 539–49.

Li, D.-Q. E. Morschhauser, H. You., and P. Dodson. 2018. "The anatomy of the syncervical of *Auroraceratops* (Ornithischia: Ceratopsia), the oldest known ceratopsian syncervical." *Journal of Vertebrate Paleontology* 38 (Suppl. 1): 69–74.

Li, Zhiheng, and Julia A. Clarke. 2015. "New insight into the anatomy of the hyolingual apparatus of *Alligator mississippiensis* and implications for reconstructing feeding in extinct archosaurs." *Journal of Anatomy* 227, no. 1: 45–61.

Li, Zhiheng, and Julia A. Clarke. 2016. "The craniolingual morphology of waterfowl (Aves, Anseriformes) and its relationship with feeding mode revealed through contrast-enhanced x-ray computed tomography and 2D morphometrics." *Evolutionary Biology* 43, no. 1: 12–25.

Lingham-Soliar, Theagarten. 1998. "Guess who's coming to dinner: a portrait of *Tyrannosaurus* as a predator." *Geology Today* 14, no. 1: 16–20.

Liyong, Jin, Chen Jun, Zan Shuqin, Richard J. Butler, and Pascal Godefroit. 2010. "Cranial anatomy of the small ornithischian dinosaur *Changchunsaurus parvus* from the Quantou Formation (Cretaceous: Aptian–Cenomanian) of Jilin Province, northeastern China." *Journal of Vertebrate Paleontology* 30, no. 1: 196–214.

Longrich, Nicholas R. 2006. "Structure and function of hindlimb feathers in *Archaeopteryx lithographica*." *Paleobiology* 32, no. 3: 417–31.

Longrich, Nicholas R., John R. Horner, Gregory M. Erickson, and Philip J. Currie. 2010. "Cannibalism in *Tyrannosaurus rex*." *PLoS One* 5, no. 10: e13419.

Longrich, Nicholas R., Ken Barnes, Scott Clark, and Larry Millar. 2013. "Caenagnathidae from the Upper Campanian Aguja Formation of west Texas, and a revision of the Caenagnathinae." *Bulletin of the Peabody Museum of Natural History* 54, no. 1: 23–49.

Lowi-Merri, Talia M., and David C. Evans. 2020. "Cranial variation in *Gryposaurus* and biostratigraphy of hadrosaurines (Ornithischia: Hadrosauridae) from the Dinosaur Park Formation of Alberta, Canada." *Canadian Journal of Earth Sciences* 57, no. 6: 765–79.

Lü, Jun-Chang, Li Xu, Hua-Li Chang, Song-Hai Jia, Ji-Ming Zhang, Dian-Song Gao, Yi-Yang Zhang, Cheng-Jun Zhang, and Fang Ding. 2018. "A new alvarezsaurid dinosaur from the Late Cretaceous Qiupa Formation of Luanchuan, Henan Province, central China." *China Geology* 1, no. 1: 28–35.

Lull, Richard S. 1908. "The cranial musculature and the origin of the frill in the ceratopsian dinosaurs." *American Journal of Science* 25, no. 149: 387.

Lull, Richard S. 1933. *A revision of the Ceratopsia or horned dinosaurs*. Vol. 3. New Haven, CT: Tuttle, Morehouse & Taylor.

Lull, Richard S., and Nelda Emelyn Wright. 1942. *Hadrosaurian dinosaurs of North America*. Vol. 40. Boulder, CO: Geological Society of America.

Lyson, Tyler R., and Nicholas R. Longrich. 2011. "Spatial niche partitioning in dinosaurs from the latest Cretaceous (Maastrichtian) of North America." *Proceedings of the Royal Society B: Biological Sciences* 278, no. 1709: 1158–64.

Ma, Waisum, Junyou Wang, Michael Pittman, Qingwei Tan, Lin Tan, Bin Guo, and Xing Xu. 2017. "Functional anatomy of a giant toothless mandible from a bird-like dinosaur: *Gigantoraptor* and the evolution of the oviraptorosaurian jaw." *Scientific Reports* 7, no. 1: 1–15.

Ma, Waisum, Michael Pittman, Stephan Lautenschlager, Luke E. Meade, and Xing Xu. 2020a. "Functional morphology of the Oviraptorosaurian and Scansoriopterygid skull." *Bulletin of the American Museum of Natural History* 440, no. 1: 229–49.

Ma, Waisum, Stephen L. Brusatte, Junchang Lü, and Manabu Sakamoto. 2020b. "The skull evolution of oviraptorosaurian dinosaurs: the role of niche partitioning in diversification." *Journal of Evolutionary Biology* 33, no. 2: 178–88.

Ma, Waisum, Michael Pittman, Richard J. Butler, and Stephan Lautenschlager. 2022. "Macroevolutionary trends in theropod dinosaur feeding mechanics." *Current Biology* 32, no. 3: 677–86.

Macaluso, Loredana, and Emanuel Tschopp. 2018. "Evolutionary changes in pubic orientation in dinosaurs are more strongly correlated with the ventilation system than with herbivory." *Palaeontology* 61, no. 5: 703–19.

MacLaren, Jamie A., Philip S. L. Anderson, Paul M. Barrett, and Emily J. Rayfield. 2017. "Herbivorous dinosaur jaw disparity and its relationship to extrinsic evolutionary drivers." *Paleobiology* 43, no. 1: 15–33.

Maidment, Susannah C. R., and Paul M. Barrett. 2011. "The locomotor musculature of basal ornithischian dinosaurs." *Journal of Vertebrate Paleontology* 31, no. 6: 1265–91.

Maidment, Susannah C. R., and Paul M. Barrett. 2014. "Osteological correlates for quadrupedality in ornithischian dinosaurs." *Acta Palaeontologica Polonica* 59, no. 1: 53–70.

Maidment, Susannah C. R., and Laura B. Porro. 2010. "Homology of the palpebral and origin of supraorbital ossifications in ornithischian dinosaurs." *Lethaia* 43, no. 1: 95–111.

Maidment, Susannah C. R., Deborah H. Linton, Paul Upchurch, and Paul M. Barrett. 2012. "Limb-bone scaling indicates diverse stance and gait in quadrupedal ornithischian dinosaurs." *PLoS One* 7, no. 5: e36904.

Maidment, Susannah C. R., Karl T. Bates, Peter L. Falkingham, Collin VanBuren, Victoria Arbour, and Paul M. Barrett. 2014a. "Locomotion in ornithischian dinosaurs: an assessment using three-dimensional computational modelling." *Biological Reviews* 89, no. 3: 588–617.

Maidment, Susannah C. R., Donald M. Henderson, and Paul M. Barrett. 2014b. "What drove reversions to quadrupedal-ity in ornithischian dinosaurs? Testing hypotheses using centre of mass modelling." *Naturwissenschaften* 101, no. 11: 989–1001.

Maiorino, Leonardo, Andrew A. Farke, Tassos Kotsakis, Luciano Teresi, and Paolo Piras. 2015. "Variation in the shape and mechanical performance of the lower jaws in ceratopsid dinosaurs (Ornithischia, Ceratopsia)." *Journal of Anatomy* 227, no. 5: 631–46.

Maiorino, Leonardo, Andrew A. Farke, Tassos Kotsakis, Pasquale Raia, and Paolo Piras. 2018. "Who is the most stressed? Morphological disparity and mechanical behavior of the feeding apparatus of ceratopsian dinosaurs (Ornithischia, Marginocephalia)." *Cretaceous Research* 84: 483–500.

Makovicky, Peter J., and Mark A. Norell. 2006. "*Yamaceratops dorngobiensis*, a new primitive ceratopsian (Dinosauria: Ornithischia) from the Cretaceous of Mongolia." *American Museum Novitates* 3530: 1–42.

Makovicky, Peter J., Brandon M. Kilbourne, Rudyard W. Sadleir, and Mark A. Norell. 2011. "A new basal ornithopod (Dinosauria, Ornithischia) from the Late Cretaceous of Mongolia." *Journal of Vertebrate Paleontology* 31, no. 3: 626–40.

Mallison, Heinrich. 2010a. "The digital Plateosaurus, I. Body mass, mass distribution, and posture assessed using CAD and CAE on a digitally mounted complete skeleton." *Palaeontologia Electronica* 13, no. 2: 8A.

Mallison, Heinrich. 2010b. "The digital Plateosaurus, II. An assessment of the range of motion of the limbs and vertebral column and of previous reconstructions using a digital skeletal mount." *Acta Palaeontologica Polonica* 55, no. 3: 433–58.

Mallison Heinrich. 2011a. "Rearing giants: kinetic-dynamic modeling of sauropod bipedal and tripodal poses." In: N. Klein, K. Remes, C. T. Gee, and P. M. Sander, eds. *Biology of the sauropod dinosaurs: understanding the life of giants*. Bloomington: Indiana University Press, 237–50.

Mallison, Heinrich. 2011b. "Defense capabilities of *Kentrosaurus aethiopicus* Hennig, 1915." *Palaeontologia Electronica* 14, no. 2: 10A.

Mallison, Heinrich. 2014. "Osteoderm distribution has low impact on the centre of mass of stegosaurs." *Fossil Record* 17, no. 1: 33–39.

Mallon, Jordan C. 2019. "Competition structured a Late Cretaceous megaherbivorous dinosaur assemblage." *Scientific Reports* 9, no. 1: 1–18.

Mallon, Jordan C., and Jason S. Anderson. 2013. "Skull ecomorphology of megaherbivorous dinosaurs from the Dinosaur Park Formation (upper Campanian) of Alberta, Canada." *PLoS One* 8, no. 7: e67182.

Mallon, Jordan C., and Jason S. Anderson. 2014a. "Implications of beak morphology for the evolutionary paleoecology of the megaherbivorous dinosaurs from the Dinosaur Park Formation (Upper Campanian) of Alberta, Canada." *Palaeogeography, Palaeoclimatology, Palaeoecology* 394: 29–41.

Mallon, Jordan C., and Jason S. Anderson. 2014b. "The functional and palaeoecological implications of tooth morphology and wear for the megaherbivorous dinosaurs from the Dinosaur Park Formation (Upper Campanian) of Alberta, Canada." *PLoS One* 9, no. 6: e98605.

Mallon, Jordan C., and Jason S. Anderson. 2015. "Jaw mechanics and evolutionary paleoecology of the megaherbivorous dinosaurs from the Dinosaur Park Formation (Upper Campanian) of Alberta, Canada." *Journal of Vertebrate Paleontology* 35, no. 2: e904323.

Mallon, Jordan C., David C. Evans, Michael J. Ryan, and Jason S. Anderson. 2012. "Megaherbivorous dinosaur turnover in the Dinosaur Park Formation (Upper Campanian) of Alberta, Canada." *Palaeogeography, Palaeoclimatology, Palaeoecology* 350: 124–38.

Mallon, Jordan C., David C. Evans, Michael J. Ryan, and Jason S. Anderson. 2013. "Feeding height stratification among the herbivorous dinosaurs from the Dinosaur Park Formation (upper Campanian) of Alberta, Canada." *BMC Ecology* 13, no. 1: 1–15.

Mallon, Jordan C., Christopher J. Ott, Peter L. Larson, Edward M. Iuliano, and David C. Evans. 2016. "*Spiclypeus shipporum* gen. et sp. Nov., a boldly audacious new chasmosaurine ceratopsid (Dinosauria: Ornithischia) from the Judith River Formation (Upper Cretaceous: Campanian) of Montana, USA." *PLoS One* 11, no. 5: e0154218.

Malone, Joshua R., Jeffrey C. Strasser, David H. Malone, Michael D. D'Emic, Lauren Brown, and John P. Craddock. 2021. "Jurassic dinosaurs on the move: gastrolith provenance and long-distance migration." *Terra Nova* 33, no. 4: 375–82.

Manning, Phillip L., David Payne, John Pennicott, Paul M. Barrett, and Roland A. Ennos. 2006. "Dinosaur killer claws or climbing crampons?" *Biology Letters* 2, no. 1: 110–12.

Manning, Phillip L., Lee Margetts, Mark R. Johnson, Philip J. Withers, William I. Sellers, Peter L. Falkingham, Paul M. Mummery, Paul M. Barrett, and David R. Raymont. 2009. "Biomechanics of dromaeosaurid dinosaur claws: application of x-ray microtomography, nanoindentation, and finite element analysis." *Anatomical Record* 292, no. 9: 1397–405.

Mannion, Philip D., Paul Upchurch, Matthew T. Carrano, and Paul M. Barrett. 2011. "Testing the effect of the rock record on diversity: a multidisciplinary approach to elucidating the generic richness of sauropodomorph dinosaurs through time." *Biological Reviews* 86, no. 1: 157–81.

Mantell, Gideon Algernon. 1841. "XI. Memoir on a portion of the lower jaw of the *Iguanodon*, and on the remains of the *Hylæosaurus* and other saurians, discovered in the strata of Tilgate Forest, in Sussex." *Philosophical Transactions of the Royal Society of London* 131: 131–51.

Mantell, Gideon Algernon. 1848. "XIII. On the structure of the jaws and teeth of the *Iguanodon*." *Philosophical Transactions of the Royal Society of London* 138: 183–202.

Marsh, Adam D., and Timothy B. Rowe. 2018. "Anatomy and systematics of the sauropodomorph *Sarahsaurus aurifontanalis* from the Early Jurassic Kayenta Formation." *PLoS One* 13, no. 10: e0204007.

Marsh, Adam D., and Timothy B. Rowe. 2020. "A comprehensive anatomical and phylogenetic evaluation of *Dilophosaurus wetherilli* (Dinosauria, Theropoda) with descriptions of new specimens from the Kayenta Formation of northern Arizona." *Journal of Paleontology* 94, no. S78: 1–103.

Marsh, Othniel Charles. 1878. "Principal characters of American Jurassic dinosaurs." *American Journal of Science* 3, no. 95: 411–16.

Marsh, Othniel Charles. 1884. "Principal characters of American Jurassic dinosaurs, VIII. The order Theropoda." *American Journal of Science* 3, no. 160: 329–40.

Marsh, Othniel Charles. 1891. "I. The Gigantic Ceratopsidæ, or Horned Dinosaurs, of North America." *Geological Magazine* 8, no. 5: 193–99.

Marsh, Othniel Charles. 1893. "II. Restorations of *Anchisaurus*, Ceratosaurus, and Claosaurus." *Geological Magazine* 10, no. 4: 150–57.

Marsh, Othniel Charles. 1895. "On the affinities and classification of the dinosaurian reptiles." *American Journal of Science* 3, no. 300: 483–98.

Martin, J. 1987. "Mobility and feeding of *Cetiosaurus*: why the long neck?" *Occasional Paper of the Tyrrell Museum of Paleontology* 3: 150–55.

Martin, J., V. Martin-Rolland, and E. Frey. 1998. "Not cranes or masts, but beams: the biomechanics of sauropod necks." *Oryctos* 1: 113–20.

Martinez, Ricardo N., Paul C. Sereno, Oscar A. Alcober, Carina E. Colombi, Paul R. Renne, Isabel P. Montañez, and Brian S. Currie. 2011. "A basal dinosaur from the dawn of the dinosaur era in southwestern Pangaea." *Science* 331, no. 6014: 206–10.

Martínez, Rubén D. F., Matthew C. Lamanna, Fernando E. Novas, Ryan C. Ridgely, Gabriel A. Casal, Javier E. Martínez, Javier R. Vita, and Lawrence M. Witmer. 2016. "A basal lithostrotian titanosaur (Dinosauria: Sauropoda) with a complete skull: implications for the evolution and paleobiology of Titanosauria." *PLoS One* 11, no. 4: e0151661.

Marugán-Lobón, Jesús, and Luis M. Chiappe. 2022. "Ontogenetic niche shifts in the Mesozoic bird *Confuciusornis sanctus*." *Current Biology* 32, no. 7: 1629–34.

Maryańska, Teresa, and Halszka Osmólska. 1974. "Pachycephalosauria, a new suborder of ornithischian dinosaurs." *Palaeontologia Polonica* 30: 45–102.

Maryańska, Teresa, Ralph E. Chapman, and David B. Weishampel. 2004. "Pachycephalosauria." In: D. Weishampel, P. Dodson, and H. Osmólska, eds. *The Dinosauria*, 2nd ed. Berkeley: University of California Press, 464–77.

Matley, C. A. 1939. "The Coprolites of Pisdura, central province." *Records of the Geological Survey of India* 74: 535–47.

Maxwell, W. Desmond, and John H. Ostrom. 1995. "Taphonomy and paleobiological implications of Tenontosaurus-Deinonychus associations." *Journal of Vertebrate Paleontology* 15, no. 4: 707–12.

Mayr, Gerald. 2017. "The early Eocene birds of the Messel fossil site: a 48 million-year-old bird community adds a temporal perspective to the evolution of tropical avifaunas." *Biological Reviews* 92, no. 2: 1174–88.

Mazzetta, Gerardo V., Richard A. Fariña, and Sergio F. Vizcaíno. 1998. "On the palaeobiology of the South American horned theropod *Carnotaurus sastrei* Bonaparte." *Gaia* 15, no. 185: 192.

Mazzetta, Gerardo V., Adrián P. Cisilino, and R. Ernesto Blanco. 2004. "Distribución de tensiones durante la mordida en la mandíbula de *Carnotaurus sastrei* Bonaparte, 1985 (Theropoda: Abelisauridae)." *Ameghiniana* 41, no. 4: 605–17.

Mazzetta, Gerardo V., Adrián P. Cisilino, R. Ernesto Blanco, and Néstor Calvo. 2009. "Cranial mechanics and functional interpretation of the horned carnivorous dinosaur *Carnotaurus sastrei*." *Journal of Vertebrate Paleontology* 29, no. 3: 822–30.

McCrea, R. T., M. G. Lockley, and C. A. Meyer. 2001. "Global distribution of purported ankylosaur track occurrences." In: K. Carpenter, ed. *The armored dinosaurs*. Bloomington: Indiana University Press, 413–54.

McGowan, C. 1982. "The wing musculature of the Brown Kiwi *Apteryx australis mantelli* and its bearing on ratite affinities." *Journal of Zoology* 197, no. 2: 173–219.

McGowan, Christopher. 1979. "The hind limb musculature of the brown kiwi, *Apteryx australis mantelli*." *Journal of Morphology* 160, no. 1: 33–73.

McHugh, Julia. 2018. "Evidence for niche partitioning among ground-height browsing sauropods from the Upper Jurassic Morrison Formation of North America." *Geology of the Intermountain West* 5: 95–103.

McKellar, Ryan C., Emma Jones, Michael S. Engel, Ralf Tappert, Alexander P. Wolfe, Karlis Muehlenbachs, Pierre Cockx, Eva B. Koppelhus, and Philip J. Currie. 2019. "A direct association between amber and dinosaur remains provides paleoecological insights." *Scientific Reports* 9, no. 1: 1–7.

McLain, Matthew A., David Nelsen, Keith Snyder, Christopher T. Griffin, Bethania Siviero, Leonard R. Brand, and Arthur V. Chadwick. 2018. "Tyrannosaur cannibalism: a case of a tooth-traced tyrannosaurid bone in the Lance Formation (Maastrichtian), Wyoming." *Palaios* 33, no. 4: 164–73.

McNab, Brian K. 2009. "Resources and energetics determined dinosaur maximal size." *Proceedings of the National Academy of Sciences* 106, no. 29: 12,184–88.

Meers, Mason B. 2002. "Maximum bite force and prey size of *Tyrannosaurus rex* and their relationships to the inference of feeding behavior." *Historical Biology* 16, no. 1: 1–12.

Melstrom, Keegan M., Luis M. Chiappe, and Nathan D. Smith. 2021. "Exceptionally simple, rapidly replaced teeth in sauropod dinosaurs demonstrate a novel evolutionary strategy for herbivory in Late Jurassic ecosystems." *BMC Ecology and Evolution* 21, no. 1: 1–12.

Meso, J. G., Z. Qin, M. Pittman, J. I. Canale, L. Salgado, and V. Díez Díaz. 2021. "Tail anatomy of the Alvarezsauria (Theropoda, Coelurosauria), and its functional and behavioural implications." *Cretaceous Research* 124: 104830.

Midgley, J. J., G. Midgley, and W. J. Bond. 2002. "Why were dinosaurs so large? A food quality hypothesis." *Evolutionary Ecology Research* 4, no. 7: 1093–95.

Miller, Case Vincent, and Michael Pittman. 2021. "The diet of early birds based on modern and fossil evidence and a new framework for its reconstruction." *Biological Reviews* 96, no. 5: 2058–112.

Miller, Case Vincent, Michael Pittman, Xiaoli Wang, Xiaoting Zheng, and Jen A. Bright. 2022. "Diet of Mesozoic toothed birds (Longipterygidae) inferred from quantitative analysis of extant avian diet proxies." *BMC Biology* 20, no. 1: 1–37.

Milner, Andrew R. C., and James I. Kirkland. 2007. "The case for fishing dinosaurs at the St. George dinosaur discovery site at Johnson Farm." *Survey Notes* 39, no. 3: 1–3.

Mocho, P., R. Royo-Torres, and F. Ortega. 2017. "New data of the Portuguese brachiosaurid *Lusotitan atalaiensis* (Sobral Formation, Upper Jurassic)." *Historical Biology* 29, no. 6: 789–817.

Mohabey, D. M. 2005. "Late Cretaceous (Maastrichtian) nests, eggs, and dung mass (coprolites) of sauropods (titanosaurs) from India." In: Virginia Tidwell and Kenneth Carpenter, eds. *Thunder-lizards: the sauropodomorph dinosaurs*. Bloomington: Indiana University Press, 466–89.

Molnar, Ralph E. 1973. "The cranial morphology and mechanics of *Tyrannosaurus rex* (Reptilia: Saurischia)." PhD diss., University of California.

Molnar, Ralph E. 1977. "Analogies in the evolution of combat and display structures in ornithopods and ungulates." *Evolutionary Theory* 3: 165–90.

Molnar, Ralph E. 1991. "The cranial morphology of *Tyrannosaurus rex*." *Palaeontographica Abteilung A: Paläozoologie, Stratigraphie* 217: 137–76.

Molnar, Ralph E. 1996. "Preliminary report a new ankylosaur from the Early Cretaceous of Queensland, Australia." *Memoirs of the Queensland Museum* 39: 653–68.

Molnar, Ralph E. 1998. "Mechanical factors in the design of the skull of *Tyrannosaurus rex* (Osborn, 1905)." *Gaia* 15: 193.

Molnar, Ralph E. 2008. "Reconstruction of jaw musculature of *Tyrannosaurus rex*." In: Peter L. Larson and Kenneth Carpenter, eds. *Tyrannosaurus rex: the tyrant king*. Bloomington: Indiana University Press, 254–81.

Molnar, Ralph E. 2013. "A comparative analysis of reconstructed jaw musculature and mechanics of some large theropods." In: R. E. Molnar and P. J. Currie, eds. *Tyrannosaurid paleobiology*. Bloomington: Indiana University Press, 177–94.

Molnar, Ralph E., and H. Trevor Clifford. 2000. "Gut contents of a small ankylosaur." *Journal of Vertebrate Paleontology* 20, no. 1: 194–96.

Molnar, Ralph E., and H. Trevor Clifford. 2001. "An ankylosaurian cololite from the Lower Cretaceous of Queensland, Australia." In: K. Carpenter, ed. *The armored dinosaurs*. Bloomington: Indiana University Press, 399–412.

Monbaron, Michel, Dale A. Russell, and Philippe Taquet. 1999. "*Atlasaurus imelakei* ng, n. sp., a brachiosaurid-like sauropod from the Middle Jurassic of Morocco alen." *Comptes Rendus de l'Académie des Sciences, Series IIA: Earth and Planetary Science* 329, no. 7: 519–26.

Monfroy, Quentin T. 2017. "Correlation between the size, shape and position of the teeth on the jaws and the bite force in Theropoda." *Historical Biology* 29, no. 8: 1089–105.

Morhardt, Ashley Caroline. 2009. "Dinosaur smiles: do the texture and morphology of the premaxilla, maxilla, and dentary bones of sauropsids provide osteological correlates for inferring extra-oral structures reliably in dinosaurs?" Master's thesis, Western Illinois University.

Morschhauser, Eric M., David J. Varricchio, Gao Chunling, Liu Jinyuan, Wang Xuri, Cheng Xiadong, and Meng Qingjin. 2009. "Anatomy of the Early Cretaceous bird *Rapaxavis pani*, a new species from Liaoning Province, China." *Journal of Vertebrate Paleontology* 29, no. 2: 545–54.

Morschhauser, Eric M., Hailu You, Daqing Li, and Peter Dodson. 2018. "Phylogenetic history of *Auroraceratops rugosus* (Ceratopsia: Ornithischia) from the Lower Cretaceous of Gansu Province, China." *Journal of Vertebrate Paleontology* 38 (Suppl. 1): 117–47.

Morris, William J. 1970. *Hadrosaurian dinosaur bills: morphology and function*. Los Angeles: Los Angeles County Museum of Natural History.

Müller, Rodrigo T. 2020. "Craniomandibular osteology of Macrocollum itaquii (Dinosauria: Sauropodomorpha) from the Late Triassic of southern Brazil." *Journal of Systematic Palaeontology* 18, no. 10: 805–41.

Müller, Rodrigo T., and Maurício S. Garcia. 2020a. "Rise of an empire: analyzing the high diversity of the earliest sauropodomorph dinosaurs through distinct hypotheses." *Historical Biology* 32, no. 10: 1334–39.

Müller, Rodrigo T., and Maurício S. Garcia. 2020b. "A paraphyletic 'Silesauridae' as an alternative hypothesis for the initial radiation of ornithischian dinosaurs." *Biology Letters* 16, no. 8: 20200417.

Müller, Rodrigo T., Max C. Langer, Mario Bronzati, Cristian P. Pacheco, Sérgio F. Cabreira, and Sérgio Dias-Da-Silva. 2018. "Early evolution of sauropodomorphs: anatomy and phylogenetic relationships of a remarkably well-preserved dinosaur from the Upper Triassic of southern Brazil." *Zoological Journal of the Linnean Society* 184, no. 4: 1187–248.

Müller, Rodrigo T., José D. Ferreira, Flávio A. Pretto, Mario Bronzati, and Leonardo Kerber. 2020. "The endocranial anatomy of *Buriolestes schultzi* (Dinosauria: Saurischia) and the early evolution of brain tissues in sauropodomorph dinosaurs." *Journal of Anatomy* 238, no. 4: 809–27.

Mustoe, George E. 2007. "Coevolution of cycads and dinosaurs." *Cycad Newsletter* 30: 6–9.

Myhrvold, Nathan P. 2016. "Dinosaur metabolism and the allometry of maximum growth rate." *PLoS One* 11, no. 11: e0163205.

Myhrvold, Nathan P., and Philip J. Currie. 1997. "Supersonic sauropods? Tail dynamics in the diplodocids." *Paleobiology* 23, no. 4: 393–409.

Myhrvold, Nathan P., Paul C. Sereno, Stephanie L. Baumgart, Kiersten K. Formoso, Daniel Vidal, Frank E. Fish, and Donald M. Henderson. 2022. "Spinosaurids as 'subaqueous foragers' undermined by selective sampling and problematic statistical inference." *bioRxiv* 1–5. https://doi.org/10.1101/2022.04.13.487781.

Nabavizadeh, Ali. 2011. "Thyreophoran jaw mechanics and the functional significance of the predentary bone." *Journal of Vertebrate Paleontology* 31, no. S3: 164A.

Nabavizadeh, Ali. 2014. "Hadrosauroid jaw mechanics and the functional significance of the predentary bone." In: David A.

Eberth and David C. Evans, eds. *Hadrosaurs*. Bloomington: Indiana University Press, 467–83.

Nabavizadeh, Ali. 2016. "Evolutionary trends in the jaw adductor mechanics of ornithischian dinosaurs." *Anatomical Record* 299, no. 3: 271–94.

Nabavizadeh, Ali. 2020a. "New reconstruction of cranial musculature in ornithischian dinosaurs: implications for feeding mechanisms and buccal anatomy." *Anatomical Record* 303, no. 2: 347–62.

Nabavizadeh, Ali. 2020b. "Cranial musculature in herbivorous dinosaurs: a survey of reconstructed anatomical diversity and feeding mechanisms." *Anatomical Record* 303, no. 4: 1104–45.

Nabavizadeh, Ali, and David B. Weishampel. 2016. "The predentary bone and its significance in the evolution of feeding mechanisms in ornithischian dinosaurs." *Anatomical Record* 299, no. 10: 1358–88.

Navalón, G. 2014. "Reconstructing the palaeobiology of *Confuciusornis* and other confuciusornithiformes." PhD diss., University of Bristol.

Nesbitt S. J. 2011. "The early evolution of Archosaurs: relationships and the origin of major clades." *Bulletin of the American Museum of Natural History* 352: 1–292.

Nesbitt, S. J., and H. D. Sues. 2021. "The osteology of the early-diverging dinosaur *Daemonosaurus chauliodus* (Archosauria: Dinosauria) from the Coelophysis Quarry (Triassic: Rhaetian) of New Mexico and its relationships to other early dinosaurs." *Zoological Journal of the Linnean Society* 191, no. 1: 150–79.

Nesbitt, Sterling J., Alan H. Turner, Gregory M. Erickson, and Mark A. Norell. 2006. "Prey choice and cannibalistic behaviour in the theropod *Coelophysis*." *Biology Letters* 2, no. 4: 611–14.

Nesbitt, Sterling J., Nathan D. Smith, Randall B. Irmis, Alan H. Turner, Alex Downs, and Mark A. Norell. 2009. "A complete skeleton of a Late Triassic saurischian and the early evolution of dinosaurs." *Science* 326, no. 5959: 1530–33.

Nesbitt, Sterling J., Robert K. Denton, Mark A. Loewen, Stephen L. Brusatte, Nathan D. Smith, Alan H. Turner, James I. Kirkland, Andrew T. McDonald, and Douglas G. Wolfe. 2019. "A mid-Cretaceous tyrannosauroid and the origin of North American end-Cretaceous dinosaur assemblages." *Nature Ecology and Evolution* 3, no. 6: 892–99.

Newman, B. H. 1970. "Stance and gait in the flesh-eating dinosaur *Tyrannosaurus*." *Biological Journal of the Linnean Society* 2, no. 2: 119–23.

Nicholls, E. L., and A. P. Russell. 1985. "Structure and function of the pectoral girdle and forelimb of *Struthiomimus altus* (Theropoda: Ornithomimidae)." *Palaeontology* 28: 643–77.

Nopcsa, F. 1900. "Dinosaurierreste aus Siebenbürgen I. Schädel von *Limnosaurus transsylvanicus* nov. Gen. Et nov. Spec. Denkschriften der königlichen Akademie der Wissenschaften." *Mathematisch-Naturwissenschaftlichen Klasse* 68: 555–91.

Nopcsa, F. 1928. "Paleontological notes on reptiles." *Acta Geologica Hungarica* 1: 1–84.

Nordén, Klara K., Thomas L. Stubbs, Albert Prieto-Márquez, and Michael J. Benton. 2018. "Multifaceted disparity approach reveals dinosaur herbivory flourished before the end-Cretaceous mass extinction." *Paleobiology* 44, no. 4: 620–37.

Norell, Mark A., and J. Makovicky Peter. 1999. "Important features of the dromaeosaur skeleton, 2. Information from newly collected specimens of Velociraptor mongoliensis." *American Museum Novitates* 3282: 1–45.

Norell, Mark A., James M. Clark, Luis M. Chiappe, and Demberelyin Dashzeveg. 1995. "A nesting dinosaur." *Nature* 378, no. 6559: 774–76.

Norell, Mark A., Peter J. Makovicky, and Philip J. Currie. 2001. "The beaks of ostrich dinosaurs." *Nature* 412, no. 6850: 873–74.

Norman, David B. 1980. *On the ornithischian dinosaur Iguanodon bernissartensis from the Lower Cretaceous of Bernissart (Belgium)*. Belgium: Institut Royal des Sciences Naturelles de Belgique.

Norman, David B. 1984. "On the cranial morphology and evolution of ornithopod dinosaurs." *Symposia of the Zoological Society of London* 52: 521–47.

Norman, David B. 1986. "On the anatomy of Iguanodon atherfieldensis (Ornithischia: Ornithopoda)." *Bulletin-Institut royal des sciences naturelles de Belgique: Sciences de la Terre* 56: 281–372.

Norman, David B. 2020a. "*Scelidosaurus harrisonii* from the Early Jurassic of Dorset, England: cranial anatomy." *Zoological Journal of the Linnean Society* 188, no. 1: 1–81.

Norman, David B. 2020b. "*Scelidosaurus harrisonii* from the Early Jurassic of Dorset, England: postcranial skeleton." *Zoological Journal of the Linnean Society* 189, no. 1: 47–157.

Norman, David B. 2020c. "*Scelidosaurus harrisonii* from the Early Jurassic of Dorset, England: the dermal skeleton." *Zoological Journal of the Linnean Society* 190, no. 1: 1–53.

Norman, David B. 2021. "*Scelidosaurus harrisonii* (Dinosauria: Ornithischia) from the Early Jurassic of Dorset, England: biology and phylogenetic relationships." *Zoological Journal of the Linnean Society* 191, no. 1: 1–86.

Norman, David B., and David B. Weishampel. 1985. "Ornithopod feeding mechanisms: their bearing on the evolution of herbivory." *American Naturalist* 126, no. 2: 151–64.

Norman, David B., David B. Weishampel, J. M. V. Rayner, and R. J. Wootton. 1991. "Feeding mechanisms in some small herbivorous dinosaurs: processes and patterns." *Biomechanics in Evolution* 36: 161–81.

Norman, David B., Alfred W. Crompton, Richard J. Butler, Laura B. Porro, and Alan J. Charig. 2011. "The Lower Jurassic ornithischian dinosaur *Heterodontosaurus tucki* Crompton & Charig, 1962: cranial anatomy, functional morphology, taxonomy, and relationships." *Zoological Journal of the Linnean Society* 163, no. 1: 182–276.

Novas, Fernando E., Leonardo Salgado, Manuel Suárez, Federico L. Agnolín, Martín D. Ezcurra, Nicolás R. Chimento, Rita de la Cruz, Marcelo P. Isasi, Alexander O. Vargas, and David Rubilar-Rogers. 2015. "An enigmatic plant-eating theropod from the Late Jurassic period of Chile." *Nature* 522, no. 7556: 331–34.

O'Connor, Jingmai K. 2009. "A systematic review of Enantiornithes (Aves: Ornithothoraces)." PhD diss., University of Southern California.

O'Connor, Jingmai K. 2019. "The trophic habits of early birds." *Palaeogeography, Palaeoclimatology, Palaeoecology* 513: 178–95.

O'Connor, Jingmai K., and Zhonghe Zhou. 2013. "A redescription of *Chaoyangia beishanensis* (Aves) and a comprehensive phylogeny of Mesozoic birds." *Journal of Systematic Palaeontology* 11, no. 7: 889–906.

O'Connor, Jingmai K., and Zhonghe Zhou. 2020. "The evolution of the modern avian digestive system: insights from paravian fossils from the Yanliao and Jehol biotas." *Palaeontology* 63, no. 1: 13–27.

O'Connor, Jingmai K., Zhonghe Zhou, and Xing Xu. 2011a. "Additional specimen of *Microraptor* provides unique evidence of dinosaurs preying on birds." *Proceedings of the National Academy of Sciences* 108, no. 49: 19,662–65.

O'Connor, Jingmai K., Luis M. Chiappe, and Alyssa Bell. 2011b. "Pre-modern birds: avian divergences in the Mesozoic." In: Gareth Dyke and Gary Kaiser, eds. *Living dinosaurs: the evolutionary history of modern birds.* Norwell, MA: Wiley-Blackwell, 39–114.

O'Connor, Jingmai, Xiaoli Wang, Corwin Sullivan, Yan Wang, Xiaoting Zheng, Han Hu, Xiaomei Zhang, and Zhonghe Zhou. 2018. "First report of gastroliths in the Early Cretaceous basal bird Jeholornis." *Cretaceous Research* 84: 200–208.

Olsen, Aaron M. 2017. "Feeding ecology is the primary driver of beak shape diversification in waterfowl." *Functional Ecology* 31, no. 10: 1985–95.

Olsen, Aaron M. 2019. "A mobility-based classification of closed kinematic chains in biomechanics and implications for motor control." *Journal of Experimental Biology* 222, no. 21: jeb195735.

Olsen, Aaron M., and Mark W. Westneat. 2016. "Linkage mechanisms in the vertebrate skull: structure and function of three-dimensional, parallel transmission systems." *Journal of Morphology* 277, no. 12: 1570–83.

Organ, Christopher Lee. 2006a. "Thoracic epaxial muscles in living archosaurs and ornithopod dinosaurs." *Anatomical Record, Part A: Discoveries in Molecular, Cellular, and Evolutionary Biology* 288, no. 7: 782–93.

Organ, Chris L. 2006b. "Biomechanics of ossified tendons in ornithopod dinosaurs." *Paleobiology* 32, no. 4: 652–65.

Osborn, H. 1924. "Three new Theropoda, *Protoceratops* zone, central Mongolia." *American Museum Novitates* 144: 1–12.

Ősi, Attila, Paul M. Barrett, Tamás Földes, and Richárd Tokai. 2014. "Wear pattern, dental function, and jaw mechanism in the Late Cretaceous ankylosaur *Hungarosaurus*." *Anatomical Record* 297, no. 7: 1165–80.

Ősi, Attila, Edina Prondvai, Jordan Mallon, and Emese Réka Bodor. 2017. "Diversity and convergences in the evolution of feeding adaptations in ankylosaurs (Dinosauria: Ornithischia)." *Historical Biology* 29, no. 4: 539–70.

Osmólska, H. "Antorbital fenestra of archosaurs and its suggested function." 1985. *Fortschritte der Zoologie* 30: 159–62.

Ostrom, John H. 1961. "Cranial morphology of the hadrosaurian dinosaurs of North America." *Bulletin of the American Museum of Natural History* 122, no 2: 39–186.

Ostrom, John H. 1964a. "A functional analysis of jaw mechanics in the dinosaur *Triceratops*." *Postilla* 88: 1–35.

Ostrom, John H. 1964b. "A reconsideration of the paleoecology of hadrosaurian dinosaurs." *American Journal of Science* 262, no. 8: 975–97.

Ostrom, John H. 1966. "Functional morphology and evolution of the ceratopsian dinosaurs." *Evolution* 20: 290–308.

Ostrom, John H. 1969. *Osteology of Deinonychus antirrhopus, an unusual theropod from the Lower Cretaceous of Montana.* New Haven, CT: Yale University Press.

Ostrom, John H. 1976. "*Archaeopteryx* and the origin of birds." *Biological Journal of the Linnean Society* 8, no. 2: 91–182.

Ostrom, John H. 1995. "Wing biomechanics and the origin of bird flight." *Neues Jahrbuch für Geologie und Paläontologie-Abhandlungen* 195: 253–66.

Ostrom, John H., and P. Wellnhofer. 1986. "The Munich specimen of *Triceratops* with a revision of the genus." *Zitteliana* 14: 111–58.

Otero, Alejandro. 2010. "The appendicular skeleton of *Neuquensaurus*, a Late Cretaceous saltasaurine sauropod from Patagonia, Argentina." *Acta Palaeontologica Polonica* 55, no. 3: 399–426.

Otero, Alejandro. 2018. "Forelimb musculature and osteological correlates in Sauropodomorpha (Dinosauria, Saurischia)." *PLoS One* 13, no. 7: e0198988.

Otero, Alejandro, Andrew R. Cuff, Vivian Allen, Lauren Sumner-Rooney, Diego Pol, and John R. Hutchinson. 2019. "Ontogenetic changes in the body plan of the sauropodomorph dinosaur *Mussaurus patagonicus* reveal shifts of locomotor stance during growth." *Scientific Reports* 9, no. 1: 1–10.

Otero, Alejandro, José L. Carballido, and Agustín Pérez Moreno. 2020. "The appendicular osteology of *Patagotitan mayorum* (Dinosauria, Sauropoda)." *Journal of Vertebrate Paleontology* 40, no. 4: e1793158.

Owen, Richard. 1840. *Odontography; or, Treatise on the comparative anatomy of the teeth; their physiological relations, mode of development, and microscopic structure in the vertebrate animal.* 2 vols. London: Hippolyte Bailliere.

Owen, Richard. 1842. "Report on British Fossil reptiles, Pt. II." *Reports of the British Association for the Advancement of Science* 11: 60–204.

Owocki, Krzysztof, Barbara Kremer, Martin Cotte, and Hervé Bocherens. 2020. "Diet preferences and climate inferred from oxygen and carbon isotopes of tooth enamel of *Tarbosaurus bataar* (Nemegt Formation, Upper Cretaceous, Mongolia)." *Palaeogeography, Palaeoclimatology, Palaeoecology* 537: 109190.

Pacheco, Cristian, Rodrigo T. Müller, Max Langer, Flávio A. Pretto, Leonardo Kerber, and Sérgio Dias da Silva. 2019. "*Gnathovorax cabreirai*: a new early dinosaur and the origin and initial radiation of predatory dinosaurs." *PeerJ* 7: e7963.

Padian, Kevin. 2022. "Why tyrannosaurid forelimbs were so short: an integrative hypothesis." *Acta Palaeontologica Polonica* 67, no. 1: 63–76.

Palma Liberona, José A., Sergio Soto-Acuña, Marco A. Mendez, and Alexander O. Vargas. 2019. "Assessment and interpretation of negative forelimb allometry in the evolution of non-avian Theropoda." *Frontiers in Zoology* 16, no. 1: 1–13.

Papp, M. J., and L. M. Witmer. 1998. "Cheeks, beaks, or freaks: a critical appraisal of buccal soft-tissue anatomy in ornithischian dinosaurs." *Journal of Vertebrate Paleontology* 18 (Suppl. 3): 69A.

Park, Jin-Young, Yuong-Nam Lee, Philip J. Currie, Yoshitsugu Kobayashi, Eva Koppelhus, Rinchen Barsbold, Octávio Mateus, Sungjin Lee, and Su-Hwan Kim. 2020. "Additional skulls of *Talarurus plicatospineus* (Dinosauria: Ankylosauridae) and implications for paleobiogeography and paleoecology of armored dinosaurs." *Cretaceous Research* 108: 104340.

Park, Jin-Young, Yuong-Nam Lee, Philip J. Currie, Michael J. Ryan, Phil Bell, Robin Sissons, Eva B. Koppelhus, Rinchen

Barsbold, Sungjin Lee, and Su-Hwan Kim. 2021. "A new ankylosaurid skeleton from the Upper Cretaceous Baruungoyot Formation of Mongolia: its implications for ankylosaurid postcranial evolution." *Scientific Reports* 11, no. 1: 1–10.

Parsons, William L., and Kristen M. Parsons. 2009. "A new ankylosaur (Dinosauria: Ankylosauria) from the Lower Cretaceous Cloverly Formation of central Montana." *Canadian Journal of Earth Sciences* 46, no. 10: 721–38.

Paul, Gregory S. 1988. "Physiological, migratorial, climatological, geophysical, survival, and evolutionary implications of Cretaceous polar dinosaurs." *Journal of Paleontology* 62, no. 4: 640–52.

Paul, Gregory S. 2017. "Restoring maximum vertical browsing reach in sauropod dinosaurs." *Anatomical Record* 300, no. 10: 1802–25.

Paul, Gregory S., and P. Christiansen. 2000. "Forelimb posture in neoceratopsian dinosaurs: implications for gait and locomotion." *Paleobiology* 26, no. 3: 450–65.

Paulina-Carabajal, Ariana, and Jorge O. Calvo. 2021. "Redescription of the braincase of the rebbachisaurid sauropod *Limaysaurus tessonei* and novel endocranial information based on CT scans." *Anais da Academia Brasileira de Ciências* 93 (Suppl. 2): 1–14.

Perle, Altangerel, Teresa Maryanska, and Halszka Osmólska. 1982. "*Goyocephale lattimorei* gen. et sp. N., a new flat-headed pachycephalosaur (Ornithischia, Dinosauria) from the Upper Cretaceous of Mongolia." *Acta Palaeontologica Polonica* 27, no. 1–4: 115–127.

Perle, Altangerel, Mark A. Norell, Luis M. Chiappe, and James M. Clark. 1993. "Flightless bird from the Cretaceous of Mongolia." *Nature* 362, no. 6421: 623–26.

Perry, George L. W. 2021. "How far might plant-eating dinosaurs have moved seeds?" *Biology Letters* 17, no. 1: 20200689.

Persons, W. Scott, IV, and Philip J. Currie. 2011a. "Dinosaur speed demon: the caudal musculature of *Carnotaurus sastrei* and implications for the evolution of South American abelisaurids." *PLoS One* 6, no. 10: e25763.

Persons, W. Scott, IV, and Philip J. Currie. 2011b. "The tail of *Tyrannosaurus*: reassessing the size and locomotive importance of the M. caudofemoralis in non-avian theropods." *Anatomical Record* 294, no. 1: 119–31.

Persons, W. Scott, IV, and Philip J. Currie. 2012. "Dragon tails: convergent caudal morphology in winged archosaurs." *Acta Geologica Sinica*, Eng. ed., 86, no. 6: 1402–12.

Persons, W. Scott, IV, and Philip J. Currie. 2014. "Duckbills on the run: the cursorial abilities of hadrosaurs and implications for tyrannosaur-avoidance strategies." In: David A.

Eberth and David C. Evans, eds. *Hadrosaurs*. Bloomington: Indiana University Press, 449.

Persons, W. Scott, IV, and Philip J. Currie. 2016. "An approach to scoring cursorial limb proportions in carnivorous dinosaurs and an attempt to account for allometry." *Scientific Reports* 6, no. 1: 1–12.

Persons, W. Scott, IV, and Philip J. Currie. 2020. "The anatomical and functional evolution of the femoral fourth trochanter in ornithischian dinosaurs." *Anatomical Record* 303, no. 4: 1146–57.

Peterson, Joseph E., and Karsen N. Daus. 2019. "Feeding traces attributable to juvenile *Tyrannosaurus rex* offer insight into ontogenetic dietary trends." *PeerJ* 7: e6573.

Peterson, Joseph E., Michael D. Henderson, Reed P. Scherer, and Christopher P. Vittore. 2009. "Face biting on a juvenile tyrannosaurid and behavioral implications." *Palaios* 24, no. 11: 780–84.

Peterson, Joseph E., Z. Jack Tseng, and Shannon Brink. 2021. "Bite force estimates in juvenile *Tyrannosaurus rex* based on simulated puncture marks." *PeerJ* 9: e11450.

Piperno, Dolores R., and Hans-Dieter Sues. 2005. "Dinosaurs dined on grass." *Science* 310, no. 5751: 1126–28.

Pittman, Michael, and Xing Xu. 2020. "Pennaraptoran theropod dinosaurs past progress and new frontiers." *Bulletin of the American Museum of Natural History* 440, no. 1: 1–355.

Podlesnov, A. V. 2018. "Morphology of the craniovertebral joint in *Psittacosaurus sibiricus* (Ornithischia: Ceratopsia)." *Paleontological Journal* 52, no. 6: 664–76.

Poropat, Stephen F., Philip D. Mannion, Paul Upchurch, Travis R. Tischler, Trish Sloan, George H. K. Sinapius, Judy A. Elliott, and David A. Elliott. 2020. "Osteology of the wide-hipped titanosaurian sauropod dinosaur *Savannasaurus elliottorum* from the Upper Cretaceous Winton Formation of Queensland, Australia." *Journal of Vertebrate Paleontology* 40, no. 3: e1786836.

Poropat, Stephen F., Martin Kundrát, Philip D. Mannion, Paul Upchurch, Travis R. Tischler, and David A. Elliott. 2021. "Second specimen of the Late Cretaceous Australian sauropod dinosaur *Diamantinasaurus matildae* provides new anatomical information on the skull and neck of early titanosaurs." *Zoological Journal of the Linnean Society* 192, no. 2: 610–74.

Porro, Laura B. 2007. "Feeding and jaw mechanism in *Heterodontosaurus tucki* using finite element analysis." *Journal of Vertebrate Paleontology* 27 (Suppl. 3): 131A.

Porro, Laura B. 2009. "Cranial biomechanics in the early dinosaur *Heterodontosaurus*." PhD diss., University of Cambridge.

Porro, Laura B., Richard J. Butler, Paul M. Barrett, Scott Moore-Fay, and Richard L. Abel. 2010. "New heterodontosaurid specimens from the Lower Jurassic of southern Africa and the early ornithischian dinosaur radiation." *Earth and Environmental Science Transactions of the Royal Society of Edinburgh* 101, no. 3–4: 351–66.

Porro, Laura B., C. M. Holliday, F. Anapol, L. C. Ontiveros, L. T. Ontiveros, and C. F. Ross. 2011. "Free body analysis, beam mechanics, and finite element modeling of the mandible of *Alligator mississippiensis*." *Journal of Morphology* 272: 910–37.

Porro, Laura B., Lawrence M. Witmer, and Paul M. Barrett. 2015. "Digital preparation and osteology of the skull of *Lesothosaurus diagnosticus* (Ornithischia: Dinosauria)." *PeerJ* 3: e1494.

Porter, William Ruger, and Lawrence M. Witmer. 2020. "Vascular patterns in the heads of dinosaurs: evidence for blood vessels, sites of thermal exchange, and their role in physiological thermoregulatory strategies." *Anatomical Record* 303, no. 4: 1075–103.

Powers, Mark James, Corwin Sullivan, and Philip John Currie. 2020. "Re-examining ratio based premaxillary and maxillary characters in Eudromaeosauria (Dinosauria: Theropoda): divergent trends in snout morphology between Asian and North American taxa." *Palaeogeography, Palaeoclimatology, Palaeoecology* 547: 109704.

Prasad, Vandana, Caroline A. E. Strömberg, Habib Alimohammadian, and Ashok Sahni. 2005. "Dinosaur coprolites and the early evolution of grasses and grazers." *Science* 310, no. 5751: 1177–80.

Preuschoft, Holger, and Nicole Klein. 2013. "Torsion and bending in the neck and tail of sauropod dinosaurs and the function of cervical ribs: insights from functional morphology and biomechanics." *PLoS One* 8, no. 10: e78574.

Prieto-Márquez, A. 2010. "Global phylogeny of Hadrosauridae (Dinosauria: Ornithopoda) using parsimony and Bayesian methods." *Zoological Journal of the Linnean Society* 159: 435–502.

Prieto-Márquez, A., Jonathan R. Wagner, and Thomas Lehman. 2020a. "An unusual 'shovel-billed' dinosaur with trophic specializations from the early Campanian of Trans-Pecos Texas, and the ancestral hadrosaurian crest." *Journal of Systematic Palaeontology* 18, no. 6: 461–98.

Prieto-Márquez, A., Joan Garcia-Porta, Shantanu H. Joshi, Mark A. Norell, and Peter J. Makovicky. 2020b. "Modularity and heterochrony in the evolution of the ceratopsian dinosaur frill." *Ecology and Evolution* 10, no. 13: 6288–309.

Raath, M. A. 1974. "Fossil vertebrate studies in Rhodesia: further evidence of gastroliths in prosauropod dinosaurs." *Arnoldia* 7, no. 5, 1–7.

Radermacher, V. J., V. Fernandez, E. R. Schachner, R. J. Butler, E. M. Bordy, M. N. Hudgins, W. J. de Klerk, K. E. J. Chapelle, and J. N. Choiniere. 2021. "A new *Heterodontosaurus* specimen elucidates the unique ventilatory macroevolution of ornithischian dinosaurs." *eLife* 10: e66036.

Rauhut, Oliver W. M., and Diego Pol. 2019. "Probable basal allosauroid from the early Middle Jurassic Cañadón Asfalto Formation of Argentina highlights phylogenetic uncertainty in tetanuran theropod dinosaurs." *Scientific Reports* 9, no. 1: 1–9.

Rauhut, Oliver W. M., Christian Foth, and Helmut Tischlinger. 2018. "The oldest *Archaeopteryx* (Theropoda: Avialiae): a new specimen from the Kimmeridgian/Tithonian boundary of Schamhaupten, Bavaria." *PeerJ* 6: e4191.

Rayfield, Emily J. 2004. "Cranial mechanics and feeding in *Tyrannosaurus rex*." *Proceedings of the Royal Society B: Biological Sciences* 271, no. 1547: 1451–59.

Rayfield, Emily J. 2005a. "Aspects of comparative cranial mechanics in the theropod dinosaurs *Coelophysis*, *Allosaurus* and *Tyrannosaurus*." *Zoological Journal of the Linnean Society* 144, no. 3: 309–16.

Rayfield, Emily J. 2005b. "Using finite-element analysis to investigate suture morphology: a case study using large carnivorous dinosaurs." *Anatomical Record, Part A: Discoveries in Molecular, Cellular, and Evolutionary Biology* 283, no. 2: 349–65.

Rayfield, Emily J. 2007. "Finite element analysis and understanding the biomechanics and evolution of living and fossil organisms." *Annual Review of Earth and Planetary Sciences* 35: 541–76.

Rayfield, Emily J. 2011. "Structural performance of tetanuran theropod skulls, with emphasis on the Megalosauridae, Spinosauridae and Carcharodontosauridae." *Special Papers in Palaeontology* 86: 241–53.

Rayfield, Emily J., David B. Norman, Celeste C. Horner, John R. Horner, Paula May Smith, Jeffrey J. Thomason, and Paul Upchurch. 2001. "Cranial design and function in a large theropod dinosaur." *Nature* 409, no. 6823: 1033–37.

Rayfield, Emily J., David B. Norman, Celeste C. Horner, John R. Horner, Paula May Smith, Jeffrey J. Thomason, and Paul Upchurch. 2002. Reply to: Frazzetta, T. H., and Kenneth V. Kardong. "Prey attack by a large theropod dinosaur." *Nature* 416, no. 6879: 388.

Rayfield, Emily J., Angela C. Milner, Viet Bui Xuan, and Philippe G. Young. 2007. "Functional morphology of spinosaur 'crocodile-mimic' dinosaurs." *Journal of Vertebrate Paleontology* 27, no. 4: 892–901.

Rees, Jan, and Johan Lindgren. 2005. "Aquatic birds from the Upper Cretaceous (Lower Campanian) of Sweden and the biology and distribution of hesperornithiforms." *Palaeontology* 48, no. 6: 1321–29.

Rega, Elizabeth, R. B. Holmes, and Alex Tirabasso. 2010. "Habitual locomotor behavior inferred from manual pathology in two Late Cretaceous chasmosaurine ceratopsid dinosaurs, *Chasmosaurus irvinensis* (CMN 41357) and *Chasmosaurus belli* (ROM 843)." In: Michael J. Ryan, Brenda J. Chinnery-Allgeier, and David E. Eberth, eds. *New perspectives on horned dinosaurs: the Royal Tyrrell Museum Ceratopsian Symposium*. Bloomington: Indiana University Press, 340–54.

Reichel, Miriam. 2010a. "The heterodonty of *Albertosaurus sarcophagus* and *Tyrannosaurus rex*: biomechanical implications inferred through 3-D models." *Canadian Journal of Earth Sciences* 47, no. 9: 1253–61.

Reichel, Miriam. 2010b. "A model for the bite mechanics in the herbivorous dinosaur *Stegosaurus* (Ornithischia, Stegosauridae)." *Swiss Journal of Geosciences* 103, no. 2: 235–40.

Reiss, Stefan, and Heinrich Mallison. 2014. "Motion range of the manus of *Plateosaurus engelhardti* von Meyer, 1837." *Palaeontologia Electronica* 17, no. 1: 1–19.

Reisz, Robert R., Diane Scott, Hans-Dieter Sues, David C. Evans, and Michael A. Raath. 2005. "Embryos of an Early Jurassic prosauropod dinosaur and their evolutionary significance." *Science* 309, no. 5735: 761–64.

Reisz, Robert R., Aaron R. H. LeBlanc, Hillary C. Maddin, Thomas W. Dudgeon, Diane Scott, Timothy Huang, Jun Chen, Chuan-Mu Chen, and Shiming Zhong. 2020. "Early Jurassic dinosaur fetal dental development and its significance for the evolution of sauropod dentition." *Nature Communications* 11, no. 1: 1–9.

Rhodes, Matthew M., and Philip J. Currie. 2020. "The homology, form, and function of the microraptorine lateral pubic tubercle." *Journal of Vertebrate Paleontology* 40, no. 1: e1755866.

Rhodes, Matthew M., Gregory F. Funston, and Philip J. Currie. 2020. "New material reveals the pelvic morphology of Caenagnathidae (Theropoda, Oviraptorosauria)." *Cretaceous Research* 114: 104521.

Rhodes, Matthew M., Donald M. Henderson, and Philip J. Currie. 2021. "Maniraptoran pelvic musculature highlights evolutionary patterns in theropod locomotion on the line to birds." *PeerJ* 9: e10855.

Rivera-Sylva, Héctor E., Christina I. Barrón-Ortízb, Rosalba Lizbeth Nava Rodríguez, José Rubén Guzmán-Gutiérreza, Fernando Cabral Valdez, and Claudio de León Dávila. 2019. "Preliminary assessment of hadrosaur dental microwear from the Cerro del Pueblo Formation (Upper Cretaceous: Campanian) of Coahuila, northeastern Mexico." *Paleontología Mexicana* 8, no. 1: 17–28.

Roach, Brian T., and Daniel L. Brinkman. 2007. "A reevaluation of cooperative pack hunting and gregariousness in *Deinonychus antirrhopus* and other nonavian theropod dinosaurs." *Bulletin of the Peabody Museum of Natural History* 48, no. 1: 103–38.

Rogers, Raymond R., David W. Krause, and Kristina Curry Rogers. 2003. "Cannibalism in the Madagascan dinosaur *Majungatholus atopus*." *Nature* 422, no. 6931: 515–18.

Romer, Alfred S. 1923. "Crocodilian pelvic muscles and their avian and reptilian homologues." *Bulletin of the American Museum of Natural History* 48: 533–52.

Romer, Alfred S. 1927. "The pelvic musculature of ornithischian dinosaurs." *Acta Zoologica* 8, nos. 2-3: 225–75.

Rowe, Andre J., and Eric Snively. 2021. "Biomechanics of juvenile tyrannosaurid mandibles and their implications for bite force: evolutionary biology." *Anatomical Record* 305, no. 2: 373–92.

Rowe, T., and J. Gauthier. 1990. "Ceratosauria." In: D. T. Weishampel, P. Dodson, and H. Osmólska, eds. *The dinosauria*, 2nd ed. Berkeley: University of California Press, 151–68.

Russell, Dale A. 1970. "A skeletal reconstruction of *Leptoceratops gracilis* from the upper Edmonton Formation (Cretaceous) of Alberta." *Canadian Journal of Earth Sciences* 7, no. 1: 181–84.

Russell, Dale A. 1972. "Ostrich dinosaurs from the Late Cretaceous of western Canada." *Canadian Journal of Earth Sciences* 9, no. 4: 375–402.

Russell, L. S. 1935. "Musculature and function in the Ceratopsia." *Natural Museum of Canada Bulletin* 77: 39–48.

Russell, L. S. 1940. *Edmontonia rugosidens (Gilmore): an armoured dinosaur from the Belly River series of Alberta.* Toronto: University of Toronto Press.

Rybczynski, N., and M. K. Vickaryous. 2001. "Evidence of complex jaw movement in the Late Cretaceous." In: K. Carpenter, ed. *The armored dinosaurs*. Bloomington: Indiana University Press, 299–317.

Rybczynski, N., Alex Tirabasso, Paul Bloskie, Robin Cuthbertson, and Casey Holliday. 2008. "A three-dimensional animation model of *Edmontosaurus* (Hadrosauridae) for testing chewing hypotheses." *Palaeontologia Electronica* 11, no. 2: 9A.

Sakagami, Rina, and Soichiro Kawabe. 2020. "Endocranial anatomy of the ceratopsid dinosaur *Triceratops* and interpretations of sensory and motor function." *PeerJ* 8: e9888.

Sakamoto, Manabu. 2010. "Jaw biomechanics and the evolution of biting performance in theropod dinosaurs." *Proceedings of the Royal Society B: Biological Sciences* 277, no. 1698: 3327–33.

Sakamoto, Manabu. 2021. "Assessing bite force estimates in extinct mammals and archosaurs using phylogenetic predictions." *Palaeontology* 64: 743–53.

Salgado, Leonardo, and Jorge O. Calvo. 1992. "Cranial osteology of *Amargasaurus cazaui* Salgado and Bonaparte (Sauropoda, Dicraeosauridae) from the Neocomian of Patagonia." *Ameghiniana* 29, no. 4: 337–46.

Salgado, Leonardo, Rodolfo A. Coria, and Jorge O. Calvo. 1997. "Evolution of titanosaurid sauropods: phytogenetic analysis based on the postcranial evidence." *Ameghiniana* 34, no. 1: 3–32.

Samman, Tanya. 2013. "Tyrannosaurid craniocervical mobility: a preliminary qualitative assessment." In: *Tyrannosaurid palaeobiology*. Bloomington: Indiana University Press, 195–210.

Sampson, Scott D. 1993. "Cranial ornamentations in ceratopsid dinosaurs: systematic, behavioural, and evolutionary implications." PhD diss., University of Toronto.

Sampson, Scott D., and Lawrence M. Witmer. 2007. "Craniofacial anatomy of *Majungasaurus crenatissimus* (Theropoda: Abelisauridae) from the late Cretaceous of Madagascar." *Journal of Vertebrate Paleontology* 27, no. S2: 32–104.

Sampson, Scott D., Matthew T. Carrano, and Catherine A. Forster. 2001. "A bizarre predatory dinosaur from the Late Cretaceous of Madagascar." *Nature* 409, no. 6819: 504–6.

Sampson, Scott D., M. A. Loewen, A. A. Farke, E. M. Roberts, C. A. Forster, et al. 2010. "New horned dinosaurs from Utah provide evidence for intracontinental dinosaur endemism." *PLoS One* 5, no. 9: e12292.

Sander, P. M. 1999. "Life history of the Tendaguru sauropods as inferred from long bone histology." *Mitteilungen aus dem Museum für Naturkunde der Humboldt-Universität Berlin, Geowissenschaftliche Reihe* 2: 103–12.

Sander, P. Martin, Carole T. Gee, Jürgen Hummel, and Marcus Clauss. 2010. "Mesozoic plants and dinosaur herbivory." In: Carol T. Gee, ed. *Plants in Mesozoic time: morphological innovations, phylogeny, ecosystems*. Bloomington: Indiana University Press, 331–59.

Sander, P. Martin, Andreas Christian, Marcus Clauss, Regina Fechner, Carole T. Gee, et al. 2011. "Biology of the sauropod dinosaurs: the evolution of gigantism." *Biological Reviews* 86, no. 1: 117–55.

Sanders, F., K. Manley, and K. Carpenter. 2001. "Gastroliths from the Lower Cretaceous sauropod *Cedarosaurus weiskopfae*." In: D. H. Tanke and K. Carpenter, eds. *Mesozoic vertebrate life*. Bloomington: Indiana University Press, 166–80.

Sattler, Franziska, and Daniela Schwarz. 2021. "Tooth replacement in a specimen of *Tyrannosaurus rex* (Dinosauria,

Theropoda) from the Hell Creek Formation (Maastrichtian), Montana." *Historical Biology* 33, no. 7: 949–72.

Schachner, Emma R., Tyler R. Lyson, and Peter Dodson. 2009. "Evolution of the respiratory system in nonavian theropods: evidence from rib and vertebral morphology." *Anatomical Record* 292, no. 9: 1501–13.

Schade, Marco, Oliver W. M. Rauhut, and Serjoscha W. Evers. 2020. "Neuroanatomy of the spinosaurid *Irritator challengeri* (Dinosauria: Theropoda) indicates potential adaptations for piscivory." *Scientific Reports* 10, no. 1: 1–9.

Schaeffer, Joep, Michael J. Benton, Emily J. Rayfield, and Thomas L. Stubbs. 2020. "Morphological disparity in theropod jaws: comparing discrete characters and geometric morphometrics." *Palaeontology* 63, no. 2: 283–99.

Schroeder, Katlin, S. Kathleen Lyons, and Felisa A. Smith. 2021. "The influence of juvenile dinosaurs on community structure and diversity." *Science* 371, no. 6532: 941–44.

Schubert, Blaine W., and Peter S. Ungar. 2005. "Wear facets and enamel spalling in tyrannosaurid dinosaurs." *Acta Palaeontologica Polonica* 50, no. 1.

Schumacher, G. H. 1973. "The head muscles and hyolaryngeal skeleton of turtles and crocodilians. In: C. Gans, ed. *Biology of the reptilia*. Vol. 4. London: Academic Press.

Schwarz, Daniela, Eberhard Frey, and Christian A. Meyer. 2007a. "Pneumaticity and soft-tissue reconstructions in the neck of diplodocid and dicraeosaurid sauropods." *Acta Palaeontologica Polonica* 52, no. 1.

Schwarz, Daniela, Eberhard Frey, and Christian A. Meyer. 2007b. "Novel reconstruction of the orientation of the pectoral girdle in sauropods." *Anatomical Record* 290, no. 1: 32–47.

Schwarz-Wings, Daniela. 2009. "Reconstruction of the thoracic epaxial musculature of diplodocid and dicraeosaurid sauropods." *Journal of Vertebrate Paleontology* 29, no. 2: 517–34.

Schwarz-Wings, Daniela, and Eberhard Frey. 2008. "Is there an option for a pneumatic stabilization of sauropod necks? An experimental and anatomical approach." *Palaeontologia Electronica* 11, no. 3: 17A.

Schwarz-Wings, Daniela, C. A. Meyer, E. Frey, H. R. Manz-Steiner, and R. Schumacher. 2010. Mechanical implications of pneumatic neck vertebrae in sauropod dinosaurs. *Proceedings of the Royal Society B: Biological Sciences* 277, no. 1678: 11–17. https://doi.org/10.1098/rspb.2009.1275.

Sciscio, Lara, Fabien Knoll, Emese M. Bordy, Michiel O. de Kock, and Ragna Redelstorff. 2017. "Digital reconstruction of the mandible of an adult Lesothosaurus diagnosticus with insight into the tooth replacement process and diet." *PeerJ* 5: e3054.

Seebacher, Frank. 2001. "A new method to calculate allometric length-mass relationships of dinosaurs." *Journal of Vertebrate Paleontology* 21, no. 1: 51–60.

Sellers, William Irvin, and Phillip Lars Manning. 2007. "Estimating dinosaur maximum running speeds using evolutionary robotics." *Proceedings of the Royal Society B: Biological Sciences* 274, no. 1626: 2711–16.

Sellers, William Irvin, Phillip Lars Manning, T. Lyson, K. Stevens, and L. Margetts. 2009. "Virtual palaeontology: gait reconstruction of extinct vertebrates using high performance computing." *Palaeontologia Electronica* 12: 11A.

Sellers, William Irvin, Lee Margetts, Rodolfo Anibal Coria, and Phillip Lars Manning. 2013. "March of the titans: the locomotor capabilities of sauropod dinosaurs." *PLoS One* 8, no. 10: e78733.

Sellers, William Irvin, Stuart B. Pond, Charlotte A. Brassey, Philip L. Manning, and Karl T. Bates. 2017. "Investigating the running abilities of *Tyrannosaurus rex* using stress-constrained multibody dynamic analysis." *PeerJ* 5: e3420.

Senter, Phil. 2006. "Comparison of forelimb function between *Deinonychus* and *Bambiraptor* (Theropoda: Dromaeosauridae)." *Journal of Vertebrate Paleontology* 26, no. 4: 897–906.

Senter, Phil. 2007. "Necks for sex: sexual selection as an explanation for sauropod dinosaur neck elongation." *Journal of Zoology* 271, no. 1: 45–53.

Senter, Phil, and Sara L. Juengst. 2016. "Record-breaking pain: the largest number and variety of forelimb bone maladies in a theropod dinosaur." *PLoS One* 11, no. 2: e0149140.

Senter, Phil, and James H. Robins. 2005. Range of motion in the forelimb of the theropod dinosaur *Acrocanthosaurus atokensis*, and implications for predatory behaviour. *Journal of Zoology* 266: 307–18.

Senter, Phil, and Corwin Sullivan. 2019. "Forelimbs of the theropod dinosaur *Dilophosaurus wetherilli*: range of motion, influence of paleopathology and soft tissues, and description of a distal carpal bone." *Palaeontologia Electronica* 22, no. 2: 30.

Sereno, Paul C. 1986. "Phylogeny of the bird-hipped dinosaurs (Order Ornithischia)." *National Geographic Research* 2: 234–56.

Sereno, Paul C. 1987. "The ornithischian dinosaur *Psittacosaurus* from the Lower Cretaceous of Asia and the relationships of the Ceratopsia." PhD diss., Columbia University.

Sereno, Paul C. 1991. "*Lesothosaurus*, 'fabrosaurids,' and the early evolution of Ornithischia." *Journal of Vertebrate Paleontology* 11, no. 2: 168–97.

Sereno, Paul C. 1997. "The origin and evolution of dinosaurs." *Annual Review of Earth and Planetary Sciences* 25, no. 1: 435–89.

Sereno, Paul C. 2012. "Taxonomy, morphology, masticatory

function and phylogeny of heterodontosaurid dinosaurs." *ZooKeys* 226: 1–225.

Sereno, Paul C., and Fernando E. Novas. 1993. "The skull and neck of the basal theropod *Herrerasaurus ischigualastensis.*" *Journal of Vertebrate Paleontology* 13, no. 4: 451–76.

Sereno, Paul C., and Jeffrey A. Wilson. 2005. "Structure and evolution of a sauropod tooth battery." In: Kristina Curry Rogers and Jeffrey Wilson, eds. *The sauropods: evolution and paleobiology.* Berkeley: University of California Press.

Sereno, Paul C., and Dong Zhimin. 1992. "The skull of the basal stegosaur *Huayangosaurus taibaii* and a cladistic diagnosis of Stegosauria." *Journal of Vertebrate Paleontology* 12, no. 3: 318–43.

Sereno, Paul C., Catherine A. Forster, Raymond R. Rogers, and Alfredo M. Monetta. 1993. "Primitive dinosaur skeleton from Argentina and the early evolution of Dinosauria." *Nature* 361, no. 6407: 64–66.

Sereno, Paul C., Allison L. Beck, Didier B. Dutheil, Boubacar Gado, Hans C. E. Larsson, et al. 1998. "A long-snouted predatory dinosaur from Africa and the evolution of spinosaurids." *Science* 282, no. 5392: 1298–302.

Sereno, Paul C., Jeffrey A. Wilson, Lawrence M. Witmer, John A. Whitlock, Abdoulaye Maga, Oumarou Ide, and Timothy A. Rowe. 2007. "Structural extremes in a Cretaceous dinosaur." *PLoS One* 2, no. 11: e1230.

Sereno, Paul C., Zhao Xijin, and Tan Lin. 2010. "A new psittacosaur from inner Mongolia and the parrot-like structure and function of the psittacosaur skull." *Proceedings of the Royal Society B: Biological Sciences* 277, no. 1679: 199–209.

Sereno, Paul C., Ricardo N. Martínez, and Oscar A. Alcober. 2013. "Osteology of *Eoraptor lunensis* (Dinosauria, Sauropodomorpha)." *Journal of Vertebrate Paleontology* 32, no. S1: 83–179.

Sereno, Paul C., Nathan Myhrvold, Donald M. Henderson, Frank E. Fish, Daniel Vidal, Stephanie L. Baumgart, Tyler M. Keillor, Kiersten K. Formoso, and Lauren L. Conroy. 2022. "*Spinosaurus* is not an aquatic dinosaur." *bioRxiv* 1–47. https://doi.org/10.1101/2022.05.25.493395.

Serrano, Francisco José, and Luis María Chiappe. 2017. "Aerodynamic modelling of a Cretaceous bird reveals thermal soaring capabilities during early avian evolution." *Journal of the Royal Society Interface* 14, no. 132: 20170182.

Seymour, Roger S. 2009. "Raising the sauropod neck: it costs more to get less." *Biology Letters* 5, no. 3: 317–19.

Seymour, Roger S., and Harvey B. Lillywhite. 2000. "Hearts, neck posture and metabolic intensity of sauropod dinosaurs." *Proceedings of the Royal Society B: Biological Sciences* 267, no. 1455: 1883–87.

Singh, Suresh A., Armin Elsler, Thomas L. Stubbs, Russell

Bond, Emily J. Rayfield, and Michael J. Benton. 2021. "Niche partitioning shaped herbivore macroevolution through the early Mesozoic." *Nature Communications* 12, no. 1: 1–13.

Siviero, Bethania C. T., Elizabeth Rega, William K. Hayes, Allen M. Cooper, Leonard R. Brand, and Art V. Chadwick. 2020. "Skeletal trauma with implications for intratail mobility in *Edmontosaurus annectens* from a monodominant bonebed, Lance Formation (Maastrichtian), Wyoming USA." *Palaios* 35, no. 4: 201–14.

Skutschas, Pavel P., Vera A. Gvozdkova, Alexander O. Averianov, Alexey V. Lopatin, Thomas Martin, et al. 2021. "Wear patterns and dental functioning in an Early Cretaceous stegosaur from Yakutia, eastern Russia." *PLoS One* 16, no. 3: e0248163.

Słowiak, Justyna, Victor S. Tereshchenko, and Łucja Fostowicz-Frelik. 2019. "Appendicular skeleton of *Protoceratops andrewsi* (Dinosauria, Ornithischia): comparative morphology, ontogenetic changes, and the implications for non-ceratopsid ceratopsian locomotion." *PeerJ* 7: e7324.

Słowiak, Justyna, Tomasz Szczygielski, Michał Ginter, and Łucja Fostowicz-Frelik. 2020. "Uninterrupted growth in a non-polar hadrosaur explains the gigantism among duck-billed dinosaurs." *Palaeontology* 63, no. 4: 579–99.

Smith, David K. 1990. "Osteology of *Oviraptor philoceratops*, a possible herbivorous theropod from the Upper Cretaceous of Mongolia." *Journal of Vertebrate Paleontology* 10, no. S3: 42A.

Smith, David K. 1992. "The type specimen of *Oviraptor philoceratops*, a theropod dinosaur from the Upper Cretaceous of Mongolia." *Neues Jahrbuch für Geologie und Paläontologie, Abhandlungen* 186: 365–88.

Smith, David K. 1993. "The type specimen of *Oviraptor philoceratops*, a theropod dinosaur from the Upper Cretaceous of Mongolia." *Neues Jahrbuch für Geologie und Paläontologie Abhandlungen* 186, no. 3: 365–88.

Smith, David K. 2015. "Craniocervical myology and functional morphology of the small-headed therizinosaurian theropods *Falcarius utahensis* and *Nothronychus mckinleyi.*" *PLoS One* 10, no. 2: e0117281.

Smith, David K. 2021a. "Forelimb musculature and function in the therizinosaur Nothronychus (Maniraptora, Theropoda)." *Journal of Anatomy* 239, no. 2: 307–35.

Smith, David K. 2021b. "Hind limb muscle reconstruction in the incipiently opisthopubic large therizinosaur *Nothronychus* (Theropoda; Maniraptora)." *Journal of Anatomy* 238, no. 6: 1404–24.

Smith, Joshua B. 2005. "Heterodonty in *Tyrannosaurus rex*: implications for the taxonomic and systematic utility of theropod dentitions." *Journal of Vertebrate Paleontology* 25, no. 4: 865–87.

Smithwick, Fiann M., Robert Nicholls, Innes C. Cuthill, and Jakob Vinther. 2017. "Countershading and stripes in the theropod dinosaur *Sinosauropteryx* reveal heterogeneous habitats in the Early Cretaceous Jehol Biota." *Current Biology* 27, no. 21: 3337–43.

Snively, Eric, and Anthony P. Russell. 2003. "Kinematic model of tyrannosaurid (Dinosauria: Theropoda) arctometatarsus function." *Journal of Morphology* 255, no. 2: 215–27.

Snively, Eric, and Anthony P. Russell. 2007a. "Functional variation of neck muscles and their relation to feeding style in Tyrannosauridae and other large theropod dinosaurs." *Anatomical Record* 290, no. 8: 934–57.

Snively, Eric, and Anthony P. Russell. 2007b. "Functional morphology of neck musculature in the Tyrannosauridae (Dinosauria, Theropoda) as determined via a hierarchical inferential approach." *Zoological Journal of the Linnean Society* 151, no. 4: 759–808.

Snively, Eric, and Anthony P. Russell. 2007c. "Craniocervical feeding dynamics of *Tyrannosaurus rex*." *Paleobiology* 33, no. 4: 610–38.

Snively, Eric, Anthony P. Russell, and G. L. Powell. 2004. "Evolutionary morphology of the coelurosaurian arctometatarsus: descriptive, morphometric and phylogenetic approaches." *Zoological Journal of the Linnean Society* 142: 525–53.

Snively, Eric, Donald M. Henderson, and Doug S. Phillips. 2006. "Fused and vaulted nasals of tyrannosaurid dinosaurs: implications for cranial strength and feeding mechanics." *Acta Palaeontologica Polonica* 51, no. 3: 435–54.

Snively, Eric, John R. Cotton, Ryan Ridgely, and Lawrence M. Witmer. 2013. "Multibody dynamics model of head and neck function in *Allosaurus* (Dinosauria, Theropoda)." *Palaeontologia Electronica* 16, no. 2: 11A.

Snively, Eric, A. P. Russell, G. L. Powell, J. M. Theodor, and M. J. Ryan. 2014. "The role of the neck in the feeding behaviour of the Tyrannosauridae: inference based on kinematics and muscle function of extant avians." *Journal of Zoology* 292, no. 4: 290–303.

Snively, Eric, Haley O'Brien, Donald M. Henderson, Heinrich Mallison, Lara A. Surring, et al. 2019. "Lower rotational inertia and larger leg muscles indicate more rapid turns in tyrannosaurids than in other large theropods." *PeerJ* 7: e6432.

Söderblom, Fredrik. 2017. "Morphological variation in the hadrosauroid dentary." Master's thesis, Uppsala University.

Solounias, Nikos, and Sonja M. C. Moelleken. 1992. "Dietary adaptations of two goat ancestors and evolutionary considerations." *Geobios* 25, no. 6: 797–809.

Solounias, Nikos, and Sonja M. C. Moelleken. 1993. "Dietary adaptation of some extinct ruminants determined by premaxillary shape." *Journal of Mammalogy* 74, no. 4: 1059–971.

Solounias, Nikos, Mark Teaford, and Alan Walker. 1988. "Interpreting the diet of extinct ruminants: the case of a non-browsing giraffid." *Paleobiology* 14, no. 3: 287–300.

Sonkusare, Hemant, Bandana Samant, and D. M. Mohabey. 2017. "Microflora from sauropod coprolites and associated sediments of Late Cretaceous (Maastrichtian) Lameta Formation of Nand-Dongargaon basin, Maharashtra." *Journal of the Geological Society of India* 89, no. 4: 391–97.

Starck, D., and A. Barnikol. 1954. "Beitrage zur Morphologie der Trigeminusmuskulatur der Vogel (besonders der Accipitres, Cathartidae, Striges und Anseres)." *Morphologisches Jahrbuch* 94: 1–64.

Stevens, Kent A. 2006. "Binocular vision in theropod dinosaurs." *Journal of Vertebrate Paleontology* 26, no. 2: 321–30.

Stevens, Kent A. 2013. "The articulation of sauropod necks: methodology and mythology." *PLoS One* 8, no. 10: e78572.

Stevens, Kent A., and J. Michael Parrish. 1999. "Neck posture and feeding habits of two Jurassic sauropod dinosaurs." *Science* 284, no. 5415: 798–800.

Stevens, Kent A., and J. Michael Parrish. 2000. Response to: Upchurch, P. "Neck posture in sauropod dinosaurs." *Science* 287, no. 5453: 547.

Stevens, Kent A., and J. Michael Parrish. 2005a. "Neck posture, dentition, and feeding strategies in Jurassic sauropod dinosaurs." In: Virginia Tidwell and Kenneth Carpenter, eds. *Thunder-lizards: the sauropodomorph dinosaurs.* Bloomington: Indiana University Press, 212–32.

Stevens, Kent A., and J. Michael Parrish. 2005b. "Digital reconstructions of sauropod dinosaurs and implications for feeding." In: Kristina Curry Rogers and Jeffrey Wilson, eds. *The sauropods: evolution and paleobiology.* Berkeley: University of California Press, 178–200.

Stokosa, Kathy. 2005. "Enamel microstructure variation within." In: K. Carpenter, ed. *The carnivorous dinosaurs.* Bloomington: Indiana University Press, 163.

Straight, William H., Jonathan D. Karr, Julia E. Cox, and Reese E. Barrick. 2004. "Stable oxygen isotopes from theropod dinosaur tooth enamel: interlaboratory comparison of results and analytical interference by reference standards." *Rapid Communications in Mass Spectrometry* 18, no. 23: 2897–903.

Strickson, Edward, Albert Prieto-Márquez, Michael J. Benton, and Thomas L. Stubbs. 2016. "Dynamics of dental evolution in ornithopod dinosaurs." *Scientific Reports* 6, no. 1: 1–11.

Stubbs, Thomas L., Michael J. Benton, Armin Elsler, and Albert Prieto-Márquez. 2019. "Morphological innovation and the evolution of hadrosaurid dinosaurs." *Paleobiology* 45, no. 2: 347–62.

Suarez, Celina A., Hai-Lu You, Marina B. Suarez, Da-Qing Li, and J. B. Trieschmann. 2017. "Stable isotopes reveal rapid enamel elongation (amelogenesis) rates for the Early Cretaceous iguanodontian dinosaur *Lanzhousaurus magnidens*." *Scientific Reports* 7, no. 1: 1–8.

Sues, Hans-Dieter. 1980. "Anatomy and relationships of a new hypsilophodontid dinosaur from the Lower Cretaceous of North America." *Palaeontographica* A169, nos. 1–3: 51–72.

Sues, Hans-Dieter, and Peter M. Galton. 1987. "Anatomy and classification of the North American Pachycephalosauria (Dinosauria: Ornithischia)." *Palaeontographica, Abteilung A* 178: 1–40.

Sues, Hans-Dieter, Eberhard Frey, David M. Martill, and Diane M. Scott. 2002. "*Irritator challengeri*, a spinosaurid (Dinosauria: Theropoda) from the Lower Cretaceous of Brazil." *Journal of Vertebrate Paleontology* 22, no. 3: 535–47.

Sullivan, Corwin, and Xing Xu. 2017. "Morphological diversity and evolution of the jugal in dinosaurs." *Anatomical Record* 300, no. 1: 30–48.

Taggart, R. E., A. T. Cross, D. L. Wolberg, E. Stump, and G. Rosenberg. 1997. "The relationship between land plant diversity and productivity and patterns of dinosaur herbivory." In *Dinofest international: proceedings of a symposium sponsored by Arizona State University*, vol. 403. Philadelphia: Academy of Natural Sciences, 416.

Tahara, Rui, and Hans C. E. Larsson. 2011. "Cranial pneumatic anatomy of *Ornithomimus edmontonicus* (Ornithomimidae: Theropoda)." *Journal of Vertebrate Paleontology* 31, no. 1: 127–43.

Tait, John, and Barnum Brown. 1928. *How the Ceratopsia carried and used their head*. Montreal: McGill University Press.

Tanoue, Kyo, Hai-Lu You, and Peter Dodson. 2009a. "Comparative anatomy of selected basal ceratopsian dentitions." *Canadian Journal of Earth Sciences* 46, no. 6: 425–39.

Tanoue, Kyo, Barbara S. Grandstaff, Hai-Lu You, and Peter Dodson. 2009b. "Jaw mechanics in basal Ceratopsia (Ornithischia, Dinosauria)." *Anatomical Record* 292, no. 9: 1352–69.

Tanoue, Kyo, H. L. You, and Peter Dodson. 2010. "Mandibular anatomy in basal Ceratopsia." In *New perspectives on horned dinosaurs: the Royal Tyrell Museum Ceratopsian Symposium*. Bloomington: Indiana University Press, 234–50.

Taquet, P. 1976. "Ostéologie d'Ouranosaurus nigeriensis, Iguanodontide du Crétacé Inférieur du Niger." In: *Géologie et Paléontologie du Gisement de Gadoufaoua (Aptien du Niger)*. Paris: Editions Du Centre National De La Recherche Scientifique, 57–168.

Taylor, Adam C., Stephan Lautenschlager, Zhao Qi, and Emily J. Rayfield. 2017. "Biomechanical evaluation of different musculoskeletal arrangements in *Psittacosaurus* and implications for cranial function." *Anatomical Record* 300, no. 1: 49–61.

Taylor, Michael P. 2014. "Quantifying the effect of intervertebral cartilage on neutral posture in the necks of sauropod dinosaurs." *PeerJ* 2: e712.

Taylor, Michael P., and Mathew J. Wedel. 2013a. "Why sauropods had long necks; and why giraffes have short necks." *PeerJ* 1: e36.

Taylor, Michael P., and Mathew J. Wedel. 2013b. "The effect of intervertebral cartilage on neutral posture and range of motion in the necks of sauropod dinosaurs." *PLoS One* 8, no. 10: e78214.

Taylor, Michael P., Mathew J. Wedel, and Darren Naish. 2009. "Head and neck posture in sauropod dinosaurs inferred from extant animals." *Acta Palaeontologica Polonica* 54, no. 2: 213–20.

Teaford, Mark F. 1994. "Dental microwear and dental function." *Evolutionary Anthropology: Issues, News, and Reviews* 3, no. 1: 17–30.

Teaford, Mark F. 1988. "A review of dental microwear and diet in modern mammals." *Scanning Microscopy* 2, no. 2: 1149–66.

Tereshchenko, V. S. 1996. "A reconstruction of the locomotion of *Protoceratops*." *Paleontological Journal C/C of Paleontologicheskii Zhurnal* 30: 232–45.

Tereschenko, V. S. 2004. "On the heterocelous vertebrae in horned dinosaurs (Protoceratopidae, Neoceratopsia)." *Paleontological Journal* 38, no. 2: 200–205.

Tereschenko, V. S. 2008. "Adaptive features of protoceratopoids (Ornithischia: Neoceratopsia)." *Paleontological Journal* 42, no. 3: 273–86.

Therrien, F., D. M. Henderson, and C. B. Ruff. 2005. "Bite me: biomechanical models of theropod mandibles and implications for feeding behavior." In: K. Carpenter, ed. *The carnivorous dinosaurs*, 179–237. Bloomington: Indiana University Press.

Therrien, F., Darla K. Zelenitsky, Jared T. Voris, and Kohei Tanaka. 2021. "Mandibular force profiles and tooth morphology in growth series of *Albertosaurus sarcophagus* and *Gorgosaurus libratus* (Tyrannosauridae: Albertosaurinae) provide evidence for an ontogenetic dietary shift in tyrannosaurids." *Canadian Journal of Earth Sciences* 58, no. 9: 812–28.

Thompson, Stefan, and Robert Holmes. 2007. "Forelimb stance and step cycle in *Chasmosaurus irvinensis* (Dinosauria: Neoceratopsia)." *Palaeontologia Electronica* 10, no. 1: 1–17.

Thulborn, Richard A. 1971. "Tooth wear and jaw action in the Triassic ornithischian dinosaur Fabrosaums." *Journal of Zoology* 164, no. 2: 165–79.

Thulborn, Richard A. 1972. "The post-cranial skeleton of the Triassic ornithischian dinosaur *Fabrosaurus australis*." *Palaeontology* 15: 1–32.

Thulborn, Richard A. 1974. "Thegosis in herbivorous dinosaurs." *Nature* 250, no. 5469: 729–31.

Thulborn, Richard A. 1978. "Aestivation among ornithopod dinosaurs of the African Trias." *Lethaia* 11, no. 3: 185–98.

Thulborn, Richard A. 1991. "Morphology, preservation and palaeobiological significance of dinosaur copralites." *Palaeogeography, Palaeoclimatology, Palaeoecology* 83, no. 4: 341–66.

Tiffney, Bruce H. 1992. "The role of vertebrate herbivory in the evolution of land plants." *Palaeobotanist* 41: 87–97.

Tiffney, Bruce H. 1997. "Land plants as food and habitat." In: Brett-Surman, Michael K., Thomas R. Holtz, and James O. Farlow, eds. *The complete dinosaur*. Bloomington: Indiana University Press, 352–70.

Tiffney, Bruce H. 2004. "Vertebrate dispersal of seed plants through time." *Annual Review of Ecology, Evolution, and Systematics* 35: 1–29.

Torices, Angelica, Ryan Wilkinson, Victoria M. Arbour, Jose Ignacio Ruiz-Omeñaca, and Philip J. Currie. 2018. "Puncture-and-pull biomechanics in the teeth of predatory coelurosaurian dinosaurs." *Current Biology* 28, no. 9: 1467–74.

Tsai, Henry P., and Casey M. Holliday. 2015. "Articular soft tissue anatomy of the archosaur hip joint: structural homology and functional implications." *Journal of Morphology* 276: 601–30.

Tsai, Henry P., Kevin M. Middleton, John R. Hutchinson, and Casey M. Holliday. 2018. "Hip joint articular soft tissues of non-dinosaurian Dinosauromorpha and early Dinosauria: evolutionary and biomechanical implications for Saurischia." *Journal of Vertebrate Paleontology* 38, no. 1: e1427593.

Tsai, Henry P., Kevin M. Middleton, John R. Hutchinson, and Casey M. Holliday. 2020. "More than one way to be a giant: convergence and disparity in the hip joints of saurischian dinosaurs." *Evolution* 74, no. 8: 1654–81.

Tschopp, Emanuel, Octávio Mateus, and Mark Norell. 2018. "Complex overlapping joints between facial bones allowing limited anterior sliding movements of the snout in diplodocid sauropods." *American Museum Novitates* 3911: 1–16.

Tsuihiji, Takanobu. 2004. "The ligament system in the neck of *Rhea americana* and its implication for the bifurcated neural spines of sauropod dinosaurs." *Journal of Vertebrate Paleontology* 24, no. 1: 165–72.

Tsuihiji, Takanobu. 2010. "Reconstructions of the axial muscle insertions in the occipital region of dinosaurs: evaluations of past hypotheses on Marginocephalia and Tyrannosauridae using the Extant Phylogenetic Bracket approach." *Anatomical Record* 293, no. 8: 1360–86.

Tumanova, T. A. 1983. "The first ankylosaur from the Lower Cretaceous of Mongolia." *Transactions of the Joint Soviet-Mongolian Paleontological Expedition* 24: 110–20.

Tumanova, T. A. 1985. "Skull morphology of the ankylosaur *Shamosaurus scutatus* from the Lower Cretaceous of Mongolia." In: P. Taquet and C. Sudre, eds. *Les Dinosaures de la Chine à la France*. Toulouse: Muséum d'Histoire Naturelle de Toulouse et Muséum d'Histoire Naturelle, Chonqing, 73–79.

Tütken, T. 2011. "The diet of sauropod dinosaurs." In: N. Klein, K. Remes, C. T. Gee, and P. M. Sander, eds. *Biology of the sauropod dinosaurs: understanding the life of giants*. Bloomington: Indiana University Press.

Tweet, Justin S., Karen Chin, Dennis R. Braman, and Nate L. Murphy. 2008. "Probable gut contents within a specimen of *Brachylophosaurus canadensis* (Dinosauria: Hadrosauridae) from the Upper Cretaceous Judith River Formation of Montana." *Palaios* 23, no. 9: 624–35.

Ullmann, Paul V., Matthew F. Bonnan, and Kenneth J. Lacovara. 2017. "Characterizing the evolution of wide-gauge features in stylopodial limb elements of titanosauriform sauropods via geometric morphometrics." *Anatomical Record* 300, no. 9: 1618–35.

Unwin, D. M., Altangerel Perle, and C. N. G. Trueman. 1995. "*Protoceratops* and *Velociraptor* preserved in association: evidence for predatory behaviour in dromaeosaurid dinosaurs?" *Journal of Vertebrate Paleontology* 15: 57–57.

Upchurch, P. 2000. "Neck posture in sauropod dinosaurs." *Science* 287: 547b.

Upchurch, P., and P. M. Barrett. 2000. "The evolution of sauropod feeding mechanisms." In: H.-D. Sues, ed. *The evolution of herbivory in terrestrial vertebrates: perspectives from the fossil record*. Cambridge: Cambridge University Press, 42–78.

Upchurch, P., P. M. Barrett, and P. Dodson. 2004. "Sauropoda." In: D. B. Weishampel, P. Dodson, and H. Osmólska, eds. *The dinosauria*, 2nd ed. Berkeley: University of California Press, 259–322.

Upchurch, P., P. M. Barrett, and Peter M. Galton. 2007. "A phylogenetic analysis of basal sauropodomorph relationships: implications for the origin of sauropod dinosaurs." *Special Papers in Palaeontology* 77: 57.

VanBuren, Collin S., Nicolás E. Campione, and David C. Evans. 2015. "Head size, weaponry, and cervical adaptation: testing craniocervical evolutionary hypotheses in Ceratopsia." *Evolution* 69, no. 7: 1728–44.

Van Valkenburgh, Blaire, and Ralph E. Molnar. 2002. "Dinosaurian and mammalian predators compared." *Paleobiology* 28, no. 4: 527–43.

Varriale, Frank J. 2004. "Dental microwear in *Triceratops* and *Chasmosaurus* and its implication for jaw mechanics in Ceratopsidae." *Journal of Vertebrate Paleontology* 24, no. S3: 124A–125A.

Varriale, Frank J. 2008. "Dental microwear and jaw mechanics in basal neoceratopsians." *Journal of Vertebrate Paleontology* 28 (Suppl. 3): 156A.

Varriale, Frank J. 2011. "Dental microwear and the evolution of mastication in ceratopsian dinosaurs." PhD diss., Johns Hopkins University.

Varriale, Frank J. 2016. "Dental microwear reveals mammal-like chewing in the neoceratopsian dinosaur *Leptoceratops gracilis*." *PeerJ* 4: e2132.

Varricchio, David J. 2001. "Gut contents from a Cretaceous tyrannosaurid: implications for theropod dinosaur digestive tracts." *Journal of Paleontology* 75, no. 2: 401–6.

Versluys, J. 1910. "Streptostylie bei Dinosauriern." *Zoologische Jahrbücher Abteilung für Anatomie und Ontogenie der Tiere* 30: 175–260.

Versluys, J. 1912. "Das Streptostylie-Problem und die Bewegung im Schadel bei Sauropsida." *Zoologische Jahrbücher* (Suppl.) 15: 545–716.

Versluys, J. 1923. "Der Schadel des Skeletts von *Trachodon annectus* im Senckenberg-Museum." *Abhandlungen der Senckenbergischen Naturforschenden Gesellschaft* 38:1–19.

Vickaryous, Matthew K. 2006. "New information on the cranial anatomy of *Edmontonia rugosidens* Gilmore, a Late Cretaceous nodosaurid dinosaur from Dinosaur Provincial Park, Alberta." *Journal of Vertebrate Paleontology* 26, no. 4: 1011–13.

Vickaryous, Matthew K., and Anthony P. Russell. 2003. "A redescription of the skull of *Euoplocephalus tutus* (Archosauria: Ornithischia): a foundation for comparative and systematic studies of ankylosaurian dinosaurs." *Zoological Journal of the Linnean Society* 137, no. 1: 157–86.

Vickaryous, Matthew K., Anthony P. Russell, Philip J. Currie, and Xi-Jin Zhao. 2001. "A new ankylosaurid (Dinosauria: Ankylosauria) from the Lower Cretaceous of China, with comments on ankylosaurian relationships." *Canadian Journal of Earth Sciences* 38, no. 12: 1767–80.

Vidal, Daniel, and Veronica Diez Diaz. 2017. "Reconstructing hypothetical sauropod tails by means of 3D digitization: *Lirainosaurus astibiae* as case study." *Journal of Iberian Geology* 43, no. 2: 293–305.

Vidal, Daniel, Pedro Mocho, Adrián Páramo, José Luis Sanz, and Francisco Ortega. 2020a. "Ontogenetic similarities between giraffe and sauropod neck osteological mobility." *PLoS One* 15, no. 1: e0227537.

Vidal, Daniel, P. Mocho, A. Aberasturi, J. L. Sanz, and F. Ortega.

2020b. "High browsing skeletal adaptations in *Spinophorosaurus* reveal an evolutionary innovation in sauropod dinosaurs." *Scientific Reports* 10, no. 1: 1–10.

Vidal, Luciano da Silva, Paulo Victor Luiz Gomes da Costa Pereira, Sandra Tavares, Stephen L. Brusatte, Lílian Paglarelli Bergqvist, and Carlos Roberto dos Anjos Candeiro. 2020c. "Investigating the enigmatic Aeolosaurini clade: the caudal biomechanics of *Aeolosaurus maximus* (Aeolosaurini/Sauropoda) using the neutral pose method and the first case of protonic tail condition in Sauropoda." *Historical Biology* 3, no. 9: 1836–56.

Virág, Attila, and Attila Ősi. 2017. "Morphometry, microstructure, and wear pattern of neornithischian dinosaur teeth from the Upper Cretaceous Iharkut locality (Hungary)." *Anatomical Record* 300, no. 8: 1439–63.

Voegele, Kristyn K., Paul V. Ullmann, Matthew C. Lamanna, and Kenneth J. Lacovara. 2020. "Appendicular myological reconstruction of the forelimb of the giant titanosaurian sauropod dinosaur *Dreadnoughtus schrani*." *Journal of Anatomy* 237, no. 1: 133–54.

Voegele, Kristyn K., Paul V. Ullmann, Matthew C. Lamanna, and Kenneth J. Lacovara. 2021. "Myological reconstruction of the pelvic girdle and hind limb of the giant titanosaurian sauropod dinosaur *Dreadnoughtus schrani*." *Journal of Anatomy* 238, no. 3: 576–97.

Vullo, Romain, Ronan Allain, and Lionel Cavin. 2016. "Convergent evolution of jaws between spinosaurid dinosaurs and pike conger eels." *Acta Palaeontologica Polonica* 61, no. 4: 825–28.

Walker, C. 1981. "New subclass of birds from the Cretaceous of South America." *Nature* 292: 51–53.

Walker, J. D., J. W. Geissman, S. A. Bowring, and L. E. Babcock. 2018. *Geologic time scale 414*, v.5.0. Boulder, CO: Geological Society of America.

Wang, Chun-Chieh, Yen-Fang Song, Sheng-Rong Song, Qiang Ji, Cheng-Cheng Chiang, et al. 2015. "Evolution and function of dinosaur teeth at ultramicrostructural level revealed using synchrotron transmission x-ray microscopy." *Scientific Reports* 5, no. 1: 1–11.

Wang, Min, Jingmai K. O'Connor, Xing Xu, and Zhonghe Zhou. 2019. "A new Jurassic scansoriopterygid and the loss of membranous wings in theropod dinosaurs." *Nature* 569, no. 7755: 256–59.

Wang, Shuo, Shukang Zhang, Corwin Sullivan, and Xing Xu. 2016. "Elongatoolithid eggs containing oviraptorid (Theropoda, Oviraptorosauria) embryos from the Upper Cretaceous of southern China." *BMC Evolutionary Biology* 16, no. 1: 67.

Wang, Shuo, Josef Stiegler, Romain Amiot, Xu Wang, Guo-hao

Du, James M. Clark, and Xing Xu. 2017. "Extreme ontogenetic changes in a ceratosaurian theropod." *Current Biology* 27, no. 1: 144–48.

Wang, Shuo, Qiyue Zhang, and Rui Yang. 2018. "Reevaluation of the dentary structures of caenagnathid oviraptorosaurs (Dinosauria, Theropoda)." *Scientific Reports* 8, no. 1: 391.

Wang, Xiaoli, Jingmai K. O'Connor, Xiaoting Zheng, Min Wang, Han Hu, and Zhonghe Zhou. 2014. "Insights into the evolution of rachis dominated tail feathers from a new basal enantiornithine (Aves: Ornithothoraces)." *Biological Journal of the Linnean Society* 113, no. 3: 805–19.

Wang, Xuri, Jingmai K. O'Connor, Bo Zhao, Luis M. Chiappe, Chunling Gao, and Xiaodong Cheng. 2010. "New species of *enantiornithes* (Aves: Ornithothoraces) from the Qiaotou Formation in northern Hebei, China." *Acta Geologica Sinica*, Eng. ed. 84, no. 2: 247–56.

Watanabe, Akinobu, Paul M. Gignac, Amy M. Balanoff, Todd L. Green, Nathan J. Kley, and Mark A. Norell. 2019. "Are endocasts good proxies for brain size and shape in archosaurs throughout ontogeny?" *Journal of Anatomy* 234, no. 3: 291–305.

Wedel, Mathew J. 2003a. "The evolution of vertebral pneumaticity in sauropod dinosaurs." *Journal of Vertebrate Paleontology* 23, no. 2: 344–57.

Wedel, Mathew J. 2003b. "Vertebral pneumaticity, air sacs, and the physiology of sauropod dinosaurs." *Paleobiology* 29, no. 2: 243–55.

Wedel, Mathew J. 2005. "Postcranial skeletal pneumaticity in sauropods and its implications for mass estimates." In: Kristina Curry Rogers and Jeffrey Wilson, eds. *The Sauropods: evolution and paleobiology*. Berkeley: University of California Press, 201–28.

Wedel, Mathew J. 2006. "Origin of postcranial skeletal pneumaticity in dinosaurs." *Integrative Zoology* 1, no. 2: 80–85.

Wedel, Mathew J. 2007. "What pneumaticity tells us about 'prosauropods,' and vice versa." *Special Papers in Palaeontology* 77: 207–22.

Wedel, Mathew J. 2009. "Evidence for bird-like air sacs in saurischian dinosaurs." *Journal of Experimental Zoology Part A: Ecological Genetics and Physiology* 311, no. 8: 611–28.

Wedel, Mathew J., R. Kent Sanders, and Frank P. Cuozzo. 2002. *Osteological correlates of cervical musculature in Aves and Sauropoda (Dinosauria: Saurischia), with comments on the cervical ribs of Apatosaurus*. Berkeley: Museum of Paleontology, University of California.

Wegweiser, M., B. Breithaupt, and R. Chapman. 2004. "Attack behavior of tyrannosaurid dinosaur(s): Cretaceous crime scenes, really old evidence, and 'smoking guns.'" *Journal of Vertebrate Paleontology* 24, no. S3: 127A.

Weishampel, David B. 1983. "Hadrosaurid jaw mechanics." *Acta Palaeontologica Polonica* 28, nos. 1–2: 271–80.

Weishampel, David B. 1984a. "Evolution of jaw mechanisms in ornithopod dinosaurs." *Advances in Anatomy, Embryology and Cell Biology* 87: 1–109.

Weishampel, David B. 1984b. "Interactions between Mesozoic plants and vertebrates: fructifications and seed predation." *Neues Jahrbuch für Geologie und Paläontologie* 167, no. 2: 224–50.

Weishampel, David B. 1993. "Beams and machines: modeling approaches to the analysis of skull form and function." In: James Hankens and Brian K. Hall, eds. *The skull*. Chicago: University of Chicago Press.

Weishampel, David B. 1995. "Fossils, function, and phylogeny." In: James Hankens and Brian K. Hall, eds. *The skull*. Vol. 3, *Functional and evolutionary mechanisms*. Chicago: University of Chicago Press, 303–44.

Weishampel, David B. 2004. "Ornithischia." In: D. B. Weishampel, P. Dodson, and H. Osmólska, eds. *The Dinosauria*, 2nd ed. Berkeley: University of California Press, 323–24.

Weishampel, David B., and Ronald E. Heinrich. 1992. "Systematics of hypsilophodontidae and basal Iguanodontia (Dinosauria: Ornithopoda)." *Historical Biology* 6, no. 3: 159–84.

Weishampel, David B., and Coralia-Maria Jianu. 2000. "Plant-eaters and ghost lineages: dinosaurian herbivory revisited." In: H.-D. Sues, ed. *Evolution of herbivory in terrestrial vertebrates: perspectives from the fossil record*. Cambridge: Cambridge University Press, 123–43.

Weishampel, David B., and Coralia-Maria Jianu. 2011. *Transylvanian dinosaurs*. Baltimore: Johns Hopkins University Press.

Weishampel, David B., and D. B. Norman. 1989. "Vertebrate herbivory in the Mesozoic: jaws, plants, and evolutionary metrics." *Special Paper of the Geological Society of America* 238: 87–100.

Weishampel, David B., Peter Dodson, and Halszka Osmólska, eds. 1990. *The Dinosauria*. Berkeley: University of California Press.

Weishampel, David B., Peter Dodson, and Halszka Osmólska, eds. 2004. *The Dinosauria*, 2nd ed. Berkeley: University of California Press.

Welles, Samuel P. 1984. "*Dilophosaurus wetherilli* (Dinosauria, Theropoda): osteology and comparisons." *Palaeontographica. Abteilung A, Paläozoologie, Stratigraphie* 185, no. 4–6: 85–180.

Wellnhofer, Peter. 2008. "*Archaeopteryx*." *Spektrum der Wissenschaft* 8: 100.

White, Paul D., David E. Fastovsky, and Peter M. Sheehan. 1998. "Taphonomy and suggested structure of the dinosaurian assemblage of the Hell Creek Formation (Maastrichtian), eastern Montana and western North Dakota." *Palaios* 13, no. 1: 41–51.

White, Theodore E. 1958. "The braincase of *Camarasaurus lentus* (Marsh)." *Journal of Paleontology* 32, no. 3: 477–94.

Whitlock, John A. 2011. "Inferences of diplodocoid (Sauropoda: Dinosauria) feeding behavior from snout shape and microwear analyses." *PLoS One* 6, no. 4: e18304.

Whitlock, John A. 2017. "Was *Diplodocus* (Diplodocoidea, Sauropoda) capable of propalinal jaw motion?" *Journal of Vertebrate Paleontology* 37, no. 2: e1296457.

Whitlock, John A., Jeffrey A. Wilson, and Matthew C. Lamanna. 2010. "Description of a nearly complete juvenile skull of *Diplodocus* (Sauropoda: Diplodocoidea) from the Late Jurassic of North America." *Journal of Vertebrate Paleontology* 30, no. 2: 442–57.

Whitney, M. R., A. R. H. LeBlanc, A. R. Reynolds, and K. S. Brink. 2020. "Convergent dental adaptations in the serrations of hypercarnivorous synapsids and dinosaurs." *Biology Letters* 16, no. 12: 20200750.

Wieland, George Reber. 1906. "Dinosaurian gastroliths." *Science* 23, no. 595: 819–82.

Wiersma, Kayleigh, and P. Martin Sander. 2017. "The dentition of a well-preserved specimen of *Camarasaurus* sp.: implications for function, tooth replacement, soft part reconstruction, and food intake." *PalZ* 91, no. 1: 145–61.

Wiley, Edward Orlando, and Bruce S. Lieberman. 2011. *Phylogenetics: theory and practice of phylogenetic systematics*. New York: John Wiley & Sons.

Wilkinson, David M., and Graeme D. Ruxton. 2013. "High C/N ratio (not low-energy content) of vegetation may have driven gigantism in sauropod dinosaurs and perhaps omnivory and/or endothermy in their juveniles." *Functional Ecology* 27, no. 1: 131–35.

Williams, Vincent S., Paul M. Barrett, and Mark A. Purnell. 2009. "Quantitative analysis of dental microwear in hadrosaurid dinosaurs, and the implications for hypotheses of jaw mechanics and feeding." *Proceedings of the National Academy of Sciences* 106, no. 27: 11,194–99.

Wilson, Jeffrey A., and Matthew T. Carrano. 1999. "Titanosaurs and the origin of 'wide-gauge' trackways: a biomechanical and systematic perspective on sauropod locomotion." *Paleobiology* 25, no. 2: 252–67.

Wing, Scott L., and Hans-Dieter Sues. 1992. "Mesozoic and early Cenozoic terrestrial ecosystems." In: Anna K. Behrensmeyer, John D. Damuth, William A. DiMichele, et al., eds. *Terrestrial ecosystems through time: evolutionary paleoecology of terrestrial plants and animals*. Chicago: University of Chicago Press.

Wing, Scott L., and Bruce H. Tiffney. 1987. "The reciprocal interaction of angiosperm evolution and tetrapod herbivory." *Review of Palaeobotany and Palynology* 50, nos. 1–2: 179–210.

Wings, Oliver. 2007. "A review of gastrolith function with implications for fossil vertebrates and a revised classification." *Acta Palaeontologica Polonica* 52, no. 1.

Wings, Oliver, and P. Martin Sander. 2007. "No gastric mill in sauropod dinosaurs: new evidence from analysis of gastrolith mass and function in ostriches." *Proceedings of the Royal Society B: Biological Sciences* 274, no. 1610: 635–40.

Wings, Oliver, Thomas Tütken, Denver W. Fowler, Thomas Martin, Hans-Ulrich Pfretzschner, and Ge Sun. 2015. "Dinosaur teeth from the Jurassic Qigu and Shishugou Formations of the Junggar Basin (Xinjiang/China) and their paleoecologic implications." *Paläontologische Zeitschrift* 89, no. 3: 485–502.

Witmer, Lawrence M. 1987. "The nature of the antorbital fossa of archosaurs: shifting the null hypothesis." In: P. J. Currie and E. H. Koster, eds. *Fourth symposium on Mesozoic terrestrial ecosystems, short papers*. Occasional Paper No. 3. Drumheller, AB: Tyrrell Museum of Palaeontology, 230–35.

Witmer, Lawrence M. 1995. "The extant phylogenetic bracket and the importance of reconstructing soft tissues in fossils." In: J. J. Thomason, ed. *Functional morphology in vertebrate paleontology*. Cambridge: Cambridge University Press, 19–33.

Witmer, Lawrence M. 1997. "The evolution of the antorbital cavity of archosaurs: a study in soft-tissue reconstruction in the fossil record with an analysis of the function of pneumaticity." *Journal of Vertebrate Paleontology* 17, no. S1: 1–76.

Witmer, Lawrence M. 2001. "Nostril position in dinosaurs and other vertebrates and its significance for nasal function." *Science* 293, no. 5531: 850–53.

Witmer, Lawrence M., and Ryan C. Ridgely. 2008. "The paranasal air sinuses of predatory and armored dinosaurs (Archosauria: Theropoda and Ankylosauria) and their contribution to cephalic structure." *Anatomical Record* 291, no. 11: 1362–88.

Witmer, Lawrence M., and Ryan C. Ridgely. 2009. "New insights into the brain, braincase, and ear region of tyrannosaurs (Dinosauria, Theropoda), with implications for sensory organization and behavior." *Anatomical Record* 292, no. 9: 1266–96.

Woodruff, D. Cary. 2017. "Nuchal ligament reconstructions in diplodocid sauropods support horizontal neck feeding postures." *Historical Biology* 29, no. 3: 308–19.

Woodruff, D. Cary, Thomas D. Carr, Glenn W. Storrs, Katja Waskow, John B. Scannella, Klara K. Nordén, and John P. Wilson. 2018. "The smallest diplodocid skull reveals cranial ontogeny and growth-related dietary changes in the largest dinosaurs." *Scientific Reports* 8, no. 1: 1–12.

Woodruff, D. Cary, David Trexler, and Susannah CR Maidment. 2019. "Two new stegosaur specimens from the Upper Jurassic Morrison Formation of Montana, USA." *Acta Palaeontologica Polonica* 64, no. 3: 461–80.

Wosik, Mateusz, Mark B. Goodwin, and David C. Evans. 2019. "Nestling-sized hadrosaurine cranial material from the Hell Creek Formation of northeastern Montana, USA, with an analysis of cranial ontogeny in *Edmontosaurus annectens*." *Paleobios* 36: 1–18.

Wright, J. L. 1999. "Ichnological evidence for the use of the forelimb in iguanodontid locomotion." *Special Papers in Palaeontology* 60: 209–19.

Wyenberg-Henzler, Taia. 2022. "Ecomorphospace occupation of large herbivorous dinosaurs from Late Jurassic through to Late Cretaceous time in North America." *PeerJ* 10: e13174.

Wyenberg-Henzler, Taia, R. Timothy Patterson, and Jordan C. Mallon. 2021. "Size-mediated competition and community structure in a Late Cretaceous herbivorous dinosaur assemblage." *Historical Biology* 1–11.

Wyenberg-Henzler, Taia, R. Timothy Patterson, and Jordan C. Mallon. 2022. "Ontogenetic dietary shifts in North American hadrosaurids (Dinosauria: Ornithischia)." *Cretaceous Research* 135: 105177.

Xing, Li-Da, Jerald D. Harris, Gerard D. Gierliński, Murray K. Gingras, Julien D. Divay, Yong-Gang Tang, and Philip J. Currie. 2012a. "Early Cretaceous pterosaur tracks from a "buried" dinosaur tracksite in Shandong Province, China." *Palaeoworld* 21, no. 1: 50–58.

Xing, Li-Da, Phil R. Bell, W. Scott Persons IV, Shuan Ji, Tetsuto Miyashita, Michael E. Burns, Qiang Ji, and Philip J. Currie. 2012b. "Abdominal contents from two large Early Cretaceous compsognathids (Dinosauria: Theropoda) demonstrate feeding on confuciusornithids and dromaeosaurids." *PLoS One* 7, no. 8: e44012.

Xing, Li-Da, W. Scott Persons IV, Phil R. Bell, Xing Xu, Jianping Zhang, Tetsuto Miyashita, Fengping Wang, and Philip J. Currie. 2013. "Piscivory in the feathered dinosaur Microraptor." *Evolution* 67, no. 8: 2441–45.

Xing, Li-Da, Bruce M. Rothschild, Patrick S. Randolph-Quinney, Yi Wang, Alexander H. Parkinson, and Hao Ran. 2018. "Possible bite-induced abscess and osteomyelitis in *Lufengosaurus* (Dinosauria: sauropodomorph) from the Lower Jurassic of the Yimen Basin, China." *Scientific Reports* 8, no. 1: 1–8.

Xu, Xing. 1997. "A new psittacosaur (*Psittacosaurus mazongshanensis* sp. nov.) from Mazongshan area, Gansu Province, China." In: Z.-M. Dong, ed. *Sino-Japanese Silk Road Dinosaur Expedition*. Beijing: China Ocean Press, 48–67.

Xu, Xing, Peter J. Makovicky, Xiao-Lin Wang, Mark A. Norell, and Hai-Lu You. 2002. "A ceratopsian dinosaur from China and the early evolution of Ceratopsia." *Nature* 416, no. 6878: 314–17.

Xu, Xing, James M. Clark, Jinyou Mo, et al. 2009. "A Jurassic ceratosaur from China helps clarify avian digital homologies." *Nature* 459, no. 7249: 940–44.

Xu, Xing, Hailu You, Kai Du, and Fenglu Han. 2011. "An *Archaeopteryx*-like theropod from China and the origin of Avialae." *Nature* 475, no. 7357: 465–70.

Xu, Xing, Xiaoting Zheng, Corwin Sullivan, Xiaoli Wang, Lida Xing, Yan Wang, Xiaomei Zhang, Jingmai K. O'Connor, Fucheng Zhang, and Yanhong Pan. 2015. "A bizarre Jurassic maniraptoran theropod with preserved evidence of membranous wings." *Nature* 521, no. 7550: 70–73.

Yang, Yuqing, Wenhao Wu, Paul-Emile Dieudonné, and Pascal Godefroit. 2020. "A new basal ornithopod dinosaur from the Lower Cretaceous of China." *PeerJ* 8: e9832.

Yates, Adam M., and James W. Kitching. 2003. "The earliest known sauropod dinosaur and the first steps towards sauropod locomotion." *Proceedings of the Royal Society B: Biological Sciences* 270, no. 1525: 1753–58.

Yates, Adam M., Matthew F. Bonnan, Johann Neveling, Anusuya Chinsamy, and Marc G. Blackbeard. 2010. "A new transitional sauropodomorph dinosaur from the Early Jurassic of South Africa and the evolution of sauropod feeding and quadrupedalism." *Proceedings of the Royal Society B: Biological Sciences* 277, no. 1682: 787–94.

You, Hailu, Daqing Li, Qiang Ji, Matthew C. Lamanna, and Peter Dodson. 2005. "On a new genus of basal neoceratopsian dinosaur from the Early Cretaceous of Gansu Province, China." *Acta Geologica Sinica* 79, no. 5: 593–97.

You, Hai-Lu, Yoichi Azuma, Tao Wang, Ya-Ming Wang, Zhi-Ming Dong, and H. L. You. 2014. "The first well-preserved coelophysoid theropod dinosaur from Asia." *Zootaxa* 3873, no. 3: 233–49.

Young, Mark T., and Matthew D. Larvan. 2010. "Macroevolutionary trends in the skull of sauropodomorph dinosaurs—the largest terrestrial animals to have ever lived." In: Ashraf M. T. Elewa, ed. *Morphometrics for nonmorphometricians*. Berlin: Springer, 259–69.

Young, Mark T., Emily J. Rayfield, Casey M. Holliday, Lawrence M. Witmer, David J. Button, Paul Upchurch, and Paul M. Barrett. 2012. "Cranial biomechanics of *Diplodocus* (Dinosauria, Sauropoda): testing hypotheses of feeding behaviour in an extinct megaherbivore." *Naturwissenschaften* 99, no. 8: 637–43.

Zaher, H., D. Pol, B. Navarro, R. Delcourt, and A. Carvalho. 2020. "An Early Cretaceous theropod dinosaur from Brazil sheds light on the cranial evolution of the Abelisauridae." *Comptes Rendus Palevol* 19: 101–15.

Zanno, Lindsay E., and Peter J. Makovicky. 2011. "Herbivorous ecomorphology and specialization patterns in theropod dinosaur evolution." *Proceedings of the National Academy of Sciences* 108, no. 1: 232–37.

Zanno, Lindsay E., and Peter J. Makovicky. 2013. "No evidence for directional evolution of body mass in herbivorous theropod dinosaurs." *Proceedings of the Royal Society B: Biological Sciences* 280, no. 1751: 20122526.

Zanno, Lindsay E., David D. Gillette, L. Barry Albright, and Alan L. Titus. 2009. "A new North American therizinosaurid and the role of herbivory in 'predatory' dinosaur evolution." *Proceedings of the Royal Society B: Biological Sciences* 276, no. 1672: 3505–11.

Zanno, Lindsay E., Khishigjav Tsogtbaatar, Tsogtbaatar Chinzorig, and Terry A. Gates. 2016. "Specializations of the mandibular anatomy and dentition of *Segnosaurus galbinensis* (Theropoda: Therizinosauria)." *PeerJ* 4: e1885.

Zanno, Lindsay E., Ryan T. Tucker, Aurore Canoville, Haviv M. Avrahami, Terry A. Gates, and Peter J. Makovicky. 2019. "Diminutive fleet-footed tyrannosauroid narrows the 70-million-year gap in the North American fossil record." *Communications Biology* 2, no. 1: 1–12.

Zhang, Fucheng, Per G. P. Ericson, and Zhonghe Zhou. 2004. "Description of a new enantiornithine bird from the Early Cretaceous of Hebei, northern China." *Canadian Journal of Earth Sciences* 41, no. 9: 1097–107.

Zhang, Y. 1988. *Sauropod dinosaurs (I) Shunosaurus: the Middle Jurassic dinosaur fauna from Dashanpu, Zigong, Sichuan.* Vol. 1. Chengdu: Sichuan Scientific and Technological Publishing House, 42–67.

Zhao, X., D. Li, G. Han, H. Zhao, F. Liu, L. Li, and X. Fang. 2007. "*Zhuchengosaurus maximus* from Shandong Province." *Acta Geoscientica Sinica* 28: 111–22.

Zhao, Xi-Jin, and Philip J. Currie. 1993. "A large crested theropod from the Jurassic of Xinjiang, People's Republic of China." *Canadian Journal of Earth Sciences* 30, no. 10: 2027–36.

Zheng, Xiaoting, Hai-Lu You, Xing Xu, and Zhi-Ming Dong. 2009. "An Early Cretaceous heterodontosaurid dinosaur with filamentous integumentary structures." *Nature* 458, no. 7236: 333–36.

Zheng, Xiaoting, Larry D. Martin, Zhonghe Zhou, David A. Burnham, Fucheng Zhang, and Desui Miao. 2011. "Fossil evidence of avian crops from the Early Cretaceous of China." *Proceedings of the National Academy of Sciences* 108, no. 38: 15,904–7.

Zheng, Xiaoting, Jingmai K. O'Connor, Fritz Huchzermeyer, Xiaoli Wang, Yan Wang, Xiaomei Zhang, and Zhonghe Zhou. 2014. "New specimens of *Yanornis* indicate a piscivorous diet and modern alimentary canal." *PLoS One* 9, no. 4: e95036.

Zheng, Xiaoting, Xiaoli Wang, Corwin Sullivan, Xiaomei Zhang, Fucheng Zhang, Yan Wang, Feng Li, and Xing Xu. 2018. "Exceptional dinosaur fossils reveal early origin of avian-style digestion." *Scientific Reports* 8, no. 1: 1–8.

Zhou, Zhonghe, and Larry D. Martin. 2011. "Distribution of the predentary bone in Mesozoic ornithurine birds." *Journal of Systematic Palaeontology* 9, no. 1: 25–31.

Zhou, Zhonghe, and Fucheng Zhang. 2002. "A long-tailed, seed-eating bird from the Early Cretaceous of China." *Nature* 418, no. 6896: 405–9.

Zhou, Zhonghe, and Fucheng Zhang. 2005. "Discovery of an ornithurine bird and its implication for Early Cretaceous avian radiation." *Proceedings of the National Academy of Sciences* 102, no. 52: 18,998–19,002.

Zhou, Zhonghe, Julia Clarke, Fucheng Zhang, and Oliver Wings. 2004. "Gastroliths in *Yanornis*: an indication of the earliest radical diet-switching and gizzard plasticity in the lineage leading to living birds?" *Naturwissenschaften* 91, no. 12: 571–74.

Zhou, Zhonghe, Julia Clarke, and Fucheng Zhang. 2008. "Insight into diversity, body size and morphological evolution from the largest Early Cretaceous enantiornithine bird." *Journal of Anatomy* 212, no. 5: 565–77.

Zusi, Richard L. 1984. *A functional and evolutionary analysis of rhynchokinesis in birds.* Smithsonian Contributions to Zoology 395. Washington, DC: Smithsonian Institution.

Zweers, Gart. 1991. "Transformation of avian feeding mechanisms: a deductive method." *Acta Biotheoretica* 39, no. 1: 15–36.

Index

Abelisauridae: classification of, 55, 67; cranial musculature in, 68, 71; dentition in, 68; dietary ecology in, 70, 73; feeding biomechanics in, 70–72; phylogenetic position of, 54; postcranial adaptations in, 72–73; skull morphology in, 66–68; temporal predator–prey relationships of, 292–294

Abelisaurus, temporal predator–prey relationships of, 294. *See also* Abelisauridae

Abrictosaurus. See Heterodontosauridae

Abydosaurus. See Macronaria

Acrocanthosaurus: skeleton of, 91; skull of, 85; temporal predator–prey relationships of, 293. *See also* Carcharodontosauridae

acrodonty, 23

adductor chamber, 40

Africa, 10–11

Afrovenator, temporal predator–prey relationships of, 292. *See also* Megalosauridae

Agilisaurus, temporal predator–prey relationships of, 292. *See also* Neornithischia, non-cerapodan

air sacs, 53, 153, 158

Ajkaceratops. See Ceratopsia, non-ceratopsid

Alamosaurus, temporal predator–prey relationships of, 295. *See also* Macronaria

Albertosaurinae: classification of, 95; phylogenetic position of, 94. *See also* Tyrannosauroidea

Albertosaurus: cranial musculature in, 103; reconstructed head of, 96; skeleton of, 105; skull of, 96; temporal predator–prey relationships of, 295. *See also* Tyrannosauroidea

Alioramus, temporal predator–prey relationships of, 295. *See also* Tyrannosauroidea

Allosauridae: classification of, 74, 82, 84; cervical musculature in, 87, 89; cranial musculature in, 86–87; dentition in, 37, 84,

86; dietary ecology in, 91–92; feeding biomechanics in, 86–90; neck function in, 87, 90; phylogenetic position of, 75; postcranial adaptations, 90–92; skull morphology in, 82–84, 86

Allosauroidea: classification of, 82 84; phylogenetic relationships within, 75; temporal predator–prey relationships of, 292–295. *See also* Allosauridae; Carcharodontosauridae; Metriacanthosauridae; Neovenatoridae

Allosaurus, 8; cervical musculature in, 46, 47, 87; cranial musculature in, 87; reconstructed head of, 8, 83, 293; skeleton of, 91; skull of, 30, 83; temporal predator–prey relationships of, 292–293; tooth of, 38, 86. *See also* Allosauridae

alveolus, 22

Alverezsauridae: classification of, 110, 120; dietary ecology in, 120–121; phylogenetic position of, 109; postcranial adaptations, 120–121; skull morphology in, 120

Alxasaurus. See Therizinosauria

Amargasaurus: cranial musculature in, 171; reconstructed head of, 167. *See also* Diplodocoidea

Ambopteryx. See Scansoriopterygidae

anatomical directionality, 32

anatomy: comparative, 15–18, 53; qualitative analysis of, 15, 26; quantitative analysis of, 15, 26. *See also* biomechanical analysis

Anatosaurus, as a synonym of *Edmontosaurus*, 6

Anchiceratops. See Ceratopsidae

Anchiornis. See Troodontidae

Anchisauridae (Anchisauria): classification of, 145–146; phylogenetic position of, 146. *See also* "Prosauropoda"

Anchisaurus. See "Prosauropoda"

Andes Mountains, 11

angiosperms, 12–13, 299–301

Ankylosauria: classification of, 4, 213, 216; cranial musculature in, 219, 221–223; dentition in, 9, 22, 37, 217–220; dietary ecology in, 218–219; 224–225, 297–300; early feeding studies in, 4, 9; feeding mechanisms and biomechanics in, 219–223; phylogenetic relationships of, 206; postcranial adaptations in, 223–224; skull morphology in, 214–219; temporal predator–prey relationships of, 292, 295–296

Ankylosauridae: classification of, 216; phylogenetic position of, 206. *See also* Ankylosauria

Ankylosaurus: skull of, 215; temporal predator–prey relationships of, 295. *See also* Ankylosauria

Antarctica, 11

antorbital fenestra, 34–35

Anzu: cranial musculature in, 127; reconstructed head of, 124; skeleton of, 128; skull of, 124. *See also* Oviraptorosauria

Apatosaurus, temporal predator–prey relationships of, 292. *See also* Diplodocoidea

appendicular skeleton. *See* pectoral limb skeleton; pelvic limb skeleton

Archaeoceratops: cranial musculature in, 277; coronoid eminence of, 273; reconstructed head of, 266; skull of, 266. *See also* Ceratopsia, non-ceratopsid

Archaeopteryx, 2, 139–140

Archosauria, 10

Argentinosaurus, temporal predator–prey relationships of, 295. *See also* Macronaria

Arrhinoceratops. See Ceratopsidae

Asfaultovenator, temporal predator–prey relationships of, 292. *See also* Allosauroidea

Asia, 11

Asteriornis, 143

LIFE SCIENCE BOOKS FROM HOPKINS PRESS

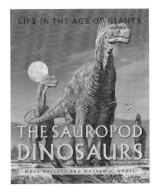

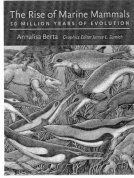

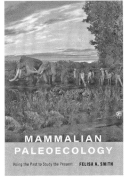

The Sauropod Dinosaurs
Life in the Age of Giants

Mark Hallett and Mathew J. Wedel

The best illustrated and most comprehensive book ever published on the largest land animals the world has ever known.

The Rise of Reptiles
320 Million Years of Evolution

Hans-Dieter Sues

The defining masterwork on the evolution of reptiles.

The Rise of Marine Mammals
50 Million Years of Evolution

Annalisa Berta
graphics editor James L. Sumich

"An excellent addition to paleontology collections."
—*Choice*

Birds of Stone
Chinese Avian Fossils from the Age of Dinosaurs

Luis M. Chiappe and Meng Qinqin

Captivating photographs of the world's most detailed bird fossils illuminate the early diversity of avifauna.

Mammalian Paleoecology
Using the Past to Study the Present

Felisa A. Smith

"A tour de force. One of the world's most eminent evolutionary biologists, Felisa Smith synthesizes paleontology and ecology to tell the story of mammal evolution and put today's environmental crisis into the perspective of Earth history."—Steve Brusatte, University of Edinburgh

 @JohnsHopkinsUniversityPress

 @HopkinsPress

 @JHUPress